ADAPTING TO CLIMATE CHANGE
Thresholds, Values, Governance

Adapting to climate change is one of the most challenging problems facing humanity. The time for adaptation action to ongoing and future climate change is now upon us. Living with climate change involves reconsidering our lifestyles and goals for the future, which are linked to our actions as individuals, societies and governments worldwide. This book presents the latest science and social science research on how and whether the world can adapt to climate change. Written by some of the world's leading experts, both academics and practitioners, on governance, ecosystem services and human interactions, the book examines the nature of the risks to ecosystems and the thresholds of change. It demonstrates how values, culture and the constraining forces of governance can act as significant barriers and limits to action. Adaptation will not be costless, indeed it will be painful for many. As both an extensive state-of-the-art review of science and as a holistic assessment of adaptation options, this book is essential reading for all those concerned with responses to climate change, especially researchers, policy-makers, practitioners and graduate students.

The main features include:

- Historical, contemporary and future insights into adaptation to climate change
- Wide-ranging coverage of adaptation issues from different perspectives: climate science, hydrology, engineering, ecology, economics, human geography, anthropology and political science
- Contributions from leading researchers and practitioners from around the world.

W. NEIL ADGER is Professor of Environmental Economics in the School of Environmental Sciences at the University of East Anglia, Norwich, UK. He has led the programme on adaptation in the Tyndall Centre for Climate Change Research at the University of East Anglia since its inception in 2000. He served as a Lead Author in the Millennium Ecosystem Assessment and as a Convening Lead Author for the Fourth Assessment Report of the Intergovernmental Panel on Climate Change. He is a co-recipient as a member of the IPCC of the Nobel Peace Prize 2007. He was awarded a Philip Leverhulme Prize from the Leverhulme Trust in 2001 for his research achievements.

IRENE LORENZONI is Lecturer in Environmental Politics and Governance at the School of Environmental Sciences at the University of East Anglia, Norwich, UK.

Her interest is in the understanding of, and engagement with, climate change and energy. She is deputy leader of the adaptation programme of the Tyndall Centre for Climate Change Research, and a contributing author for the Fourth Assessment Report of the IPCC on barriers to adaptation.

KAREN L. O'BRIEN is a Professor in the Department of Sociology and Human Geography at the University of Oslo, Norway and Scientific Chair of the Global Environmental Change and Human Security (GECHS) project of the International Human Dimensions Programme on Global Environmental Change (IHDP). Her current research focuses on climate change adaptation as a social process, and on the role that values and worldviews play in responding to environmental change. She was a Lead Author on the Fourth Assessment Report of the IPCC.

ADAPTING TO CLIMATE CHANGE
Thresholds, Values, Governance

Edited by

W. NEIL ADGER
University of East Anglia

IRENE LORENZONI
University of East Anglia

KAREN L. O'BRIEN
University of Oslo

CAMBRIDGE
UNIVERSITY PRESS

CAMBRIDGE UNIVERSITY PRESS
Cambridge, New York, Melbourne, Madrid, Cape Town, Singapore,
São Paulo, Delhi, Dubai, Tokyo, Mexico City

Cambridge University Press
The Edinburgh Building, Cambridge CB2 8RU, UK

Published in the United States of America by Cambridge University Press, New York

www.cambridge.org
Information on this title: www.cambridge.org/9780521764858

First published 2009
First paperback edition 2010

A catalogue record for this publication is available from the British Library

ISBN 978-0-521-76485-8 Hardback

Contents

v

Contents vii

Contributors

Vanessa Abrahamson Research Associate in the Department of Epidemiology and Public Health, University College London, UK.

W. Neil Adger Professor in the Tyndall Centre for Climate Change Research, School of Environmental Sciences, University of East Anglia, Norwich, UK.

Arun Agrawal Associate Professor in the School of Natural Resources and Environment, University of Michigan, USA.

John M. Anderies Assistant Professor at the School of Human Evolution and Social Change and School of Sustainability, Arizona State University, USA.

Nigel W. Arnell Professor and Director of the Walker Institute for Climate System Research, University of Reading, UK.

Iulie Aslaksen Senior Research Fellow in the Research Department of Statistics, Oslo, Norway.

Roberta Balstad Senior Research Scientist and Senior Fellow with Center for International Earth Science Information Network, Earth Institute, Columbia University, New York, USA.

Rachel Berger Climate Change Policy Advisor at Practical Action, Rugby, UK.

Maria Brockhaus Researcher, Centre for International Forestry Research, Bogor, Indondesia.

Nick Brooks Visiting Research Fellow at the Tyndall Centre for Climate Change Research, School of Environmental Sciences, University of East Anglia, Norwich, UK.

Andrew F. Casely Research Fellow in the Institute of Geography, University of Edinburgh, UK.

Matthew B. Charlton Researcher in the Walker Institute for Climate System Research, University of Reading, UK.

Sarah Coulthard Researcher at the Amsterdam Institute for Metropolitan and International Development Studies, University of Amsterdam, the Netherlands.

Suraje Dessai Lecturer in the Department of Geography, University of Exeter, UK.

Alena Drieschova PhD Candidate in Department of Political Science, University of Toronto, Canada.

Andrew J. Dugmore Professor of Geosciences, University of Edinburgh, UK.

Hallie Eakin Associate Professor in the School of Sustainability, Arizona State University, Tempe, Arizona, USA.

Per Ove Eikeland Research Fellow at the Fridtjof Nansen Institute, Oslo, Norway.

Inger Marie G. Eira Linguistic Researcher in the Department of Linguistics, Nordic Sami Institute, Saami University College, Kautokeino, Norway.

Jonathan Ensor Policy Researcher at Practical Action, Rugby, UK.

Roger Few Senior Research Fellow in the School of Development Studies, University of East Anglia, Norwich, UK.

Timothy J. Finan Director of the Bureau of Applied Research in Anthropology and Professor in the Department of Anthropology both at the University of Arizona, USA.

Itay Fischhendler Lecturer in the Department of Geography, The Hebrew University of Jerusalem, Israel.

James D. Ford Postdoctoral Fellow in the Department of Geography, McGill University, Montreal, Canada.

Vladimir Gil Adjunct Associate Research Scientist at the Center for Environmental Research and Conservation, Earth Institute, Columbia University, New York, USA.

Mark Giordano Principal Researcher and Leader of Water and Society Research, International Water Management Institute, Colombo, Sri Lanka.

Marisa Goulden Senior Research Associate in the Tyndall Centre for Climate Change Research, School of Development Studies, University of East Anglia, Norwich, UK.

Thomas Heyd Sessional Lecturer in the Department of Philosophy, University of Victoria, Canada.

Mike Hulme Professor in the School of Environmental Sciences, University of East Anglia, Norwich, UK and founding Director of the Tyndall Centre for Climate Change Research.

Alistair Hunt Research Officer in the Department of Economics and International Development, University of Bath, UK.

Tor Håkon Inderberg Research Fellow in the Fridtjof Nansen Institute, Oslo, Norway.

Tori L. Jennings Postgraduate Researcher in the Department of Anthropology, University of Wisconsin–Madison, USA.

Hermann Kambiré Department of Sociology, University of Ouagadougou, Burkina Faso.

Christian Keller Professor of Archaeology, Department of Culture Studies and Oriental Languages, IKOS, University of Oslo, Norway.

Jan Erling Klausen Senior Researcher at the Norwegian Institute for Urban and Regional Research, Oslo, Norway.

Richard J. T. Klein Senior Research Fellow at the Stockholm Environment Institute, Sweden and Visiting Researcher at the Potsdam Institute for Climate Impact Research, Germany.

Robert Lempert Senior Physical Scientist at RAND Corporation and Professor at the Pardee RAND Graduate School, Virginia, USA.

Irene Lorenzoni Lecturer in the School of Environmental Sciences, University of East Anglia, Norwich, UK and deputy leader of the programme on adaptation Tyndall Centre for Climate Change Research.

Luke Lovell Hydrologist in Water Engineering and Management Skill Group, Halcrow Group Limited, UK.

Sabine Marx Associate Director at the Center for Research on Environmental Decisions and Adjunct Research Scientist at the International Research Institute for Climate and Society, Earth Institute, Columbia University, New York, USA.

Svein D. Mathiesen Professor at the Saami University College and the Norwegian School of Veterinary Science, and affiliated with the International Centre for Reindeer Husbandry, Norway.

Robert McLeman Assistant Professor in the Department of Geography, University of Ottawa, Canada.

Thomas H. McGovern Professor of Archeology, Hunter Bioarcheology Laboratory, Department of Anthropology, Hunter College, City University of New York, USA.

Susanne C. Moser Director, Susanne Moser Research and Consulting, Santa Cruz, California, USA.

Annett Möhner Independent researcher, Bonn, Germany.

Lars Otto Næss Research Fellow at the Institute of Development Studies, University of Sussex, Brighton, UK.

Donald R. Nelson Assistant Professor in the Department of Anthropology, University of Georgia, Athens, Georgia, USA.

Sophie Nicholson-Cole Senior Consultant at Climate Change and Environmental Futures Division at Atkins, Peterborough, and Senior Research Associate at the Tyndall Centre for Climate Change Research, School of Environmental Sciences, University of East Anglia, Norwich, UK.

Karen L. O'Brien Professor in the Department of Sociology and Human Geography, University of Oslo, Norway and Chair of the Global Environmental Change and Human Security Project of the IHDP.

Tim O'Riordan Emeritus Professor in the School of Environmental Sciences, University of East Anglia, Norwich, UK.

Ben Orlove Professor in Environmental Science and Policy at the University of California, Davis, USA and Adjunct Senior Research Scientist at the International Research Institute for Climate Prediction, Earth Institute, Columbia University, New York, USA.

Anthony G. Patt Research Scholar at the International Institute for Applied Systems Analysis, Laxenburg, Austria.

Nicolas Perrin Senior Social Development Specialist in the Social Development Department, World Bank, Washington, USA.

Garry Peterson Canada Research Chair and Assistant Professor in the Department of Geography in the School of Environment at McGill University, Montreal, Canada.

Roger Pielke, Jr Professor in the Environmental Studies Program and a Fellow of the Cooperative Institute for Research in Environmental Sciences, University of Colorado, Boulder, USA.

Rosalind Raine Professor in Department of Epidemiology and Public Health, University College London, UK.

Tim Reeder Regional Climate Change Programme Manager in the Thames Region of the Environment Agency, UK.

Erik S. Reinert Adjunct Professor, Saami University College, Kautokeino, Norway and Professor, Tallinn University of Technology, Estonia.

Hugo Reinert Postdoctoral Research Fellow at the Centre for Ecology and Hydrology, Edinburgh, UK.

Roly Russell Postdoctoral Research Scholar at the Earth Institute, Columbia University, New York, USA.

Inger-Lise Saglie Professor at the Norwegian University of Life Sciences and Senior Researcher at the Norwegian Institute for Urban and Regional Research, Oslo, Norway.

Konrad Smiarowski Graduate Student, Anthropology Department, City University of New York, USA.

Knut Bjørn Stokke Researcher at the Norwegian Institute for Urban and Regional Research, Oslo, Norway.

Owen Tarrant Principal Scientist in the Flood and Coastal Erosion Risk Management Science Programme, Environment Agency, UK.

Tim Taylor Research Officer in the Department of Economics and International Development, University of Bath, UK.

Emma L. Tompkins Senior Lecturer in the Sustainability Research Institute, School of Earth and Environment, University of Leeds, UK.

Ellen Inga Turi Researcher in Political Science at Saami University College, Kautokeino, Norway.

Katharine Vincent Postdoctoral Research Fellow, School of Geography, Archaeology and Environmental Studies, University of the Witwatersrand, South Africa.

Jon Wicks Chief Engineer, Halcrow Group Limited, UK.

Marte Winsvold Researcher at the Norwegian Institute for Urban and Regional Research, Oslo, Norway.

Johanna Wolf Senior Research Associate at the Tyndall Centre for Climate Change Research, School of Environmental Sciences, University of East Anglia, Norwich, UK.

Preface

Almost every day new scientific evidence suggests that the climate is changing due to human action and will continue to change over our lifetimes and those of the next generations. It would be inconceivable that humans, as the most adaptable of species, would not adapt to this challenge. But the state of science about how and whether we adapt and the cost and consequences of such adaptations is nowhere near that of science of atmospheric change. Of course, societies adapt all the time to diverse risks and challenges. So, drawing on theoretical and empirical research, we should be able to discern how adaptation to a changing climate will proceed. Herein lies the impetus for this book.

Until recently, adaptation has been somewhat sidelined, or some would say, actually tabooed, in the climate change discourse. Many argue that investing in adapting to the impacts distracts from the major task of mitigating the causes of anthropogenic climate change by reducing greenhouse gas emissions. Others are convinced that adaptation will automatically happen, once environmental changes become visible. But the time for adaptation action has arrived and the demand for information and rigorous science in this area is increasing exponentially. The funding for adaptation research is growing, and so are the questions that need to be addressed. Many of these questions are directly related to the process of adaptation, and to one overarching question: can we live with climate change?

In this book we examine whether there are real limits and constraints to adapting to the impacts of climate change that is already observed and which is projected for the future. The book is interdisciplinary because adaptation involves deliberate changes and decision-making about resources, values and priorities. The volume spans the natural sciences related to climate change and ecological change, as well as social science analysis of decision-making, involving contributions from economics, anthropology, political science and human geography. Taken together, the chapters paint an illuminating picture of historical, contemporary and future adaptation actions, and draw on work from the developing world, Europe and the Americas, as well as the international policy landscape.

We show that adaptation is necessary, but also fraught with difficulties and challenges. Adaptation in ecosystems and in society will not advance in a smooth manner in response to slow trends in climate parameters such as temperature. Rather it is a messy business involving societal decision-making, attitudes to acceptable risks, and structural constraints within society. The scale of the necessary transformations of societies and economies to face the coming climate change is enormous. Even if rigorous greenhouse gas mitigation efforts are pursued, the climatic changes that are anticipated to occur over the next decades will require changes in infrastructure, in behaviour and more fundamentally, in society's relationship with its environment. This book shows that even moderate levels of climate change can be difficult to adapt to due to constraints in physical and biological systems, due to culture and values in making collective decisions, and due to inertia in governance systems.

Questions of whether and how adaptation will occur have been central to our own research over the past five years. We have helped to steer two major research programmes: the adaptation programme of the Tyndall Centre for Climate Change Research in the UK and the project on the Potentials of and Limits to Adaptation in Norway (PLAN), funded by the Research Council of Norway. We also worked closely together in the IPCC Fourth Assessment Report on adaptation, where we became more convinced that adaptation is severely limited if it is to be the head-in-the-sand, primary response to the coming climate crisis. As we presented this work in Norway, in the UK and internationally, we realised that although adaptation occurs around the world in very different historical, cultural and policy contexts, the processes, the constraints and the ways forward for sustainable action are common to all. We convened a conference at the Royal Geographical Society in London in February 2008 on this topic. The resulting papers, selected through rigorous peer review, are the basis of this book.

We thank the funding bodies of our underlying research, and of the conference from which this book derives for support. In the UK these are the research councils who support the Tyndall Centre: Natural and Environment Research Council, Engineering and Physical Science Research Council and the Economic and Social Research Council. In Norway, the research and conference was supported by the Research Council of Norway's NORKLIMA programme. Internationally, the conference was supported by the Global Environmental Change and Human Security (GECHS) project of the IHDP. We thank colleagues and supporters of the PLAN and GECHS projects, who recognise that understanding adaptation to climate change in any particular place requires an international research endeavour. We also thank colleagues in the Tyndall Centre, particularly Nigel Arnell, Emma Tompkins,

Mike Hulme and the present adaptation team for stimulation and discussions over the years. The Tyndall Centre is unique in its spirit of scientific openness and true interdisciplinarity – traits that deserve nurturing and preservation.

We thank the participants to the conference, and especially extend our thanks to many within our organisations who contributed to the production of this book and for stimulating the scientific learning and refinement that occurred at the conference. These include Jacquie Gopal, Vanessa McGregor, Helen Adams, Lauren Roffey, Anita Wreford, Saleemul Huq, Mike Hulme, Diana Liverman, Andrew Watkinson, Bo Kjellen, Chris West, Roger Street, Siri Mittet, Asher Minns, Linda Sygna and Kirsten Ulsrud. We also thank our colleagues in the IPCC, particularly Shardul Agrawala, Monirul Mirza, Cecilie Conde, Juan Pulhin, Roger Pulwarty, Barry Smit and Kyoshi Takahashi, as well as colleagues in the Tyndall Centre for discussions which initially stimulated our thoughts towards this volume. Finally we thank Professor Bob Watson of the Tyndall Centre for his contribution to the Royal Geographical Society conference and for encouragement towards seeking the messages for policy and government from this endeavour.

1

Adaptation now

W. Neil Adger, Irene Lorenzoni and Karen L. O'Brien

Introduction

Look out the window and assess the weather. If it is hot, change into a lighter shirt. If it is raining, take an umbrella. This is adaptation to changing weather.

Adaptation to changing climate is a different matter. The climate may change either slowly or rapidly, and the changes may be irreversible and impossible to predict with any accuracy. The simple principles of adapting to changing weather begin to break down when the climate changes. In the context of climate change the options for adaptation may involve relocating homes, moving cities, changing the foods we grow and consume, seeking compensation for economic damages, and mourning the loss of our favourite place or iconic species. The difference between adapting to changing weather and adapting to a changing climate lies both in the time-frame and in the significance of the changes required. Moreover, the consequences of *not* adapting to climate change may be far more serious than not adapting to changing weather.

There are two aspects of climate change that have profound significance for adaptation. First is the growing recognition that the weather is no longer 'natural'. While the weather varies and changes seasonally as part of the natural rhythm of our lives, climate change, as it is presently observed, is now beyond all reasonable doubt driven by human activities. This induces a feeling, for some, that the world is sullied, and nature itself is at an end (McKibben, 1999). Adapting to changes that are caused by humans thus involves changing our understanding of our relationship to the climate system. The second aspect of climate change that has implications for adaptation is that it involves harm to some (now and in the future) on the basis of gain to others (in the past, present and future). Hence climate change raises

Adapting to Climate Change: Thresholds, Values, Governance, eds. W. Neil Adger, Irene Lorenzoni and Karen L. O'Brien. Published by Cambridge University Press. © Cambridge University Press 2009.

questions of justice, responsibility and obligations. If human activities are driving climate change, then adaptation involves issues such as compensation and liability. 'Blaming the weather' is no longer a benign and apolitical excuse for uncontrollable natural phenomena. Instead, adaptation to climate change is both a social and a political process.

These social and political dimensions form the backdrop for the analyses presented in this book. Here and in the following chapters we analyse how adaptation occurs, how it may be limited by unknown and sometimes non-linear responses in physical and biological systems, and how societies act both in terms of the values that they hold and the collective action that they undertake.

We already know that adaptation is necessary – the impacts of climate change are already apparent or in some cases predictable with some certainty, as discussed in numerous reports of the Intergovernmental Panel on Climate Change (IPCC) (Parry et al., 2007). While there are many uncertainties concerning present and future adaptation to a changing climate, the book presents emerging findings that have major implications for current discussions and debates about climate change.

First, the chapters in this book suggest that adaptation to climate change is, in general, a desirable outcome: adaptation will often, for example, promote other benefits that can lead to equitable and sustainable development. Building resilience and the capacity to adapt to climate change promotes flexibility, learning and protection of ecosystems from shifting into ultimately undesirable states and provides common good resources to cope with change in general as well as direct social and environmental benefits.

However, even with such ancillary benefits, this book also shows that building resilience in the face of a changing climate is not going to be costless. In fact, adaptation may involve significant transformations rather than incremental changes, some of which will be painful to those in societies reluctant to, or not able to, embrace change. International action and funding may be required to assist in promoting resilience, not only to finance adaptation projects, but also to facilitate the exchange of knowledge and practices that embrace a resilience approach to adaptation.

But if adaptation is indeed such a universally 'good thing', then why does it not occur spontaneously, for the benefit of all? Herein lies the second major contribution of this book: adaptation is limited as a response to the climate crisis. We argue that global-scale analyses of adaptation cannot capture the complexity and diversity of changes that are already taking place in response to climate change, nor can they capture the significance of the losses that are already being experienced. The limits to adaptation, as a response to climate change, depend on ecological thresholds, individual and cultural values, and institutions and governance. As these social, physical and ecological factors together will determine whether adaptation is successful, analyses that overlook one aspect may present a dangerously misleading understanding of the consequences of climate change.

Third, the analysis in this book suggests strongly that the science of adaptation to climate change cannot determine an optimal path between abating the cause of climate change (mitigating greenhouse gas emissions) and adapting to the risks of climate change, at least at the global scale. Framing the global problem of climate change as a trade-off between 'mitigation' and 'adaptation' in effect involves accepting climate change that may breach too many potential thresholds and lead to a loss of resilience, causing harm to people and places that cannot readily be compensated. Accepting, and working towards achieving, a safe level of global climate change involves judgements in the present which may easily be seen in the future (Page, 2006; Caney, 2008). Furthermore, in popular discourse, mitigation and adaptation involve actions and processes that are invariably intertwined and feed into each other; blurring, therefore, a more localised level, the distinction between the two.

Adaptation and its limits

The calculus between adaptation to climate risks and mitigative action to reduce emissions is fundamentally difficult, given the uncertainty created in the global experiment of climate change. Yet, as Gardiner (2004) portrays the distinction, the future can be characterised as a choice between either simply adapting to the consequences of unabated climate change or reducing the risks through abatement (mitigation) of climate change. In the first case the world will be adapting to 'sudden unpredictable large scale impacts which descend at random on particular individuals, communities, regions and industries and visit them with pure irrecoverable costs' (p. 574). This can be compared to mitigation-led strategies where adaptation would be 'addressing gradual, predictable, incremental impacts, phased in so as to make adaptation easier' (p. 574). Stern (2007), Dietz et al. (2008) and others argue that economics has (some of) the tools to make judgements on the trade-offs between mitigation and adaptation, or at least to make them explicit. Parry et al. (2008) and many others argue that there is indeed a globally optimal strategy between mitigation and adaptation. Other approaches suggest that multiple metrics, coupled with knowledge and judgement of unacceptable thresholds in Earth systems, can provide the necessary global scale analysis of the trade-offs between coping with the consequences of climate change or reducing them through decarbonising the economy to mitigate the risk in the first place (Schneider and Lane, 2006; Schneider et al., 2007; Lenton et al., 2008). All of these approaches rely on being able to identify a safe level or rate of change, or at least a socially acceptable level of risk to be avoided.

All of the chapters of this book analyse adaptation in the explicit recognition that adaptation is not a simple and straightforward substitute for action to prevent climate change in the first place. In focusing on what can and should be done in the face of unavoidable climate change, we are acutely aware of the dangers of 'making the case for adaptation a self-fulfilling prophecy' (Gardiner, 2004, p. 574). Much of the severe and potentially catastrophic climate change is eminently avoidable through early and sustained action to reduce emissions of greenhouse gases. Such reductions can occur through many channels – individual behaviour, the development of new technologies, government regulations and new architectures for international co-ordinated action (Barrett, 2007; Stern, 2007, 2008). Although the mechanisms and means for such mitigation measures are well known, whether the necessary mitigation actions are taken is nonetheless dependent on the ability and willingness of societies and ecosystems to cope with and adapt to climate change.

How to respond to climate change at the global scale is not, however, a simple trade-off between the economic damages of climate change impacts and the economic costs of reducing fossil-fuel dependency. The trade-offs are more complex for a number of reasons. First, as the chapters of this book point out, individual species and natural communities are directly limited in their adaptation capacity. While it is possible to envisage how ecosystem services that are of value to humans will be affected by climatic changes, many of the ecological impacts are fundamentally unknowable in terms of ecological processes and surprises.

Second, from many philosophical positions and belief systems, ecosystems have intrinsic value over and above the services they provide to humans. From these perspectives, there is a moral imperative to avoid climate change that threatens global and local extinctions of species, even if non-humans do not have explicit rights within many national and international legal frameworks. Such moral imperatives may appear to be vague and outside the domain of the politics of climate change, but they are not. The imperative to protect non-human species are embodied in law and culture throughout the world: from UN World Heritage Sites to the US Endangered Species Act, through to stewardship ethics in all major world religions. In addition, a material rationale for conservation can be justified by the emerging realisation that ecosystems provide supporting, regulating and cultural services that underpin human life and well-being (as described in the Millennium Ecosystem Assessment, 2005).

Finally, new observations of climate and the impacts of climate change and new models based on improved understanding of physical processes of climate change continually emerge, raising new and penetrating insights and potentially dire scenarios of future climate change. For example, since the IPCC Fourth Assessment Report was published in 2007, new projections from global assessments suggest that observed and projected sea level rises may in fact exceed those reported in

IPCC (Hansen, 2007; Rahmstorf, 2007) but that there are high levels of uncertainty around projections of sea level change that could rise by up to 7 metres with loss of land-based ice sheets in Greenland and Antarctica.

Similarly, new reviews suggest that aerosols from traditional pollutants continue to mask regional warming trends and that these pollutants are likely to be reduced in many countries due to their health impacts. The combined effect will be to unmask the real warming trend raising global mean temperatures above those previously estimated in stabilisation scenarios (Ramanathan and Feng, 2008). The world is, therefore, potentially already committed to 2.4 °C warming due to emissions even up to 2005. Research on ocean acidification has also introduced new questions about the future of marine ecosystems under climate change (Orr et al., 2005; Hoegh-Guldberg et al., 2007). These new findings suggest that the probability of global society being required to adapt to climate and resource states hugely different from today's is indeed high and that radical changes in where and how we live are likely to be necessary.

The challenge of adaptation

The critical and overarching challenge of climate change is how and when to act in the face of scientific evidence. As we demonstrate in this book, this is more multi-faceted than simple models suggest. First, ecosystems and social–ecological systems can absorb significant perturbations if they are resilient. But when thresholds are breached, they often undergo significant regime shifts into alternate states that may be equally resilient, yet are often undesirable from human perspectives.

Second, the impacts and consequences of climate change can be valued according to different metrics, which include but are certainly not limited to economic measures. For example, Schneider et al. (2000) identify five numeraires for judging the significance of climate change impacts, including monetary loss, loss of life, biodiversity loss, distribution and equity, and quality of life. Adaptations measures taken by individuals, communities, groups and generations may reflect one metric over another, and be closely tied to prioritised values. When it comes to decisions on whether or not to act in the face of scientific evidence about climate change, the question inevitably arises of whose values count. The values that are pursued and those that are ignored can easily become enmeshed in the politics of climate change adaptation.

Third, the implementation of adaptation is essentially a governance issue. Adaptation involves deliberate action, or inaction, taken by individuals and through collective action. Governance involves processes through which we engage with our environment and the rest of society: governance involves those activities which make a 'purposeful effort to guide, steer, control or manage sectors or facets of societies' (Kooiman, 1993, p. 2). The dilemmas of governance concern the location of power and influence within society, relating again to whose values count, and to the presumption of collective wisdom over myopic individual choices taken on the basis of self interest (Adger and Jordan, 2009). The scale of adaptation action required is enormous, yet at the same time the geopolitical systems that are in the thrall of the carbon economy creates massive inertia. Under these circumstances it is not enough to simply state that resources should be shared, adaptation should be funded through international transfers, or people and settlements should move in the face of risk. These actions will not take place. Economists label these inertias as market failure or government failure. This book shows the governance challenges of promoting necessary adaptation are significant even if they are largely assumed away in simple models of adaptation action. In reality, the governance of adaptation is likely to be complex and somewhat messy – a legacy of past modes of operating combined with the persistence of outdated paradigms that make it difficult to enact effective adaptations to an issue as complex and multifaceted as climate change.

The implications of thresholds for adaptive action

A threshold is defined as 'a level or point at which something starts or ceases to happen or come into effect' (Soanes and Stevenson, 2008, p. 1502). There are many thresholds for adaptive action, and they generally fall into two categories. First, there are thresholds at which adaptive actions first appear. These are the levels or points when responses come into effect and reduce vulnerability to the negative effects of climate change. Second, there are thresholds beyond which adaptive actions cease to be effective in reducing vulnerability. These can, in effect, be considered limits to adaptation, in that adaptation no longer represents a successful response to climate change. While the first type of threshold is important for initiating positive actions in response to climate change, the second type is of greater concern, as it defines the changes that cannot be adapted to, as well as the losses that will be incurred as a result of climate change.

Current scientific discourses on limits to adaptation focus on immutable thresholds in biological and technological parameters, or even in unaffordable economic costs. Thus 2 °C of global mean warming is regarded as a threshold of dangerous anthropogenic interference for its impacts on sensitive ecosystems such as coral reefs (Schellnuber et al., 2006). But framed another way, adaptation by humans is endogenous to the way in which society operates and hence any limits are contingent on parameters such as ethics, knowledge, attitudes to risk and cultural constraints on action (Adger et al., 2009). Meze-Hausken (2008) similarly argues that although some thresholds can be quantified (most often by experimental design where other variables are held constant, or by statistical analysis), others 'can only be defined through subjective assessments of levels of acceptable risk and impact, as well as on expectations and experience' (Meze-Hausken, 2008, p. 318). For Meze-Hausken (2008) adaptation is considered the adjustment *to* a response or impact, with the possible consequence of either increasing or reducing the threshold level. In other words, thresholds may change over time, depending on adaptive actions. This represents a third type of threshold – a dynamic threshold that is itself influenced by adaptation measures. This draws attention to the importance of assessing the implications of adaptation measures for thresholds, not only from physical or ecological perspectives, but also from social, cultural and experiential perspectives.

Reducing the vulnerability of households and communities to climate change has been identified as a key response by both the climate change and disaster risk reduction communities (Schipper and Pelling, 2006; UNISDR, 2008). Vulnerability approaches can directly address the physical risks of climate change through technological interventions, such as adjustments to infrastructure or new varieties of seeds. They may also address the underlying and systemic factors that contribute to vulnerability in the first place, such as land tenure laws, unequal access to markets or credit, or a lack of social safety nets. Finally, vulnerability approaches often focus on enhancing adaptive capacity, by improving access to education, financial resources, information such as seasonal climate forecasts or diversifying livelihoods. Together, all of these strategies can help to increase the thresholds at which climate change creates negative outcomes. Vulnerability reduction itself can be considered an adaptive response to climate change.

Yet what about the thresholds that define conditions beyond which society can successfully adapt? Schneider and Lane (2006) discuss 'imaginable surprises', such as deglaciation of Greenland or changes in the North Atlantic Thermohaline Circulation, which would present numerous and, arguably for many people and species, insurmountable barriers to adaptation. They also discuss the possibility of 'true surprises' that have yet to be imagined or taken into consideration when discussing climate change impacts, vulnerability and adaptation. These critical thresholds are sometimes referred to as 'tipping points', in that at a particular moment in time a small change can have large and long-term consequences for a system (Lenton et al., 2008).

The possibility of surpassing critical climate change thresholds has important implications for adaptive actions. First, adapting to a world beyond 'tipping points' requires foresight and investment, ideally sustained over long periods of time. Large-scale infrastructure projects, massive population resettlement schemes and changes to global food production systems will have to be planned, financed and implemented amidst tremendous uncertainty about the future. Second, surpassing climate change thresholds will lead to innumerable losses, regardless of adaptation measures. Although the potential changes will undoubtedly create unequal outcomes, the changes will be so dramatic that the real equity issues are intergenerational. Demanding future generations to adapt to changes set in motion by past and present human activities raises ethical and philosophical questions that are only beginning to be addressed (Gardiner, 2004; Adger et al., 2006; Caney, 2008; Jagers and Duus-Otterström, 2008). Third, the consequences of the tipping points described by Lenton et al. (2008), such as the collapse of the West Antarctic Ice Sheet or dieback of the Amazon rainforest, will create changes in physical systems and ecosystem services, as well as geopolitical changes and transformation of economic and social systems. The notion of 'adaptation to climate change' is unlikely to be the main concern under such scenarios, but rather the focus is likely to shift towards adaptation to complex emergencies and disasters. Finally, although much can be done to adapt to climate change thresholds, the question of what type of a world we want to live in and whose values count in deciding this must be addressed. These aspects of climate change adaptation are discussed below in relation to the chapters of the book.

Values in adaptation: whose and how they count

Adaptation, like most other changes in society and economies, involves a multitude of decisions taken by individuals acting in their own perceived interest, but with impacts and ramifications well beyond those actions both in space and time. As with all such situations, people act collectively as well as individually, and hence governments have a role in steering society towards longer-term outcomes. Adaptation actions are likely to be undertaken by individuals or businesses if they perceive early rewards or benefits from their actions, such as reduced damages from extreme weather events or cheaper insurance. In order for these actions to be economically efficient the individuals and businesses concerned must bear all the costs and receive all the benefits from their actions. There are cases however, where private actions create externalities that must be borne by others. In addition, in many cases, the incentives to act to adapt to climate change are not sufficiently strong, or there are property rights and public good aspects which hinder private action. These situations where adaptation is not efficient or optimal in any meaningful sense are not the minority of cases. Negative externalities and spillovers are, in effect, pervasive. As Hanemann (2000) suggests, economically optimal adaptations are built into economics models of adaptation, almost in the hope and expectation that they will occur rather than on evidence that they do occur.

What then are the key roles of public policy in adaptation? Some economists writing in this area suggest that much adaptation will occur spontaneously through adjustments to markets and individual behaviour (as discussed by Hanemann,

2000). But markets are, in effect, constructs of the laws, regulations and collective will of the agents and regulators involved. Governments, as an expression of collective will, influence everything. The major objectives of public policy to adapt to climate change therefore would seem to be (1) to protect vulnerable populations by reducing their vulnerability and exposure to risk; (2) to provide information for planning and stimulating adaptation, and (3) to protect important public goods (such as nature conservation) as well as to provide public goods such as human security and protection (such as coastal defence and early warnings of extreme events). In addition, a strong signal from the public sector that it is taking adaptation seriously can induce increased action in the private sector.

These principles are similar to many arenas of public policy, from social welfare to environment to health. In all of these areas, the implementation of policy is hugely contested as the values, goals and belief in policy prescriptions and instruments varies widely (Adger et al., 2009). One of the greatest problems in implementing adaptation lies in identifying who and what is vulnerable, and even in specifying who has the right and responsibility for identifying who and what is vulnerable. Understanding the wider implications of adaptation measures requires that many important normative and ethical issues be discussed and debated.

Aside from providing adaptation actions directly for nature conservation or other reasons, government can aid private actors by providing information about the likely environmental changes and impacts, and the options for adaptation. Importantly, government may become involved in private adaptations to shift the burden of the costs from the victim to the polluter. It is important to be clear who is responsible for compensation. If the polluter is to pay for adaptation, then we need to be able to pinpoint the links between who emitted greenhouse gases in the past and the impacts that we are now suffering. Under widely established principles of law, polluters should compensate victims by at least the value of the harm inflicted. A natural analogy to the climate change case is the set of those compensations claims under tort law for harm caused by toxic substances (so-called toxic torts) (Farber, 2007). Clearly legal processes are not in such a position to directly point the finger of blame for climate change at present. But there is a strong possibility that advances in environmental sciences to attribute proportions of individual extreme weather events to human-induced climate change will bring the issue of liability to the fore in the next decade (Allen and Lord, 2004; Allen et al., 2007).

Such discussions of public policy in adaptation largely assume, of course, that governments act in a far-sighted manner to promote their citizens towards equitable and sustainable outcomes. Importantly it also assumes that governments, as the agents of collective action, have the necessary means and knowledge to implement that vision. The chapters of this book point to the limits of such assumptions. Most often governments act in the interest of the most vociferous and influential actors in society, to the detriment of others who are less powerful and influential. So while some adaptations to climate change may be efficient, they may leave behind the most vulnerable. There are, as discussed in Chapter 13 by Hallie Eakin and colleagues, therefore inherent trade-offs between public policy interventions based on the dominant policy paradigms of efficiency and those based on minimising vulnerability, or even building resilience. Vulnerability approaches suggest that some risks are unacceptable and should be avoided at all costs, an approach consistent with Rawlsian accounts of equity and justice (Dow et al., 2006; Paavola and Adger, 2006). Resilience approaches suggest that system resilience and learning can come at the expense of loss to individuals. Hence the objectives of adaptation, and how governments act on underlying values, makes a huge difference in terms of outcome.

Making adaptation happen for the common good

Adaptation has always taken place, and is likely to continue doing so. Human beings have been able to adapt to changing environments and societies, surviving and flourishing overall. However, if we hold a lens to the adaptation process and analyse it further in detail, it becomes clear that environmental and social change does not affect everyone equally. Less resilient communities – and more vulnerable individuals – can be severely affected by change, thus limiting their opportunities for adaptation.

The prospect of climatic changes of greater magnitude and frequency than those experienced throughout most of human history beg the question of whether adaptation is possible and how adaptation to present and future changes may be facilitated. In very simple terms, adaptation entails an adjustment to changing conditions. On a social level, this can be interpreted as some form of cognitive or behavioural response at individual and collective levels, both being invariably entwined. Understanding adaptation in the context of climate change requires careful consideration of two dimensions: scale (Who is responding where, to what?) and purpose (Why are we responding? What are the aims of adaptation?).

Let us consider these in turn. Adaptation occurs at different but related levels. Policies shaped by national and international circumstances set objectives to be achieved at local and regional levels. Individuals and organisations however do not operate in isolation. Interpretation of information and its translation into decisions and behaviours are affected by social context, individual characteristics and direct experiences. In other words, adaptation is a multi-scalar process of multi-level governance, concerned with the interaction of individual and collective behaviours acting from the bottom–up and the top–down in response to changing circumstances (see Pelling et al., 2008; Urwin and Jordan, 2008). Given, however, that any response to changing circumstances is in part shaped by entrenched practices, beliefs and perspectives, adaptation in itself can result in imperfect outcomes, or even maladaptation. Existing procedures and processes can themselves hinder a process of adaptation that has to now aim, it has been argued, to achieve much more than incremental change.

Regarding our second point about the aims of adaptation, it is important to note that some normative perspectives argue that the imperative lies not only in ensuring humankind's survival in the long term, but guaranteeing a certain degree of individual and social welfare in the present as well as the future. The adaptation literature has in recent years focused on the differential impacts of climate change on populations, communities and nations worldwide, distinguishing and assessing winners and losers from future changes (O'Brien and Leichenko, 2003). The debate has again shifted, philosophically underpinned by considerations about how adaptation must happen for the common good, coupled with assisting the most vulnerable.

Taking these two dimensions into account together begs the question of how successful adaptation may take place despite (or within) the governance constraints we have acknowledged exist and persist in enacting change? Can – and if so, how – governance structures and processes facilitate a transition towards a future under more pronounced climate change, ensuring collective and individual human welfare?

It has been argued that at the international level, standards of responsibility and accountability tend to be defined by prevailing ideological paradigms, hampering drives to create institutions for global environmental governance based on shared ethics, justice and equity considerations (Okereke, 2008). It appears that although there are diverse forms of governance that could be combined into novel hybridised forms, adaptable and flexible to context-specific needs and changing circumstances, the prevalent drive appears to denote an incremental change of the status quo (Backstrand, 2008).

Part of the challenge in enacting governance for successful adaptation are the different scales of decision-making and the incongruence between aspirations, aims, priorities and interests. Historical legacies manifested in struggles of power and interests shape our world today. Nevertheless there are demonstrations of how these can be overcome in order to enable more sensitive, useful and malleable forms of governance for adaptation. Embedding the space for constructive reflection and learning within existing processes, can result in a better appreciation and management of existing knowledge, within the boundaries of scientific and socio-economic uncertainty. Engendering greater interaction between individuals and collectives has also been variedly explored and proposed as means to enhance the capacity to adapt (e.g. Pelling et al., 2008).

Others have emphasised the need to widen the adaptation focus beyond climate change to encompass sustainability. Justice, equity and well-being are goals within debates about poverty reduction and development; although these may be driven by concerns other than environmental, these can at times be complementary and related to climate change adaptation (Eriksen and O'Brien, 2007; Wilbanks, 2007).

Governance for adaptation therefore appears as a worthwhile yet elusive set of processes, defined by issues of scale, context, understanding and interactions between different levels. It is invariably entwined with uncertainty, knowledge, perceptions, goals, priorities, transparency, responsibility and accountability. The evolving characteristics of the future, under the influence of climate change, suggests that ensuring the common good with a view to supporting the most vulnerable may entail reflexively revising and reviewing the effectiveness of current governance structures and processes, ensuring their flexibility and suitability to evolving circumstances and understandings.

Contribution of this book

The book is divided into three sections, each of which covers one of the key themes identified above. Part I looks at the role of various types of thresholds in adaptation. These thresholds are influenced not only by physical factors, but also by social, cultural and cognitive ones. Several of the chapters consider the ways that thresholds can be changed over time through adaptive actions. The main argument that emerges from these chapters is that developing resilience in both ecosystems and society should be considered an essential foundation for adaptation to future changes and uncertainty.

The section starts with a discussion in Chapter 2 by Garry Peterson on the ecological limits to adaptation. Peterson argues that human modification and simplification of ecosystem services has reduced the ability of ecosystems to self-regulate, thereby increasing the possibilities for abrupt changes in ecological functioning. This decline has important consequences for the ability of people to adapt to climate change, which Peterson exemplifies by discussing potential agricultural-water regime shifts. He emphasises that declines in ecosystem services may lead to abrupt and surprising changes that pose new challenges to adaptation. Restoration or enhancement of ecosystem services offers one mechanism for building resilience to climate change, and should thus be considered as an important adaptation strategy.

Climate change impacts involve both the ecological impacts highlighted by Garry Peterson and physical impacts such as on water resources, infrastructure and coasts. Chapters 3 and 4 examine in detail two issues where engineering solutions and their social context are relevant. In Chapter 3 Nigel Arnell and Matt Charlton illustrate both supply- and demand-side adaptations to increasing water scarcity in southern England. They show that technologically feasible water-supply solutions, such as building new reservoirs, are constrained by environmental regulation and scarcity of land. Yet even demand-side interventions, such as using price mechanisms to encourage wise use of water, are themselves limited by social regulators. Hence adaptation is socially constrained even where technology could overcome limits to change. In Chapter 4, Tim Reeder and colleagues present a study on adaptation measures that are needed to protect London and the Thames Estuary, which has one of the best tidal flood defence systems in the world. The Thames Estuary 2100 project examines flood risk management options that protect against incremental sea level rise and different magnitudes of storm surge events in a future under climate change. By identifying thresholds of future flood risk, proactive measures can be designed to adapt the system and manage the new level of risk. This includes identification of the point at which it is considered impractical to intervene further to manage flood risk through engineering. One of the important lessons from this study is that the lead time needed to make decisions to adapt to change is often long, and one cannot simply wait until an adaptation threshold has been reached or exceeded to respond.

In Chapter 5, Suraje Dessai and colleagues examine the role of uncertainty in climate change adaptation, and question the assumption that improved climate predictions are a prerequisite for adaptive responses to climate change. They emphasise that since climate is only one among many uncertain processes that influence society, climate predictions should not be a central tool to guide adaptation. Instead, strategies that are robust against a wide range of plausible climate futures should be used as a basis for decisions about adaptation. In other words, the limits to climate prediction should not be interpreted as a limit to adaptation. Improved predictions based on more accurate and precise computer models may come at the expense of improved understandings of the vulnerability of adaptive decisions to large and irreducible uncertainty.

In Chapter 6, Anthony Patt looks at the potential role of seasonal climate forecasts in building adaptive capacity. These forecasts, developed to address interannual climate variability, can be considered a valuable first step towards climate change adaptation, which first and foremost requires flexibility of responses. Nonetheless, Patt points out that although a lack of flexibility can be culturally embedded and difficult to change, one real benefit of forecasts is that they build linkages between scientists, public institutions and end-users of the information. In other words, they not only can help to improve coping with climate variability, but they also establish a foundation for successful climate change adaptation in the future. A key point here is that it is not only the technological aspects of forecasts that matter, but the social and institutional connections that are made between people. When it comes to adaptation, collaboration and partnerships will matter.

In Chapter 7, Andrew Dugmore and colleagues examine historical evidence from the archaeological record and present some sobering conclusions about climate change adaptation among the Norse settlements of Greenland. They show that although the Norse communities successfully adapted their livelihoods to changing climate conditions, they nonetheless made decisions that reduced their resilience to natural and cultural changes, and as a result the settlements collapsed and disappeared. Facing multiple stressors, including the combined challenges of economic change, cultural contact with the Inuit and unanticipated changes in the climate, they were unable to successfully respond. One lesson that can be drawn from this example is that successful adaptation along a particular pathway of development may at the same time decrease resilience, and eventually lead to crisis.

A more optimistic picture of adaptation is given by James Ford in Chapter 8, based on observed, empirical evidence from the Arctic community of Igloolik in the Canadian territory of Nunavut. In exploring whether there are limits to adaptation to sea ice change, Ford found signs of increasing adaptability, and evidence of social learning through trial and error, which has led to the evolution of Inuit traditional knowledge. However, although a combination of risk management, avoidance and sharing strategies, facilitated by Inuit knowledge, enabled the increased physical risks of reduced sea ice to be moderated, Ford's study also suggests that limits to adaptation are most likely to be cultural in nature, whereby the trade-offs necessary to maintain food security may compromise the social and cultural values of the Inuit.

Part II of the book comprises theoretical and empirical analysis of the implications of diverse values and worldviews in implementing adaptation. The chapters examine how these values are manifest in observed behaviour and how these values are captured in models and theories of value. Diverse values are apparent, for example, in attitudes to risk among elderly people faced with risks from heat wave, and from self-identified locals dealing with flood risk compared to those of outsiders and outside agencies. The description and analysis of values and underlying cultural traits is a central, if not the central problem, of the social sciences. Hence chapters explore how values are described and made tractable within economic cost–benefit analysis, with vulnerability analysis and analysis of motivations and outcomes of demographic change. This part of the book illustrates that global scale analysis of adaptation pathways cannot easily capture the diversity and often conflictual nature of values and value change.

Ben Orlove in Chapter 9 demonstrates how adaptation is used by government and other agencies dealing with climate change and discusses the values inherent in that use. He argues that the use of the term, evolving over the past three decades within the scientific communities, reinforces the drive for interventionist development, often to the detriment of priorities and perceptions of those farmers and rural residents in developing countries who actually face the risks. Karen O'Brien in Chapter 10 brings social theory to bear on the key issue of changing values. Even if social science can capture values in decision-making, these values (such as what people care about and their relationship to places they live) are likely to change over time, which can influence the ways that adaptation measures are viewed by future generations. Many Norwegians, for example, hold traditional, modern or post-modern values that influence understandings of what it is to be Norwegian and to be a citizen. Some of these values, including values related to snow cover, are challenged by a changing climate. Discussions about the economic aspects of the potential loss of ski days may therefore be irrelevant to people who value snow cover as something far more intrinsic to Norwegian identity. Similar insights are offered by Johanna Wolf and colleagues in Chapter 11 who examine differences in perceptions of risk to heat wave in elderly populations who are ostensibly at risk from heat waves due to their health status. They show that some individuals define themselves by their independence and hence are resistant to any external assistance to reduce their exposure to risk, even by those in their social networks. This study raises important questions concerning how effective government interventions, in this case in public health where vulnerabilities are persistent and entrenched in deeply held values.

Clearly there are limits to how economics deals with social values that do not appear in markets. This limitation is widely recognised, not least in the Stern Review (Stern, 2007) and the commentaries on that analysis (Neumayer, 2007). Alistair Hunt and Tim Taylor in Chapter 12 argue that economics continues to make progress towards incorporating such non-market values into cost and risk assessments. They used stated preference techniques, for example, to elicit values for historic and valued cultural assets (a church and a brewery in Sussex) at risk from flooding and show both how incorporation of preferences and acknowledgement of the time discounting of such values affects outcomes of decisions made on purely efficiency grounds. They suggest, rightly, that such techniques need to be complemented by other analytical tools to incorporate values into decision-making. But Eakin and colleagues in Chapter 13 question more directly the basis of efficiency as a guiding principle in adaptation decision-making. They show that there are inherent trade-offs between efficiency-driven and vulnerability- or resilience-led approaches to adaptation. Vulnerability approaches, in particular, deploy a radically different view of ethics in appraising values, whereby loss to some cannot be compensated by gain to others (as in utilitarian framings). Jonathan Ensor and Rachel Berger in Chapter 14 describe the principles of decision-making for collective community action based on notions of communitarianism and social capital that contrast with utilitarian notions of individual aggregation of preferences and utility. Their illustrations of community-based adaptation to climate change demonstrate that culture and the collective good are in fact central to how people value their own well-being. But even with a community focus, the techniques to elicit and reflect values are fraught with difficulty. Across these methodological chapters the key issue of the partial nature of all methods in handling values comes through.

The implications of diverse values are further explored in the chapters by Tori Jennings (Chapter 15), Sarah Coulthard (Chapter 16) and Thomas Heyd and Nick Brooks (Chapter 17). Each of these demonstrates how values held by those that are less powerful, or whose values do not easily fit into established analytical tools, are subordinated or simply ignored. Jennings demonstrates how locals and outsiders construct and manage their landscape of risk. In the case of the village of Boscastle in south-west England, the implementation of flood risk measures after a major flood event in 2004 excluded local knowledge and practice to the detriment of inclusive planning and engagement with the risks involved. Sarah Coulthard argues that fishing communities and those people engaged in fishing as their source of livelihood worldwide have already 'earned the right to be considered as expert adapters', dealing continuously with fluctuating and variable resources. She shows how common embedded values in fishing communities, in this case in South Asia, focus on fantastically innovative means to keep on fishing. But since their identities are bound up with this activity, it will be extremely difficult and painful for fishers to 'hang up their nets' – this is an adaptation too far for the value systems of these people and places. This common finding, reflected in the discussion by Heyd and Brooks, reflects a key observation of this book – that adaptation to climate change (of the transformative type likely to be necessary) will be painful for many.

An important frontier of research on adaptation is how values relating to people's knowledge of and sense of place, relate to their adaptation decision-making. Chapters by Roberta Balstad and colleagues (Chapter 18) and by Robert McLeman (Chapter 19) address this issue through application of insights from psychology, cultural geography and demography. Balstad and colleagues demonstrate, through analysis of the historical case of dealing with drought in the Great Plains of the USA in the 1930s, that different adaptation strategies undertaken by different proportions of the population (settlers and more established farmers) show that they process the same information in different ways due to their prior experiences and become locked into single strategies and ways out of the problem. Thus the settlers of eastern Dakota came up against a climate extreme they had never experienced and many decided that exiting from farming was the only option. One of the most significant transformative adaptation decisions that can be taken by any individual is to move location. Demographic theory offers insights into how such decisions are taken and with what motivations. McLeman, again drawing on the experience of rural residents of the US Great Plains, shows how networks, social capital and underlying sense of place are important determinants of how such adaptation is actually carried out.

Part III of the book addresses governance issues in the context of adaptation spanning a range of analytical scales, from the international to the local, with theoretical and empirical contributions from Europe, the Middle East, Africa and North and South America. Susanne Moser in Chapter 20 offers a systematic overview of the pivotal role of both governance structures and processes in decision-making. Guided by key questions, and informing theoretical perspectives with practical bottom–up considerations, Moser appraises decision-making in action and whether this results in successful adaptation. She stresses our limited current understanding of the relationships between theoretical views of governance structures and processes, and the complex social dynamics that result in practical change. Moser's final words caution against the illusionary prospect of pursuing a 'perfect governance approach that promises perfect adaptation', given that governance – in the form of different structures, processes and mechanisms – can both inhibit and encourage adaptation.

Constraints to adaptation in the form of different governance structures are underlined by Sophie Nicholson-Cole and Tim O'Riordan (Chapter 23), as their analysis of coastal zone management in England reveals the mismatch between national policies and strategies on the one hand, and differing preferences for adaptation at regional and local scales on the other. The contrast between national strategies and local strongly held preferences creates a state of dysfunctional, rather than adaptive, governance, challenging the possibilities of a sustainable future coastline. How can governance mechanisms overcome these structural barriers to action? Chapter 25 by Maria Brockhaus and Hermann Kambiré outlines how a decentralisation process cannot result in effective adaptation governance unless underpinned structural, behavioural and resource limitations are overcome through the use of local knowledge and institutional responsiveness to local realities. The inadequacy of some governance goals, driven by particular paradigms, is further illustrated at cross-national scales by Erik Reinert and colleagues (Chapter 26) with regards to the mismatch of traditional local pastoralist knowledge and practices with international and national market economy priorities. The message here, once again, is not all doom and gloom – recent developments in governance structures have enabled the recognition of traditional ways of knowing and doing, gradually overcoming ingrained constraints to sustainable adaptation to environmental change.

Sometimes, however, even with transparent and appropriate governance structures, resources are not sufficient for providing the social protection and public goods required in the face of climate change risks. This is certainly the case for many developing countries. Richard Klein and Annett Möhner (Chapter 29) challenge the responsiveness of international adaptation funds to developing countries' needs, proposing a reduction in administrative complexity, integration of themes and priorities across overseeing institutions as well as promoting procedural and guidance transparency. The shifting goalposts created by variability in resource availability are explored by Alena Drieschova and colleagues who in Chapter 24 underline the mechanisms available and used to manage water. They emphasise how differing priorities and circumstances shape governance responses.

Chapters 21, 22 and 28 in this section outline how governance processes, structures and mechanisms – to continue using Moser's words – can be modified and developed to overcome some of the observed reinforcing dynamics which inhibit adaptive capacity. Timothy Finan and Donald Nelson's work (Chapter 21) in the drought-stricken region of Ceará (north-eastern Brazil) illustrates how a methodological innovation – integrating community participatory research tools with a GIS-based mapping platform – may reform the persistent structural inequity created by century-old procedures resulting in proactive and pre-emptive local planning. Both Marisa Goulden and colleagues (Chapter 28), as well as Arun Agrawal and Nicolas Perrin (Chapter 22), examine the role of local institutions in the adaptation of rural groups, within the context of emerging national adaptation strategies. Agrawal and Perrin's chapter reveals the lack of attention dedicated by national governments to the institutional structures which would enable the poor to take action. They point out the criticality of focusing on, and integrating, institutional processes and structures in order that they may translate into more successful adaptation interventions and investment. Although Goulden and co-authors focus on different mechanisms of adaptation, comparison of societal adaptations in three different locations also supports the view that the success of adaptation strategies rests on understanding and recognition of multiple drivers of vulnerability through appropriate governance structures, accompanied by adequate access to resources.

Tor Inderberg and Per Ove Eikeland (Chapter 27) develop an institutional approach and apply it to exploring constraints to adaptive capacity in a national energy system, suggesting that institutional factors may hinder climate change adaptation, due to strong path dependencies, resonating with the findings and argument by Reinert and colleagues. Marte Winsvold and co-authors (Chapter 30) explore coordination, under different modes of governance, and their influence on adaptation, by relating these to theories on organizational learning, identifying how different forms of coordination may interact with actors' characteristics and responses.

Overall, Part III on governance illustrates the long-term consequences of getting it wrong may be dire, resulting in a decrease of future adaptive capacity. The pending question therefore is how – under the pressures of increasing environmental change – we may harness individual and social abilities and ingenuity to enable successful adaptation through effective governance.

In the concluding chapter, Donald Nelson weaves together key messages in the book and considers them in a cultural context. He points out that changes in our material behaviour influence our level of adaptedness at any point in time, whether we are conscious of this or not. This is a particularly important point, given the current financial crisis and its potential implications for climate change responses. Nelson also reminds us that novel climate regimes do not signify the end of the world – they may, however, signify an age in which we have to radically reassess our understanding of the world.

Conclusions

This book reveals many important facets of adaptation that have not yet been included or addressed in mainstream discourses on climate change. The issues raised in this book present some real reasons for concern. If humans must learn to live with climate change, a situation that is increasingly recognised as imminent among both science and policy communities, then some key questions must be openly debated at all levels of governance, and by individuals and groups with differing values and belief systems. The chapters in this book debate the ecological, social, cultural and cognitive thresholds for adaptation and the question of how much change can we live with. They raise issues about societies' willingness, and ability, to adapt, given that it is partly limited by subjective thresholds. Entwined with these are considerations about what will be lost if, or when, we cannot adapt and the consideration of which losses we may be prepared to accept if this is the case. A major part of the book focuses on the decision-making processes underpinning adaptation, with a specific focus on governance. Some of the chapters tackle the major issue of the role of governance in the midst of climatic changes that will deeply affect human well-being, possibly in the context of undesirable outcomes. Some authors suggest there may be means of reconciling individual and group interests with adaptation for the common good, including the well-being of future generations, whereas others point to the overlooked and often opaque processes of decision-making and policy, which confuse and cloud adaptation.

Adaptation is a social process with implications for ecosystem services, economic and political stability, and culture, among many other things. Yet the science of adaptation has not yet progressed to the point where we have a solid understanding of what is actually involved in adapting to dramatic changes and uncertainties that are both predicted and increasingly observed. Bob Watson has suggested that, given the new scientific evidence on emissions trajectories and climate sensitivity referred to earlier in the chapter, the UK should plan for the effects of a 4 °C global average temperature increase compared to pre-industrial levels (Randerson, 2008). This book shows that, in effect, there is no science on how we are going to adapt to 4 °C warming. And although further research on adaptation to climate change can provide the knowledge base and insights about what we can live with, and what we cannot live with, whether we choose to act upon this knowledge remains to be seen.

The chapters in this book demonstrate that adaptation to date is an imperfect process, driven by our limited understanding and ability to act. One of the underlying messages of this book is that adaptation to changing future circumstances, including the climate, will and should take place in the form of both win–win options and actions that will only in retrospect be able to be assessed in terms of their success. Acknowledging the drivers and goals of decision-making is crucial. Ultimately the adaptation challenge is whether society has the capacity to both adapt sustainably to a changing climate and at the same time create an alternative future that limits the amount of change that we impose on the planet and on future society.

References

Adger, W. N. and Jordan, A. (eds.) 2009. *Governing Sustainability*. Cambridge: Cambridge University Press.

Adger, W. N., Paavola, J., Huq, S. and Mace, M. J. (eds.) 2006. *Fairness in Adaptation to Climate Change*. Cambridge: MIT Press.

Adger, W. N., Dessai, S., Goulden, M., Hulme, M., Lorenzoni, I., Nelson, D. R., Naess, L. O., Wolf, J. and Wreford, A. 2009. 'Are there social limits to adaptation to climate change?' *Climatic Change* **93**: 335–354.

Allen, M., Pall, P., Stone, D., Stott, P., Frame, D., Min, S.-K., Nozawa, T. and Yukimoto, S. 2007. 'Scientific challenges in the attribution of harm to human influence on climate', *University of Pennsylvania Law Review* **155**: 1353–1399.

Allen, M. R. and Lord, R. 2004. 'The blame game', *Nature* **432**: 551–552.

Backstrand, K. 2008. 'Accountability of networked climate governance: the rise of transnational climate partnerships', *Global Environmental Politics* **8**(3): 74–102.

Barrett, S. 2007. 'Proposal for a new climate change treaty system', *Economists' Voice* **4**(3): 6. Available at www.bepress.com/ev/vol4/iss3/art6.

Caney, S. 2008. 'Human rights, climate change and discounting', *Environmental Politics* **17**: 536–555.

Dietz, S., Hepburn, C. and Stern, N. 2008. 'Economics, ethics and climate change', in Basu, K. and Kanbur, R. (eds.) *Arguments for a Better World: Essays in Honour of Amartya Sen, vol. 2, Society, Institutions and Development*. Oxford: Oxford University Press, pp. 365–386.

Dow, K., Kasperson, R. E. and Bohn, H. 2006. 'Exploring the social justice implications of adaptation and vulnerability', in Adger, W. N., Paavola, J., Huq, S. and Mace, M. J. (eds.) *Fairness in Adaptation to Climate Change*. Cambridge: MIT Press, pp. 79–96.

Eriksen, S. H. and O'Brien, K. 2007. 'Vulnerability, poverty and the need for sustainable adaptation measures', *Climate Policy* **7**: 337–352.

Farber, D. A. 2007. 'Basic compensation for victims of climate change', *University of Pennsylvania Law Review* **155**: 1605–1634.

Gardiner, S. M. 2004. 'Ethics and global climate change', *Ethics* **114**: 555–600.

Hanemann, W. M. 2000. 'Adaptation and its measurement', *Climatic Change* **45**: 571–581.

Hansen, J. E. 2007. 'Scientific reticence and sea level rise', *Environmental Research Letters* **2**: doi 10.1088/1748–9326/2/2/024002.

Hoegh-Guldberg, O., Mumby, P. J., Hooten, A. J., Steneck, R. S., Greenfield, P., Gomez, E., Harvell, C. D., Sale, P. F., Edwards, A. J., Caldeira, K., Knowlton, N., Eakin, C. M., Iglesias-Prieto, R., Muthiga, N., Bradbury, R. H., Dubi, A. and Hatziolos, M. E. 2007. 'Coral reefs under rapid climate change and ocean acidification', *Science* **318**: 1737–1742.

Jagers, S. C. and Duus-Otterström, G. 2008. 'Dual climate change responsibility: on moral divergences between mitigation and adaptation', *Environmental Politics* **17**: 576–591.

Kooiman, J. (ed.) 1993. *Modern Governance*. London: Sage.

Lenton, T. M., Held, H., Kriegler, E., Hall, J. W., Lucht, W., Rahmstorf, S. and Schellnhuber, H. J. 2008. 'Tipping elements in the Earth's climate system', *Proceedings of the National Academy of Sciences of the USA* **105**: 1786–1793.

McKibben, B. 1999. *The End of Nature*, 2nd edn. London: Penguin.

Meze-Hausken, E. 2008. 'On the (im-)possibilities of defining human climate thresholds', *Climatic Change* **89**: 299–324.

Millennium Ecosystem Assessment 2005. *Ecosystems and Human Well-Being: Synthesis*. Washington, DC: Island Press.

Neumayer, E. 2007. 'A missed opportunity: the Stern Review on climate change fails to tackle the issue of non-substitutable loss of natural capital', *Global Environmental Change* **17**: 297–301.

O'Brien, K. and Leichenko, R. 2003. 'Winners and losers in the context of global change', *Annals of the Association of American Geographers* **93**: 99–113.

Okereke, C. 2008. 'Equity norms in global environmental governance', *Global Environmental Politics* **8**(3): 25–50.

Orr, J. C., Fabry, V. J., Aumont, O., Bopp, L., Doney, S. C., Feely, R. A., Gnanadesikan, A., Gruber, N., Ishida, A., Joos, F., Key, R. M., Lindsay, K., Maier-Reimer, E., Matear, R., Monfray, P., Mouchet, A., Najjar, R. G., Plattner, G.-K., Rodgers, K. B., Sabine, C. L., Sarmiento, J. L., Schlitzer, R., Slater, R. D., Totterdell, I. J., Weirig, M.-F., Yamanaka, Y. and Yool, A. 2005. 'Anthropogenic ocean acidification over the twenty-first century and its impact on calcifying organisms', *Nature* **437**: 681–686.

Paavola, J. and Adger, W. N. 2006. 'Fair adaptation to climate change', *Ecological Economics* **56**: 594–609.

Page, E. A. 2006. *Climate Change, Justice and Future Generations*. Cheltenham: Elgar.

Parry, M. L., Canziani, O. F., Palutikof, J. P., Hanson, C. E. and van der Linden, P. J. (eds.) 2007. *Climate Change 2007: Impacts, Adaptation and Vulnerability. Contribution of Working Group II to the Fourth Assessment Report of the Intergovernmental Panel on Climate Change*. Cambridge: Cambridge University Press.

Parry, M., Palutikof, J., Hansen, C. and Lowe, J. 2008. 'Squaring up to reality', *Nature Climate Change Reports* **2**: 68–70.

Pelling, M., High, C., Dearing, J. and Smith, D. 2008. 'Shadow spaces for social learning: a relational understanding of adaptive capacity to climate change within organisations', *Environment and Planning A* **40**: 867–884.

Rahmstorf, S. 2007. 'A semi-empirical approach to projecting future sea-level rise', *Science* **315**: 368–371.

Ramanathan, V. and Feng, Y. 2008. 'On avoiding dangerous anthropogenic interference with the climate system: formidable challenges ahead', *Proceedings of the National Academy of Sciences of the USA* **105**: 14 245–14 250.

Randerson, J. 2008. 'Climate change: prepare for global temperature rise of 4 °C, warns top scientist', *Guardian*: 7 August 2008.

Schellnhuber, H. J., Cramer, W., Nakicenovic, N., Wigley, T. and Yohe, G. (eds.) 2006. *Avoiding Dangerous Climate Change*. Cambridge: Cambridge University Press.

Schipper, L. and Pelling, M. 2006. 'Disaster risk, climate change and international development: scope for, and challenges to, integration', *Disasters* **30**: 19–38.

Schneider, S. H. and Lane, J. 2006. 'Dangers and thresholds in climate change and the implications for justice', in Adger, W. N., Paavola, J., Huq, S. and Mace, M. J. (eds.) *Fairness in Adaptation to Climate Change*. Cambridge: MIT Press, pp. 23–51.

Schneider, S. H., Kuntz-Duriseti, K. and Azar, C. 2000. 'Costing nonlinearities, surprises, and irreversible events', *Pacific and Asian Journal of Energy* **10**: 81–106.

Schneider, S. H., Semenov, S., Patwardhan, A., Burton, I., Magadza, C. H. D., Oppenheimer, M., Pittock, A. B., Rahman, A., Smith, J. B., Suarez, A. and Yamin, F. 2007. 'Assessing key vulnerabilities and the risk from climate change', in M. L. Parry, O. F. Canziani, J. P. Palutikof, C. E. Hanson and P. J. van der Linden (eds.) *Climate Change 2007: Impacts, Adaptation and Vulnerability. Contribution of Working Group II to the Fourth Assessment Report of the Intergovernmental Panel on Climate Change*. Cambridge: Cambridge University Press, pp. 779–810.

Soanes, C. and Stevenson, A. (eds.) 2008. *Concise Oxford English Dictionary*, 11th Edn, revised. Oxford: Oxford University Press.

Stern, N. 2007. *Economics of Climate Change: The Stern Review*. Cambridge: Cambridge University Press.

Stern, N. 2008. 'The economics of climate change', *American Economic Review: Papers and Proceedings* **98**(2): 1–37.

UN International Strategy for Disaster Reduction (UNISDR) 2008. *Links between Disaster Risk Reduction, Development and Climate Change*. Report prepared for the Commission on Climate Change and Development, Sweden.

Urwin, K. and Jordan, A. 2008. 'Does public policy support or undermine climate change adaptation? Exploring policy interplay across different scales of governance', *Global Environmental Change* **18**: 180–191.

Wilbanks, T. J. 2007. 'Scale and sustainability', *Climate Policy* **7**: 278–287.

Part I

Adapting to thresholds in physical
and ecological systems

2

Ecological limits of adaptation to climate change

Garry Peterson

Introduction

The human domination of Earth's ecosystems imposes ecological limits to the ability of humanity to adapt to climate change. Humanity already uses a substantial proportion of Earth's ecosystem services, and there are limits to the extent that humanity can increase this use further, particularly in the context of climate change. There are two reasons for this: first, human modification of ecosystems is decreasing the supply and undermining the reliability of many of these services, and climatic change is likely to amplify these changes. Second, the simplification of Earth's ecosystems has reduced the ability of ecosystems to self-regulate, which increases the possibilities for abrupt changes in ecological functioning. Abrupt changes are much more difficult to adapt to than gradual changes. In this chapter, I review evidence for regime shifts in agricultural ecosystems, and discuss how climate change could alter these regime shifts. Based on these examples, I argue that adaptation policies should consider and aim to reduce the ecological limits to adaptation by focusing on building ecological resilience in combination with climate change mitigation and adaptation.

Living in the Anthropocene

Climate change is occurring on a planet that is already dominated by humans. Humanity's modification of the Earth is the product of both intentional activities, such as converting wild ecosystems to agricultural ones to increase the supply of food, and unintentional activities, such as the production of climate change via the burning of fossil fuels. Human action has shifted biogeochemical and water flows,

Adapting to Climate Change: Thresholds, Values, Governance, eds. W. Neil Adger, Irene Lorenzoni and Karen L. O'Brien. Published by Cambridge University Press. © Cambridge University Press 2009.

25

modified land cover, increased erosion, homogenized biota, and introduced new chemical compounds into the biosphere. Scientists have proposed that this new human-dominated geological era should be named the Anthropocene, in recognition of human dominance in shaping the functioning of the Earth system (Steffen et al., 2004).

Estimating human domination

The scope of humanity's transformations of the biosphere is multifaceted, partially understood and difficult to summarize. Ecological footprint analysis and human appropriation of the ecological production have been used to assess human use of flows from ecosystems. For example, estimates of humanity's ecosystem footprint show that it exceeds the productive capacity of Earth and suggests that human civilization is unsustainably depleting the Earth's ecosystems (Wackernagel et al., 2002). Estimates of human appropriation of the products of photosynthesis suggest that humanity appropriates about 24% of the Earth's productivity (Haberl et al., 2007), and that this appropriation supports about half the metabolism of human civilization (most of the rest comes from fossil fuels) (Vitousek et al., 1986; Haberl, 2006). These estimates differ in some details, and they do not assess qualitative changes in ecosystems. Nonetheless, they show that humans currently control a large part of global ecological flows, and that there is simply not enough free ecological space for global civilization to greatly expand its use of ecosystems.

The reliance of human well-being on the continued supply of ecosystem services constrains future growth of humanity's impact on the biosphere. It is projected that between 2000 and 2050 the world's economy will grow between 3-fold and 6-fold (Millennium Ecosystem Assessment, 2005). However, such an increase in the use of the biosphere is not likely to be possible, due to lack of ecosystems into which humans can expand. For example, increasing human use of biomass energy while avoiding large CO_2 emissions or reducing food production requires that biomass be produced on abandoned agricultural lands. However, the productivity of these lands is low, suggesting that they could probably supply no more than 5% of current global energy needs (Field et al., 2008). Furthermore, humanity would find it difficult to deal with declines in the productivity of ecosystems. While in theory technology may be able to decouple human well-being from ecosystems, in practice improvements in technology have been used to extend human dominion over the biosphere. Although technological and social innovation have greatly increased the human benefits derived from the use of ecosystems, these gains in efficiency have expanded the size of the economy and resulted in increasing rather than decreasing use of global ecosystems (Millennium Ecosystem Assessment, 2005).

Consequently, reducing the human use of ecosystems could result in a substantial decline in human well-being.

Changes in ecological quality

Humanity's domination of the biosphere has changed its qualitative nature, in addition to appropriating a significant proportion of it. These changes have reduced the availability of multiple benefits that people receive from nature. The Millennium Ecosystem Assessment (MA), the first comprehensive global assessment of the state and possible futures of the benefits people receive from nature, assessed ecosystem services in four categories: *provisioning services,* the material that people extract from ecosystems such as food, water and forest products; *regulating services*, which modulate changes in climate and regulate floods, disease, wastes and water quality; *cultural services*, which consist of recreational, aesthetic and spiritual benefits; and *supporting services*, such as soil formation, photosynthesis and nutrient cycling, which underpin all these services (Millennium Ecosystem Assessment, 2005).

The MA found that the supply of ecosystem services is decreasing, at the same time that the demand for them is increasing. Of the ecosystem services assessed by the MA, 60% were declining, while the demands for over 80% of the services were increasing (Table 2.1). Part of the reason for these declines is that human modification of the biosphere to increase the supply of services that people receive from agro-ecosystems has led to declines in ecosystem services produced by *other* ecosystems. In particular, the increase in the provisioning services produced by human-dominated systems has been accompanied by declines in regulating ecosystem services. These declines influence not only resources available to people, but they also affect human well-being and security.

Ecosystem resilience and regime shifts

Regulating ecosystem services are important for maintaining the resilience of human-dominated ecosystems, for example by moderating the effects of extreme weather events, and similarly for maintaining the reliable production of ecosystem services in those and other ecosystems. The decline in these services not only exposes people to more environmental fluctuations, but also decreases the resilience of ecosystems. The decline in regulating ecosystem services has important consequences for the ability of humans to adapt to climate change.

Climate change, along with other forms of global change, is likely to increase the shocks and disturbances that ecosystems are exposed to in the future, while at the same time it is creating new ecological arrangements that influence the ability

Table 2.1 *Trends in supply and demand for ecosystem services. Supply is declining for most ecosystems, but mixed or increasing for some, in particular food production. Demand for all assessed ecosystem services are increasing except for services in italics that show mixed increase and decrease in demand, and those in bold italics declining demand*

	Provisioning ecosystem services	Regulating ecosystem services	Cultural ecosystem services
Increasing supply	Crops, livestock, aquaculture	Global climate regulation	
Mixed supply	Timber, *cotton*	Water flow regulation, disease control	Recreation and ecotourism
Declining supply	*Fuelwood*, genetic resources, biochemical resources, fresh water, **capture fisheries, wild foods**	Local climate regulation, erosion control, water quality regulation, pest control, pollination, natural hazard regulation	Spiritual and religious values, aesthetic values

Source: Adopted from Millennium Ecosystem Assessment (2005).

of ecosystems to respond. In other words, human simplification of the world's ecosystems is reducing the ability of ecosystems to continue to function reliably in the face of shocks and changes. The MA found that declines of regulating ecosystem services have been substantially driven by human modifications of ecosystems to increase agricultural ecosystem services (Millennium Ecosystem Assessment, 2005). Furthermore, agriculture is expected to expand due to increases in world population, meat consumption and demand for the use of biofuels. This expansion is expected to occur while climate change is altering precipitation, soil moisture and runoff. The combination of increases in agricultural intensity and extent combined with a shift in global hydrology due to climate change increases the possibility of surprising change being produced in agro-ecosystems. Understanding the consequences of these changes requires understanding ecological resilience.

Ecological resilience is the ability of an ecosystem to persist despite disruption and change (Holling, 1973; Peterson et al., 1998). Ecosystem dynamics are defined by both internal dynamics, such as vegetation growth, and external forces, such as precipitation. Regime shifts occur when the external forces exceed the resilience of a system, or when gradual internal changes decrease a system's resilience to the extent that it reorganizes, shifting from being organized around one set of mutually reinforcing processes to another. These shifts

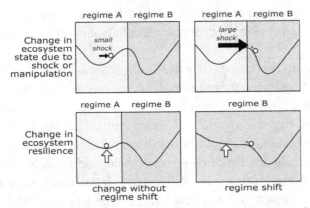

Figure 2.1 Regime shifts can be caused by either shocks or changes in ecosystem resilience.

can occur from either changes in the structure of a system or changes in the way it operates. The differences are illustrated in Figure 2.1, which shows how the feedbacks that maintain a system can be represented as the shape of a landscape, with the condition or state of a system represented as a ball. Multiple valleys in the landscape represent the potential for alternative regimes. A system can be pushed into another regime by an external shock, such as a flood, or by a change in ecological structure, such as deforestation. Such changes push the state of the system from one regime into another. The system will remain in this regime unless another large shock pushes it back.

Alternatively, a system can move into another regime if the forces structuring the system also change, for example through changes in the ability of soil to hold moisture. If these changes eliminate the feedback processes that define a regime, then the system will move to another regime. These changes are often distinguished in the resilience literature as changes in either fast or slow variables: shocks occur much faster than the structural changes in how the system operates. In nature these two types of shifts are not so neatly separated. As the resilience of a regime declines, ever-smaller shocks can push it into another regime. Therefore, changes in slow variables can make it easier for changes in fast variables to produce a shift in ecological regimes.

Climate change and regime shifts

Climate change researchers have focused on *climate* regime shifts – the possibilities of abrupt climate change due to positive feedback processes involving processes such as arctic sea ice, tundra vegetation or thermohaline circulation (Lenton et al., 2008). However, many of the impacts of climate change will be realized through their impacts on ecosystem services that people rely upon. Consequently climate change and ecological changes can individually or in combination produce regime shifts.

The relationship between climate and ecological change varies, depending upon the dynamics of both climate and ecological systems. Human modification of eco-systems does not have a simple relationship to resilience to climate change. To simplify things I consider gradual and abrupt climate change, as well as ecological change that tracks non-linear shifts in climate change versus ecological change. If ecosystems exhibit resilience they will respond non-linearly to climate change. A small amount of climate change will result in little change to an ecosystem, but once a threshold is passed the system will undergo a regime shift and reorganize into a new state. When climate change is abrupt, it can drive a regime shift in a system that responds gradually to climate change. However, local ecological fac-tors shape the resilience of a system to climate change. They may either increase or decrease resilience, depending upon the situation. What is important here is that an abrupt change in an ecosystem can arise from either ecological dynamics or an abrupt climate change (Figure 2.2).

Ecological change, or human ecological engineering, can alter an ecosys-tem's response to climate change. In particular, human actions can increase or decrease the resilience of an ecosystem to climate change. Consequently, human interventions in the biosphere can shape the ability of ecosystems to adapt to climate change. Regulating ecosystem services, such as pollination or climate moderation, help maintain the resilience of ecosystems, and these services are declining worldwide. Consequently, one would expect, in general, the resilience of global ecosystems to be decreasing. Below, I explore this interaction using the example of how agriculture's modifications of hydrological flows alters ecological resilience.

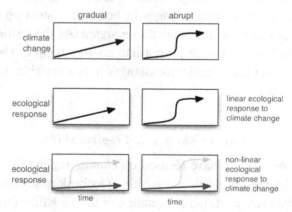

Figure 2.2 Possible relationships between climate change and ecological change.

Climate change and agricultural regime shifts

In terms of limiting the ability of humans to adapt to climate change, it is the trans-formation of much of the Earth's terrestrial surface to agricultural lands that is likely to be the most substantial. Not only does it represent a massive modification of Earth's ecological functioning, but its continuation is considered to be essential for human well-being. Agricultural ecosystems cover an estimated 40% of Earth's surface, but they also create impacts on other ecosystems. One of the major ways that agriculture affects distant ecosystems is through its modification of global water flows. Agriculture does this in some obvious ways. About two-thirds of the water removed from rivers is used for irrigation (Scanlon et al., 2007), and the water that flows from agricultural lands into rivers and lakes carries with it agri-cultural fertilizers that reduce water quality in aquatic ecosystems (Bennett et al., 2001; Galloway et al., 2004). However, less obviously, agriculture alters atmos-pheric flows of water due to the impacts of irrigation and deforestation on global evapotranspiration (Gordon et al., 2005). It is via both direct and indirect impacts on other ecosystems that agriculture has increased the supply of desired ecosystem services, such as food and fibre, but at the same time led to unintended declines in non-agricultural ecosystem services, such as fisheries, flood regulation and down-stream recreational opportunities (Millennium Ecosystem Assessment, 2005).

Managing trade-offs is difficult due to the social and ecological complexities involved, and managing them will be made even more difficult in a changing cli-mate. However, while these changes would be difficult to cope with even if gradual and predictable, ecological research suggests that declines in ecosystem services may also be abrupt and surprising – and difficult to reverse. Abrupt changes in eco-system services can occur due to shifts between different ecosystem regimes, and this presents a substantial challenge to ecosystem management and development goals (Gunderson and Holling, 2002; Folke et al., 2004).

Three parts of the hydrological cycle where agriculture can trigger regime shifts

There is evidence for the presence of numerous regime shifts in ecosystems, includ-ing hydrologically mediated regime shifts in agriculture (see Gordon et al., 2008 for a review of these shifts). Agriculture is strongly connected to other ecosystems due to its influence on water flows. The hydrological cycle connects different ecosys-tems because runoff, groundwater and evapotranspiration move materials among different ecosystems, and alter energy balances in landscapes. Consequently, changes in hydrological flows can produce changes that are distant from the loca-tion where the flow is changed. These changes can lead to regime shifts at different locations in the hydrological cycle. Locally, agriculture can change infiltration and soil moisture to produce terrestrial regime shifts. In watersheds, agriculture's alterations of runoff quantity and quality can produce aquatic regime shifts in downstream ecosystems. Finally, agriculture's interactions with atmospheric moisture can produce climatic regime shifts.

Locally, vegetation and soil water interact through effects on infiltration, soil water holding capacity and root water uptake to produce *vegetation patchiness, salinization* and *soil structure* regime shifts. Vegetation patchiness interacts with the flow of water to concentrate water and nutrients (Rietkerk et al., 2004; Peters et al., 2006), fire, grazing or precipitation changes can trigger regime shifts. Deforestation that causes a rise in the water table can cause salinization (Cramer and Hobbs, 2005; Anderies et al., 2006). Soil structure regime shifts can occur if soil moisture holding capacity is decreased by soil compaction and crusting to an extent that it reduces the capacity of the soil to recover its moisture holding capacity during wetter periods (Bossio et al., 2007).

Regionally, agriculturally driven change in water flows, nutrient levels and sediment loads can produce regime shifts in downstream aquatic systems. *Freshwater eutrophication* (Carpenter, 2005) and *hypoxic zones* are produced when nutrients used in agriculture accumulate and are recycled within a region (Diaz and Rosenberg, 2008). Accumulation of sediments in rivers can cause shifts in *river channel position* (Hooke, 2003).

At ecosystem scales, changes in vegetation can alter precipitation in ways that can possibly produce regime shifts (Higgins et al., 2002). Theory suggests that regime shifts can occur if vegetation cover responds non-linearly to changes in precipitation and vegetation has a sufficiently strong effect on precipitation that it can alter the amount of vegetation cover (Sternberg, 2001; Scheffer et al., 2005). These include larger-scale regime shifts such as *wet savanna systems* and *dry savanna forest* regime shifts (Hutyra et al., 2005), *cloud or fog forests* regime shifts (Dawson 1998; del-Val et al., 2006) and *shifts in monsoon behaviour* (Higgins et al., 2002; Zickfeld et al., 2005). These regime shifts are shown in Table 2.2, along with an assessment of the evidence supporting their existence (Gordon et al., 2008).

Climate change interacts with agricultural change to produce regime shifts. Regime shifts can be triggered by the interaction of changes internally in a system with changes in external drivers (Figure 2.3). Using this scheme climate and non-climate processes that alter the resilience of these agriculture–water regime shifts can be identified. In many cases, climate drivers interact with other external drivers, such as agricultural practices, to determine a system's resilience to a regime shifts. Some processes are jointly produced by agriculture practices and climate, such as erosion, fire and water balance. These processes are key to understanding adaptation to climate change in the case of agricultural regime shifts (Table 2.2).

Table 2.2 *Climate and non-climate factors that drive agricultural and water regime shifts*

Regime shift[a]	Internal slow variable	Climate shocks	Climate resilience	Strongly coupled	Non-climate shock	Non-climate resilience
Freshwater eutrophication	Sediment and watershed soil phosphorus	Extreme rainfall	Temperature	Erosion	Soil disturbance	Nutrient/soil management
Coastal hypoxic zones	Aquatic biodiversity	Extreme rainfall, storms	Temperature	Erosion	River dredging	Nutrient/soil management, fisheries management
River channel position	River channel shape	Extreme rainfall	Water balance, average precipitation	Erosion	Canal construction	River dredging, levee building
Vegetation patchiness	Vegetation pattern	Drought	Average precipitation	Fires	Shrub removal	Grazing, changes in net primary production
Salinization	Water table salt accumulation	Wet years	Average precipitation	Water balance	Irrigation?	Increased water leakage in the soil, irrigation
Soil structure	Soil organic matter	Drought, dry spells	Proportion of rainfall in extreme events	Fire	Biomass removal	Loss of fallows, soil management
Wet savanna–dry savanna	Moisture recycling, energy balance	Drought	Rainfall	Fire	Agricultural conversion	Reductions in net primary production
Monsoon circulation	Energy balance, advective moisture flows	Unknown	Temperature, precipitation, change in off-shore sea surface temperature	Evapotranspiration	Unknown	Land cover, irrigation
Forest–savanna	Moisture recycling, energy balance	Droughts	Rainfall	Fire	Agricultural conversion	Reductions in net primary production
Cloud forest	Leaf area	Drought	Fog frequency	Fire	Deforestation	Unknown

[a]Bold, italics and plain text are used to identify the different types of regime

33

Regime shifts vary in their response to climate change and ecological modification. The hydrological consequences of climate change interact with agriculturally induced changes in hydrology to shape the vulnerability of a system to regime shifts. For example, reduced soil organic matter, a critical slow variable, can lead to decreased water-holding capacity. Less water in the soil reduces capacity to cope with a high frequency of dry spells (Bossio et al., 2007; Enfors and Gordon, 2007). In the Mississippi River basin, increasing precipitation in the late autumn and spring influences nitrogen runoff, which expands the size of the hypoxic zone in the Gulf of Mexico (Donner and Scavia, 2007). In the other direction, the recent drought in the Murray–Darling basin in Australia has reduced the rate of dryland salinization expansion because the drought has kept water tables low; consequently, the return of wetter conditions could have disastrous consequences (Anderies et al., 2006).

Climatic shocks such as extreme rainfall or drought can trigger different regime shifts. The vulnerability of agriculture–water regime shifts to climate change can be summarized by identifying how regime shifts vary in their response to changes and variability in precipitation and average temperature. For example, extreme rainfall can trigger regime shifts such as river channel shifts, salinization and hypoxia. Increases in average temperature can alter the way that an ecosystem responds, for example, by reducing risk of salinization (assuming evapotranspiration also increases), or by increasing risks of hypoxia (assuming that algal growth is stimulated). Common non-climate factors that decrease regime shift resilience include shocks such as soil disturbance and land clearing, resilience-shaping activities such as local soil management, landscape management and the management of entire watersheds. The ways that regime shifts respond to shocks and management varies. However, by identifying the way that shocks and slow changes alter the resilience of a regime, the potential for different types of climate change and ecosystem management to increase or decrease the risk of a regime shift can be assessed (Figure 2.3).

Agriculture–aquatic system regime shifts occur over a wide range of scales and vary from years to millennia in their reversibility (Figure 2.4). For example, freshwater eutrophication is often irreversible, or only reversible after massive reductions of phosphorus inputs for decades or longer due to internal cycling of phosphorus within the lake system and accumulation of phosphorus in watershed soils (Carpenter, 2005). Agriculture–soil regime shifts tend to operate at the field to landscape scale with varying degrees of reversibility. Although soil structure regime shifts occur at small spatial scales, their impact can cascade across the landscape, as exemplified with the development of the Dust Bowl in the 1930s in the USA. The Dust Bowl started at the scale of individual fields and expanded non-linearly to impact the agricultural regions of the USA (Peters et al., 2004).

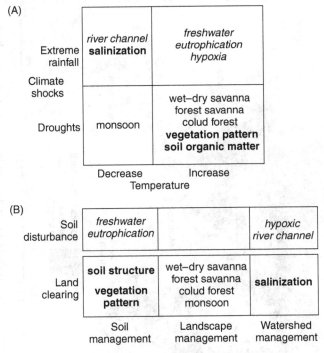

Figure 2.3 Vulnerability and resilience of regime shifts.

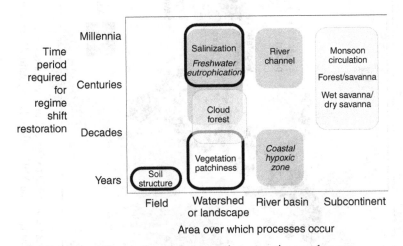

Figure 2.4 Regime shift creation and restoration over time and space.

Broad-scale weather patterns caused individual fields to become highly connected, creating massive dust storms that non-linearly aggravated the situation (Peters et al., 2004). Finally, agriculture–atmosphere regime shifts tend to operate at relatively large spatial and temporal scales, although uncertainty remains about the important

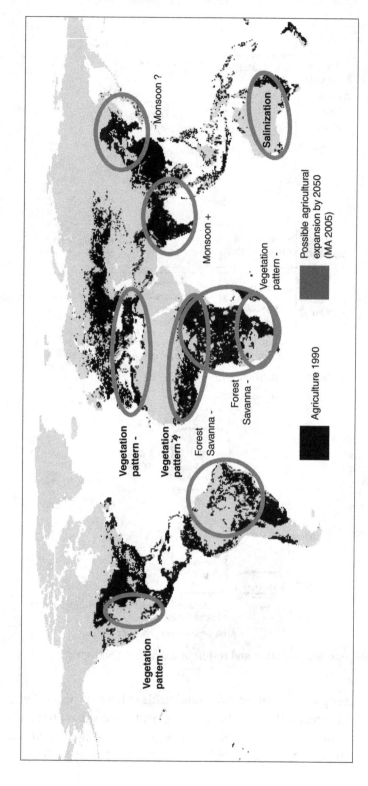

Figure 2.5 Regions vulnerable to agriculture–water regime shifts superimposed on areas of agriculture and projections of agriculture expansion by 2050 from the Millennium Ecosystem Assessment (2005) scenarios.

scales of land–atmosphere feedbacks. For example, while evapotranspiration from the forests is the main source of water for precipitation in the Amazon, patchy regional deforestation which increases landscape heterogeneity can contribute to an increase in rainfall through the establishment of anomalous convective circulations, while large-scale deforestation would substantially decrease precipitation even in very distant places (D'Almeida et al., 2007). Consequently, the activities necessary to avoid or respond to these regime shifts vary across scales, and will be influenced by different types of institutions, governance and infrastructure.

Identifying processes that strongly connect climate and the management of agricultural landscapes presents places where people can intervene to manage or monitor changes in resilience. Fire and erosion are particularly important as these processes extend across scales and connect different types of regime shifts, as mentioned in the Dust Bowl example. Fire is a key process mediating vegetation patchiness, wet–dry savanna, forest savanna, cloud forest regime shifts, whereas erosion mediates regime shifts in soil structure, river channel, freshwater eutrophication and coastal hypoxia.

Mapping the vulnerability of regions to regime shifts is difficult because processes operating at different scales regulate these shifts. However, Figure 2.5 shows a rough assessment of the areas of the world that are vulnerable to large-scale regime shifts, in a world of both agricultural expansion and climate change. This global map cannot identify the many specific areas that are vulnerable to freshwater eutrophication, or coastal hypoxia (Diaz and Rosenberg, 2008). It can, however, tentatively identify where more regional assessments could be conducted, and it draws attention to the need for better assessments of vulnerability to regime shifts.

Summary

This chapter emphasizes that the majority of the impacts of climate change on human well-being may occur through changes in ecosystem services. This situation is problematic, because scientists are not able to predict in detail the ecological consequences of global environmental change, and human civilization is rapidly degrading the ability of ecosystems to produce services, in particular the regulating services that help humans and ecosystems cope with shocks and change. However, the restoration or enhancement of ecosystem services also offer a mechanism of adapting to climate change and increasing the supply of ecosystem services.

The possibility of ecosystems undergoing abrupt regime shifts complicates adaptation to climate change. Regime shifts are often triggered by shocks, but their resilience to shocks is controlled by slow ecological processes. A resilient ecosystem can undergo a regime shift if it experiences large shocks, while a non-resilient

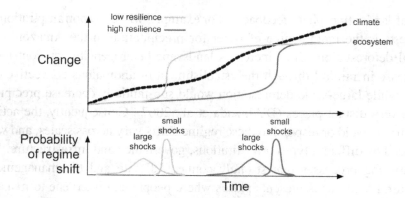

Figure 2.6 Probability of regime shifts due to climate change depends upon both the occurrence of shocks and ecosystem resilience.

ecosystem may be unable to persist if it experiences only small shocks. In terms of ecological adaptation to climate change, reducing the types of shocks that a system is exposed to as well as increasing its resilience can alter the situations under which a system experiences a regime shift. Managing these processes may enable regime shifts to be delayed or prevented altogether (Figure 2.6). Some of the key climatic and agricultural processes regulating agriculture–water regime shifts were described above, including how they alter the resilience of different types of regime shifts. It must be emphasized that the regulation or management of these processes is a place where people can intervene to increase or decrease the resilience of these systems.

In conclusion, research on climate change impacts and adaptation needs to identify potential situations of 'dangerous ecological change' and seek to better understand how to avoid these situations. Ecologists believe that there is substantial potential for ecological degradation to affect the global economy and human well-being. There is a need to better understand the places and situations in which these damages can occur, as well as the policies and practices necessary to avoid them. The potential for abrupt ecological change can, in fact, be used to improve human well-being on a changing planet. This requires analysis of human-dominated ecosystems and large-scale ecological dynamics. Creating this understanding requires better connecting ecological, climatological and social data across a range of scales. Such work has begun, but may need to be greatly accelerated if human civilization is to thrive in the twenty-first century.

Acknowledgements

This paper builds upon work conducted in collaboration with Line Gordon and Elena Bennett and was supported by a Canada Research Chairs Program.

References

Anderies, J. M., Ryan, P. and Walker, B. H. 2006. 'Loss of resilience, crisis, and institutional change: lessons from an intensive agricultural system in southeastern Australia', *Ecosystems* **9**: 865–878.

Bennett, E. M., Carpenter, S. R. and Caraco, N. F. 2001. 'Human impact on erodable phosphorus and eutrophication: a global perspective', *BioScience* **51**: 227–234.

Bossio, D., Critchley, W., Geheb, K., Van Lynden, G. and Mati, B. 2007. 'Conserving land – protecting water', in Molden, D. (ed.) *Water for Food, Water for Life: A Comprehensive Assessment of Water Management*. London: Earthscan, pp. 551–583.

Carpenter, S. R. 2005. 'Eutrophication of aquatic ecosystems: bistability and soil phosphorus', *Proceedings of the National Academy of Sciences of the USA* **102**: 10002–10005.

Cramer, V. A. and Hobbs, R. J. 2005. 'Assessing the ecological risk from secondary salinity: a framework addressing questions of scale and threshold responses', *Austral Ecology* **30**: 537–545.

D'Almeida, C., Vorosmarty, C. J., Hurtt, G. C., Marengo, J. A., Dingman, S. L. and Keim, B. D. 2007. 'The effects of deforestation on the hydrological cycle in Amazonia: a review on scale and resolution', *International Journal of Climatology* **27**: 633–647.

Dawson, T. E. 1998. 'Fog in the California redwood forest: ecosystem inputs and use by plants', *Oecologia* **117**: 476–485.

Del-Val, E., Armesto, J. J., Barbosa, O., Christie, D. A., Gutierrez, A. G., Jones, C. G., Marquet, P. A. and Weathers, K. C. 2006. 'Rain forest islands in the Chilean semiarid region: fog-dependency, ecosystem persistence and tree regeneration', *Ecosystems* **9**: 598–608.

Diaz, R. J. and Rosenberg, R. 2008. 'Spreading dead zones and consequences for marine ecosystems', *Science* **321**: 926–929.

Donner, S. D. and Scavia, D. 2007. 'How climate controls the flux of nitrogen by the Mississippi River and the development of hypoxia in the Gulf of Mexico', *Limnology and Oceanography* **52**: 856–861.

Enfors, E. I. and Gordon, L. J. 2007. 'Analysing resilience in dryland agro-ecosystems: a case study of the Makanya catchment in Tanzania over the past 50 years', *Land Degradation and Development* **18**: 680–696.

Field, C. B., Campbell, J. E. and Lobell, D. B. 2008. 'Biomass energy: the scale of the potential resource', *Trends in Ecology and Evolution* **23**: 65–72.

Folke, C., Carpenter, S., Walker, B., Scheffer, M., Elmqvist, T., Gunderson, L. and Holling, C. S. 2004. 'Regime shifts, resilience, and biodiversity in ecosystem management', *Annual Review of Ecology, Evolution and Systematics* **35**: 557–581.

Galloway, J. N., Dentener, F. J., Capone, D. G., Boyer, E. W., Howarth, R. W., Seitzinger, S. P., Asner, G. P., Cleveland, C. C., Green, P. A., Holland, E. A., Karl, D. M., Michaels, A. F., Porter, J. H., Townsend, A. R. and Vorosmarty, C. J. 2004. 'Nitrogen cycles: past, present, and future', *Biogeochemistry* **70**: 153–226.

Gordon, L. J., Steffen, W., Jonsson, B. F., Folke, C., Falkenmark, M. and Johannessen, A. 2005. 'Human modification of global water vapor flows from the land surface', *Proceedings of the National Academy of Sciences of the USA* **102**: 7612–7617.

Gordon, L. J., Peterson. G. D. and Bennett, E. M. 2008. 'Agricultural modifications of hydrological flows create ecological surprises', *Trends in Ecology and Evolution* **23**: 211–219.

Gunderson, L. and Holling, C. (eds.) 2002. *Panarchy: Understanding Transformations in Human and Natural Systems*. Washington, DC: Island Press.

Haberl, H. 2006. 'The global socioeconomic energetic metabolism as a sustainability problem', *Energy* **31**: 87–99.

Haberl, H., Erb, K. H., Krausmann, F., Gaube, V., Bondeau, A., Plutzar, C., Gingrich, S., Lucht, W. and Fischer-Kowalski, M. 2007. 'Quantifying and mapping the human appropriation of net primary production in earth's terrestrial ecosystems', *Proceedings of the National Academy of Sciences of the USA* **104**: 12942–12945.

Higgins, P. A. T., Mastrandrea, M. D. and Schneider, S. H. 2002. 'Dynamics of climate and ecosystem coupling: abrupt changes and multiple equilibria', *Philosophical Transactions of the Royal Society of London B* **357**: 647–655.

Holling, C. S. 1973. 'Resilience and stability of ecological systems', *Annual Review of Ecology and Systematics* **4**: 1–23.

Hooke, J. 2003. 'River meander behaviour and instability: a framework for analysis', *Transactions of the Institute of British Geographers* **28**: 238–253.

Hutyra, L. R., Munger, J. W., Nobre, C. A., Saleska, S. R., Vieira, S. A. and Wofsy, S. C. 2005. 'Climatic variability and vegetation vulnerability in Amazonia', *Geophysical Research Letters* **32**: L24712.

Lenton, T. M., Held, H., Kriegler, E., Hall, J. W., Lucht, W., Rahmstorf, S. and Schellnhuber, H. J. 2008. 'Tipping elements in the Earth's climate system', *Proceedings of the National Academy of Sciences of the USA* **105**: 1786–1793.

Millennium Ecosystem Assessment 2005. *Ecosystems and Human Well-Being: Synthesis*. Washington, DC: Island Press.

Peters, D. P. C., Pielke, R. A., Bestelmeyer, B. T., Allen, C. D, Munson-McGee, S. and Havstad, K. M. 2004. 'Cross-scale interactions, nonlinearities, and forecasting catastrophic events', *Proceedings of the National Academy of Sciences of the USA* **101**: 15130–15135.

Peters, D. P. C., Bestelmeyer, B. T., Herrick, J. E., Fredrickson, E. L., Monger, H. C. and Havstad, K. M. 2006. 'Disentangling complex landscapes: new insights into arid and semiarid system dynamics', *BioScience* **56**: 491–501.

Peterson, G. D., Allen, C. R. and Holling, C. S. 1998. 'Ecological resilience, biodiversity and scale', *Ecosystems* **1**: 6–18.

Rietkerk, M., Dekker, S. C., de Ruiter, P. C. and Van de Koppel, J. 2004. 'Self-organized patchiness and catastrophic shifts in ecosystems', *Science* **305**: 1926–1929.

Scanlon, B. R., Jolly, I., Sophocleous, M. and Zhang, L. 2007. 'Global impacts of conversions from natural to agricultural ecosystems on water resources: quantity versus quality', *Water Resources Research* **43**: doi 10.1029/2006WR005486.

Scheffer, M., Holmgren, M., Brovkin, V. and Claussen, M. 2005. 'Synergy between small- and large-scale feedbacks of vegetation on the water cycle', *Global Change Biology* **11**: 1003–1012.

Steffen, W., Sanderson, A., Tyson, P. D., Jager, J., Matson, P. M., Moore, I. B., Oldfield, F., Richardson, K., Schnellnhuber, H. J., Turner, B. L. and Wasson, R. J. 2004. *Global Change and the Earth System: A Planet under Pressure*. New York: Springer-Verlag.

Sternberg, L. D. L. 2001. 'Savanna-forest hysteresis in the tropics', *Global Ecology and Biogeography* **10**: 369–378.

Vitousek, P. M., Ehrlich, P. R., Ehrlich, A. H. and Matson, P. A. 1986. 'Human appropriation of the products of photosynthesis', *BioScience* **36**: 368–373.

Wackernagel, M., Schulz, N. B., Deumling, D., Linares, A. C., Jenkins, M., Kapos, V., Monfreda, C., Loh, J., Myers, N., Norgaard, R. and Randers, J. 2002. 'Tracking the ecological overshoot of the human economy', *Proceedings of the National Academy of Sciences of the USA* **99**: 9266–9271.

Zickfeld, K., Knopf, B., Petoukhov, V. and Schellnhuber, H. J. 2005. 'Is the Indian summer monsoon stable against global change?', *Geophysical Research Letters* **32**: 10.1029/2005GL022771.

3

Adapting to the effects of climate change on water supply reliability

Nigel W. Arnell and Matthew B. Charlton

Introduction

Climate change is expected to produce higher temperatures, drier summers and wetter winters across southern England. Reductions in water availability are expected as a consequence (Arnell, 2004) with direct abstractions becoming less reliable during summer and more seasonal, higher intensity rainfall producing high runoff and less water able to percolate into aquifers (Environment Agency, 2005). In an area already facing water deficits and supply challenges (Environment Agency, 2007a), and with increasing population demands, adaptation in the short term (to 2030) is necessary. With water resources in south-east England under increasing pressure, water companies and their regulators are exploring options to adapt not only to altered demands, but also to the challenge of climate change. The water supply industry in England and Wales is well aware of the challenge of climate change, and methodologies exist to both estimate the effects of climate change and support adaptation decisions (Arnell and Delaney, 2006). The industry has also identified a wide range of options for addressing the supply–demand imbalance, covering both supply-side and demand-side measures.

However, there are specific barriers to the implementation of each option, and some generic constraints on the ability of water supply companies to adapt to a changing climate. This chapter presents preliminary results from an assessment of the barriers to adaptation to water supply shortage in a case study catchment in south-east England with multiple supply companies. The investigation applies a conceptual framework, which distinguishes between generic barriers affecting the ability of supply companies to make adaptation decisions, and specific barriers to the implementation of each option. The preliminary analysis suggests that

Adapting to Climate Change: Thresholds, Values, Governance, eds. W. Neil Adger, Irene Lorenzoni and Karen L. O'Brien. Published by Cambridge University Press. © Cambridge University Press 2009.

whilst there is a widespread awareness of the challenge of climate change, and a conceptual understanding of the need for adaptation, some of the generic barriers that will affect detailed evaluations and actual adaptation decisions have yet to be approached. The analysis also shows that different individual adaptation options are assessed differently by different stakeholders, and that there are differences in the barriers to adoption between supply-side and demand-side measures. First, however, this chapter develops the general conceptual framework for the characterisation of the barriers to adaptation used in the study.

Barriers to adaptation: a conceptual framework

The proposed conceptual framework for the characterisation of the barriers to adaptation in a particular place (Figure 3.1) identifies two broad types of barrier. *Generic barriers* influence the way the adaptation challenge is defined and potential adaptation responses identified and selected. They can be considered cognitive and information/knowledge barriers and affect the capacity to acknowledge or recognise the problem and the solutions. *Specific barriers* relate to individual adaptation options and influence the capacity to carry out the solutions.

There are five generic barriers. The first relates to the identification of the need for adaptation (in organisational learning terms, the identification of a signal and the interpretation of the signal in terms of adaptation: Berkhout et al., 2006). The second influences the extent to which the need for adaptation can be specified in terms which inform adaptation decisions. This will be a function of the characteristics of available climate scenarios (the variability between scenarios, for example, and the extent to which they represent changes in relevant climate drivers), the ability to translate these scenarios into potential impacts on the system of

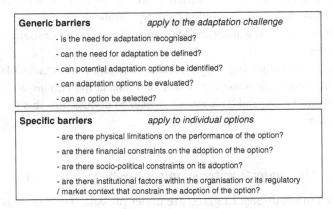

Figure 3.1 Characterisation of two types of barriers to adaptation.

interest, and local geographical circumstances. A third potential generic barrier is the identification of feasible adaptation options. Institutional competences or preferences may mean that certain options are not identified: Berkhout et al. (2006) defined the concept of 'adaptation space' to characterise the options perceived to be available to an organisation. The final two generic barriers constrain the ability of an adapting organisation first to evaluate potential options, and second to select and monitor a strategy. Evaluation and selection requires organisations to have procedures to articulate knowledge and codify practices, and for monitoring of feedback (Berkhout et al., 2006). These procedures may be internally defined, or may be imposed by external regulators; they may facilitate or constrain adaptation.

There are four types of specific barriers relating to individual adaptation options. Physical barriers are constraints on the performance of an adaptation option. There may be technical constraints, for example, to the amount of climate change that a specific measure can cope with. Financial barriers refer not only to the absolute cost of an option, but also to the ability of the organisation to raise funds to cover the costs; this will be a function of the wealth of the organisation and its access to resources. Socio-political barriers include the attitudes and reactions of stakeholders, affected parties and pressure groups to individual adaptation options. Finally, the characteristics of the individual organisation may affect its ability to implement a specific option, and the regulatory or market context may constrain specific choices. Institutional barriers may exist at both the generic and specific level and illustrate some degree of overlap between the two groups.

The research seeks to assess five propositions, drawn from the above discussion, and this chapter presents a preliminary assessment of these:

(1) The availability of credible climate scenarios is a generic barrier to adaptation.
(2) Specific barriers to the implementation of individual adaptation options in southern England are largely financial, socio-political and institutional, rather than physical.
(3) Different stakeholder groups rate different adaptation options, and barriers to their implementation, differently, reflecting their organisational objectives.
(4) The current institutional framework for water management constrains adaptation to climate change.
(5) Uncertainty in the future impacts of climate change on resource availability affects the feasibility and implementation of different adaptation options differently. Adger et al. (2009) and Dessai and colleagues (Chapter 5) discuss how this uncertainty may limit adaptation.

The context: water supply in southern England

Water resources in south-east England are under pressure from increasing demand and increasing environmental obligations. Five independent private-sector water supply companies (six until 2007, prior to a merger between two companies) provide water to the region. The companies are subject to environmental regulation by the Environment Agency (who issue and control licences to abstract water subject to regional and catchment water resources strategies) and economic regulation by Ofwat (who control prices to customers and hence determine investments). Ofwat sets company price limits every five years in its Periodic Review process, which requires companies to make five-year projections of investment requirements. The fifth Periodic Review (PR09) is currently under way and will be completed in 2009. The Periodic Review requires water companies to produce 25-year Water Resources Management Plans as the basis for their investment strategies. Industry-standard methods are used to project future resource availability and demand over this timescale, through the Environment Agency's Water Resource Planning Guidelines (Environment Agency, 2007b). Draft Water Resources Management Plans were published in April and May 2008.

A broader context for water resources planning is set by the implementation of the European Union Water Framework Directive, and by the central government Department for Environment, Food and Rural Affairs (Defra). Defra's overarching vision for English water resources in 2030 was published in 2008 (Defra, 2008), and embraces enhanced environmental quality, sustainable use of water resources, reduced greenhouse gas emissions, and embedded adaptation to climate change and other pressures. The vision also reaffirms strong government support for a twin-track approach to water resources management, combining both supply-side measures and demand-side measures. It sets out procedures and policies by which Defra will influence the actions of regulators, water companies and consumers.

Each water supply company manages its water resources via a number of 'water resource zones' (there are 15 separate 'water resource zones' in East Sussex and Kent). Some of the zones are interconnected, but the zones do not overlap. The zones do not necessarily correspond to catchment or administrative boundaries and largely reflect the historical evolution of individual supply companies. Major catchments are therefore divided amongst water resource zones managed by different supply companies.

Across south-east England as a whole, approximately 70% of public water supplies are taken from groundwater, with the remaining 30% taken either directly from rivers or from supply reservoirs. Some of these reservoirs are fed by transfers from several source catchments, and some reservoirs are used to support direct river abstractions downstream. The mix of sources varies considerably between individual water resource zones, reflecting underlying geology.

Table 3.1 summarises future resource availability in the eastern part of south-east England (Water Resources in the South East (WRSE), 2006). With no new resource developments and a medium population growth assumption, there would be a deficit

Table 3.1 *Supply–demand deficit in eastern south-east England by 2025*

	Supply–demand (Ml/d) by 2025		
	No new resources	Some resource developments	Company-proposed resource developments
Medium population growth			
Some demand management	–30	–5	+50
Much demand management	–20	+5	+60
High population growth			
Some demand management	–75	–50	+5
Much demand management	–58	–33	+22

Dry year annual average water resources: no allowance for climate change (WRSE, 2006).

of between 20 and 30 Ml/d (approximately 3–5% of current supply), depending on the assumed effect of demand management measures. If all the resource developments proposed by the water companies were implemented, this would turn into a surplus of between 50 and 60 Ml/d; a 'compromise' involving some new resource development would mean that supplies and demand were approximately in balance. A high population growth assumption obviously increases demand and the risk of a deficit.

The calculations in Table 3.1 incorporate all the 'feasible' supply-side options (as identified by water supply companies) and very optimistic assumptions about the implementation and effectiveness of demand-side measures. In practice, of course, there is much controversy and much discussion in the water industry and other stakeholders around these assumptions. The Campaign to Protect Rural England, for example, complains of 'a disproportionate emphasis on the creation of additional reservoir capacity' (Warren, 2007), whilst the WRSE report itself notes that 'some of the water efficiency scenarios considered ... are very challenging' (WRSE, 2006). Despite such differences, it is important to note that, even without climate change, it is acknowledged that matching supply to demand will be a great challenge and water companies are addressing the changes and uncertainties through their Water Resources Management Plans.

Table 3.1 does not include explicitly the effects of climate change on supply (it is included in the effects on demand). A reduction in reliable supplies of 5% corresponds to a reduction of around 30 Ml/d, increasing still further the deficits in Table 3.1.

The case study: methods

The study is focused in one catchment – the Medway in Kent – which is covered by water resources zones operated by three water supply companies (prior to a merger in 2007, the water resources zones were operated by four companies). Water is exported out of the catchment to other parts of the south-east, and transferred within zones within the catchment. Sixty per cent of public water supplies for the catchment are taken from surface water sources, including rivers regulated by upstream reservoirs (Environment Agency, 2005); agriculture abstractions are very small and industrial abstractions are almost entirely withdrawn directly from groundwater. The Environment Agency assesses rivers and groundwater units within the catchment as having no additional water available for abstraction during low flows (Environment Agency, 2005). The Medway catchment was chosen because it has known water resources pressures and a variety of potential adaptation options, and is served by several supply companies.

The study design involves three stages. The first is to make credible simulations of the effect of climate change on resource availability, and prepare narrative descriptions of future resource availability under 'central', 'wet' and 'dry' scenarios. The second stage identifies realistic adaptation options from the literature and existing resource plans, and characterises the advantages, disadvantages and potential barriers to each option. The third stage involves structured discussions with water management stakeholders – regulators, water supply companies, environmental groups, councils etc. – to explore and assess options and their barriers.

Preliminary results

Impact of climate change on resources in the Medway

Flows in the Medway have been simulated using the Mac-PDM hydrological model (Arnell, 1999). Model parameters were optimised in a three-stage tuning process using the observed flow data (from gauging station 40003 at Teston) for the period 1980–1983 and validated over the period 1984–1989. In the first stage the land cover and soil classes were determined. Next, the soil parameters were optimised by generating parameter sets consisting of combinations of values of field capacity, saturation capacity and a parameter describing the distribution of soil moisture capacity between a range of ±75% of the initial value at 25% increments, producing 245 parameter combinations (once all instances of field capacity exceeding saturation capacity are removed). This was followed by an additional 49 runs to optimise the flow routing parameters.

Climate scenarios characterising change in mean monthly rainfall, temperature and potential evaporation were created from the UKCIP02 scenarios (Hulme

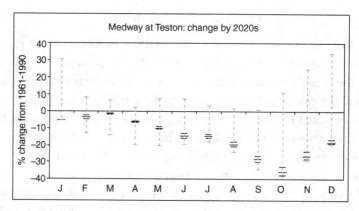

Figure 3.2 Change in mean monthly runoff in the Medway catchment, by the 2020s.

et al., 2002) and five additional climate models (ECHAM4/OPYC, CGCM2, CSIRO MKII, GFDC_R30 and CCSR/NIES2). Figure 3.2 shows the change in mean monthly runoff by the 2020s in the Medway catchment under the UKCIP02 scenarios. The scenarios from the other climate models demonstrate a range of impacts including substantial increases and decreases in flows, reflecting sensitivity to climate model uncertainty. By the 2020s, average annual runoff in the Medway catchment falls by between 11% and 13% under the UKCIP02 scenarios; under the other scenarios, the change in average annual runoff varies from a decrease of 18% to an increase of 14%. The effects on deployable output of the supply systems in the Medway catchment will depend on operating rules, but as a first approximation, a reduction of 11–13% (approximately 28–34 Ml/d across the Medway catchment), in line with the change in average annual runoff, is feasible.

Characterisation of adaptation options

Table 3.2 summarises adaptation options that have been identified by water companies, the Environment Agency, pressure groups and local councils as potentially feasible in the Medway catchment. Some of these are specific resource schemes (which will also serve other catchments), whilst others are options applicable to water resources more generally. Many of these options have been incorporated into the Water Resources Management Plans of the water companies responsible for water resources in the Medway catchment. The complex responsibility for water resources in the catchment means it is necessary to consider schemes across the Kent region. The table provides indicative estimates of the potential contribution of each option, where these are available (estimates are in many cases very generalised, and not to be taken too literally).

Each of the options has been proposed to deal with a future imbalance between supply and demand, and none has been developed specifically with climate change in mind. The supply-side options are generally well established with clearly definable properties. The demand-side options, however, are generally less well established, and their effectiveness is highly uncertain. It is highly unlikely that one option alone will be sufficient to meet future water resource requirements.

Characterising the barriers to adaptation

As shown in Figure 3.1 there are five potential generic barriers to adaptation. The first potential barrier is clearly no obstacle in the Medway catchment. The regulators, the water supply companies, local councils and pressure groups all identify climate change as a challenge to the future security of water resources. The second barrier also does not appear to be a major obstacle, at least at the strategic level of assessment that has been undertaken so far. An established methodology exists for incorporating the effects of climate change into strategic resource assessments (see Arnell and Delaney, 2006), and uncertainty over the potential magnitude of climate change effects on supply reliability has not deterred the search for adaptation options. Between them, the water supply companies, regulators and pressure groups have identified a very large number of potential adaptation options (the adaptation space is wide), although as will be shown below the attitudes towards these different options vary considerably between different organisations. None of the proposed options is specific to climate change. One water supply company specifically asked consultees to its 2008 draft Water Resources Management Plan to suggest additional strategic options.

The fourth potential barrier – ability to evaluate adaptation options – has not yet been seriously approached as assessments are currently in their early stages. However, this evaluation will require a more sophisticated set of scenarios and methodologies than are currently available to water supply companies. In particular, evaluation is likely to be based on risk analyses, using multiple scenarios. Whilst such scenarios are currently being produced (for example for the UKCIP09 scenario set), there are as yet no practical guidelines for applying risk analyses to the assessment of adaptation options in the water industry. The fifth potential barrier – option selection – has also not yet been approached. In practice, water supply companies will have to select and implement options that have been agreed by the environmental regulator (the Environment Agency), the economic regulator (Ofwat) and, where relevant, planning and building control authorities. A consulting mechanism is in place, but has not yet been tested.

Table 3.2 gives a preliminary assessment of the specific barriers to the identified adaptation options. This assessment is based on reviews of documents produced

Table 3.2 *Key specific barriers to potential adaptation options in the Medway catchment*

Option	Details[a]	Potential contribution[b] (Ml/d)	Physical	Financial	Socio-political	Institutional
Supply side						
Bulk transfers	Within region and from outside region (e.g. from Thames)	?	Environmental impacts Network capacity constraints High energy use	High unit costs		Ability to strike deals
Effluent re-use	Kent recycling scheme Re-use of water from Margate–Broadstairs	Up to 20	Moderate energy use	Moderate unit costs	Public acceptability	
Aquifer storage and recovery		5	Limited capacity	Moderate unit costs		
Desalination		?	Environmental impacts High energy use	High unit costs	Pressure group objections (strong)	
Local resources	Enlarge Bewl Bridge Reservoir	20	Availability of water	High unit costs	Pressure group objections (moderate)	
	New reservoir at Broadoak	40	Availability of water Environmental impacts	High unit costs	Pressure group objections (strong)	
	New reservoir at Clay Hill	18	Availability of water Environmental impacts	High unit costs	Pressure group objections (strong)	
	New groundwater sources	9	Limited capacity			
	Use of winter flood storages	?	Uncertainty over effectiveness			Multi-purpose management

Increased connectivity	Transfers between zones	15	Network capacity constraints			Ability to strike deals
Demand side						
Reduce distribution leakage	Across region	29	Uncertainty over effectiveness	High unit costs Not funded through capital streams		
Metering and tariff structures	New houses and retrofitting	10–12% reduction in per capita demand; volume effect depends on growth rates	Uncertainty over effectiveness	Not funded through capital streams	Customer willingness to reduce usage	Political support for widespread metering
Water efficiency	New houses and retrofitting	8–21% reduction in per capita demand; volume effect depends on growth rates	Uncertainty over effectiveness	Costs incurred by customers	Customer willingness to install efficient devices	Lack of measures to encourage adoption
License trading	Transfer of surplus licenses between organisations	?	Uncertainty over effectiveness		Abstractors willingness to sell licenses	Lack of market history
Public education		?	Uncertainty over effectiveness		Customer willingness to reduce usage	
Curb population growth in the catchment		?		Impact on local economy	Local council willingness to deter growth	Mechanisms for curbing growth

[a] Specific schemes where appropriate.
[b] Increase in supply or reduction in demand.

51

by local councils, water companies, the Environment Agency and some pressure groups, and will be reviewed with stakeholders in the final stage of the research. It is possible to draw four key preliminary conclusions. First, there are physical barriers to most of the supply-side options, relating partly to the constraints posed by environmental obligations and partly to uncertainty over whether there would be enough water to sustain the options (particularly filling reservoirs). Second, the physical barriers to most of the demand-side options relate to uncertainty over the magnitude of their contribution to reducing the supply–demand deficit. Third, there are significant pressure group objections to many of the supply-side options – largely on environmental grounds. Finally, there are potential customer barriers to the implementation of many demand-side measures.

Whilst this is a preliminary assessment, it appears that there will be some significant challenges in adapting to the effects of climate change on water resources in southern England.

Conclusions

This chapter presents a preliminary assessment of the barriers to adaptation to water supply shortage due to climate change in a catchment in southern England. The assessment has identified a number of generic barriers, relating to the challenge of adaptation as a whole, and specific barriers relating to individual adaptation options. The next stage of the project is to explore these barriers in more detail with stakeholders in the catchment. At the generic level, the availability of credible scenarios has not yet hindered adaptation, although is likely to have a greater effect when detailed plans are developed. Different stakeholders clearly value different adaptation options differently, and there is a clear difference in the characteristics of the barriers between supply-side and demand-side options.

Acknowledgements

This research was funded by the Tyndall Centre, contributing to the research programme on 'Building resilience: what are the limits to adaptation?'

References

Adger, W. N., Dessai, S., Goulden, M., Hulme, M., Lorenzoni, I., Nelson, D., Naess, L., Wolf, J., Wreford, A. 2009. 'Are there social limits to adaptation to climate change?', *Climatic Change* **93**: 335–354.

Arnell, N. W. 1999. 'A simple water balance model for the simulation of streamflow over a large geographic domain', *Journal of Hydrology* **217**: 314–335.

Arnell, N. W. 2004. 'Climate change impacts on river flows in Britain: the UKCIP02 scenarios', *Journal of the Chartered Institute of Water and Environmental Management* **18**: 112–117.

Arnell, N. W. and Delaney, E. K. 2006. 'Adapting to climate change: public water supply in England and Wales', *Climatic Change* **78**: 227–255.

Berkhout, F., Hertin, J. and Gann, D. 2006. 'Learning to adapt: organisational adaptation to climate change impacts', *Climatic Change* **78**: 135–156.

Defra 2008. *Future Water: the Government's Water Strategy for England*, Cm7319. London: The Stationery Office.

Environment Agency 2005. *Medway Catchment Abstraction Management Strategy.* Bristol: Environment Agency.

Environment Agency 2007a. *Identifying Areas of Water Stress.* Bristol: Environment Agency.

Environment Agency 2007b. *Water Resources Planning Guideline*, April 2007 and updates. Bristol: Environment Agency.

Hulme, M., Jenkins, G. J., Lu, X., Turnpennt, J. R., Mitchell, T. D., Jones, R. G., Lowe, J., Murphy, J. M., Hassell, D., Boorman, P., McDonald, R. and Hill, S. 2002. *Climate Change Scenarios for the United Kingdom: The UKCIP02 Scientific Report.* Norwich: Tyndall Centre for Climate Change Research, School of Environmental Sciences, University of East Anglia.

Warren, G. 2007. *A Water Resource Strategy for the South East of England.* Ashford: Campaign to Protect Rural England (CPRE) South East.

Water Resources in the South East 2006. *WRSE Report on the Latest South East Plan Housing Provision and Distribution.* Bristol: Environment Agency.

4

Protecting London from tidal flooding: limits to engineering adaptation

Tim Reeder, Jon Wicks, Luke Lovell and Owen Tarrant

Introduction

London and the Thames Estuary have one of the best tidal flood defence systems in the world, which offers a standard of protection in excess of a 1 in 1000 year flood (up to at least 2030). However potential drivers such as climate change, socio-economic change and asset deterioration will continue to increase the level of flood risk into the future. Given the long lead times required to implement large-scale infrastructure projects, now is the right time to start planning for the future. Thames Estuary 2100 (TE2100) is an initiative by Anglian, Southern and Thames Regions of the Environment Agency (2008) to develop a plan for flood risk management in the estuary for the next 100 years. The development of this plan is based on a phased programme of study and consultation. This chapter describes one particular work element of TE2100, entitled 'Limits to adaptation' (Halcrow for the Environment Agency, 2006).

The 'Limits to adaptation' study was initiated by the TE2100 team to gain an early appreciation of the likely limits of large-scale 'hard engineering-biased' flood risk management options against incremental sea level rise and different magnitudes of storm surge event in the future. The study has provided TE2100 with a rich insight into the hydraulic performance and possible design considerations of 'hard' engineering biased options, such as an outer estuary barrage, in order to aid in the development of different sets of options. Furthermore, this study has allowed the team to explore the outer limit of the TE2100 climate scenarios against recently emerging climate science, most notably, the findings reported at the Exeter Conference on Avoiding Dangerous Climate Change (2005), where the collapse of the Western Antarctic ice shelf and the melting of the Greenland ice sheet were

Adapting to Climate Change: Thresholds, Values, Governance, eds. W. Neil Adger, Irene Lorenzoni and Karen L. O'Brien. Published by Cambridge University Press. © Cambridge University Press 2009.

postulated with a subsequent eventual 16 m Above Ordnance Datum (AOD) + rise in Mean Sea Level (MSL). The study was conceptualised, at least in part, on the basis of reports published by the ATLANTIS project (Tol et al., 2005).

Decision pathways and defining adaptation thresholds

It has been clear from the outset of TE2100 that to deal with uncertainty associated with the likely effects of climate change there was a need to move away from reactive flood defence towards the proactive adaptive management of future flood risk. Historically, London's flood defences have been raised and improved in the aftermath of various flood catastrophes – typically to a height just above that of the flood that had just been experienced. These incremental raisings can be readily seen in many of the flood walls flanking the River Thames, whose present-day crest height was largely defined by the 1953 flood event.

The proactive management of risk promoted within TE2100 sees a series of timed interventions seeking to manage flood risk within an acceptable zone. This vision recognises that if the risk were to be left unmanaged, it would increase in the future as the impacts of climate change along with development pressures on the floodplain become more acute and as the asset base deteriorates with time (Figure 4.1). However, through the implementation of risk management responses at different points in the century, this risk can be managed within the appropriate bounds. The appropriate bound for flood risk is largely determined through interpretation of the government's guidance on flood risk management and will include an element of cost–benefit analysis.

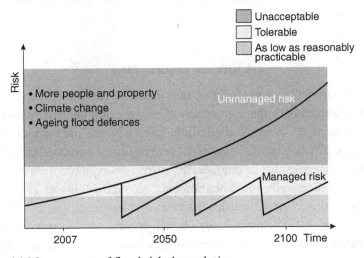

Figure 4.1 Management of flood risk through time.

This timeline of risk management interventions underpin the concept of the *decision pathway* (Donovan et al., 2006). Given an understanding of the likely future trajectory of risk, along with a target level or *threshold* of future flood risk, the decision to adapt the system to manage the new level of risk can be taken in a timely fashion. Different assumptions about the future drivers of change, different adaptation thresholds, or different aspirations for flood risk management, can all generate different decision pathways. These completed decision pathways then can be appraised in terms of their robustness to future uncertainty in addition to other key sustainability criteria.

One of the main aims of the study was to try to define the various points, in terms of sea level rise, at which various engineering responses would face a critical threshold which would force a further adaptive change in the system. The study thus defined a number of key adaptation thresholds, which are listed as follows:

Threshold 1: The point at which the freeboard allowance within the existing flood defences is eroded by a given surge event, or by a future spring tide.

Threshold 2: The point at when the height of existing downriver defences and the crest level of the existing Thames Barrier would need to be raised.

Threshold 3: The point at which the existing Thames Barrier and associated walls and embankments cannot be adapted further, leading to the possible move to an outer estuary barrier (for example at Southend).

Threshold 4: The point at which it is necessary to modify the structure at Southend into a barrage (i.e. there is limited movement of tidal flow upstream of Southend).

Threshold 5: The point at which it is considered impractical to further intervene to manage flood risk through engineering (i.e. the overall engineering limit to adaptation).

Approach to modelling

To explore a number of potential engineering responses to sea level rise at the broad scale required both the application of engineering judgement and extensive one-dimensional hydrodynamic modelling. Each engineering response was tested against a range of sea level rise scenarios ranging up to a maximum of 8 m above the current mean sea level. The engineering responses explored were constrained to (1) raising of defence walls and embankments; (2) modifying the Thames Barrier; and (3) construction of new throttles, barriers and barrages. The overall approach is shown in Figure 4.2. The figure shows how hydraulic models representing different future flood risk management responses are run with a series of extreme tides

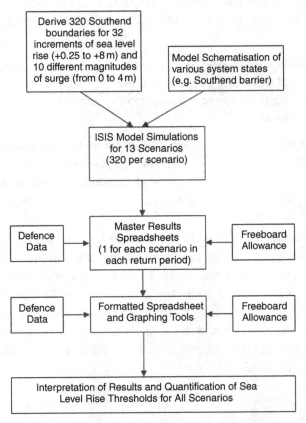

Figure 4.2 Schematic flowchart modelling different flood management responses at Southend.

at Southend to calculate in-bank water levels at a number of locations along the tidal Thames. These water levels are then compared to defence data (i.e. information on current and future crest levels), with an allowance for freeboard also taken into account. In effect, and in this study, the freeboard level sits below the defence crest level and accounts for the need to recognise various uncertainties associated with modelling and estimating in-river water levels (for example uncertainty in the model parameters used, the shape of the tide and statistical errors).

Boundaries

Since the principal objective of the study was to assess the potential impacts of climate change on various tidal defence system states, it was necessary to consider how various climate change scenarios might affect the downriver boundary of the tidal Thames model, situated at Southend-on-Sea. It had previously been determined that climate change has the potential to affect peak water levels at Southend in the following ways:

- Through increasing sea levels as a result of both thermal expansion of the oceans and through melting of freshwater glaciers and ice caps.
- Through increasing the frequency and magnitude – and changing the storm tracks – of extreme surge events.

The study kept separate any absolute rise in sea level from potential increases in surge height. This is important in defining an appropriate tide shape at Southend, as it is known that different tide shapes (with the same peak water level) will propagate differently along the estuary and can result in significantly different water levels in places.

Figure 4.3 illustrates how the change in surge magnitude affects the peak water level and volume of water under the curve. In this case, the surge acts on the fourth and fifth tides. The more extreme surges cause higher peak water levels and will push a greater volume of water into the estuary. Figure 4.4 illustrates the change in the hydrograph shape resulting from different ratios of sea level rise and surge contribution for the same peak water level. It can be seen that a surge peak has a steeper flood limb than the sea level rise peak and this will influence peak water levels further up the estuary.

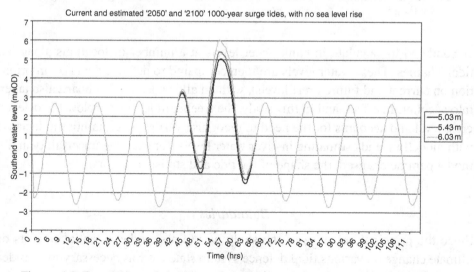

Figure 4.3 Translation of the downriver boundary for three definitions of the 1000-year surge event.

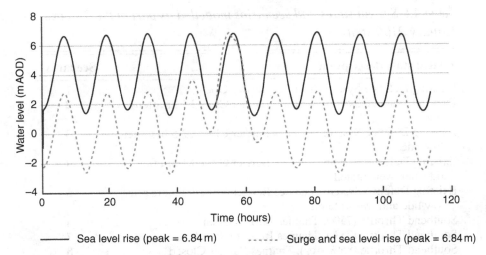

Figure 4.4 Change in hydrograph shape for different ratios of sea level rise and surge contribution for a peak water level of 6.84 m.

Responses

In all, a total of 17 engineering adaptation responses to rising sea levels were modelled, including the existing system of defences and modification to the way this is operated. For each of the 17 engineering responses, it was also possible to assess the effect of raising defences. Table 4.1 provides a brief description of the 17 scenarios of responses used in the model.

Model simulations

Model simulations were undertaken using the ISIS one-dimensional hydrodynamic river modelling software (www.halcrow.com/isis). The models were schematised as in-bank, meaning water did not spill out onto the floodplain. This meant that the results for each scenario could be reused to define overtopping thresholds for different defence levels. Once the simulations were completed, the maximum water levels from each design event were extracted and stored in a spreadsheet, which was then used to facilitate analysis. Various overtopping thresholds were then calculated for each response, both with defences at current levels, and with defences raised by 1 m.

Assumptions

The large number of engineering adaptation responses, combined with the many potential future extreme sea levels, necessitated that a series of assumptions be made in order to keep the study to a manageable size. The more important assumptions are listed below.

Table 4.1 *System states and scenarios modelled as part of the Thames Estuary 2100 study*

System state description	Scenario
Existing Defence System, Thames Barrier Open	1
Existing Defence System, Thames Barrier Closed 1 hr after low tide	2
Existing Defence System, Thames Barrier Closed 2.5 hrs after low tide	3
Existing Defence System, Thames Barrier Closed 1 hr after low tide and over-rotated	4
Existing Defence System, Thames Barrier Closed 2.5 hrs after low tide and over-rotated	5
Southend Throttle (75%), Thames Barrier Open	6
Southend Throttle (75%), Thames Barrier Closed	7
Southend Throttle (85% area), Thames Barrier Closed	8
Southend Throttle (85% width), Thames Barrier Closed	9
Tilbury Barrier (closing 1 hr after low tide), Thames Barrier Closed (1 hr)	10
Southend Barrier (closing 2.5 hrs after low tide), Thames Barrier Open	11
Southend Barrier (closing 2.5 hrs after low tide), Thames Barrier Closed (2.5 hrs)	12
Southend Barrier (closing 2.5 hrs after low tide), Thames Barrier Closed (1 hr)	13
Southend Barrier (closing 1 hr after low tide), Thames Barrier Open	14
Southend Barrier (closing 1 hr after low tide), Thames Barrier Closed (1 hr)	15
Southend Barrage (8.2 m), Thames Barrier Open	16
Southend Barrage (10 m), Thames Barrier Open	17

Boundary conditions: core assumptions

- The potential for climate change to increase rainfall/runoff rates was ignored as the study was tasked with investigating the impacts arising from sea level rise.
- The surge was assumed as a 'double peaked' 1953-type surge with 3-hour offset. It is one of several possible shapes for a surge and each shape will propagate along the estuary differently with varying local peak water levels. As long as the surge type and convolution remain consistent, then relative comparisons will be valid.

Model setup: core assumptions

- The closure rules of the Southend barrier and the Thames Barrier are defined relative to time of low tide, with two options considered by this study. The barriers either close

2.5 hours after low tide or 1 hour after low tide. In reality, the time at which the barrier will be closed is much less certain (due to forecasting storm surges).

- The use of an in-bank model may lead to elevated water levels, compared to those that might be observed if the model allowed water to spill into the floodplain once it had overtopped defences. However, the effect of this assumption on the overall results is likely to be small.

Defence and freeboard assumptions

- Only one indicative value of freeboard allowance was assigned to each embayment. This may not be entirely representative, particularly if the embayment frontage is very long and composed of several different defence types.

Results assumptions

- For each embayment, only one water level has been used to represent the water level for the whole embayment. Earlier investigation suggests that for the embayments with the longest frontage, the water level difference between the upriver node of an embayment and the downriver node is bounded within a range of $\pm 150\,$mm. This is considered to an acceptable range of uncertainty for a study at this broad scale.

Results

Figure 4.5 illustrates a very small selection of results from the study, presenting a timeline of potential engineering adaptation against increases in sea levels. It assumes that a 1000-year standard of protection must be sustained. As well as increasing mean sea level, it allows for an increase of 0.4 m in the 1000-year surge magnitude (note that the thresholds occur at different levels when other surge allowances are analysed). The figure essentially shows one potential 'decision pathway' for the Thames Estuary and highlights the 'failure' point, in terms of sea level rise, for several adaptive flood risk management responses.

Assuming that the requirement is to maintain a 1 in 1000-year standard of protection for the urbanised embayments along the Thames Estuary, then the absolute maximum rise in mean sea level that the potential engineering adaptations tested in this study could accommodate is:

- 5.25 m (for a 1 m increase in surge magnitude); or
- 5.75 m (with a 0.4 m increase in surge magnitude); or
- 6.0 m (with no increase in surge magnitude).

Note that the adaptation threshold used to define this limit is the future sea level resulting in overtopping of the raised (by 1 m) flood defences upriver of a

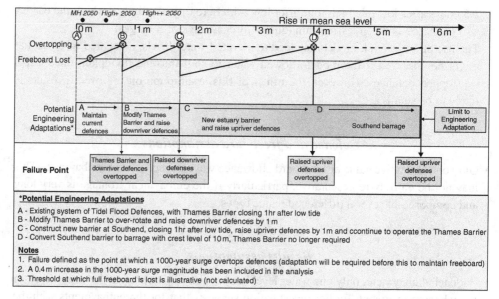

Figure 4.5 A timeline of potential engineering adaptation to longer-term climate change.

Southend barrage. Fortunately, a sea level rise of 5.25 m by 2100 is higher than that postulated for any of the climate change scenarios currently in use by the TE2100 project (for example, the High++ scenario allows for 3.2 m of mean sea level rise combined with 1 m increase in 1 in 1000-year surge by 2100).

Looking at the central TE2100 climate change scenario (the UKCIP Medium High scenario at 2050), the following adaptation thresholds can be defined:

• The existing nominal freeboard allowance upriver of the Thames Barrier would be eroded through the assumed 0.4 m increase in the 1000-year surge magnitude (and no increase in mean sea level is needed).

• The nominal freeboard allowance downriver of the Thames Barrier would be eroded by a 0.25 m increase in sea level (along with the 0.4 m increase in surge magnitude).

• The Thames Barrier, along with many of the downriver defences, would be overtopped by a future 1 in 1000-year event (i.e. with a 0.4 m increase in surge magnitude) if sea levels rose by 0.75 m. To maintain a 1 in 1000-year standard of protection along the estuary beyond this point would necessitate raising both the Thames Barrier and the downriver defences.

• The downriver defences, raised by 1 m, would be overtopped in a 1000-year event if sea levels rise by 1.75 m. To maintain a 1000-year standard of protection, an outer estuary barrier would need to be constructed (assumed to be at Southend).

• With a crest level of 8.2 m AOD, an outer estuary barrier would fail by overtopping of upriver defences if sea levels rise by 4 m. To maintain a 1000-year standard of protection, the outer estuary barrier would need to be converted to, or replaced by, a barrage (i.e. a cross-estuary structure which is normally closed).

- An outer estuary barrage with a crest level of 10 m AOD fails to offer a 1000-year standard of protection when sea levels rise by 5.75 m. This is considered to be the end point with respect to engineering adaptations to sea level rise.

Conclusions

It is clearly evident that to be truly proactive in our approach to flood risk management we cannot wait until one of these adaptation thresholds has been reached or indeed exceeded. The lead time needed to make decisions to adapt to change is often long – for example, it took 30 years from the 1953 flood event for the present-day flood risk management system (with the Thames Barrier as its centrepiece) to be fully operational. This implies the need for careful monitoring of the trajectory of risk and the drivers of change to ensure that the decision to adapt is taken in a timely fashion.

It should also be noted, that this work has not explored the economic, social and environmental 'costs' associated with the implementation of the responses. The 'costs' of some of the response explored above may indeed be prohibitive, thus narrowing the envelope of sea level rise we could adapt to with engineering or structural responses alone. It is likely that a complete portfolio of responses – both structural and non-structural – will have to be employed in the Thames Estuary as we adapt to future flood risk.

References

Avoiding Dangerous Climate Change 2005. *International Symposium on the Stabilisation of Greenhouse Gas Concentrations*, Report of the International Scientific Steering Committee. Available at www.metoffice.gov.uk/climatechange/

Donovan, B., Von Lany, P., Wells, T. and Hall, J. 2006. 'Introducing the concept of decision pathways to help make adaptive and robust flood risk management decisions under uncertainty', paper presented at the *Flood and Coastal Erosion Risk Management Conference*, Defra, London.

Environment Agency 2008. *Thames Estuary 2100*. Available at www.environment-agency.gov.uk/te2100/ (accessed 21 July 2008).

Halcrow for the Environment Agency 2006. *Thames Estuary 2100 Limits to Adaptation: Final Modelling Report*. Bristol: Environment Agency.

Tol, R. S. J., Bohn, M., Downing, T. E., Guillerminet, M.-L., Hizsnyik, E., Kasperson, R., Lonsdale, K., Mays, C., Nicholls, R. J., Olsthoorn, A. A., Pfeifle, G., Poumadere, M., Toth, F. L., Vafeidis, N., Van der Werff, P. E. and Hakan Yetkiner, I. 2005. *Adaptation to 5 Metres of Sea Level Rise: ATLANTIS – Adaptation to Worst Imaginable Sea Level Rise*. Available at www.uni-hamburg.de/Wiss/FB/15/Sustainability/annex8.pdf (accessed 21 July 2008).

5

Climate prediction: a limit to adaptation?

Suraje Dessai, Mike Hulme, Robert Lempert
and Roger Pielke, Jr

Introduction

Projections of future climate and its impacts on society and the environment have
been crucial for the emergence of climate change as a global problem for public
policy and decision-making. Climate projections are based on a variety of scenarios, models and simulations which contain a number of embedded assumptions.
Central to much of the discussion surrounding adaptation to climate change is
the claim – explicit or implicit – that decision-makers need accurate, and increasingly precise, assessments of the future impacts of climate change in order to
adapt successfully. According to Füssel (2007), 'the effectiveness of pro-active
adaptation to climate change often depends on the accuracy of regional climate
and impact projections, which are subject to substantial uncertainty'. Similarly,
Gagnon-Lebrun and Agrawala (2006) note that the level of certainty associated with climate change and impact projections is often key to determining the
extent to which such information can be used to formulate appropriate adaptation
responses. If true, these claims place a high premium on accurate and precise climate predictions at a range of geographical and temporal scales. But is effective
adaptation tied to the ability of the scientific enterprise to predict future climate
with accuracy and precision?

This chapter addresses this important question by investigating whether or not
the lack of accurate climate predictions represents a limit – or perceived limit – to
adaptation. We examine the arguments implicit in the various claims made about
climate prediction and adaptation, and suggest that an approach focused on robust
decision-making is less likely to be constrained by epistemological limits and
therefore more likely to succeed than an approach focused on optimal decision-
making predicated on the predictive accuracy of climate models.

Adapting to Climate Change: Thresholds, Values, Governance, eds. W. Neil Adger, Irene Lorenzoni and Karen L.
O'Brien. Published by Cambridge University Press. © Cambridge University Press 2009.

The chapter is organized in five sections, including this introduction. The following section provides evidence of claims that accurate climate prediction on timescales of years to decades at regional and finer spatial scales is necessary for decision-making related to adaptation. This evidence is drawn from peer-reviewed literature and from published science funding strategies and government policy documents. The third section discusses the challenges to accurate climate prediction and why science will consistently be unable to provide reliable and precise predictions of future climate at the regional and local scales that are claimed to be relevant for adaptation. The section that follows explores alternatives to climate prediction, with a focus on robust decision-making. The latter captures a variety of approaches that differ from traditional optimum expected utility analysis in that they characterize uncertainty with multiple representations of the future rather than a single set of probability distributions. They use robustness, rather than optimality, as a decision criterion. The final section draws together some conclusions and implications for climate and science policy.

Climate prediction for adaptation decision-making

Scientific understandings of phenomena are often tested via predictions that are compared against observations. For example, weather forecasters evaluate the skill of their forecasts by comparing predicted weather against actual weather events. Decision-makers also make predictions about the relationship of actions and outcomes when they choose one course of action over another. Such predictions involve some expectation of the consequences of action and the desirability of those consequences. Lasswell and Kaplan (1950) explain: 'decision making is forward looking, formulating alternative courses of action extending into the future, and selecting among alternatives by expectations about how things will turn out.'

There is therefore a natural tendency for policy-makers to look to scientists to aid decision making by providing insight on how the future will turn out. In many cases, science has provided enormous benefits to decision-makers, either by providing an accurate forecast of future events, such as knowledge of an approaching storm, or by enabling technological innovation that helps decision-makers consciously steer the future toward desired outcomes, such as with the invention of vaccines that improve public health. But there are other circumstances where an improper reliance on scientific prediction to enable decision-making does not have such positive outcomes; policy responses to earthquakes are a notable example (see Sarewitz and Pielke Jr, 1999).

Climate science has proven to be enormously valuable in detecting and attributing recent changes in the climate system. Science has shown that the climate system is undergoing unprecedented changes that cannot be explained solely by

Table 5.1 *Statements about climate prediction and adaptation from the peer-reviewed and grey literature*

We must be able to predict more accurately the climatic effect of increased levels of atmospheric carbon dioxide. This is now the major uncertainty in assessing environmental impact ... We must learn to anticipate the ... consequences of climatic change. (Cooper, 1978) – scientist perspective

In planning the rational use and distribution of ... resources, reliable predictions of the climatic future are ... absolutely essential. (Kelly, 1979) – scientist perspective

It is ... essential that GCM [global climate model] predictions are accompanied by quantitative estimates of the associated uncertainty in order to render them usable in planning mitigation and adaptation strategies. (Murphy et al., 2004) – scientist perspective

It is ... vital that more detailed regional climate change predictions are made available both in the UK and internationally so that cost-effective adaptation and appropriate mitigation action can be planned. Met Office Hadley Centre (MOHC, 2007) – scientist perspective

NERC-funded science must play a leading role in the development of risk-based predictions of the future state of the climate – on regional and local scales, spanning days to decades. Advances in climate science ... are necessary to develop the high-resolution regional predictions needed by decision makers. New scientific knowledge will enable policy-makers to develop adaptation and mitigation strategies. NERC Strategy 2007–2012 (NERC, 2007) – science funding agency perspective

Policy needs robust climate science. Societies need robust infrastructures to deal with extreme weather conditions. Such measures will rely on scientific understanding and accurate predictions of regional climate change ... (Patrinos and Bamzai, 2005) – decision-maker perspective

Plans will only be effective to the extent that climate science can provide ... agencies with climate scenarios that describe a range of possible future climates that California may experience, at a scale useful for regional planning. Reducing uncertainty in projections of future climates is critical to progress ... (Hickox and Nichols, 2003) – decision-maker perspective

Increased acceptance that some degree of climate change is inevitable is now coupled with increasing demand from communities, industry and government for reliable climate information at high resolution and with accurate extremes. There must, therefore, be development in regionalizing climate information, principally through downscaling. World Meteorological Organization (WMO, 2008) – international organisation perspective

The climate models will, as in the past, play an important, and perhaps central, role in guiding the trillion dollar decisions that the peoples, governments and industries of the world will be making to cope with the consequences of changing climate ... adaptation strategies require more accurate and reliable predictions of regional weather and climate extreme events than are possible with the current generation of climate models. World Modelling Summit for Climate Prediction, ECMWF – Reading (UK), 6–9 May 2008 – scientist perspective

Predicting the effects of climate change on hydrological and ecological processes is crucial to avoid future conflicts over water and to conserve biodiversity ... downscaling climate predictions and assessing their impact on mountain environments is an exciting scientific challenge that may allow us to protect the livelihoods of millions of people. NERC PhD studentship at the University of Bristol http://www.ggy.bris.ac.uk/PGadmissions/projects/buytaert-phd2.pdf – scientist perspective natural factors. Unless both natural and anthropogenic forcings are included, climate model simulations cannot mimic the observed continental- and global-scale changes in surface temperature, and other climate-related biogeophysical phenomena, of the last 100 years. Under scenarios of increasing greenhouse gas emissions, climate models estimate that the climate system will continue to change for many more decades and longer.

The ability of climate models to reproduce the time-evolution of observed global mean temperature (within an uncertainty range) has given them much credibility. Advances in scientific understanding and in computational resources have increased the credibility of climate models when projecting into the future using scenarios of greenhouse gas emissions and other climate-forcing agents. Many climate scientists, science funding agencies and decision-makers have argued that quantifying the uncertainty and providing more accuracy and precision in assessments of future climate change is crucial to devise adaptation strategies. The quotes in Table 5.1 exemplify some of these voices. Table 5.1 includes two quotes from the late 1970s to show that this sort of thinking has been around for at least 30 years.

If such claims are true, then they place a high premium on accurate and precise climate predictions at a range of geographical and temporal scales as a key element of decision-making related to climate adaptation. Under this line of reasoning, such predictions become indispensable, and indeed a prerequisite for, effective adaptation decision-making. According to these views, adaptation would be limited by the uncertainties and imprecision that afflicts climate prediction. The next section briefly assesses the state of climate prediction from an adaptation perspective and asks whether indeed accurate and precise predictions of future climate can (ever) be delivered.

Are there limits to climate prediction?

The accuracy of climate predictions is limited by fundamental, irreducible uncertainties. Uncertainty means that more than one outcome is consistent with expectations. For climate prediction, uncertainties can arise from limitations in knowledge (for example, cloud physics), from randomness (for example, due to the chaotic nature of the climate system), and also from intentionality, as decisions made by people can have significant effects on future climate and on future vulnerability (for example, future greenhouse gas emissions, population, economic growth, development etc.). Some of these uncertainties can be quantified, but many simply cannot, meaning that there is some level of irreducible ignorance in our understandings of future climate (Dessai and Hulme, 2004).

A 'cascade' or 'explosion' of uncertainty arises when conducting climate change impact assessments for the purposes of making national and local adaptation decisions (Jones, 2000). In climate projections used for the development of long-term adaptation strategies, uncertainties from the various levels of the assessment accumulate. For example, there are uncertainties associated with future emissions of greenhouse gases and aerosol precursors, uncertainties about the response of the climate system to these changes (due to structural, parameter and initial conditions uncertainty) and uncertainties about impact modelling and the spatial and temporal distributions of impacts. Wilby (2005) has shown that the uncertainty associated with impact models (in his case a water resources model) arising from the choice of model calibration period, model structure, and non-uniqueness of model parameter sets, can be substantial and comparable in magnitude to the uncertainty in greenhouse gas emissions.

Recent increases in computational power have allowed the partial quantification of model uncertainty in climate projections using techniques such as perturbed-physics ensembles (Stainforth et al., 2005), multi-model ensembles (Tebaldi and Knutti, 2007), statistical emulators (Rougier and Sexton, 2007) and other techniques. This has partially moved the science from deterministic climate projections to probabilistic climate projections, but the interpretation of the latter are much disputed (Stainforth et al., 2007). Most of this work is done with GCMs of coarse resolution (for example 300–500km grids), but ensembles of regional climate model simulations (for example 25–100km grids) are also being developed (Murphy et al., 2007, which includes the next set of national UK climate scenarios, UKCIP09). Studies that have propagated these various uncertainties for the purposes of adaptation assessments (sometimes called end-to-end analysis) have found large uncertainty ranges in climate impacts (Whitehead et al., 2006; Wilby and Harris, 2006; Dessai and Hulme, 2007; New et al., 2007). They have also found that the impacts are highly conditional on assumptions made in the assessment, for example with respect to weightings of GCMs (according to some criteria, such as performance against past observations) or to the combination of GCMs used. Some have cautioned that the use of probabilistic climate information may misrepresent uncertainty and therefore lead to bad adaptation decisions (Hall, 2007). Hall (2007) warns that improper consideration of the residual uncertainties of probabilistic climate information (which is always incomplete and conditional) in optimization exercises, could lead to maladaptation and be far from optimal.

Future prospects for reducing these large uncertainties are limited for several reasons. Only part of the modelled uncertainty space has been explored up to now (due to computational expense) so uncertainty in predictions is likely to increase even as computational power increases. It has proved elusive to find 'objective' constraints with which to reduce the uncertainty in predictions (see Allen and Frame, 2007; Roe and Baker, 2007, in the context of climate sensitivity). The problem of equifinality (sometimes also called the problem of 'model identifiability' or 'non-uniqueness') – that many different model structures and many different parameter sets of a model can produce similar observed behaviour of the system under study – has rarely been addressed in climate change studies except in some impact sectors such as water resources (see, for example, Wilby, 2005).

It is also important to recognize that when considering adaptation, climate is only one of many processes that influence outcomes, sometimes important in certain decision contexts, other times not (Adger et al., 2007). Many of the other processes (for example, globalization, economic priorities, regulation, cultural preferences etc.) are not considered to be amenable to prediction. This raises the question of why climate should be treated differently, or why accuracy in one element of a complex and dynamic system would be of benefit given that other important elements are fundamentally unpredictable. One answer is that we currently live in a society with a strong emphasis on science- and evidence-based policy-making. This has led predictive scientific modelling to be elevated above other evidence base because it can be measured and because of its claimed predictive power (Evans, 2008).

The quotes in Table 5.1 imply that more accurate (i.e. reduced uncertainty) and more precise (i.e. higher resolution) regional climate change predictions will help to solve the challenge of adaptation by providing a more faithful description of the future. However, Bankes (1993) notes that such efforts fall prey to false reductionism: 'The belief that the more details a model contains the more accurate it will be. This reductionism is false in that no amount of detail can provide validation, only the illusion of realism.' This mindset is visible in the climate science community with many efforts geared towards increasing the spatial resolution of climate models and adding further components to the model structure. Furthermore, there appears to be confusion amongst users about the relationship between accuracy and precision. Higher precision, in the form of higher spatial (for example, 25 km grids) and temporal (for example, sub-daily estimates) resolution, is often equated with greater realism (i.e. higher accuracy), but that is not necessarily the case. High precision can have low accuracy and high accuracy can have low precision. For example, the statement that 'global mean temperature is projected to increase between 1.4 and 5.8 °C by the end of the century' may prove to have high accuracy but low precision. Correspondingly, the statement that 'maximum summer temperature is projected to increase by 5.7 °C by the end of the century in the London area' may prove to have high precision but low accuracy. According to the *Oxford English Dictionary*, accurate means 'correct in all details', while precise contains a notion of trying to specify a detail exactly.

We have discussed accuracy and precision in the context of spatial and temporal resolution, but as climate projections move into the probabilistic realm there are interesting trade-offs between accuracy and precision. Figure 5.1 shows two probability density functions (PDFs), where the dotted PDF is less precise than the full PDF, but the dotted PDF is more accurate than the full PDF. In this case, precision can be characterized as the standard deviation of the measurements. The larger the standard deviation the lower

S. Dessai et al.

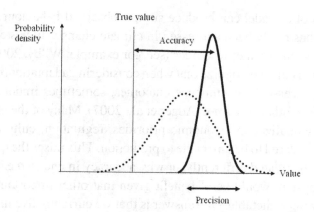

Figure 5.1 Accuracy and precision for two probability density functions.

the precision. Accuracy relates to the difference between the true value and the PDF in question. The higher the difference the lower the accuracy. Extremely wide PDFs have low precision but may be accurate; they may also make it difficult to make decisions (at least under an optimization paradigm). On the other hand, narrow PDFs with high precision may lead to inaccurate results and therefore to maladaptation (false negatives and false positives). We expect that climate scientists will provide users with wide PDFs over the next few years and probably decades for regional and local climate projections. This is likely to be accompanied by a user demand for further precision (i.e. narrower PDFs).

There are also fundamental reasons why climate prediction may fail to fulfil the mission expected of it by the advocates quoted above. For some scholars (see Ravetz, 2003), complex models of open systems are best viewed as heuristic tools which help our understanding of what we can observe, measure or estimate, rather than 'truth machines' which determine our future. Oreskes et al. (1994) argue that verification and validation of numerical models in the Earth sciences is impossible; models can only be evaluated in relative terms, making their predictive value open to question. In the context of complex climate models, Stainforth et al. (2007) have reiterated this point: 'statements about future climate relate to a never before experienced state of the system; thus it is impossible to either calibrate the model for the forecast regime of interest or confirm the usefulness of the forecasting process.'

Based on ten case studies (from weather to earthquake prediction and many others), Pielke Jr et al. (2000) came up with five conditions that are needed for prediction to be useful for decision-making:

(1) Predictive skill is known

In other words, decision-makers have a basis for calibrating the expected accuracy of the prediction. Government weather forecast agencies issue many millions of forecasts every year, providing a rich basis of experience for evaluating predictive performance. In a situation where the forecast is *sui generis*, an evaluation of expected accuracy is necessarily based on factors other than actual performance.

(2) Decision-makers have experience with understanding and using predictions

When decision-makers have experience with using a particular forecast they develop the ability to calibrate its strengths and weaknesses. Research on the use and value of seasonal climate forecasts has indicated that decision-makers often fail to understand the forecasts in the context of the decision environment, and because seasonal climate anomalies, such as El Niño Southern Oscillation, occur only every several years, it is difficult to acquire enough experience for the forecast to become meaningful.

(3) The characteristic time of the predicted event is short

In order for feedback to take place between a forecast – a decision – and an outcome, the time period of an event being predicted needs to be short enough for information on the outcome associated with the decision to be evaluated and factored into the subsequent decision-making process. Predictions of events far into the future by definition cannot be verified or learned from on the timescale of decision-making.

(4) There are limited alternatives

In some situations decision-makers have alternative approaches to decision-making that do not require reliance on predictions. Earthquake policy is an example of such a situation. While some scientists hold out hope for developing predictive skill of particular earthquakes, policy-makers have chosen to focus on engineering design of structures such that buildings will withstand shaking regardless of when the event occurs. By contrast, for those who live in low-lying areas exposed to tsunamis, there is little alternative to a well-functioning early warning system to facilitate evacuation from a coming tsunami.

(5) The outcomes of various courses of action are understood in terms of well-constrained uncertainties

Decision-makers need to understand with some degree of accuracy how various alternative courses of action will relate to particular outcomes. Otherwise, there is no basis for expecting one decision to lead to desired outcomes any more than another decision. A prediction will inform effective decision-making only if it is helpful in discriminating among alternative courses of action in terms of their expected outcomes.

Unfortunately, climate prediction at the decadal to centennial scale fails to meet all these conditions. Predictive skill is unknown, and for long-term predictions cannot be known (condition 1). The accuracy at global and continental level is considered to be higher than at the regional level, but at regional to local scales accuracy is largely unknown. There is little (but slowly growing) experience of decision-makers using long-term climate predictions (2) because until the 1980s or 1990s climate was widely assumed to be stationary and long-term climate predictions were non-existent or speculative. The predictions we are considering here are long-term (3), from a decade up to a century. Alternatives to prediction exist (4) and are discussed in the next section. Finally (5), the outcomes of alternative adaptation strategies often depend little on discriminating among various climate predictions.

This section has shown that there are important limitations to our ability to predict future climate conditions for adaptation decision-making. These include widening uncertainties (as we gain more knowledge of how the climate system operates), lack of objective constraints (with which to reduce the uncertainty of predictions), irreducible uncertainties and the problem of equifinality. Furthermore, there is much evidence that shows that climate is only one of many uncertain processes that influence society and its activities. This suggests that climate prediction should not be the central tool to guide adaptation to climate change. We argue therefore that adaptation efforts should not be limited by the lack of reliable (accurate and precise) foresight about future climate conditions. The next section elaborates on alternatives to prediction.

Making decisions despite deep uncertainties

Individuals and organizations commonly take actions without accurate predictions of the future to support them. They manage the uncertainty by making decisions or establishing decision processes that produce satisfactory results in the absence of good predictions. For instance, no one expects to predict the results of scientific research. Organizations nonetheless undertake such activity. For instance, a private firm might fund multiple initial research and development projects that offer potential new products, assess their progress, and continue those few that seem most promising. Such an adaptive policy often proves a successful response to the lack of predictive ability.

In recent years, a number of researchers have begun to use climate models to provide information that can help evaluate alternative responses to climate change, without necessarily relying on accurate predictions as a key step in the assessment process. The basic concept rests on an exploratory modelling approach (Bankes, 1993) where analysts use multiple runs of one or more simulation models to systematically explore the implications of a wide range of assumptions and to make policy arguments whose prospects for achieving desired ends is unaffected by the uncertainties.

One fundamental step in such analyses is to use climate models to identify potential vulnerabilities of proposed adaptation strategies. For instance, Dessai (2005) uses information from climate models to identify potential weaknesses in strategies that water management agencies in the UK have put in place to address future climate change. This analysis does not require accurate predictions of future climate change. Rather it only requires a range of plausible representations of future climate that can be used to help the water agencies better understand where their vulnerabilities may lie. This is similar to the argument that effective responses to future earthquakes depends not on knowing when the next earthquake will occur, but simply a general sense of where earthquakes do occur. Even without accurate probabilistic information on the likelihood of identified vulnerabilities, such information can prove very useful to decision-makers.

Dessai (2005) found that the water company's water resource plan remains robust to much of the uncertainty space sampled. However, this was in part due to the fact that the company used among the driest available climate model (HadCM3) and the large supply options considered. The criterion upon which robustness was assessed in Dessai (2005) was security of supply. If the analysis had been done on the basis of financial considerations (i.e. minimizing costs and maximizing benefits) the water company's plan could not be considered robust as it would be over-investing. A combination of high greenhouse gas emissions in the near future, low aerosol forcing and large precipitation decreases would require further investment by the water company. Using a similar analytic approach in a very different policy area, Dixon et al. (2007) showed that the current United States government program that offers federal subsidies to encourage private sector provision of insurance against terrorism actually saves the US taxpayer money over a very wide range (over an order of magnitude) of assumptions about the likelihood of future terrorist attacks. This result, based on consideration of thousands of simulation-model-generated scenarios without any claim to predictive skill led to a concise, policy-relevant result invoked by an important senator on the floor of the US Senate (Congressional Record, Nov 16, 2007, Sen. Dodd).

Non-predictive information from climate models can also help decision-makers identify and assess actions that may reduce their vulnerabilities to future climate change. Such approaches generally fall under the heading of robust decision-making (Lempert et al., 2006). The IPCC defines robustness as 'strength; degree to which a system is not given to influence'. Lempert and Schlesinger (2000) propose that society should seek strategies that are robust against a wide range of plausible climate change futures. For these authors, robust strategies perform well (though not necessarily optimally) compared to the alternatives over a wide range of assumptions about the future. In this sense, robust strategies are 'insensitive' to the resolution of the uncertainties. In general, there can be a trade-off between optimality and robustness such that a robust strategy may sacrifice some optimal performance in order to achieve less sensitivity to violated assumptions (Lempert and Collins, 2007).

A variety of analytic approaches have been proposed to identify and assess robust strategies. For instance, information-gap (info-gap) decision theory (Ben-Haim, 2006) has been applied to climate impact related areas such as flood management (Hine and Hall, 2006) and conservation management (Regan et al., 2005). An info-gap is the disparity between what is known and what needs to be known in order to make a well-founded decision. Info-gap decision theory is a non-probabilistic decision theory seeking to optimize robustness to failure, or opportunity of wind-fall. This differs from classical decision theory, which typically maximizes the expected utility.

The RAND group recently worked with Southern California's Inland Empire Utilities Agency (IEUA) to help identify vulnerabilities due to climate change and other uncertainties in the agency's long-range water management plans and to assess additional actions the agency might take to reduce those vulnerabilities (Groves et al., 2008a). They combined downscaled climate projections for the IEUA region with a simulation of the agency's system and hydrology, used the resulting model to create roughly 1000 scenarios, and identified the key factors that would cause the IEUA to suffer significant shortages. The analysis suggested that under its current investment and management plan IEUA was likely to suffer such shortages only if precipitation declines were large, the agency failed to meet its ambitious recycling goals, and the amount of rainwater percolating into the groundwater declined. The analysis shows that all three factors would need to occur simultaneously for future IEUA shortages to become likely. This information, which the agency and its stakeholders found very useful, required a wide range of plausible climate projections but did not require accurate probabilistic estimates of which of these plausible projections were most likely.

The analysis also evaluated a range of adaptation options for IEUA (Groves et al., 2008b). Each option has a particular combination of early actions and actions that can be taken at a later date if groundwater supplies run too low. Testing each option over the 1000 scenarios helped IEUA understand the extent to which early action could reduce future climate-related and other vulnerabilities and the extent to which adaptation, that is responding to future observations of impending shortages, could also address these vulnerabilities. Without requiring accurate probabilistic predictions, this analysis helped IEUA understand its most attractive adaptation options.

This section has shown that there are alternatives to basing adaptation decisions on claims of being able to predict future climate (with accuracy and precision). These alternatives may use plausible scenarios derived from climate models, but they do not require accurate and precise predictions of future climate change, and in fact operate under the assumption that such predictive abilities will not be forthcoming. Central to such approaches is the identification of strategies that work well across a wide range of uncertainties. This ethos is particularly appropriate for adaptation to climate change since many of the non-climatic processes that influence effective adaptation (for example, economic growth, policy regulation, human behaviour) are generally accepted as not being amenable to prediction.

Conclusions

Given the deep uncertainties involved in climate prediction (and even more so in the prediction of climate impacts) and given that climate is usually only one factor in decisions aimed at climate adaptation, we conclude that the 'predict and provide' approach to science in support of climate change adaptation is significantly flawed. Other areas of public policy have come up with similar conclusions (for example, earthquake risk, national security, public health). We therefore argue that the epistemological limits to climate prediction should not be interpreted as a limit to adaptation, despite the widespread belief that it is. By avoiding an approach that places climate prediction (and consequent risk assessment) at its heart, successful adaptation strategies can be developed in the face of this deep uncertainty. We suggest that decision-makers systematically examine the performance of their adaptation strategies/policies/activities over a wide range of plausible futures driven by uncertainty about the future state of climate and many other economic, political and cultural factors. They should choose a strategy that they find sufficiently robust across these alternative futures. Such an approach can identify successful adaptation strategies without accurate and precise predictions of future climate.

These findings have significant implications for science policies as well. At a time when government expects decisions to be based on the best possible science (evidence-based policy-making), we have shown that the science of climate prediction is unlikely to fulfil the expectations of decision-makers. Overprecise climate predictions can potentially lead to bad decisions if misinterpreted or used incorrectly. From a science policy perspective it is worth reflecting on where science funding agencies should focus their efforts if one of the goals is to maximize the societal benefit of science in society. The recent World Modelling Summit for Climate Prediction called for a substantial increase in computing power (an

increase by a factor of 1000) in order to provide better information at the local
level. We believe, however, that society will benefit much more from a greater
understanding of the vulnerability of climate-influenced decisions to large irreducible uncertainties than in seeking to increase the accuracy and precision of the next
generation of climate models.

Acknowledgements

Dessai was supported by funding from the Tyndall Centre core contract with the
NERC, EPSRC and ESRC, by the EPSRC funded project 'Simplicity, Complexity
and Modelling' (EP/E018173/1) and by an ESRC–SSRC Collaborative Visiting
Fellowship. Lempert and Pielke, Jr were supported by the National Science
Foundation (Grants No. 0345925 and 0345604 respectively).

References

Adger, W. N., Agrawala, S., Mirza, M., Conde, C., O'Brien, K., Pulhin, J., Pulwarty,
    R. S., Smit, B. and Takahashi, K. 2007. 'Assessment of adaptation practices,
    options, constraints and capacity', *Climate Change 2007: Impacts, Adaptation and
    Vulnerability. Contribution of Working Group II to the Fourth Assessment Report of
    the Intergovernmental Panel on Climate Change.* Cambridge: Cambridge University
    Press, pp. 717–743.
Allen, M. R. and Frame, D. J. 2007. 'Call off the quest', *Science* **318**: 582–583.
Bankes, S. 1993. 'Exploratory modeling for policy analysis', *Operations Research* **41**:
    435–449.
Ben-Haim, Y. 2006. *Info-Gap Decision Theory: Decisions under Severe Uncertainty*,
    2nd edn. Oxford: Academic Press.
Cooper, C. F. 1978. 'What might man-induced climate change mean', *Foreign Affairs* **56**:
    500–520.
Dessai, S. 2005. 'Robust adaptation decisions amid climate change uncertainties',
    unpublished, PhD thesis, Norwich: University of East Anglia.
Dessai, S. and Hulme, M. 2004. 'Does climate adaptation policy need probabilities?',
    Climate Policy **4**: 107–128.
Dessai, S. and Hulme, M. 2007. 'Assessing the robustness of adaptation decisions to
    climate change uncertainties: a case study on water resources management in the
    East of England', *Global Environmental Change* **17**: 59–72.
Dixon, L., Lempert, R. J., LaTourrette, T. and Reville, R. T. 2007. *The Federal Role in
    Terrorism Insurance: Evaluating Alternatives in an Uncertain World*, MG-679-
    CTRMP. Santa Monica: RAND Corporation.
Evans, S. A. 2008. 'A new look at the interaction of scientific models and policymaking',
    workshop report, 13 February 2008, Policy Foresight Programme, James Martin
    Institute, Oxford University, Oxford.
Füssel, H. M. 2007. 'Vulnerability: a generally applicable conceptual framework for
    climate change research', *Global Environmental Change* **17**: 155–167.
Gagnon-Lebrun, F. and Agrawala, S. 2006. *Progress on Adaptation to Climate Change
    in Developed Countries: An Analysis of Broad Trends.* ENV/EPOC/GSP(2006)1/
    FINAL. Paris: Organization for Economic Cooperation and Development.

Groves, D., Knopman, D., Lempert, R., Berry, S. and Wainfan, L. 2008a. *Presenting Uncertainty about Climate Change to Water Resource Managers*, RAND TR-505-NSF. Santa Monica: RAND Corporation.

Groves, D., Lempert, R., Knopman, D. and Berry, S. 2008b. *Preparing for an Uncertain Climate Future Climate in the Inland Empire: Identifying Robust Water Management Strategies*, RAND DB-550-NSF. Santa Monica: RAND Corporation.

Hall, J. 2007. 'Probabilistic climate scenarios may misrepresent uncertainty and lead to bad adaptation decisions', *Hydrological Processes* **21**: 1127–1129.

Hickox, W. H. and Nichols, M. D. 2003. 'Climate research', *Issues in Science and Technology* **19**: 6–7.

Hine, D. J. and Hall, J. W. 2006. 'Convex analysis of flood inundation model uncertainties and info-gap flood management decisions', in Vrijling, K. et al. (eds.) *Stochastic Hydraulics '05, Proc. 9th Int. Symp. on Stochastic Hydraulics*, Nijmegen, Netherlands.

Jones, R. N. 2000. 'Managing uncertainty in climate change projections: issues for impact assessment', *Climatic Change* **45**: 403–419.

Kelly, P. M. 1979. 'Towards the prediction of climate', *Endeavour* **3**: 176–182.

Lasswell, H. D. and Kaplan, A. D. H. 1950. *Power and Society: A Framework for Political Inquiry*. New Haven: Yale University Press.

Lempert, R. J. and Collins, M. T. 2007. 'Managing the risk of uncertain threshold response: comparison of robust, optimum, and precautionary approaches', *Risk Analysis* **27**: 1009–1026.

Lempert, R. J. and Schlesinger, M. E. 2000. 'Robust strategies for abating climate change', *Climatic Change* **45**: 387–401.

Lempert, R. J., Groves, D. G., Popper, S. W. and Bankes, S. C. 2006. 'A general, analytic method for generating robust strategies and narrative scenarios', *Management Science* **52**: 514–528.

MOHC 2007. *Climate Research at the Met Office Hadley Centre: Informing Government Policy into the Future*. Exeter: Met Office Hadley Centre.

Murphy, J. M., Sexton, D. M. H., Barnett, D. N., Jones, G. S., Webb, M. J., Collins, M. and Stainforth, D. A. 2004. 'Quantifying uncertainties in climate change from a large ensemble of general circulation model predictions', *Nature* **430**: 768–772.

Murphy, J. M., Booth, B. B. B., Collins, M., Harris, G. R., Sexton, D. M. H. and Webb, M. J. 2007. 'A methodology for probabilistic predictions of regional climate change from perturbed physics ensembles', *Philosophical Transactions of the Royal Society of London A* **365**: 1993–2028.

NERC 2007. *Next Generation Science for Planet Earth: NERC Strategy 2007–2012*. Swindon: Natural Environment Research Council.

New, M., Lopez, A., Dessai, S. and Wilby, R. 2007. 'Challenges in using probabilistic climate change information for impact assessments: an example from the water sector', *Philosophical Transactions of the Royal Society of London A* **365**: 2117–2131.

Oreskes, N., Shraderfrechette, K. and Belitz, K. 1994. 'Verification, validation, and confirmation of numerical models in the Earth sciences', *Science* **263**: 641–646.

Patrinos, A. and Bamzai, A. 2005. 'Policy needs robust climate science', *Nature* **438**: 285.

Pielke Jr, R. A., Sarewitz, D. and Byerly Jr, R. 2000. 'Decision making and the future of nature: understanding and using predictions', in Sarewitz, D., Pielke Jr, R. A. and Byerly Jr, R. (eds.) *Prediction: Science, Decision Making and the Future of Nature*. Washington, DC: Island Press, pp. 361–387.

Ravetz, J.R. 2003. 'Models as metaphors', in Kasemir, B., Jager, J., Jaeger, C.C. and Gardner, M.T. (eds.) *Public Participation in Sustainability Science.* Cambridge: Cambridge University Press, pp. 62–78.

Regan, H.M., Ben-Haim, Y., Langford, B., Wilson, W.G., Lundberg, P., Andelman, S.J. and Burgman, M.A. 2005. 'Robust decision-making under severe uncertainty for conservation management', *Ecological Applications* **15**: 1471–1477.

Roe, G.H. and Baker, M.B. 2007. 'Why is climate sensitivity so unpredictable?', *Science* **318**: 629–632.

Rougier, J. and Sexton, D.M.H. 2007. 'Inference in ensemble experiments', *Philosophical Transactions of the Royal Society of London A* **365**: 2133–2143.

Sarewitz, D. and Pielke Jr, R.A. 1999. 'Prediction in science and policy', *Technology in Society* **21**: 121–133.

Stainforth, D.A., Aina, T., Christensen, C., Collins, M., Faull, N., Frame, D.J., Kettleborough, J.A., Knight, S.A., Martin, A., Murphy, J.M., Piani, C., Sexton, D., Smith, L.A., Spicer, R.A., Thorpe, A.J. and Allen, M.R. 2005. 'Uncertainty in predictions of the climate response to rising levels of greenhouse gases', *Nature* **433**: 403–406.

Stainforth, D.A., Allen, M.R., Tredger, E.R. and Smith, L.A. 2007. 'Confidence, uncertainty and decision-support relevance in climate predictions', *Philosophical Transactions of the Royal Society of London A* **365**: 2145–2161.

Tebaldi, C. and Knutti, R. 2007. 'The use of the multi-model ensemble in probabilistic climate projections', *Philosophical Transactions of the Royal Society of London A* **365**: 2053–2075.

Whitehead, P.G., Wilby, R.L., Butterfield, D. and Wade, A.J. 2006. 'Impacts of climate change on in-stream nitrogen in a lowland chalk stream: an appraisal of adaptation strategies', *Science of the Total Environment* **365**: 260–273.

Wilby, R.L. 2005. 'Uncertainty in water resource model parameters used for climate change impact assessment', *Hydrological Processes* **19**: 3201–3219.

Wilby, R.L. and Harris, I. 2006. 'A framework for assessing uncertainties in climate change impacts: low-flow scenarios for the River Thames, UK', *Water Resources Research* **42**: doi 02410.01029/02005WR004065.

WMO 2008. *Future Climate Change Research and Observations: GCOS, WCRP and IGBP Learning from the IPCC Fourth Assessment Report*, GCOS-117, WCRP-127, IGBP Report No. 58, WMO/TD No. 1418. Geneva: World Meteorological Organization.

6

Learning to crawl: how to use seasonal climate forecasts to build adaptive capacity

Anthony G. Patt

Introduction: climate variability and climate change

The Pacific Ocean covers almost half the Earth. Its east–west axis is longest near the Equator, and it is here that the related processes of El Niño and the Southern Oscillation, together known as ENSO, take place. El Niño refers to the periodic warming of the surface waters in the eastern tropical Pacific, while the Southern Oscillation refers to the fluctuation in air pressure differential between Darwin, Australia and Tahiti. What determines the periodicity of ENSO is the time it takes for pressure waves to cross from Indonesia to South America, and then bounce back again. Because the ocean is so wide, that process takes several years. Because there is so much water there, and water holds a lot of energy, ENSO phases can alter weather patterns around the world.

Inter-annual climate variability of this sort has always existed. In terms of human experience, it is likely that people have and will continue to experience climate change not as a gradual rise in temperature, but rather as a shift in the frequency and intensity of particular weather events. Climatic risks and climate variability are a substantial drain on the economies of least developed countries, and indeed the effects of climate variability on society are significantly greater than the effects of climate change probably will be, at least for the next 30 years (Hulme et al., 1999). Given that climate variability is not new, people have been coping with it as long as there have been people. Indeed, one can view people and cultures that have successful coping strategies in place as being well adapted to their climate. Coping strategies change and improve over time. Increased and improved reliance on many of these coping strategies may also be one of the primary ways that people adapt to climate change.

Adapting to Climate Change: Thresholds, Values, Governance, eds. W. Neil Adger, Irene Lorenzoni and Karen L. O'Brien. Published by Cambridge University Press. © Cambridge University Press 2009.

One of the major technological advances of the last 20 years that improves people's capacity to cope with climate variability is a better ability to predict seasonal climate, especially in regions where ENSO is one of the major drivers of inter-annual variability (Golnaraghi and Kaul, 1995). The use of seasonal forecasts is especially important in Africa, because African economies are so dependent on rainfed agriculture in semi-arid regions, and because there is a clear link between ENSO and seasonal climate over much of the continent. An important question, however, is whether the use of forecasts is also an effective strategy to begin adapting to climate change, or at least building the capacity to do so in the future. From a strict cost–benefit perspective, this question is meaningless: if the use of forecasts delivers benefits that outweigh their costs, then applying them makes sense, whether or not the net benefits are tied to climate change. From a programmatic perspective, however, the question is important. The amount of money that will become available to assist adaptation efforts in the next few years – such as from the Adaptation Fund of the Kyoto Protocol to the United Nations Framework Convention on Climate Change (UNFCCC), which is meant to address only anthropogenic climate change and not background variability – dwarfs the funding that has so far gone into developing and applying seasonal climate forecasts for developing countries. Resolving whether forecast application is an effective adaptation strategy could make a large difference both for deciding how to spend adaptation funds, and for influencing the resources available for forecast applications projects.

In this chapter I address the question of whether forecast application is valuable for longer-term climate adaptation, and if so, how. I go further than past analyses of this question by incorporating the lessons learned from recent research and assessment efforts in Africa. First, I draw on a recent case study of adaptation options in Mozambique, which highlights the difficulties encountered in promoting community-level adaptation. Second, I draw off of two recent assessments of the effectiveness of forecast application in Africa, which have collected detailed evidence of where seasonal climate forecasts have delivered actual benefits in Africa, and identified those factors that have led seasonal forecasts to be used or ignored. I pull the two threads together, in order to argue why particular aspects of seasonal forecast application are especially relevant for building adaptive capacity, and to identify criteria for the funding of projects.

Adaptation to climate change

Adaptation means to change to fit the environment, or, as the IPCC defines it: 'actual adjustments, or changes in decision environments, which might ultimately enhance resilience or reduce vulnerability to observed or expected changes in climate' (Adger et al., 2007, p. 720). One way to adapt is to insulate against harsh conditions, and indeed this is the tactic underlying what is probably the most common example given for climate adaptation: building protective barriers to guard against sea level rise. It also underlies the approach of making new development 'climate proof,' coping with expanded tropical disease vectors, and ensuring access to water supply. It is a top–down approach, in that investments to protect against climate change are made by public agencies. A recent background paper prepared by the UNFCCC secretariat estimated the need for annual global public expenditure of US$46–182 billion by 2030 for this kind of adaptation (UNFCCC, 2007).

A very different way to adapt, symbolized by planting drought-tolerant seeds to protect against lessened rainfall, is to modify patterns of production and consumption in ways that better suit the climate. This kind of adaptation is bottom–up, because it is driven by the decisions of private actors. Technological improvement means that this kind of adaptation can occur constantly, and the presence of climate change simply increases its potential net benefits and importance. The diffusion of any new technology or practice ultimately hinges on its acceptance and use by private actors, but public institutions can accelerate that diffusion, especially when there are high start-up costs. It is hard to estimate the costs of this type of adaptation, and efforts that have been made have tended to focus on the public component of that cost. For example, the UNFCCC has estimated the need for an additional US$3 billion per year for agricultural research and extension by 2030 (UNFCCC, 2007); this cost does not include the cost to private farmers, who would then have to experiment with new crop varieties, many of which may be more drought tolerant, but also lower yielding than what they now plant.

Flexibility and adaptive capacity

It is reasonable to assume that the most adaptive societies – in the second, bottom–up sense of the word – are those with private actors in place with the capacity to experiment with new technologies and practices, and public institutions in place with the capacity to help them. A concern for least developed countries is that the climate-sensitive poor lack this capacity required of private actors. A study by Yohe and Tol (2002) examined economic losses and loss of life in the face of climate hazards, and suggested that income is an important factor in determining the flexibility of a society to respond to climate change, both to avoid the negative consequences, and to take advantage of new opportunities. Another study, by Brooks et al. (2005), examined national-level data to see whether it is income itself that provides the adaptive capacity, or the features of society that are associated with greater flexibility, and hence higher incomes. They examined 46 variables that they suspected would be correlated with high vulnerability, and found 11 correlations significant at the 90% confidence level. These 11 all fell within the domains of governance, education and health status. These findings support a model of adaptive capacity that is oriented around flexibility: good governments can respond to changing circumstances, and educated people can participate in such changes, while healthy bodies can weather temporary harsh conditions. Once these factors were taken into account, Brooks et al. (2005) found income variables (including both average income and income equality) became insignificant predictors of vulnerability.

Historical studies have also suggested that adaptive capacity can be linked to flexibility. Diamond (2004) suggested that the society of the Norse Greenlanders collapsed because they persisted in attempting to farm in Greenland as if its climate and soils were those of Norway, and whose Christian beliefs led them to shun well-adapted technologies of their Inuit co-inhabitants. Dugmore et al. (2007; see Chapter 7 in this volume) suggest a slightly different story. The Norse settlers did adapt to the climate they found on Greenland, and for many generations maintained a stable society. But the result of their adaptations was to make it possible for them to continue to rely on agriculture even when conditions for agriculture were quite harsh. This made it possible to avoid one major difference distinguishing them from the Inuit – being tied to a particular settlement versus being nomadic – which proved to be the key to surviving the cumulative effects of climate change as it grew more severe.

Case study: adaptation in the Limpopo River Valley of Mozambique

An example from a recent World Bank study in Mozambique highlights both how a lack of flexibility can be culturally imbedded, and how difficult this can be to change. The Limpopo River floodplain was devastated by flooding in 2000, but a more frequent concern is water scarcity. Climate change probably will exacerbate both problems, increasing the magnitude of floods, and decreasing the average amount of rainfall (Parry et al., 2007). In the aftermath of the 2000 floods, the state attempted to move people out of the floodplain, but they resisted. To understand why, the World Bank surveyed both the smallholder farmers living there, as well as policy-makers for the region (Patt and Schröter, 2008). They found that while the policy-makers perceived climate change as a serious problem, farmers did not. The farmers viewed climate-related risks such as drought, floods and disease as less serious than other risks they faced, such as crime and economic difficulties. Moreover, relative to the policy-makers, the farmers viewed the climate-related risks as growing less serious over time, and the non-climate risks as growing more serious. While they did complain of erratic rainfall and low crop yields, they attributed these primarily to their failure to follow their established farming practices, and the consequent dissatisfaction of their ancestors' spirits. In short, despite falling yields, these farmers did not perceive a need to adapt to climate change, and indeed viewed their own changed actions as part of the problem, not the solution. Patt and Schröter (2008) concluded that this set of beliefs about climate made them quite unlikely to want to participate in government- or NGO-sponsored adaptation programmes, or to engage in their own autonomous adaptation.

Concurrent with this study on risk perception and attribution, the Mozambique Red Cross was attempting to develop a programme to integrate climate change adaptation concerns into its activities in the same communities along the Limpopo River, as part of a grant they had received from the Netherlands Red Cross. Given the lack of clarity over exactly what climate changes would occur in the district, and over what timescales, the Mozambique national office and Gaza Province regional office appeared to be having difficulty deciding what actually to do. One activity that they did engage in was a set of training workshops on climate change in several communities. These training workshops had explained to the farmers attending them the believed causes of climate change, what the long-term effects would likely be globally and for Mozambique, and what some of the long-term adaptation strategies would likely include. As part of its study, the World Bank surveyed farmers in these communities, allowing a comparison between those farmers who had attended the Red Cross workshops, and those who had not. In all the areas covered by the workshops, the survey revealed no significant difference in knowledge. In this particular case, people had either not believed, or not remembered, the information they had received. Patt and Schröter (2008) found this not to be surprising, given that what was presented at the workshops did not contain information immediately salient to the farmers, information on which they could immediately act.

Applying seasonal climate forecasts for adaptation

One type of information that can be salient is a seasonal climate forecast. Beginning in the mid-1980s, climatologists began to issue experimental long-lead-time ENSO forecasts (Cane et al., 1986). A few years later, in 1994, Cane et al. (1994) published a study linking maize production in Zimbabwe with ENSO; they demonstrated that variations in ENSO explained over 60% of the variance in annual harvests. This suggested that it could be extremely valuable to make use of ENSO-based forecasts in order to guide actual decisions. For example, the Famine Early Warning System, which had been operational in Africa for well over a decade, issued food security alerts based on estimates of food reserves, recent harvests, and – prior to harvest – actual rainfall. But forecasts could allow such agencies to start planning several months even further in advance, based on pre-season estimates of rainfall, and this could prove important. In Zimbabwe in 1991–1992, for example, there was a devastating drought associated with a strong El Niño. Prior to the rainy season, the government had large stockpiles of maize. They sold much of this stockpile in late 1991, before it became apparent that there was a major drought under way. Several months later, after a catastrophic harvest, they bought grain back to cover their food shortfall, but at a much higher price. If there had been a seasonal forecast available, and if the government had relied on it, they might have decided not to sell off their stockpile in the first place (Stewart et al., 1996).

Potential benefits for coping and adaptation

Just as with food security planning, such forecasts could allow people in various sectors – agriculture, public health, dam management, disaster prevention – to fine-tune their decisions to the coming year's rainfall. In theory, farmers could plant different crop varieties, hospitals could stockpile malaria drugs, dam managers could enter into different contracts to buy or sell electricity, and civil protection agencies could develop flooding contingency plans. In practice, a variety of barriers stood in the way of the forecasts' use, including the fact that there were often competing forecasts, that their probabilistic character made them difficult to interpret,

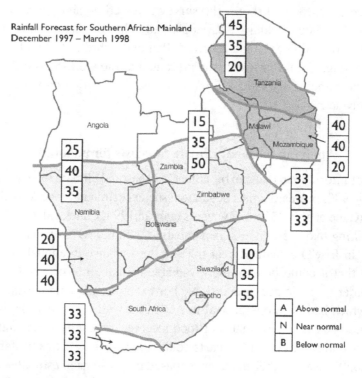

Figure 6.1 Seasonal forecast issued at the Kadoma COF, September 1997.

that their skill was often quite low, and that they were a new piece of information that people were not in the habit of using. The Southern African Regional Climate Outlook Forum (COF), held in 1997, initiated the effort to develop a single consensus forecast (shown in Figure 6.1), imbued with the legitimacy of all of the region's national meteorological and hydrological services (NMHSs), and to communicate that forecast to national-level planners (National Oceanic and Atmospheric Administration, 1999). With the COFs, which became annual or biannual events across several regions of Africa, Latin America and Asia, was a concomitant set of training workshops designed to build the capacity of NMHSs to issue seasonal forecasts, and local-level pilot projects aimed at increasing the capacity of decision-makers to use them.

One long-standing justification for investing in forecast application is that it could assist in efforts at climate adaptation. In November 2003, NOAA convened a workshop focusing specifically on learning from forecast application to inform climate adaptation, commissioning twenty-eight separate background papers looking at different lessons that research on climate variability could offer for climate change (National Oceanic and Atmospheric Administration, 2003). Washington et al. (2006) have put together what is perhaps the most cogent argument that the best first step toward adapting to climate change, specifically in Africa, is to improve coping capacity for climate variability. Their argument rests on three points: the especially low capacity in Africa both to climate science and to relate the results of that work to society; the potential of forecast applications to build that capacity; and, the results of an agent-based model showing that farmers who apply forecasts also perform better under a regime of climate change. Their second point is really the core of their argument: that applying forecasts in Africa is already building the capacity of African climatologists and climatological organizations, and building the linkages between these scientists and societal users of climate information in such a way that can strengthen the 'foundation' for successful climate adaptation in the future. To support these points, they first cite a number of studies in Africa that had examined the use of forecasts, many of which had identified institutional factors such as poor communication channels (for example, Tarhule and Lamb, 2003) or low forecast credibility (for example, Patt and Gwata, 2002), and which, the authors claim, suggest that forecasts are already bringing real value (Ingram et al., 2002; Phillips et al., 2002; Amissah-Arthur, 2003; Luseno et al., 2003; Thomson et al., 2003; Boone et al., 2004; Hansen and Indeje, 2004; Ziervogel and Downing, 2004; Morse et al., 2005; Ziervogel et al., 2005). Second, they show that current climate models have a difficult time delivering a clear and consistent message about future climate changes for Africa (McCarthy et al., 2001); hence, it is difficult to know how to prepare for long-term climate change, and improved coping with climate variability provides the better option.

Case study: recent evaluations of forecast value in Africa

Since Washington et al. (2006) argued these points, NOAA commissioned an assessment of climate forecast applications in Africa, the results of which add an important degree of texture to the arguments raised. What this assessment did was to collect both the published and unpublished findings concerning forecast use in Africa, and for each to evaluate (a) whether the study in question had identified potential benefits from forecast application, (b) whether the study had observed actual benefits accruing to users, and (c) what factors had led the potential benefits to translate into actual benefits (Patt and Winkler, 2007). At the same time, the International Research Institute for Climate and Society (IRI) published a report highlighting a series of success stories of climate forecast applications (Hellmuth et al., 2007). Together with the director of the major forecast applications centre for East Africa, the coordinators of the two assessments pooled their results to support a single analysis (Patt et al., 2007). What they found was that in the vast majority of cases, including all of those that Washington et al. (2006) cite as evidence of forecasts' benefits, researchers had identified potential benefits of using forecasts. In very few cases, however, had researchers observed actual benefits to real people. These included three cases of use by farmers (Patt et al., 2005; Patt, 2006; Diarra and Kangah, 2007), one case of use by dam managers (Axel and Céron, 2007), one case of use for malaria control (Connor et al., 2007) and two cases of use for emergency management (Erkineh, 2007; Lucio et al., 2007). Details on these cases appear in Table 6.1.

What distinguished these success stories was the presence of three related factors: the close and formal cooperation of forecasters and representatives of end-users to identify and respond to users' needs; from that cooperation, the development of a forecast that matched the exact needs of the particular user, and hence could easily be operationalized; and, the careful communication of that forecast to users in an interactive setting, where users could have their questions and concerns addressed by the forecasters. For example, in both the highlands of East Africa (Githeko and Ndegwa, 2001; Zhou et al., 2004) and the semi-arid region of southern Africa (Thomson et al., 2005, 2006), researchers had shown that seasonal and shorter-term climate forecasts could be used to predict malaria outbreaks, in turn suggesting that public health agencies could apply the forecasts to their decisions of where to focus attention, such as stockpiling anti-malaria drugs, distributing insecticide-treated bed nets and spraying for mosquitoes (World Health Organization, 2001). In both regions, malaria experts and representatives from public health ministries had been participating in the COFs, in theory able to use the information to improve decisions. However, it was in southern Africa that the World Health Organization (WHO) regional office joined forces with the IRI and the regional Drought Monitoring Centre (DMC) to organize a different meeting, the Malaria

Table 6.1 Cases of observed benefits of forecast use in Africa

Country, year	Sector	Activity	Benefits observed	Reference
Ethiopia, 2002–03	Food security	NMHS, Disaster Preparedness Agency, NGOs and UN organizations developed drought contingency plan based on forecast	Early action that ensured availability of food where and when needed	Erkineh (2007)
Mali, 2003–04	Subsistence farming	NMHS, extension service, rural development agencies and media cooperated to support farmers' experimental plot management	Increased yields of 10–80%, depending of crop type, on experimental plots	Diarra and Kangah (2007)
Mauritania, Mali and Senegal, 2001–04	Dam management	MétéoFrance and dam management authority developed operational water release model for Manantali Dam based on downscaled forecast	Up to 4% increase in power production and availability of water for irrigation	Axel and Céron (2007)
Mozambique, 1999–2000	Disaster management	NMHS and disaster management agency developed flood contingency plan based on forecast	Response to unprecedented flooding was faster than it otherwise would have been	Lucio et al. (2007)
Southern African Region, 2005–06	Malaria control	WHO, DMC, NMHSs, Ministries of Health and IRI collaborated to implement a Malaria Early Warning System	Up to 90% reduction in malaria cases compared to analog year	Connor et al. (2007)
Zimbabwe, 1997	Commercial farming	Commercial Farmers' Union, NMHS, seed company and independent forecasters developed El Niño response strategy	Maize yields reported to be higher compared to analog years	Patt (2006)
Zimbabwe, 2003–04	Subsistence farming	Extension service, academic researchers and local schools organized community forecast application workshops	Increased yields of 3–17%, depending on year, among farmers who reported using forecasts	Patt et al. (2005)

Outlook Forum (MALOF), separate from the COF (Connor et al., 2007). While the COFs in both East and southern Africa continued to prepare probabilistic rainfall forecasts for the entire region like those in Figure 6.1, the MALOF set out to identify the exact information that users needed, and to deliver it to them. This included more specific pre-season maps of where rainfall would be of sufficient quantity to promote the development of the mosquito vector, and putting in place mechanisms to communicate weekly rainfall predictions and observations during the rainy season to the appropriate decision-makers. These decision-makers were themselves at the MALOF meeting, and so they were able to develop strategies for translating this information into actual actions. The result was that in southern Africa, during an especially wet 2005–06 rainy season, morbidity and mortality were kept to levels far below what would likely have been the case in the absence of the climate information (Connor et al., 2007).

These common features of the cases where benefits have been observed from forecast application adds further support to the arguments that Washington et al. (2006) make, namely that forecast application can be the context to build the linkages between climatologists and end-users of their information that will be necessary for effective adaptation. But one has to be explicit: while these linkages seem to be a necessary condition to obtaining real value from forecasts, many projects where people have attempted to apply forecasts have failed to build the necessary linkages (Patt and Winkler, 2007). For example, in both Ethiopia and Kenya there have been efforts to develop and implement forecasts for dam management; in neither country are forecasts being used, because the necessary partnerships between the dam managers and climatologists never materialized (Babu and Korecha, 2001; Oludhe, 2003). Projects that show potential but not actual value to users may not be undertaking the task of establishing institutional connections, which may be much harder and more time-consuming than programming a model or interviewing stakeholders.

The assessments of African forecast applications also demonstrate that the ways in which people derive value from forecasts themselves may in fact be quite different from how they will adapt to climate change in the longer term. This suggests that in some cases the *only* benefits for adaptation of forecast application may be the partnerships they promote, and not the actual learning that derives from those partnerships. Three examples illustrate this. First, in Ethiopia, the primary use that people have made of forecasts has been to prepare for drought and resulting food insecurity, and yet there is reason to believe that climate change will make drought less of a problem in the future than it has been in the past (Indeje et al., 2000; IPCC, 2007). Second, in Zimbabwe, both modelling results (Phillips et al., 1998) and the field study with subsistence farmers (Patt et al., 2005) demonstrated that forecasts can be most valuable not when they predict drought, but rather when they predict normal to above normal rains. Forecasts of favourable growing conditions allow farmers to depart from generally risk-averse practices, and take advantage of much higher yielding longer-season varieties. The effect of climate change, however, is likely to be decreased rainfall over most of southern Africa, and fewer years when the risky strategy of planting longer-season maize is justified by a forecast (IPCC, 2007). Indeed, the areas of Zimbabwe where these studies took place may, in fact, become unsuitable for rainfed agriculture. Third, in the public health sector, forecasts are most useful to prepare for periodic malaria epidemics, which are most common on the fringes of those regions where, because of temperature and rainfall, malaria is endemic (Thomson et al., 2006). The effect of climate change will be to change the locations of those endemic areas substantially. For example, with increased warming, the East African highlands may change from a place experiencing intermittent epidemic malaria to one experiencing endemic malaria; with increased drying, the southern African semi-arid regions may change from a place experiencing epidemic malaria to one that is malaria-free (Martens et al., 1997; Ebi et al., 2005). As a result, the benefits that a particular region may derive from a malaria early warning system may be transitory.

Discussion

Applying seasonal climate forecasts is like a baby learning to crawl. Eventually what a baby will need to do is learn to walk, and indeed many babies learn to walk without ever having gone through the crawling stage. But crawling can help to develop the necessary coordination that walking will later demand. And crawling can offer its own rewards; a crawling baby can never carry anything, but at least she can look around further than when she is just sitting there. So too can seasonal forecast application be a valuable way to improve adaptive capacity and accelerate future adaptation. Forecast application can also offer immediate rewards, even if the actions taken today are quite different from what will be taken in the future.

The case study from Mozambique demonstrates some of the challenges facing adaptation planners. Just as scientific and technical capacity for climate adaptation may be low in many developing countries, as Washington et al. (2006) demonstrate for Africa, so too can cultural capacity also be a constraint. People probably do need information about climate change, and need to learn how to use that information to guide their immediate decisions and long-term planning. The difficulty is that science tends to influence policies and decisions only when it is salient, credible and legitimate (Mitchell et al., 2006). Climate change is a slow process, to which few decisions that people take now are likely to be sensitive, and hence scenarios of long-term climate change are unlikely to be salient to many actors. This creates a hurdle to overcome to begin considering climate information to guide decisions, and yet only when this hurdle is overcome can scientists and end-users begin to work together in a manner that builds, over time, the credibility and legitimacy of their information.

It is hard to develop effective partnerships between climatologists and users in the absence of a problem to be solved, and hard to maintain them unless the problem is persistent or repetitive. Even if the presence of funding means that people show up to work together, they require a pressing challenge to focus their attention. Seasonal climate forecasts offer a way around this impasse, because they do offer information that is potentially valuable, year after year. What makes forecasts relevant for adaptation is thus not their use per se, but rather in the necessary condition to that use: real collaboration between climatologists, government agencies and end-users, to solve pressing challenges. What makes forecast application uniquely valuable for the development of these networks and partnerships is that their use is an annual event, and can be tested, evaluated and improved upon. The application of seasonal climate forecasts is a valuable way, maybe the best way, for climatologists, government agencies and private actors to learn to work together in a way that can develop the cultural and institutional flexibility that underpin adaptive capacity.

From this, one can identify three criteria that should be applied when funding new programmes or projects to apply seasonal climate forecasts in order to improve adaptive capacity. First, projects need to incorporate the conditions identified as essential for forecast value: collaboration between climatologists and sectoral experts, the development of user-specific forecasts, and participatory communication with end-users. The most beneficial projects, for the purposes of enhancing adaptive capacity, are those that go into environments where collaboration and communication are currently poorest. For example, Patt (2006) showed that commercial farmers in Zimbabwe in 1997 were able to develop links, through their Commercial Farmers' Union, with climatologists, and use those links to develop and communicate their own crop-specific forecasts; subsistence farmers, by contrast, had little opportunity to communicate, either directly or indirectly via the agricultural extension service, with climatologists. A project to help commercial farmers use forecasts would probably generate greater immediate benefits than a project to help subsistence farmers, but the latter would do far more to enhance adaptive capacity.

Second, projects need to be of a duration that is long enough to begin to develop collaborative partnerships, establish trust and realize benefits. The minimum time for this is probably three to five years, not even taking into account the time necessary before and after to identify the necessary actors, plan the interventions and report on the results. Projects of a shorter duration may be useful for teams of researchers to develop models and decision-support tools to help apply forecasts, and these may be quite valuable for sustainable development purposes. But these tasks will not create the institutional linkages and communication channels necessary for improving adaptive capacity.

Third, projects need to include evaluation of whether benefits are actually being obtained from the use of the forecasts. Ideally, they should be designed in such a way as to test this rigorously, such as through the types of controlled studies reported by Axel and Céron (2007), Diarra and Kangah (2007) and Patt et al. (2005). Such an evaluation is an essential check that the necessary collaboration and communication are taking place. If such an evaluation indicates that no benefits are being received, it could be that the necessary collaborations have failed to materialize, or it could be that forecast skill in that location is not high enough to be valuable to the actors involved. Either case would be a good reason for discontinuing the project. Even if it is one that is enhancing institutional linkages, these are not linkages that will persist once it becomes apparent that they are not generating immediate value. On the other hand, if the evaluation provides evidence that participants are receiving value from the forecasts, then this could provide the stimulus for the project to attract more participants into its network. For example, the project reported on by Diarra and Kangah (2007) started in 1982 with a group of 16 farmers, with funding from the Swiss Agency for Development and Cooperation guaranteed for five years. As the initial evaluations were positive, farmers from neighbouring communities asked to be able to join, and the project began to grow. By 2007 there were more than 2,000 farmers participating.

Seasonal climate forecasts are an important technology to help people cope with inter-annual climate variability, and their application can be an important tool to build adaptive capacity. So far, countries in Africa and elsewhere have only touched the surface in making forecasts useful to decision-makers. There is a great potential for new projects in this area, and they represent an ideal way to direct adaptation funding. Those projects that are most worthy of such funding are those that can demonstrate that they develop new institutional linkages between climatologists, sector-specific experts and agencies, and end-users. These will be the projects that help to enhance society's flexibility in the face of climate change.

Acknowledgements

Funding for this research came from the European Union Sixth Framework Project ADAM – Adaptation and mitigation strategies, supporting European climate policy. I would like to thank Karen O'Brien for comments on an earlier draft, and Pablo Suarez, Joanne Bayer and Molly Hellmuth for helpful discussions. All remaining mistakes are my own.

References

Adger, W.N., Agrawala, S., Mirza, M.M.Q., Conde, C., O'Brien, K., Pulhin, J., Pulwarty, R., Smit, B. and Takahashi, K. 2007. 'Assessment of adaptation practices, options, constraints and capacity', in Parry, M.L., Canziani, O.F., Palutikof, J., Van der Linden, P. and Hanson, C. (eds.) *Climate Change 2007: Impacts, Adaptation and Vulnerability. Contribution of Working Group II to the Fourth Assessment Report of the Intergovernmental Panel on Climate Change*. Cambridge: Cambridge University Press, pp. 717–743.

Amissah-Arthur, A. 2003. 'Targeting forecasts for agricultural applications in sub-Saharan Africa: situating farmers in user-space', *Climatic Change* **58**: 73–92.

Axel, J. and Céron, J.P. 2007. 'Climate forecasts and the Manantali Dam', in Griffiths, J. (ed.) *Elements for Life*. London: Tudor Rose, pp. 70–71.

Babu, A.D. and Korecha, D. 2001. *Evaluation of Economic Contributions of Seasonal Outlooks for the Power Industry in Ethiopia*. Washington, DC: National Oceanic and Atmospheric Administration.

Boone, R., Galvin, K.A., Coughenour, M.B., Hudson, J.W., Weisberg, P.J., Vogel, C.H. and Ellis, J.E. 2004. 'Ecosystem modeling adds value to a South African climate forecast', *Climatic Change* **64**: 317–341.

Brooks, N., Adger, W.N. and Kelly, P.M. 2005. 'The determinants of vulnerability and adaptive capacity at the national level and the implications for adaptation', *Global Environmental Change* **15**: 151–163.

Cane, M.A., Zebiak, S. and Dolan, S. 1986. 'Experimental forecasts of El Niño', *Nature* **321**: 827–832.

Cane, M., Eshel, G. and Buckland, R. 1994. 'Forecasting Zimbabwean maize yield using eastern equatorial Pacific sea surface temperatures', *Nature* **370**: 204–205.

Connor, S.J., Da Silva, J. and Katikiti, S. 2007. 'Malaria control in southern Africa', in Hellmuth, M., Moorhead, A., Thomson, M.C. and Williams, J. (eds.) *Climate Risk Management in Africa: Learning from Practice*. New York: International Research Institute for Climate and Society, pp. 45–57.

Diamond, J. 2004. *Collapse: How Societies Choose to Fail or Succeed*. New York: Viking.

Diarra, D. and Kangah, P.D. 2007. 'Agriculture in Mali', in Hellmuth, M., Moorhead, A., Thomson, M.C. and Williams, J. (eds.) *Climate Risk Management in Africa: Learning from Practice*. New York: International Research Institute for Climate and Society, pp. 59–74.

Dugmore, A., Keller, C. and McGovern, T. 2007. 'Reflections on climate change, trade, and the contrasting fates of human settlements in the North Atlantic islands', *Arctic Anthropology* **44**: 12–36.

Ebi, K.L., Hartman, J., Chan, N., McConnell, J., Schlesinger, M. and Weyant, J. 2005. 'Climate suitability for stable malaria transmission in Zimbabwe under different climate change scenarios', *Climatic Change* **73**: 375–393.

Erkineh, T. 2007. 'Food security in Ethiopia', in Hellmuth, M., Moorhead, A., Thomson, M.C. and Williams, J. (eds.) *Climate Risk Management in Africa: Learning from Practice*. New York: International Research Institute for Climate and Society, pp. 31–44.

Githeko, A. and Ndegwa, W. 2001. 'Predicting malaria epidemics in the Kenyan highlands using climate data: a tool for decision-makers', *Global Change and Human Health* **2**: 54–63.

Golnaraghi, M. and Kaul, R. 1995. 'The science and policymaking: responding to ENSO', *Environment* **37**: 38–44.

Hansen, J. W. and Indeje, M. 2004. 'Linking dynamic seasonal climate forecasts with crop simulation for maize yield prediction in semi-arid Kenya', *Agricultural and Forest Meteorology* **125**: 143–157.

Hellmuth, M., Moorhead, A., Thomson, M. C. and Williams, J. (eds.) 2007. *Climate Risk Management in Africa: Learning from Practice.* New York: International Research Institute for Climate and Society.

Hulme, M., Barrow, E. M., Arnell, N. W., Harrison, P. A., Johns, T. C. and Downing, T. E. 1999. 'Relative impacts of human-induced climate change and natural climate variability', *Nature* **397**: 688–691.

Indeje, M., Semazzi, F. and Ogallo, L. 2000. 'ENSO signals in East African rainfall seasons', *International Journal of Climatology* **20**: 19–46.

Ingram, K., Roncoli, C. and Kirshen, P. 2002. 'Opportunities and constraints for farmers of west Africa to use seasonal precipitation forecasts with Burkina Faso as a case study', *Agricultural Systems* **74**: 331–349.

IPCC 2007. *Climate Change 2007: The Physical Science Basis. Contribution of Working Group I to the Fourth Assessment Report of the Intergovernmental Panel on Climate Change.* Cambridge: Cambridge University Press.

Lucio, F., Muianga, A. and Muller, M. 2007. 'Flood management in Mozambique', in Hellmuth, M., Moorhead, A., Thomson, M. C. and Williams, J. (eds.) *Climate Risk Management in Africa: Learning from Practice.* New York: International Research Institute for Climate and Society, pp. 15–30.

Luseno, W., McPeak, J., Barrett, C., Little, P. and Gebru, G. 2003. 'Assessing the value of climate forecast information for pastoralists: evidence from southern Ethiopia and northern Kenya', *World Development* **31**: 1477–1494.

Martens, W. J., Jetten, T. H. and Focks, D. A. 1997. 'Sensitivity of malaria, schistosomiasis and dengue to global warming', *Climatic Change* **35**: 145–156.

McCarthy, J. J., Canziani, O. F., Leary, N. A., Dokken, D. J . and White, K. S. (eds.) 2001. *Climate Change 2001: Impacts, Adaptation, and Vulnerability. Contribution of Working Group II to the Third Assessment Report of the Intergovernmental Panel on Climate Change.* Cambridge: Cambridge University Press.

Mitchell, R., Clark, W., Cash, D. and Dickson, N. (eds.) 2006. *Global Environmental Assessments: Information and Influence.* Cambridge: MIT Press.

Morse, A. P., Doblas-Reyes, F. J., Hoshen, M., Hagedorn, R. and Palmer, T. N. 2005. 'A forecast quality assessment of an end-to-end probabilistic multi-model seasonal forecast system using a malaria model', *Tellus A* **57**: 464–475.

National Oceanic, and Atmospheric Administration 1999. *An Experiment in the Application of Climate Forecasts: NOAA–OGP Activities Related to the 1997–98 El Niño Event.* Washington, DC: NOAA Office of Global Programs, US Department of Commerce.

National Oceanic, and Atmospheric Administration 2003. *Insights and Tools for Adaptation: Learning from Climate Variability.* Washington, DC: NOAA Office of Global Programs, US Department of Commerce.

Oludhe, C. 2003. *Capacity Building and the Development of Tools for Enhanced Utilization of Climate Information and Prediction Products for the Planning and Management of Hydropower Resources.* Washington, DC: National Oceanic and Atmospheric Administration.

Parry, M. L., Canziani, O. F., Palutikof, J., van der Linden, P. and Hanson, C. (eds.) 2007. *Climate Change 2007: Impacts, Adaptation and Vulnerability. Contribution of Working Group II to the Fourth Assessment Report of the Intergovernmental Panel on Climate Change.* Cambridge: Cambridge University Press.

Patt, A. G. 2006. 'Trust, respect, patience, and sea surface temperatures: useful climate forecasting in Zimbabwe', in Mitchell, R., Clark, W., Cash, D. and Dickson, N. (eds.) *Global Environmental Assessments: Information and Influence*. Cambridge: MIT Press, pp. 241–269.

Patt, A. G. and Gwata, C. 2002. 'Effective seasonal climate forecast applications: examining constraints for subsistence farmers in Zimbabwe', *Global Environmental Change* **12**: 185–195.

Patt, A. G. and Schröter, D. 2008. 'Perceptions of climate risk in Mozambique: implications for the success of adaptation and coping strategies', *Global Environmental Change* **18**: 458–467.

Patt, A. G. and Winkler, J. 2007. *Applying Climate Forecast Information in Africa: An Assessment of Current Knowledge*. Washington, DC: National Oceanic and Atmospheric Administration.

Patt, A. G., Suarez, P. and Gwata, C. 2005. 'Effects of seasonal climate forecasts and participatory workshops among subsistence farmers in Zimbabwe', *Proceedings of the National Academy of Sciences of the USA* **102**: 12 623–12 628.

Patt, A. G., Ogallo, L. and Hellmuth, M. 2007. 'Learning from 10 years of Climate Outlook Forums in Africa', *Science* **318**: 49–50.

Phillips, J., Cane, M. A. and Rosenzweig, C. 1998. 'ENSO, seaonal rainfall patterns and simulated maize yield variability in Zimbabwe', *Agricultural and Forest Meteorology* **90**: 39–50.

Phillips, J., Deane, D., Unganai, L. and Chimeli, A. 2002. 'Implications of farm-level responses to seasonal climate forecasts for aggregate grain production in Zimbabwe', *Agricultural Systems* **74**: 351–369.

Stewart, M., Clark, C., Thompson, B., Lancaster, S. and Manco, L. (eds.) 1996. *Workshop on Reducing Climate-Related Vulnerability in Southern Africa*. Washington, DC: National Oceanic and Atmospheric Administration.

Tarhule, A. and Lamb, P. 2003. 'Climate research and seasonal forecasting for West Africans', *Bulletin of the American Meteorological Society* **84**: 1741–1759.

Thomson, M. C., Indeje, M., Connor, S. J., Dilley, M. and Ward, N. 2003. 'Malaria early warning in Kenya and seasonal climate forecasts', *Lancet* **362**: 580

Thomson, M. C., Mason, S., Phindela, T. and Connor, S. J. 2005. 'Use of rainfall and sea surface temperature for malaria early warning in Botswana', *American Journal of Tropical Medicine and Hygiene* **73**: 214–221.

Thomson, M. C., Doblas-Reyes, F. J., Mason, S. J., Hagedorn, R., Connor, S. J., Phindela, T., Morse, A. P. and Palmer, T. N. 2006. 'Malaria early warnings based on seasonal climate forecasts from multi-model ensembles', *Nature* **439**: 576–579.

UNFCCC 2007. *Analysis of Existing and Planned Investment and Financial Flows Relevant to the Development of Effective and Appropriate International Response to Climate Change*. Bonn: United Nations Framework Convention on Climate Change Secretariat.

Washington, R., Harrison, M., Conway, D., Black, E., Challinor, A., Grimes, D., Jones, R., Morse, A., Kay, G. and Todd, M. 2006. 'African climate change: taking the shorter route', *Bulletin of the American Meteorological Society* **87**: 1355–1366.

World Health Organization 2001. *Malaria Early Warning Systems: A Framework for Field Research in Africa*. Geneva: World Health Organization Roll Back Malaria Cabinet Project.

Yohe, G. and Tol, R. S. J. 2002. 'Indicators for social and economic coping capacity: moving toward a working definition of adaptive capacity', *Global Environmental Change* **12**: 25–40.

Zhou, G., Minakawa, N., Githeko, A. and Yan, G. 2004. 'Association between climate variability and malaria epidemics in the East African highlands', *Proceedings of the National Academy of Sciences of the USA* **101**: 2375–2380.

Ziervogel, G. and Downing, T. 2004. 'Stakeholder networks: improving seasonal climate forecasts', *Climatic Change* **65**: 73–101.

Ziervogel, G., Bithell, M., Washington, R. and Downing, T. 2005. 'Agent-based social simulation: a method for assessing the impact of seasonal climate forecast applications among smallholder farmers', *Agricultural Systems* **83**: 1–26.

7

Norse Greenland settlement and limits to adaptation

Andrew J. Dugmore, Christian Keller, Thomas H. McGovern,
Andrew F. Casely and Konrad Smiarowski

Introduction

The end of Norse Greenland sometime in the mid to late fifteenth century AD is an iconic example of settlement desertion commonly attributed to the climate changes of the 'Little Ice Age' combined with a generalized failure to adapt (for example, Diamond, 2005). The idea of chronic Norse adaptive failure has been widely accepted, in part because other peoples in Greenland (the Thule Inuit) survived through the period of Norse extinction. Human settlement of Greenland was definitely possible through the climate fluctuations of the thirteenth to seventeenth centuries AD, despite increasingly well-documented changes in temperature, probable growing season, sea ice, storminess and sea level. The Inuit achieved sustainability during this period of instability and change, but the Norse did not. It is assumed there must have been some degree of Norse maladaptation or more constrained limits to their adaptations than those of the Inuit, and the Norse are seen to have 'chosen extinction'. We suggest that the picture emerging from recent and current research is far more complex, and propose that the Norse had achieved a locally successful adaptation to new Greenlandic resources but that their very success may have reduced the long-term resilience of the small community when confronted by a conjuncture of culture contact, climate change and new patterns of international trade.

The reasons for the final collapse of Norse Greenland are still incompletely understood, but new data from Greenland and across the North Atlantic, combined with changing ideas and developing cognitive frameworks, are refining and deepening our understanding on both adaptation and its limits (Dugmore et al., 2007a; McGovern et al., 2007).

Adapting to Climate Change: Thresholds, Values, Governance, eds. W. Neil Adger, Irene Lorenzoni and Karen L. O'Brien. Published by Cambridge University Press. © Cambridge University Press 2009.

It is apparent that the Norse in Greenland did adapt to changing conditions, in particular through the increasing utilization of marine mammals (Arneborg et al., 1999). That these adaptations were insufficient to ensure the survival of society may be inferred from the final collapse of Norse settlement, but their limited ultimate effectiveness may be best understood in terms of a failure of *resilience*, which refers to the ability of a system to maintain its structure in the face of disturbance and to absorb and utilize change (Van der Leeuw, 1994). One possibility is that the initial Norse colonization and settlement of Greenland was followed by a rising level of connection, intensification and investment in fixed resource spaces, social and material infrastructure which increased the effectiveness of adaptation, but at a cost of reduced resilience in the face of variation.

In this chapter we explore the idea that the Norse were initially well adapted to life in Greenland, with a subsistence economy based upon the seasonal coordination of the labour of dispersed households and with the ability to regulate wild resource exploitation to avoid overuse. The key failure of the Norse settlers in Greenland would thus have been not because of a clumsy or ineffective initial adaptation of European farming and hunting patterns to their new home, but later on in their limited ability to rapidly reconfigure well-established and effective mechanisms for adaptation to meet the combined challenges of economic change, culture contact with the Inuit and unanticipated climate change.

New perspectives: Norse adaptation and sustainable practices

Initial adaptation

In Greenland it is apparent that the Norse did not simply apply proven subsistence strategies based on their prior experience in Iceland or Norway. The early Greenlandic colonists did import cattle, sheep, goats, pigs and horses and set up farms ultimately tied to the pockets of inner fjord pasture vegetation in the Eastern and Western Settlements (Figure 7.1) in a dispersed pattern of farms clearly tied to pasture resources. Despite this terrestrial base, zooarchaeology and isotopic study of human bones reveal the importance of marine mammals from the first stages of settlement. Norse archaeofauna in Greenland from the earliest phases have far more seal bones than appear in any of the other North Atlantic bone collections known from the Northern Isles, Hebrides, Faroes or Iceland (Perdikaris and McGovern, in press). Migrating harp (*Phoca groenlandica*) and hooded seals (*Cystophora cristata*) encountered in Greenland were rare or absent in the rest of the Viking Age North Atlantic, and their huge populations presented a far richer resource less likely to be depleted by large-scale exploitation than the small resident harbour/common seal (*Phoca vitulina*) and grey seal (*Halichoerus gryphus*) colonies of Iceland or the British Isles. While the Norse middens do not contain

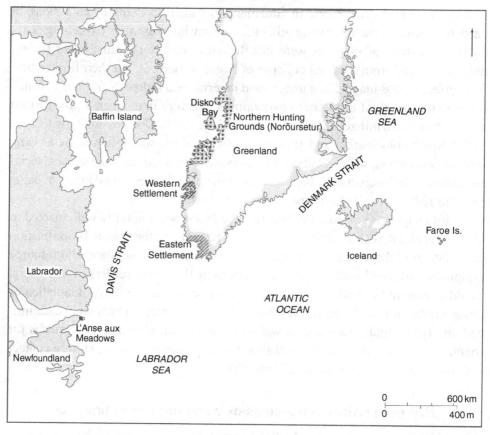

Figure 7.1 Map of Norse settlement established on the west coast of Greenland. The larger Eastern Settlement in the south has *c*. 400 farms, the Western Settlement has *c*. 80 farms. The Norðursetur (Northern Hunting Grounds) were in the Disko Bay area, *c*. 800 km north of the Eastern Settlement. Substantial, well-furnished churches with stained glass and church bells were built *c*. 1150–1300. No more churches were constructed after 1300, the Western Settlement was abandoned *c*. 1350, the last recorded contact was in 1408 and the Eastern Settlement was probably deserted by *c*. 1450.

harpoons or the ringed seals (*Phoca hispida*) usually taken with such gear, the Norse Greenlanders were certainly highly competent seal hunters who made use of nets and probably communal boat drives aimed at the millions of migrating seals arriving along the coast of West Greenland in spring. Utilization of the migrating seals may be seen as a key adaptation to provisioning Greenland settlement as the spring seal migration came at the annual low point when other stored food was probably becoming scarce. The seal hunt was critical to the entire community, and archaeofauna from middens in inland sites many hours' walk from the nearest salt water are as rich in seal bones as coastal farm middens. While the spatial organization of Norse seal hunting in Greenland is now under fresh cooperative study, it is already apparent that the spring hunt of the migratory seals was communally organized, probably drawing on the labour of whole districts for intensive mass hunts timed to catch and widely redistribute this rich seasonal peak in marine resources. It seems likely that the best modern analogue to this pattern may be found in the still-active Faroese practice of communal drive hunting (*grind*) of the pilot whale. In the Faroes, every step of the *grind* hunt (from first spotting to final division of meat) is closely regulated by tradition and written law codes extending back to the medieval period. The *grind* today is seen as so critical to community coherence and Faroese social identity as to be totally impervious to outside pressure from influential international anti-whaling groups. Norse seasonal sealing in Greenland certainly played an even more central role in year-to-year survival, and the annual hunt must have played at least as central a role in reinforcing and enhancing community solidarity.

While there is thus far limited evidence for Icelandic-style large-scale consumption of marine fish on inland farms, a few marine fish bones have been recovered far inland, as have the bones of seabirds (Enghoff, 2003; McGovern, 1985). The seabirds are almost all guillemots (*Cepphus grille*) and murres (*Uria* spp.), who nest communally on cliff sides in several parts of the outer fjords of both settlement areas. The distribution of their bones on distant farm middens again suggests some sort of communal hunting and redistribution, this time probably occurring in late summer when the moulting colonies are most vulnerable.

Walrus hunting mainly took place in the Norðursetur district far to the north of the two settlement areas and seems to have involved weeks-long voyages in both directions during the summer months (Dugmore et al., 2007a). The zooarchaeological evidence suggests that this also was a communal activity, as the fragments of bone from around the walrus tusk root chipped off during final finishing (perhaps during the long winter) are found on inland as well as coastal farms in both Eastern and Western Settlements. It would appear that this Norðursetur hunt took boats and active young people out of the settlement areas most summers. Walrus hunting again apparently represents a community-scale effort; this time aimed at securing goods for overseas trade. Fragments of Norðursetur poetry, lost saga references, and the widespread finds of walrus and polar bear amulets carved from walrus postcanines again suggest that this communal effort was embedded in a rich and well-developed cultural matrix.

Sustainable resource management and the maintenance of flexibility

There is growing evidence from Iceland and the Faroes for successful Viking Age and Norse community-level management of seabirds, waterfowl, freshwater fishing, common grazing and woodland (Simpson et al., 2002, 2003, 2004; Church et al.,

2005; Dugmore et al., 2006, 2007b; McGovern et al., 2007). As we learn more about economy in the North Atlantic, older ideas of widespread and heedless depletion of all forms of natural capital by Vikings–medieval Norse (for example McGovern et al., 1988) are being replaced by notions of more sophisticated and successful management by well-integrated communities capable of regulating access to and drawdown of potentially vulnerable resources (McGovern et al., 2007). Recent results of environmental archaeology and palaeoecology have served to underline the historical evidence for conscious and well-developed structures for management of communal and private resources. It is found in surviving medieval Icelandic law codes such as Grágás (Dennis et al., 2000), which set limits on hunting of seabirds, seals and eider ducks, and regulated use of stranded whales and driftwood. Similar law codes and a multi-tiered court *(thing)* system were also set up in Greenland soon after colonization, and while we no longer know the details, a special set of Greenlandic laws is known to have existed to regulate hunting and trips to the Norðursetur. By AD 1300 several centuries of adjustment certainly had produced a complex and well-developed legal structure regulating communal labour deployment and the distribution of catches. Contemporary Icelandic sources note special aspects of Greenlandic law codes (unfortunately without providing details: Gad, 1970). By AD 1300, Norse Greenland had become a small and somewhat isolated corner of medieval Europe which possessed literacy (in both Latin and Norse), law codes, a resident bishop or bishop's steward, a Norwegian royal representative, a monastery, a nunnery and churches equipped with imported bells and stained glass. While full-scale feudalism probably never became established in the Norse North Atlantic, society was certainly stratified at the state level. Rank and precedent played a significant role in the organization of society and economy, as in the rest of medieval Scandinavia. This modest hierarchical structure has been blamed for Norse adaptive failures (Diamond, 2005), but in many cases the ability to authoritatively regulate and manage communal resource use probably contributed to adaptive successes, particularly in conserving caribou and non-migratory seal populations.

Caribou (*Rangifer tarandus*) were utilized by the Norse throughout their settlement of Greenland. Along the long and deeply fjord-cut coastline of western Greenland where inlets frequently meet the inland ice, caribou have historically tended to fragment into localized breeding populations subject to different crash–boom cycles driven by climate and modified by differing hunting pressures (Meldgaard, 1965). In the Western Settlement area, caribou benefit from more closely interconnected grazing areas and were probably less subject to deadly range-icing in winter than caribou in the Eastern Settlement area (Vibe, 1967). Western Settlement area caribou have also proven resilient in the face of sustained human hunting as they survive in substantial numbers today. By contrast, in the entire Eastern Settlement region caribou were driven to extinction by Inuit hunters in the early nineteenth century.

The medieval Norse settlers certainly had the capacity to place heavy pressure on the Eastern Settlement caribou herds. They maintained large hunting dogs and probably employed drive systems, as well as keeping substantial numbers of competing sheep and goats in the modern summer caribou grazing areas (Degerbøl, 1934, 1941; McGovern and Jordan, 1982; McGovern, 1985). Caribou bones make up a consistent proportion of 2–5% NISP (a simple bone fragment count) of archaeofaunal assemblages from the Eastern Settlement (McGovern and Pálsdóttir, 2006). The Western Settlement range is higher at 5–27% NISP, and the stratified collections indicate no decline in caribou taken through time, despite the climatic variability seen during over 400 years of Norse occupation. This pattern suggests that the medieval Norse settlers in the large Eastern Settlement area were in fact more capable of sustainably managing their inherently more vulnerable local caribou population on the century scale than were the egalitarian Inuit hunters who succeeded them.

The utilization of common seals (*Phoca vitulina*) throughout the Norse occupation at Sandnes in the Western Settlement provides another likely example of a sustainable Norse approach to the management of wild resources in Greenland (Figure 7.2). Common seal populations tend to be localized, and over-exploitation results in the extinction of particular pods or forces them to relocate to less accessible hauling out locations. As a result, the long-term (century-scale) utilization of common seals at Sandnes and neighboring farms suggests sustainable exploitation. The notably contrasting decline in common seal utilization in the Eastern Settlement is unlikely to be a result of over-hunting, but more probably a consequence of climate changes. Common seal pups do not thrive in ice filled waters, and the presence of persistent summer sea ice tends to reduce common seal populations (Woollett et al., 2000). Thus the late thirteenth century transition to modern conditions of increased summer drift ice in southern Greenland that affected the Eastern Settlement area but not the Western, could have forced the reduction in common seal observed in the archaeofauna (Jennings and Weiner, 1996; Jennings et al., 2001; McGovern and Pálsdóttir, 2006).

Details of these Norse strategies of caribou and common seal conservation in Greenland are unclear but we may speculate that it was part of a conscious effort to conserve resilience and flexibility by underwriting the farming economy based on imported stock and the long-distance Norðursetur hunt for trade goods. In late medieval and early modern Iceland elements of resilience thinking in a context of recurring labour shortage may be embedded in restrictions on sea fishing. In order to undertake fishing, farm ownership was required, and the development of fishing camps unsupported by agriculture was legally discouraged. The rationale was that

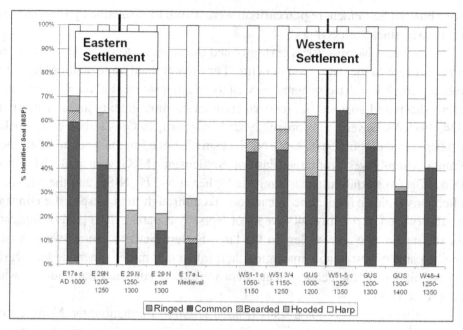

Figure 7.2 Analyses of stratified seal bone collections from Norse settlement areas. The abundance of common seals declines in the Eastern Settlement after the later thirteenth century. As increases in summer drift ice differentially affect this area it is probable that the change in the archaeofauna is driven by climate. (McGovern et al. (1993), McGovern et al. (1996), Data from Enghoff (2003), McGovern and Pálsdóttir (2006)).

fishing alone could not ensure a certain livelihood, and when it failed there would be a burden of poor relief on the wider community if those fishing did not have farming as well. In other words: specialization that could produce a greater short-term yield was discouraged because it could potentially compromise the resilience of the wider community through burdens of support during times of fishery failure. Similar rationale may have stood behind the apparent absence of specialized sealing stations in Norse Greenland and the failure to develop substantial fisheries. Despite the growing role of seals in subsistence, even small farms maintained substantial flocks of sheep and goats and at least a few cows – this was a multi-stranded economy which spread risks and coordinated labour on a community scale rather than a specialized, individualized subsistence system. Its major limitations were in its inability to accumulate multi-year surplus in the absence of cereal agriculture and the resulting recurring problem of matching high seasonal communal labour requirements with year-round provisioning limits. Even more than in Iceland, medieval Greenlanders faced a year-by-year zero sum game of allocating scarce adult labour, irreplaceable boats, and a short growing/navigation season among the demands of sealing, birding, caribou hunting, the Norðursetur voyages and farming. Long winters (chess sets are common) were balanced by heavily scheduled summers, and individual and household survival was closely connected to community cohesion and effective deployment of community resources. While a retrospective view of Norse Greenlandic economy and society will inevitably be coloured by its final extinction, it is worth emphasizing the extended period of over 400 years during which the Greenlanders successfully achieved their annual balancing act. It seems unlikely that any Norse Greenlander around the year AD 1300 would have sensed changes to come, and unlikely that any modern cultural ecologist would write up the Norse North Atlantic society as a whole as an environmental success story, especially given the scale of soil erosion in contemporary Iceland. Management based upon multi-generational incrementally adjusted legal codes and adjudicated through community court systems backed by top–down secular and religious authority need not be seen as an impediment to effective adaptation or long-term resource conservation, and indeed some combination of these environmental management tools are regularly suggested as ways forward today (for example Lovelock, 2006). There is probably no need to model the Norse Greenlanders as an arctic peasantry oppressed by Eurocentric elites or as poorly adapted to their available resources. Indeed, their problem after 1300 may have been precisely that they had achieved a complex set of well-regulated communal adaptations to their arctic homeland.

Adaptation and the long dureé

Isotopic evidence from human remains and a general tendency for the relative percentages of seal bones at archaeological sites to increase through time indicate an increasing marine component in Norse diet in Greenland (McGovern et al., 1996, 2001; Arneborg et al., 1999; McGovern and Pálsdóttir, 2006). These changes broadly reflect changes in cumulative measures of climate (Figures 7.3a and b) suggesting that the progressively more vital role in subsistence played by the seal hunt and other marine foraging could have been motivated by climate changes. Norse migratory seal hunting in Greenland was thus both highly productive and capable of expansion and intensification in response to climate changes. The ultimate cause for failure could however be as a consequence of this increasing utilization and dependence upon marine mammals. When it became the dominant source of subsistence, failure could be catastrophic and the ability to deal with a failure of the seal hunt could be minimal. In absolute terms, boosting alternative sources of subsistence to make good the deficit caused by a failure to adequately harvest the migrating seals could have been extremely difficult and the timing of the shortfall in early spring could have made the situation even worse. If a fishery fails, other options may be available; fishing gear, nets, boats and crews may be redeployed at other times to target other fish or other fishing grounds. Failure to effectively exploit the hooded and harp seals is different. If the spring cull failed because of environmental changes, conflict with

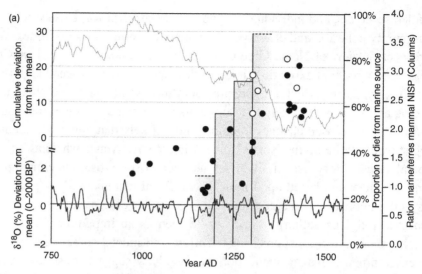

Figure 7.3 (a) The δ¹⁸O time series, a proxy temperature record, shown against human isotopic data.The δ¹⁸O time series is from GISP2 (LHS scale); the human isotopic data from RHS scale showing proportion of diet from marine sources (Mayewski et al., 1993; Arneborg et al., 1999; Mayewski and White, 2002; data from Greenland Ice Sheet Project 2www.gisp2.sr.unh.edu/). The lower line is the 5-year running mean of deviations from the long-term mean. The upper line is the same data presented as cumulative deviations (Dugmore et al., 2007a). The points represent isotopic data on Norse Greenlanders showing the proportion of marine food consumed. The histograms show the relative proportion of marine and terrestrial bones form the recent excavation at E29N (Brattahlið, Eastern Settlement) (McGovern and Pálsdóttir, 2006). The change in the cumulative deviation – marking a shift to cooler conditions – also marks the time of a distinct shift in the isotopic records of diet and changes in the archaeofauna. Closed circles represent data from the Eastern Settlement, open circles data from the Western Settlement. (b) Sea salt concentration time series, a proxy record of storminess, shown against human isotopic data. The Na⁺ concentration (sea salt) time series is from GISP2; the human isotopic data from lower RHS scale showing proportion of diet from marine sources. The histograms show the relative proportion of marine and terrestrial bones from the recent excavation at E29N (Brattahlið, Eastern Settlement) (Arneborg et al., 1999; McGovern and Pálsdóttir, 2006; Meeker and Mayewski, 2002; data from Greenland Ice Sheet Project 2www.gisp2.sr.unh.edu/). The Na⁺ in the Greenland ice cap derives principally from sea salt in the North Atlantic (Meeker and Mayewski, 2002), so temporal variations in sea salt concentrations in the GISP2 record represent a proxy record for winter storminess in the North Atlantic. The upper line (top LHS scale) is the 5-year running mean of deviations from the long-term mean. The lower line (bottom LHS scale) is the same data presented as cumulative deviations (Dugmore et al., 2007a). Storminess is an indicator of regional circulation change in the North Atlantic, and the most significant shift occurs within a short time of the final extinction of the Norse settlement. Closed circles represent data from the Eastern Settlement, open circles data from the Western Settlement.

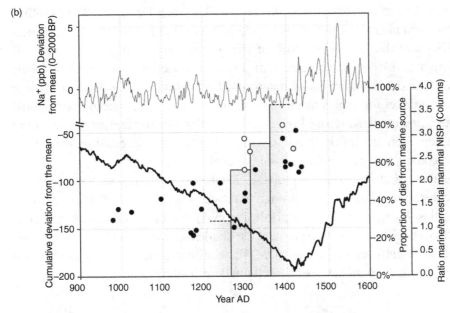

Figure 7.3 (cont.)

the Inuit, or a lack of Norse labour, the making up of the shortfall would be a great challenge as there would be no other seal migration the same year. A switch to hunting the harbour seal, which could utilized the same gear as used to hunt hooded and harp seals, while possible to some degree in the Western Settlement, would not have been possible in the Eastern Settlement (Figure 7.2).

Settlement decline

Changing trade

A key motivation for the settlement of Greenland by the Norse was to gain access to walrus ivory and furs, which are characteristic items of early Viking low-bulk, high-value trade in prestige goods. In the late tenth to early eleventh centuries when the Norse Greenland settlements were becoming established the European market for ivory and fur was buoyant and favourable. Economies changed. The development and expansion of the trade in dried Atlantic cod around AD 1100 was to have widespread economic impacts throughout Europe, which probably did not work to the advantage of the Norse Greenlanders who did little if any fishing. In the Middle Ages Hansa merchants in collaboration with Novgorod and other Russian city states developed the fur trade from the Baltic region northwards into the White Sea; elephant ivory from Africa began to provide unbeatable competition to walrus ivory in European markets and, perhaps most importantly, religious art increasingly moved away from the use of ivory (Roesdahl, 2005). The Black Death of AD 1347–1351 and subsequent plagues heavily depleted the population in Europe (Gottfried, 1983), and in Norway the loss of 30–50% of the population led to an economic collapse. Other developments could have also further eroded the trade position of the Greenlanders: the development of hemp ropes, for example, may have effectively replaced a market for cables made from walrus hide. Add increasing operational difficulties in Greenland (caused by colder, stormier weather and more sea ice) to a fundamental erosion of the Norse Greenlanders' economic position, then, their situation could have become dire (Dugmore et al., 2007a).

Under these circumstances it is probable that the limits to adaptation were defined by the constraints of adaptation through enhancement and intensification of existing activities. In other words, more effort into making established practice better or more efficient could not on its own meet the challenges faced by the Norse in fourteenth and fifteenth century Greenland. Worse still, even their ability to carry on 'business as usual' could have been fatally undermined by population decline.

Limits to adaptation: questions of resilience

Any attempt to devise an integrated model for the range of natural and human transformations affecting Norse Greenland in the fourteenth and fifteenth centuries faces a number of challenges, most importantly, how to combine the interactions of fast and slow variables operating over both large and small spatial scales. These variables are both human (including migration and colonization, settlement patterns, subsistence choices, social and economic organization, trade and kinship connections) and physical (including snowfall and sea ice distribution, storminess, growing season, the populations of marine mammals and their movements). For both human and natural systems the ways in which incremental changes may build through time and their consequences, the nature of cross-scale interactions, the occurrence of meta-stable states and the circumstances under which threshold crossing changes may occur are all important factors.

The concept of *resilience* is an idea that adds depth to ideas of adaptation and is a potentially useful aid to understanding the end of Norse Greenland. Gunderson and Holling (2002) have used the *adaptive cycle metaphor* to characterize the behaviour of far-from-equilibrium systems (Figure 7.4a). Different adaptive cycles can be used to illustrate the behaviour of fast and slow variables operating over varied spatial scales, and they can be grouped in a *panarchy* to explore the interactions between them (Figure 7.4b) Some aspects of the Greenland story appear to fit the looping heuristic framework of the panarchic formulation (Figure 7.4b) but other aspects do not.

Figure 7.4a provides a useful summary of the Norse settlement, and one that has a series of bold implications framed within it. Although the cycle is broken within the α (reorganization) and Ω (release) phases, the ribbon form of the diagram implies a

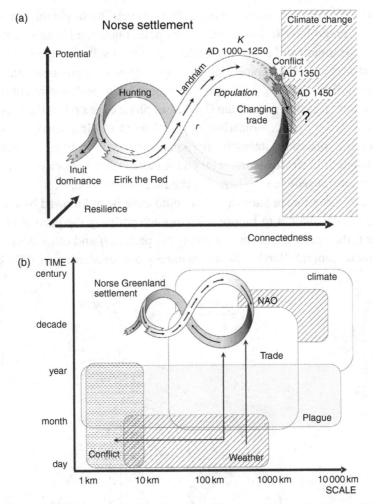

Figure 7.4 (a) The *panarchy* metatheory (Gunderson and Holling, 2002). It hypothesizes cycles of exploitation (*r* dominated) to conservation (*K* dominated) followed by collapse/release (Ω) and reorganization (α): 'connectedness' of the system increases from left to right, 'potential' from bottom to top and 'resilience' reduces into the third dimension (into the page); phases *r* to *K* are conceived to involve reduced resilience. (b) Overall pattern of change in Greenland between the tenth and sixteenth centuries AD described as a result of the interplay between fast and slow variables operating at different spatial scales that may be conceived as a 'panarchy'. NAO, North Atlantic Oscillation.

pathway of development with an inevitable crisis; on one level this is a simply reflection of the actual record of change through time – there has been just one sequence of actual events. For this case study to have wider implications a key question is what general processes may be identified that may tell us about how transformations occur elsewhere and what are the potential limits to human adaptations to climate change. Here the very constraints of the 'adaptive cycle' highlight the circumstances where events could have followed a different path. Once the Norse established settlements in the inner fjord areas and decided to base a substantial part of their provisioning strategy on the exploitation of migrating seal populations, then a series of other possibilities became harder, if not impossible to realize without substantial and far-reaching transformations (an Ω (release) phase). As a result the choices made during the initial Norse colonization and settlement of Greenland, followed by a rising level of connection, intensification and investment in fixed resource spaces, social and material infrastructure, might have increased the effectiveness of adaptation but at a cost of reduced resilience in the face of variation.

The possible alternative pathways that could have been followed by the Norse in Greenland are considered in Figure 7.5. For each pathway the pace of change will be different (the speed of movement along the pathway) and other adaptive cycles within a more general North Atlantic panarchy are interacting in different ways.

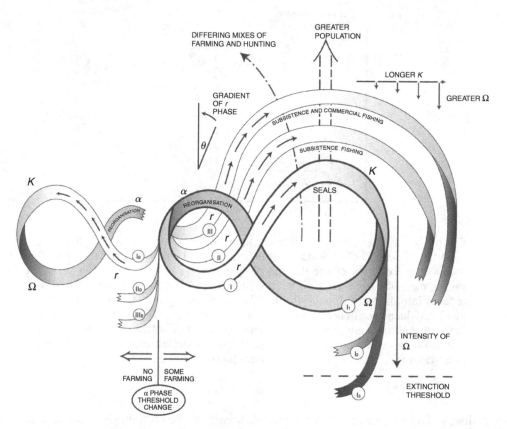

Figure 7.5 Choices made by the Norse during the initial settlement of Greenland. These lead to the sequence of events described in Figure 7.4a and shown here as r I and Ω I_3 pathway. Alternative pathways were possible the 'non-farming' route being the choice of the Inuit, and alternative mixes of hunting and fishing being followed by the Norse elsewhere in the North Atlantic.

For example, changes in both regional climate and trade will affect different *r*-phase pathways in different ways. Cod fishing in Iceland, for example, provided both a means of subsistence and also a commodity to trade, and as a result fishing provided an engine for economic growth in Iceland not available in Norse Greenland.

The possible relation between system resilience and adaptation in the face of climate change for Norse Greenland is explored further in Figure 7.6. Here a key assumption is that the crucially successful adaptive 'tool' of the Inuit is not their toggle harpoon, but their mobility, and the adaptive failure of the Norse was a loss of external trade with the decline in the market for prestige goods such as walrus ivory. Constraints are imposed progressively from the bottom LHS of the figure reducing resilience and forcing adaptation: these may be the result of either climate change or conflict that results in access being blocked to resources such as seal migration routes. Lower populations are more susceptible to catastrophic threshold-crossing events, and have fewer pathways available for avoiding them. Population decline for settlements lacking mobility or connectedness could lead to catastrophic threshold crossing events (Ω-phase transformations) as communally based provisioning systems become less viable.

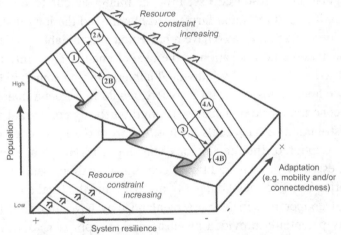

Figure 7.6 Illustration of resilience and adaptation with population and resource constraints. This figure combines ideas of resilience (*x*-axis) and adaptation (*z*-axis) with population (*y*-axis) and resource constraints (defined by diagonal lines). Non-linear, threshold changes are defined by catastrophe cusps in the upper three-dimensional surface. In the example (1) increasing resource constraint (caused by climate change or conflict) may result in adaptive change and a maintenance of population (route 1–2A) – this could be achieved by greater mobility or increased connectedness (trading links). If these adaptive changes are not possible, population may be forced to decline (route 1–2B). Population decline in settlements lacking mobility or connectedness may result in threshold crossing events of a progressively more severe nature. These thresholds may be localized and avoided (route 1–2B) or they may be catastrophic (route 3–4B).

Conclusions

In Norse Greenland successful subsistence strategies were developed and underpinned a well-integrated settlement. Ultimate failure may be attributed to limited resilience and the interplay of cultural, economic and environmental changes at local, regional and continental scales compounded by hostile cultural contacts.

Applications of adaptive cycle thinking at a range of different spatial and temporal scales, and the organization of these cycles in a panarchy provides a bold framework with which to explore limits to adaptation.

The choices made during the initial Norse colonization and settlement of Greenland, followed by a rising level of connection, intensification and investment in fixed resource spaces, social and material infrastructure, increased the effectiveness of adaptation but at a cost of reduced resilience in the face of variation. We propose that the limits of the Norse Greenlanders ability to adapt to climate change was caused by a series of interrelated factors; primarily that the success of the provisioning strategy based on seal hunting was such that initial changes in climate could be met with an intensification of existing practice (but at a hidden cost of reduced resilience).

The rich web of community connection and centuries of legal tradition which provided regulation of resource use, brought inland labour to coastal resource zones and marine food to inland farms, and maintained the long-distance walrus hunt in the Norðursetur, may also have represented a formidable barrier to individual experimentation and innovation in the face of change. Certainly the adoption of winter ringed seal hunting by individual hunters whose catch could only provision their own households, as in Inuit society, would represent a major tear in the social and economic fabric of a society centred on spring communal seal hunting and wide distribution of harp seals to those far from the fjord-side. The barriers to the ready adoption of Inuit technology may not necessarily have been imposed from above, especially when traditional communal sealing methods still proved apparently effective. The social danger of undermining the legal and ritual aspects of communal cooperation in Norse Greenland may have appeared to outweigh the flexibility potentially provided by 'just another' approach to seal hunting. If problems were perceived, one rational choice might be to avoid measures undermining the very communal solidarity that had seen Norse Greenland through so many hard times in the past, and certainly any widespread collapse of the social underpinnings of the richly interwoven subsistence network would be correctly perceived as the most immediate threat to the society as a whole.

Unpredictable shifts in local climate and sea ice could have compromised both terrestrial resources and raised the costs and hazards of the utilization of marine mammals. The progressive Inuit expansion from north-western Greenland into the coastal areas of western and south-western Greenland resulted in new cultural contacts and the potential for conflict to disrupt key Norse activities such as access to the Northern Hunting grounds and the coastal spring seal migrations. The communal organization of seal hunting and probably many other subsistence activities in Norse Greenland effectively buffered individual households against shortfall and occasional loss of life at sea, but carried with it the potential for an extreme disaster producing cascading labour shortages encompassing the whole community. Given the small size of the Norse community and the multiple demands upon labour and boats, any set of factors increasing cost and risk had the potential for exceeding buffering limits. The lack of effective multi-year food storage, the recurring shortage of active adult labour, the demands of the (increasingly unprofitable) Norðursetur and the diffuse or direct competition from the locationally flexible Inuit would certainly have posed daunting challenges to the Norse annual managerial balancing act. If even a few ambitious and able Greenlanders did emigrate back across the Atlantic their absence may have been disproportionately felt by the dwindling community left behind, and imported disease need not have been terribly virulent to have damaged this fine balance of producer and consumer.

When faced with rapid changes in a combination of both natural and human factors the limitations of the pathway chosen were probably too great and social collapse was the result. Certainly any reduction of the total population past a minimum threshold needed to carry out effective communal seal hunting would have triggered a terminal subsistence crisis, and any widespread breakdown of law and community cohesion would probably have been equally fatal. Transition to Ω phase may have been very rapid, and it is possible that no credible management strategy could have averted extinction past the tipping point. The wider lessons of Norse Greenland and the limits to human adaptation to climate change may thus be more complex than we once believed. The Norse Greenlanders did not perish because they were foolishly unwilling to adapt to arctic conditions or because of irrational economic choices. Their real lesson may be far broader and far more frightening in the modern context. It is possible to creatively adapt to new environments, build up centuries of community-based managerial expertise, wisely conserve fragile resources for communal benefit, codify the results, maintain century-scale sustainable patterns of life and society – and yet still face ultimate collapse and extinction.

Acknowledgements

We would like to acknowledge support from the Leverhulme Trust (Footsteps of the Edge of Thule) and funding from the US National Science Foundation Office of Polar Programs Arctic Social Sciences under grant number 0732327 as part of the International Polar Year Humans in the Polar Regions project 'IPY: Long term human ecodynamics in the Norse North Atlantic: cases of sustainability, survival, and collapse'. This publication is a product of the North Atlantic Biocultural Organization (NABO) cooperative.

References

Arneborg, J., Heinemeier, J., Lynnerup, N., Nielsen, H. L., Rud, N. and Sveinbjornsdóttir, A. E. 1999. 'Change of diet of the Greenland vikings determined from stable carbon isotope analysis and C14 dating of their bones', *Radiocarbon* **41**: 157–168.

Church, M. J., Arge, S. V., Brewington, S., McGovern, T. H., Woollett, J., Perdikaris, S., Lawson, I. T., Cook, G. T., Amundsen, C., Harrison R., Krivogorskaya, K. and Dunbar, E. 2005. 'Puffins, pigs, cod, and barley: palaeoeconomy at Undir Junkarinsfløtti, Sandoy, Faroe Islands', *Environmental Archaeology* **10**: 179–197.

Degerbøl, M. 1934. 'Animal bones from the Norse ruins at Brattahlið', *Meddelelser om Grønland* **88**: 149–155.

Degerbøl, M. 1941. 'The osseous material from Austmannadal and Tungmeralik', *Meddelelser om Grønland* **89**: 345–354.

Dennis, A., Foote, P. and Perkins, R. 2000. *Laws of Early Iceland: Grágás II – The Codex Regius of Grágás with Material from Other Manuscripts*. Winnipeg: University of Manitoba Press.

Diamond, J. 2005. *Collapse: How Societies Choose to Fail or Survive*. London: Allen Lane.

Dugmore, A. J., Church, M. J., Mairs, K.-A., McGovern, T. H., Newton, A. J. and Sveinbjarnardóttir, G. 2006. 'An over-optimistic pioneer fringe? Environmental perspectives on medieval settlement abandonment in Thorsmork, south Iceland', in Arneborg, J. and Grønnow, B. (eds.) *The Dynamics of Northern Societies*, Studies in Archaeology and History No. 10. Copenhagen: National Museum, pp. 333–344.

Dugmore, A. J., Keller, C. and McGovern, T. H. 2007a. 'Reflections on climate change, trade, and the contrasting fates of human settlements in the North Atlantic islands', *Arctic Anthropology* **44**: 12–36.

Dugmore, A. J., Church, M. J., Mairs, K.-A., McGovern, T. H., Perdikaris, S. and Vesteinsson, O. 2007b. 'Abandoned farms, volcanic impacts and woodland management: revisting Thorjsardalur, the Pompeii of Iceland', *Arctic Anthropology* **44**: 1–11.

Enghoff, I. B. 2003. *Hunting, Fishing, and Animal Husbandry at the Farm Beneath the Sand, Western Greenland: An Archaeozoological Analysis of a Norse Farm in the Western Settlement*. Copenhagen: Danish Polar Centers.

Gad, F. 1970. *The History of Greenland, vol. 1*. London: G. Hurst.

Gottfried, R. S. 1983. *The Black Death: Natural and Human Disaster in Medieval Europe*. New York: Free Press.

Gunderson, L. H. and Holling, C. S. 2002. *Panarchy: Understanding Transformations in Human and Natural Systems*. Washington, DC: Island Press.

Jennings, A. E. and Weiner, N. J. 1996. 'Environmental change in eastern Greenland during the last 1300 years: evidence from foraminifera and lithofacies in Nansen Fjord 68N', *The Holocene* **6**: 179–191.

Jennings, A. E., Hagen, S., Harðardóttir, J., Stein, R., Ogilvie, A. E. J. and Jónsdóttir, I. 2001. 'Oceanographic change and terrestrial human impacts in a post AD 1400 sediment record from the Southwest Iceland Shelf', in Ogilvie, A. E. J. and Jónsson, T.

(eds.) *The Iceberg in the Mist: Northern Research in Pursuit of a 'Little Ice Age'*. London: Kluwer, pp. 83–100.

Lovelock, J. E. 2006. *The Revenge of Gaia*. London: Allen Lane.

McGovern, T. H. 1985. 'Contributions to the paleoeconomy of Norse Greenland', *Acta Archaeologica* **54**: 73–122.

McGovern, T. H. and Jordan, R. H. 1982. 'Settlement and land use in the inner fjords of Godthaab District, West Greenland', *Arctic Anthropology* **19**: 63–80.

McGovern, T. H. and Pálsdóttir, A. 2006. 'Preliminary report of a medieval Norse archaeofauna from Brattahlið North Farm (KNK 2629), Qassiarsuk, Greenland', *NORSEC Zooarchaeology Laboratory Report* **34**: 1–22.

McGovern, T. H., Bigelow, G. F., Amorosi, T. and Russell, D. 1988. 'Northern Islands, human error and environmental degradation: a view of social and ecological change in the Medieval North Atlantic', *Human Ecology* **16**: 225–270.

McGovern, T. H., Amorosi, T., Perdikaris, S. and Woollett, J. W. 1996. 'Zooarchaeology of Sandnes V51: economic change at a chieftain's farm in West Greenland', *Arctic Anthropology* **33**: 94–122.

McGovern, T. H., Perdikaris, S. and Tinsley, C. 2001. 'Economy of Landnám: the evidence of zooarchaeology', in Wawn, A. and Sigurdardottir, T. (eds.) *Approaches to Vinland*. Reykjavik: Sigurdur Nordal Institute, pp. 154–165.

McGovern, T. H., Vésteinsson, O., Friðriksson, A., Church, M. J., Lawson, I. T., Simpson, I. A., Einarsson, A., Dugmore, A. J., Cook, G. T., Perdikaris, S., Edwards, K. J., Thomson, A. M., Adderley, W. P., Newton, A. J., Lucas, G., Edvardsson, R., Aldred, O. and Dunbar, E. 2007. 'Settlement, sustainability, and environmental catastrophe in Northern Iceland', *American Anthropologist* **109**: 27–51.

Meldgaard, J. 1965. *Nordboerne i Grønland*. Copenhagen: Munksgaard.

Perdikaris, S. and McGovern, T. H. In press. 'Codfish and kings, seals and subsistence: Norse marine resource use in the North Atlantic', in Rick, T. and Erlandson, J. (eds.) *Human Impacts on Ancient Marine*. Berkeley: University of California Press *Ecosystems*.

Roesdahl, E. 2005. 'Walrus ivory: demand, supply, workshops, and Greenland', in Mortensen, A. and Arge, S. (eds.) *Viking and Norse in the North Atlantic: Select Papers from the Proceedings of the 14th Viking Congress, Tórshavn 2001*. Tóshavn, Faroe Islands: Societas Scientarium Faeroensis, pp. 182–192.

Simpson, I. A., Adderley, W. P., Guðmundsson, G., Hallsdóttir, M., Sigurgeirsson, M. Á. and Snæsdóttir, M. 2002. 'Land management for surplus grain production in early Iceland', *Human Ecology* **30**: 423–443.

Simpson, I. A., Vésteinsson, O., Adderley, W. P. and McGovern, T. H. 2003. 'Fuel resources in landscapes of settlement', *Journal of Archaeological Science* **30**: 1401–1420.

Simpson, I. A., Guðmundsson, G., Thomson, A. M. and Cluett, J. 2004. 'Assessing the role of winter grazing in historic land degradation, Mývatnssveit, north-east Iceland', *Geoarchaeology* **19**: 471–503.

Van der Leeuw, S. 1994. 'Social and environmental change', *Cambridge Archaeological Journal* **4**: 130–139.

Vibe, C. 1967. 'Arctic animals in relation to climatic fluctuations', *Meddelelser om Grønland* **170**: 1–227.

Woollett, J. W., Henshaw, A. and Wake, C. 2000. 'Palaeoecological implications of archaeological seal bone assemblages: case studies from Labrador and Baffin Island', *Arctic* **53**: 395–413.

8

Sea ice change in Arctic Canada: are there limits to Inuit adaptation?

James D. Ford

Introduction

The impacts of climate change have been particularly profound in Arctic regions (ACIA, 2005; IPCC, 2007), with changes in the sea ice standing out (Kerr, 2007). For the Arctic as a whole, ice thickness and extent are decreasing, the ocean is freezing up later in the year and breaking up earlier, and the ice-free open water period is extending (Holland et al., 2006; Overland and Wang, 2007). Similar trends have been documented in the Canadian Arctic (Barber and Hanesiak, 2004; Barber and Iacozza, 2004; Nickels et al., 2006; Furgal and Prowse, 2008; Laidler and Ikummaq, 2008). Anomalous ice conditions are concentrated in recent years of the record, particularly 2002–2007 (Stroeve et al., 2007). Sea ice change is occurring in the context of other changes in the Arctic, and has been attributed to greenhouse gas emissions (IPCC, 2007).

Changing sea ice conditions have already had negative impacts on the livelihoods of the Arctic's Inuit population, many of whom rely on the frozen ocean surface for seasonal transportation between communities and as a platform for culturally important hunting activities (Correll, 2006; Nickels et al., 2006; Ford, 2008a; Ford et al., 2008b). Climate models predict sea ice change to continue into the foreseeable future (IPCC, 2007), with recent research ranking sea ice as the global system at greatest threat to crossing a 'tipping point' with climate change (Lenton et al., 2008). Studies characterizing the processes shaping community vulnerability to sea ice change, and their relation to climatic and non-climatic determinants, however, have only recently been the focus of emerging interest in the literature. Few studies have explored whether there are limits to adaptation to sea ice change, beyond which negative social, economic, and cultural trade-offs have to be made and community well-being will be compromised.

Adapting to Climate Change: Thresholds, Values, Governance, eds. W. Neil Adger, Irene Lorenzoni and Karen L. O'Brien. Published by Cambridge University Press. © Cambridge University Press 2009.

An established way of examining vulnerability to future climate change is through the identification and examination of cases with historically analogous conditions (Ford et al., 2006a; Smit and Wandel, 2006; McLeman et al., 2007; Ford, 2008a; Van Aalst et al., 2008). Examining experience and response to climate variability, change and extremes provides an empirical foundation for characterizing how communities manage and experience climate-related risks; identifying processes and conditions which determine the efficacy, availability and success of adaptations; developing greater understanding of how social and biophysical processes shape vulnerability; establishing a range of possible societal responses to future change; and helping identify and characterize limits to adaptation (Ford and Smit, 2004). This chapter examines how Inuit in the community of Igloolik, in the Canadian territory of Nunavut, experienced and responded to anomalous ice conditions in 2006, focusing on use of the ice for hunting and travel. The case study provides an opportunity to assess the processes shaping vulnerability based on empirical analysis, with limits and barriers to adaptation explored using real-time observation and community discussion. The insights are relevant to small Inuit communities across the Canadian North.

Study area

Igloolik is a coastal Inuit community of approximately 1500 people located on Igloolik Island in northern Foxe Basin in the Canadian territory of Nunavut (Figure 8.1). Located off the east coast of Melville Peninsula, the island and the mainland have a flat topography and a polar tundra climate. Sea ice dominates the surrounding waters for much of the year, with the ocean freezing in mid to late October and the ice breaking up at the end of July (Laidler and Ikummaq, 2008). The settlement has expanded rapidly since the 1960s, with both the wage economy and traditional resource harvesting sector important in the community. The harvesting of marine and terrestrial mammals is widely practised, as is common in most Inuit communities, with 'country foods' (traditional foods harvested by Inuit) contributing a significant portion of the community's nutritional intake (Ford et al., 2007). The ice acts as an essential hunting platform, from which walrus, ringed seal and polar bear are harvested (Figure 8.2). The frozen ocean surface also provides an important transportation medium, allowing seasonal access to the mainland to the south (Melville Peninsula) and Baffin Island to the north, which comprise important caribou hunting grounds, Arctic char fishing lakes, and connections to other communities including Hall Beach 100 kilometres to the south (Laidler et al., 2009). The community is representative of small Inuit communities across the Canadian North: it is remote, coastal and largely Inuit, and the harvesting of renewable resources has cultural, social and economic importance to community members (Ford et al., 2008a).

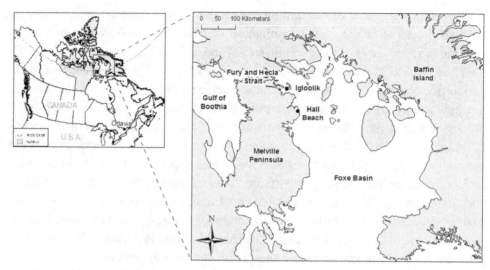

Figure 8.1 Igloolik, Nunavut.

Figure 8.2 Harvesting is an important activity for Inuit: an Igloolik hunter waits
at the ice edge for ringed seals.

Methods

Semi-structured interviews and focus groups with community members were conducted in 2006 and 2007 with the aim of: (1) documenting Inuit knowledge on sea
ice conditions and vulnerability; (2) characterizing change in sea ice conditions
over time; (3) identifying and characterizing conditions and processes shaping

vulnerability to sea ice change; and (4) identifying limits and barriers to adapting to future climate change. Inuit populations possess detailed, location-specific knowledge of the sea ice, built up through personal observation and experience, and from shared experience of members of the community (Laidler, 2006). It is therefore particularly appropriate to use Inuit knowledge to document the physical nature of ice extremes and change over time, and to identify and characterize community vulnerability and adaptability to sea ice risks (Ford et al., 2006a). Moreover, working closely with individuals and households who will be affected by climate change is particularly important when identifying limits to adaptation. In many cases, limits are likely to be socio-cultural in nature, and as such can be best specified by local people.

To provide insights into vulnerability and change in sea ice conditions, this chapter also draws upon previous research in Igloolik by Ford et al. (2006a; 2008a; 2008b) and Ford (2008a). The research also utilizes interviews contained within the Igloolik Oral History Project (IOHP), a database containing over 500 interviews with local residents on a variety of topics that was started in 1986. The information in the IOHP captures lived experience and oral history spanning the twentieth century. In conjunction with the interviews and focus groups, the IOHP interviews provide a baseline from which to evaluate and compare sea ice conditions and a context for specifying barriers and limits to adaptation.

Sea ice extremes, vulnerability and adaptation: lessons from 2006

Inuit participants described the nature of the sea ice in 2006 as significantly departing from long-term norms (Ford et al., 2009). As highlighted in Table 8.1, the ocean froze and became usable three to four weeks later than normal, with little remnant ice during the summer. Inuit observations are largely consistent with instrumental sea ice data for the region (Ford et al., 2009). In many respects, ice conditions in 2006 are representative of projected sea ice changes for the Igloolik region by mid-century (Dumas et al., 2006). This chapter uses community experience and response to sea ice conditions in 2006 to explore vulnerability and limits to adaptation.

The food system

Sea ice conditions and country foods

Focus group participants identified sea ice conditions in the summer of 2006 as constraining the ability to procure traditional foods (Ford, 2008a). It was noted

Table 8.1 *Deviation of ice conditions in the Igloolik region in 2006 compared to the long-term norm, based on Inuit traditional knowledge (TK) and instrumental data (ID)*

Time of year	Long-term norms	2006
Autumn (September–December)	Ocean freezes mid October (TK, ID). Ice thickens rapidly, usable by mid to late October (TK, ID) 1969–2005: trend of later freeze-up of 1 week per decade (ID)	Freeze-up on 5 November but then deteriorates. Not usable until late November (TK, ID) Latest freeze-up in living memory (TK) 3rd latest since 1969 (ID). Very slow to reach thickness at which it can be used (TK)
Spring (May–July)	Break-up end of July (TK, ID) 1982–2005: trend of 6 days earlier per decade (ID)	Ice breaks up end of July (TK, ID)
Summer (August, September)	Open water in Foxe Basin with significant floating ice (ice that is not attached to the land and is constantly moved by ocean currents and the wind) (ID, TK)	Large areas of open water without floating ice (TK, ID). Unprecedented in instrumental data (ID) and according to Inuit elders (TK)

Source: Based on Ford et al. (2009).

that rapid disappearance of the ice after break-up and almost complete absence of floating ice (walrus habitat) significantly reduced the ability to hunt walrus in northern Foxe Basin. Participants described walrus as located further south and beyond the range of many local hunters. Those walrus that remained in northern Foxe Basin congregated along the shoreline making them difficult to find. This affected the food system by decreasing the availability of walrus meat. It was particularly problematic for elders and more mature community members for whom walrus is an important food source, providing a source of nutrients, vitamin A and protein. Fermented and aged walrus meat (*igunak*), meanwhile, is considered a local delicacy. Limited walrus meat, especially during early summer when the meat is cached for ageing, resulted in limited production of *igunak*. Young Inuit in this study and in previous research, however, admitted to avoiding walrus due its strong and acquired taste, and hard work required to harvest the animal. Therefore, young Inuit were less affected by limited availability.

During fall, the sea ice is used for travel to caribou hunting grounds and char fishing lakes on Melville Peninsula, and is used as a platform for hunting ringed seals. Ringed seals are hunted at small pockets of open water that remain as the ice is freezing and also in bays and points of land where cracks open up (Laidler and Ikummaq, 2008). Participants noted that the slow freeze-up in 2006 limited the ability to travel to the mainland because the ice was too thin and dangerous to use until late November, except for short period at the beginning of the month. This reduced the ability to hunt seals on the ice and to travel to hunt caribou and char, reducing the availability of food from these animals. The appeal and importance of these species in the local diet magnified the implications of limited availability.

Even when the ice did become safe to use in late November, it wasn't until early December that the ice reached a thickness capable of supporting direct travel to the mainland. These routes normally start being used for travel to char fishing lakes and for caribou hunting between late October and early November. The resulting detours added travel time and in some cases doubled the travel distance, and hence cost, thereby limiting access to those with time and/or sufficient financial resources.

Community response

In light of reduced access and availability of wildlife in summer and fall, Inuit identified having to purchase more store food to meet their dietary needs (Ford, 2008a). For elders, active hunters and those with a strong connection to the land-based economy, switching to store-bought foods was not considered an equal trade-off. Country foods are preferred because they are believed to be healthier, fresher, better tasting, and have cultural significance – an observation noted across Inuit communities (Kuhnlein and Receveur, 2007). Additionally, for those who rely on country foods, switching to store-bought foods was not always an option due the high cost of commercial goods in the north and limited access to financial means (Ford, 2008a). Participants mentioned that those who did not have enough money to purchase food in 2006 had to rely on family members to share store food or use the food-bank. Some reported going without food for a couple of days and having to skip meals. The additional stress placed on household income by reliance on purchased food reinforced the negative impacts of ice conditions by adding financial burden alongside the cost of other climate adaptations, including having to use more gas to hunt and travel longer distances.

In previous years the sharing of country foods has maintained access to food and nutrition in light of environmental stress (Ford et al., 2006a; Ford, 2008a). It was noted that sharing was also important in 2006, supplementing the diet of those who were not able to hunt or who were not successful in procuring country food. Compared to previous years, however, the extent to which sharing was able to maintain a supply of country food for those without access was constrained. Particularly for walrus and caribou there was too little meat to satisfy demand.

Indeed, this case study suggests that important sharing networks, particularly inter-household networks, weaken at times of severe food stress; an observation worthy of further examination (Ford, 2008a). Flexibility in hunting behaviour has also previously enabled Inuit in Igloolik to manage fluctuations in wildlife availability, and was utilized in 2006 to maximize hunting success. In summer, hunters searched the northern Foxe Basin coastline looking for walrus in areas where they are not usually found. This resulted in partial success for some hunters, although the extra gasoline costs entailed were prohibitive for many.

Multiple stressors

Other climatic conditions magnified the impacts of ice conditions in 2006 on the access and availability of country foods. Heavy, powdery snow on the mainland in November was described as making caribou difficult to hunt. Powdery snow makes it hard to use snowmobiles, especially when towing animal carcasses, and uses a lot of gas, thereby affecting the amount of harvest that can be brought back to the community. Recent years have also witnessed the caribou migrating away from the Igloolik region. This trend, part of a natural long-term cycle, increased the difficulties of access associated with powdery snow. In fall 2006, it was therefore difficult to offset reduced sea mammal harvest with caribou meat. In this way, other climate-related conditions reduced the ability to switch species hunted – a key adaptive response to climatic stress documented previously in Igloolik and in Inuit communities across the Arctic (Riewe and Oakes, 2006). Non-climatic conditions, including higher oil and commodity prices, and underlying vulnerabilities (or slow variables) which have weakened the food system in Igloolik over time, also exacerbated the implications of sea ice conditions in 2006 (Ford et al., 2009).

Safety of ice travel

Vulnerability

October and November are widely regarded as the most dangerous times of year for using the sea ice, with areas of uncertain thickness common. The dangers related to sea ice travel in early autumn are compounded if snow falls on thin ice, as it insulates the ice underneath, promoting ice melt from the heat of the ocean (Laidler and Ikummaq, 2008). As perceived locally, the late and gradual freeze-up in 2006 increased the danger of using the ice. It was reported that hunting equipment was damaged in accidents where people fell through the ice.

Increasing physical danger of using the ice was compounded by the rising price of gasoline used to power snowmobiles (Figure 8.3). In combination with increasing commodity prices in general, this affected how community members in Igloolik experienced and responded to climatic extremes. Participants noted taking

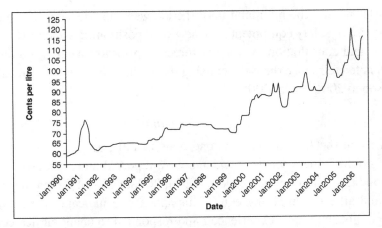

Figure 8.3 Price of regular unleaded gasoline at self-service filling stations in Yellowknife (NWT) ($Can), which can be considered indicative of prices in Nunavut. Comparable price data for Nunavut are not available. (Source: StatsCanada 2006.)

as little spare fuel as possible when making hunting or recreational trips due to the cost. Redundancy is a key feature of Inuit adaptability and in previous research in Igloolik community members noted taking more fuel than necessary to cover all eventualities (Ford et al., 2008b). Rising prices have thus reduced the ability to make extra preparations, particularly among those with a low cash flow, eroding the safety net provided by redundancy. Indeed, participants drew attention to the increasing number of people running out of gasoline while hunting in 2006. For those without adequate land skills and/or equipment, such incidents are serious and can result in loss of life.

Adaptability

In the context of increased potential danger of using the ice in 2006, participants noted that the safety implications of sea ice extremes were relatively minor compared to previous years. There were no injuries and equipment damage/loss was minimal. As documented in previous research in Igloolik – and elsewhere in Nunavut (Riewe and Oakes, 2006) – accumulated experience of adapting to sea ice variability and extremes, and the detailed knowledge of sea ice processes held by the more mature and elder hunters helped moderate the impacts of sea ice extremes. During early stages of ice formation, for example, focus groups participants explained their vigilance in evaluating the condition of the ice, using visual clues and testing using the harpoon to judge safety and avoid dangerous areas. Other reported adaptations, enabled by Inuit knowledge, included avoiding travel until December for areas which in a normal year freeze up late, and delaying travel to Baffin Island and the mainland until the ice was judged to be safe to use. The increasing use of safety equipment such as global positioning systems (GPS), satellite phones and consultation of weather forecasts prior to travelling on the ice has also been noted to reduce the risks of using the ice in a changing climate, and were widely used in 2006 (Ford, 2008b).

Emerging adaptability

Adaptability in the face of sea ice conditions in fall 2006 contrasts to previous years in Igloolik, where serious incidents were reported and attributed to dangerous ice conditions (Ford et al., 2006a, 2008a, 2008b). Given the nature of the ice in 2006 and all else remaining equal, this appears to indicate increasing adaptability. Igloolik Inuit have experienced and responded to changing ice conditions and increasing occurrence of extremes since the mid to late 1990s. Experience with stress has enabled social learning, whereby trial-and-error experience has developed and refined means of adapting. This is important in helping reduce the dangers of using the ice in a changing climate. Indeed the evolution of old – and creation of new – heuristics in response to change defines the nature of Inuit traditional knowledge, which is developed and acquired through experiential trial-and-error learning. This collective social memory, with accumulated experience responding to sea ice extremes in a changing climate, framed individual practice and decision-making in the fall of 2006 to moderate the increased risks of using the ice.

Changing sea ice conditions: are there limits to adaptation?

According to Inuit knowledge, as well as instrumental data sets, sea ice conditions in 2006 departed from the long-term mean. This provided an opportunity to identify and characterize limits to adaptation based on actual observation. The study demonstrates the adaptability of Inuit, suggesting that while barriers to adaptation may exist there are few limits to adaptation, where a limit implies an absolute barrier to adapting (Adger et al., 2009). Where limits exist, they will be differentiated by socio-economic group and activity (food consumption, hunting).

The food system and ice extremes

Limits to adaptation are apparent in the food system and are most likely to be cultural in nature. Switching to store-bought food may be considered an adaptation to sea ice extremes in the sense that individual and household access to food is maintained. It is unlikely that people will die of starvation in modern Inuit communities due to an inability to hunt or fish, a cause of death reported as late as the

1950s elsewhere in the eastern Canadian Arctic (Damas, 2002, p. 277). However, the intrinsic social and cultural importance of hunting, consuming country foods and sharing the products of the hunt is irreplaceable, representing a limit to adaptation to projected future changes in sea ice conditions if they continue to constrain food procurement at important times of the year. However, a growing population of young Inuit – the section of the population who will experience the most profound climate change effects – are more dependent on the store for their food, with country foods occasionally consumed when available but not actively sought or harvested. This trend is being noted across Inuit communities (Kuhnlein and Receveur, 2007). While country foods still remain culturally important to younger generations, for many constrained availability at certain times does not represent a limit to adaptation. Switching to reliance on store food at key times of constrained country food availability is an acceptable adaptive response for such groups.

The food system example also indicates that sea ice conditions themselves are more likely to act as a barrier to adaptation in the food system than a limit to adaptation. In 2006, for example – widely regarded as one of the most extreme years for sea ice conditions on record – harvesting was only constrained during freeze-up and summer, and Inuit community members exploited new opportunities with the longer open water period. Indeed, projections of a lengthening of the summer open water period by as much as 4 weeks by 2050 and 8 weeks by 2100 (Dumas et al., 2006), could have positive impacts for those with boat access. Summer open water is widely used for fishing and whale harvesting, with boat transportation used between communities and for the summer resupply. More open water may even create opportunities for commercial fishing which is currently limited by the length of open water. In this case, the barrier to adaptation is likely to be boat access, with fewer people having access to boats which are more expensive to purchase and maintain than snow machines which are used if there is sea ice cover. Moreover, the negative implications of sea ice conditions in 2006 were exacerbated by socio-economic conditions; determinants currently being targeted by policy programmes and areas that future adaptation policy could address (Ford et al., 2007). With subsidized gas prices in summer 2006, for example, walrus hunters would have been able to travel the extra distance to hunt successfully.

Physical risk and safety in a changing climate

Emerging adaptability with regards to the safety of hunting and travelling on the sea ice in a changing climate indicates that adaptive capacity is likely to reduce the likelihood of there being limits to adaptation. A combination of risk management, avoidance and sharing strategies, facilitated by Inuit knowledge, enabled the increased physical risks of using the ice in the autumn of 2006 to be moderated.

In contrast to previous years, it was noted that while the ice was dangerous in the autumn, there were few accidents and/or damage to equipment. This appears to be indicative of increasing adaptability over time. Sea ice conditions in 2006 need to be situated in the context of changing sea ice and other environmental conditions documented in Igloolik since the mid to late 1990s. Through trial-and-error experience of these changes, innovation, the development of new heuristics alongside memory of traditional ways, and acceptance of uncertainty, Inuit knowledge of the ice has evolved to reframe decision making in light of changed conditions. Moreover, the history of Inuit survival in the Arctic demonstrates the evolution of historical or traditional practices in response to changing conditions (Ford et al., 2006a).

Notwithstanding the apparent absence of limits to adaptation, barriers to adaptation currently exist and are likely to be important in determining vulnerability and adaptability to future change. For example, there is concern that as today's elders and experienced hunters die, many of the younger generations do not have the knowledge or land skills to promote and conduct safe and successful hunting – a concern noted across Nunavut (Ford et al., 2006a, 2006b; Gearheard et al., 2006; Furgal and Prowse, 2008). With a median age of 19 years and 41% of the population under 15 years of age (StatsCanada, 2006), many of today's children in Igloolik will be young adults and beginning to hunt and travel on the ice when the impacts of climate change become increasingly pronounced, thereby increasing exposure to climate change effects. Moreover, the ability of adaptability to 'emerge' in response to future changes will be challenged if 'land-based' knowledge and skills are not developed among today's younger generations. Identifying changing conditions, evaluating what they mean for land-based activities, and responding to reduce the risks are all contingent upon a detailed understanding of environmental conditions and 'the land' embodied in Inuit traditional knowledge. Development of cultural programmes has been advocated as a means of promoting such skills among youth and should be central to future adaptation planning initiatives (Ford, 2008b).

The economic cost of adapting also represents a barrier to adapting. In many instances, adaptation involves developing new, often longer-distance travel routes that avoid dangerous areas and utilizing new safety equipment including immersion suits, satellite phones and GPS. Access to these important technologies is unequal and depends on access to sufficient financial resources. Especially for hunting households and those with limited access to the waged economy, affording these adaptations will continue to be a challenge. A priority for future adaptation policy in Nunavut is the provision of adequate harvester assistance to enable such adaptations to be accessible to those in need (Ford et al., 2007; Ford, 2008b).

Conclusion

By focusing on the impacts, vulnerability and adaptability to sea ice conditions of Igloolik Inuit in 2006, this chapter provides insight into the potential implications of future sea ice change. More generally, the study provides insights into vulnerability and adaptation of Inuit in small communities across northern Canada. It is highlighted that vulnerability to a specific climate-related event can be exacerbated or moderated by changes in other climatic conditions and non-climatic stresses. The chapter illustrates that vulnerable groups often emerge due to the synergistic interaction of climatic and non-climatic stresses which combine to overwhelm adaptability. It is highlighted how certain groups and sectors are at greater risk than others, dependent on livelihoods and socio-economic characteristics. The food system, for instance, is susceptible to climatic extremes. The study also (re)affirms the adaptability of Inuit, with changing conditions stimulating social and institutional learning.

Community experience of and response to sea ice extremes in 2006 also provides an opportunity to empirically assess the existence and nature of limits to adaptation. *Limits* to adaptation are likely to be cultural in nature, where trade-offs necessary to maintain food security compromise social and cultural values. While these trade-offs constitute adaptation at its most basic (i.e. survival), the associated cultural impacts affect the core of how Inuit define themselves. However, if Inuit society evolves according to current trends, with increasing importance of the wage economy and store-bought food consumption, it may be that country foods have less cultural and economic importance when the worst impacts of climate change manifest themselves. Indeed, limits to adaptation are not static and evolve over time in response to climatic and non-climatic stresses. Limits to adaptation are also differentiated. For people not dependent on country foods, for instance, decreased access may be little more than an inconvenience. Barriers to adaptation are likely to be more common in creating vulnerability to sea ice change, with access to financial resources and traditional knowledge, in particular, constraining Inuit adaptive capacity. Importantly, adaptation barriers can be addressed with policy support. It is also noteworthy that it is social–cultural–economic conditions and processes that emerge as the determinants of adaptation barriers and not the physical nature of change in sea ice conditions per se. Addressing non-climatic determinants of vulnerability in the harvesting sector should be a priority of the climate change adaptation plan currently being developed by the government of Nunavut and priority for assistance from the federal government.

Acknowledgements

This research was funded by the International Polar Year CAVIAR project, ArcticNet, SSHRC and CCIAP. The author is grateful for the comments and contributions of Gita Laidler, William Gough, George Wenzel, Wayne Pollard and Karen O'Brien. Lea Berrang Ford produced Figure 8.1. The author is also grateful to community members in Igloolik who made the research possible and who gave their time to be interviewed. In particular, the contributions of Celina Irngaut, Kevin Qrunnut and John MacDonald are acknowledged.

References

ACIA 2005. *Arctic Climate Impacts Assessment*. Cambridge: Cambridge University Press.

Adger, W. N., Dessai, S., Goulden, M., Hulme, M., Lorenzoni, I., Nelson, D., Naess, L. O., Wolf, J. and Wreford, A. 2009. 'Limits and barriers to adaptation', *Climatic Change* **93**: 335–354.

Barber, D. and Iacozza, J. 2004. 'Historical analysis of sea ice conditions in M'Clintock Channel and Gulf of Boothia, Nunavut: implications for ringed seal and polar bear habitat', *Arctic* **57**: 1–14.

Barber, D. G. and Hanesiak, J. M. 2004. 'Meteorological forcing of sea ice concentrations in the southern Beaufort Sea over the period 1979 to 2000', *Journal of Geophysical Research* **109**: doi 10.1029/2003JC002027.

Correll, R. W. 2006. 'Challenges of climate change: an Arctic perspective', *Ambio* **35**: 148–152.

Damas, D. 2002. *Arctic Migrants / Arctic Villagers*. Montreal: McGill–Queens University Press.

Dumas, J., Flato, G. and Brown, R. D. 2006. 'Future projections of landfast ice thickness and duration in the Canadian Arctic', *Journal of Climate* **19**: 5175–5189.

Ford, J. D. 2008a. 'Vulnerability of Inuit food systems to food insecurity as a consequence of climate change: a case study from Igloolik, Nunavut', *Regional Environmental Change*, in press.

Ford, J. D. 2008b. 'Climate, society, and natural hazards: changing hazard exposure in two Nunavut communities', *Northern Review* **28**: 51–71.

Ford, J. D. and Smit, B. 2004. 'A framework for assessing the vulnerability of communities in the Canadian Arctic to risks associated with climate change', *Arctic* **57**: 389–400.

Ford, J. D., MacDonald, J., Smit, B. and Wandel, J. 2006a. 'Vulnerability to climate change in Igloolik, Nunavut: what we can learn from the past and present', *Polar Record* **42:** 1–12.

Ford, J. D., Smit, B. and Wandel, J. 2006b. 'Vulnerability to climate change in the Arctic: a case study from Arctic Bay, Canada', *Global Environmental Change* **16**: 145–160.

Ford, J. D., Pearce, T., Smit, B., Wandel, J., Allurut, M., Shappa, K., Ittusujurat, H. and Qrunnut, K. 2007. 'Reducing vulnerability to climate change in the Arctic: the case of Nunavut, Canada', *Arctic* **60**: 150–166.

Ford, J. D., Gough, B., Laidler, G., MacDonald, J., Qrunnut, K. and Irngaut, C. 2009. 'Sea ice, climate change, and community vulnerability in northern Foxe Basin, Canada', *Climate Research* **37**: 138–154.

Ford, J. D., Pearce, T., Gilligan, J., Smit, B. and Oakes, J. 2008a. 'Climate change and hazards associated with ice use in Northern Canada', *Arctic, Antarctic and Alpine Research* **40**: 647–659.

Ford, J. D., Smit, B., Wandel, J., Allurut, M., Shappa, K., Qrunnut, K. and Ittusujurat, H. 2008b. 'Climate change in the Arctic: current and future vulnerability in two Inuit communities in Canada', *Geographical Journal* **174**: 45–62.

Furgal, C. and Prowse, T. 2008. 'Northern Canada', in Lemmen, D., Warren, F., Bush, E. and Lacroix, J. (eds.) *From Impacts to Adaptation: Canada in a Changing Climate 2007.* Ottawa: Government of Canada, pp. 57–118.

Holland, M. M., Bitz, C. M. and Tremblay, B. 2006. 'Future abrupt reductions in the summer Arctic sea ice', *Geophysics Research Letters* **33**: 5.

IPCC 2007. *Climate Change 2007: The Physical Science Basis. Contribution of Working Group I to the Fourth Assessment Report of the Intergovernmental Panel on Climate Change.* Cambridge: Cambridge University Press.

Kerr, R. A. 2007. 'Is battered Arctic sea ice down for the count?', *Science* **318**: 33–34.

Kuhnlein, H. and Receveur, O. 2007. 'Local cultural animal food contributes high levels of nutrients for Arctic Canadian indigenous adults and children', *Journal of Nutrition* **137**: 1110–1114.

Laidler, G. 2006. 'Inuit and scientific perspectives on the relationship between sea ice and climatic change: the ideal complement?', *Climatic Change* **78**: 407–444.

Laidler, G. and Ikummaq, T. 2008 'Human geographies of sea ice: freeze/thaw processes around Igloolik, Nunavut, Canada', *Polar Record* **44**: 127–153.

Laidler, G., Ford, J., Gough, W. A. and Ikummaq, T. 2009. 'Travelling and hunting in a changing Arctic: assessing Inuit vulnerability to sea ice change in Igloolik, Nunavut', *Climatic Change*, in press.

Lenton, T. M., Held, H., Kriegler, E., Hall, J. W., Lucht, W., Rahmstorf, S. and Schellnhuber, H. J. 2008. 'Inaugural article: Tipping elements in the Earth's climate system', *Proceedings of the National Academy of Sciences of the USA* **105**: 1786–1793.

McLeman, R., Mayo, D., Strebeck, E. and Smit, B. 2007. 'Drought adaptation in rural eastern Oklahoma in the 1930s: lessons for climate change adaptation research', *Mitigation and Adaptation Strategies for Global Change* **13**: 379–400.

Nickels, S., Furgal, C., Buell, M. and Moquin, H. 2006. *Unikkaaqatigiit: Putting the Human Face on Climate Change: Perspectives from Inuit in Canada.* Ottawa: Natural Resources Canada.

Overland, J. E. and Wang, M. 2007. 'Future regional Arctic sea ice declines', *Geophysical Research Letters* **34**: L17705.

Riewe, R. and Oakes, J. 2006. *Climate Change: Linking Traditional and Scientific Knowledge.* Winnipeg: Aboriginal Issues Press.

Smit, B. and Wandel, J. 2006 'Adaptation, adaptive capacity, and vulnerability', *Global Environmental Change* **16**: 282–292.

StatsCanada 2006. 'Population counts from the 2006 Census', available at www.statcan. gc.ca (accessed 31 March 2008).

Stroeve, J., Holland, M. M., Meier, W. N., Scambos, T. and Serreze, M. 2007. 'Arctic sea ice decline: faster than forecast', *Geophysical Research Letters* **34**: L09501.

Van Aalst, M. K., Cannon, T. and Burton, I. 2008. 'Community level adaptation to climate change: the potential role of participatory community risk assessment', *Global Environmental Change* **18**: 165–179.

Part II

The role of values and culture in adaptation

9

The past, the present and some possible futures of adaptation

Ben Orlove

Introduction

Adaptation is a familiar word in the conversations of people who are concerned about climate change. They use it to describe the processes of adjusting to climate change and its impacts. It describes the actions that must be taken to reduce or eliminate harm, actions whose necessity is unquestionable once the realization strikes that no mitigation plan will be able to bring global warming to a quick halt.

This chapter calls for some reflection on these uses of the word. Though it accepts the urgency of the need to respond to climate change, it questions the naturalness of the term adaptation – the way that it is taken for granted as a key element in climate change policy – and it finds some limits to the term. This chapter suggests that the word does not always capture the full impacts of climate change and that it does not always represent accurately either the perceptions of the people affected by these impacts or the range of alternatives open to them.

This chapter develops these reservations through several sections: (1) a review of the history of the term and its use by international organizations; (2) a presentation of a local community affected by climate change, who constitute one case of the people for whom adaptations are proposed; (3) a discussion of the perceptions of climate change by this local community; (4) a review of four organizations that serve as intermediates between the international and local levels, and that have all adopted the word; and (5) a set of considerations on the way that the term operates in the relations among the international organizations, the intermediary organizations and the local community. Reduced to a single, if long, sentence, this chapter argues that the term serves the international and intermediary organizations far better than the local communities who feel the impacts most

Adapting to Climate Change: Thresholds, Values, Governance, eds. W. Neil Adger, Irene Lorenzoni and Karen L. O'Brien. Published by Cambridge University Press. © Cambridge University Press 2009.

directly; rather than transforming the great fear of a hotter planet into sustained action to address the consequences of climate change, the term can create a sense of complacency.

The history of the concept of adaptation

A brief review of the history of the word adaptation shows that the term came into use recently and in specific contexts. In turn, this history suggests that the term is accompanied by conceptual baggage that influences its meaning.

When the word first appeared in English in the early seventeenth century, it had common-sense, non-technical meanings. The *Oxford English Dictionary* offers several closely related definitions dating to this period, including a process of change ('the action or process of adapting, fitting, or suiting one thing to another') and the outcome of this process of change ('the condition or state of being adapted; adaptedness, suitableness'). By the second half of the nineteenth century, the word began to acquire specific meanings in science and other specialized areas. In *On The Origin of Species*, Darwin (1859) used the word to mean the organic modification by which an organism or species becomes adapted to its environment. A few decades later specialists in the field of optics applied it to the process by which the eye adjusts to changes in the intensity or colour of light. During this period, the word also began to refer to a copy of an object, made on purpose to fit new ends or meet new circumstances; it is this meaning that is used when one speaks of a film as being an adaptation of a novel. In the late nineteenth century, the pragmatist philosopher and psychologist John Dewey drew on Darwin's notion of adaptation to refer to the process by which individuals and groups gained knowledge of their environments to respond effectively to them and to modify them to meet their goals; his use of the concept has contributed to another common use of the term, to refer to the capacity of a person, especially a child or adolescent, to adjust to new or changing circumstances.

Loosely following the use by Darwin and other evolutionary biologists, geographers and anthropologists in the middle of the twentieth century, such as Julian Steward in the 1950s and Marvin Harris in the 1960s and 1970s, discussed culture and institutions as key features that allow human groups to make use of the natural resources in their environments, using the term adaptation for this meaning as well. Others emphasized the hazards in environments as the constraints or risks to which humans respond. An important example is Gilbert White's discussion (1945) of the social and political rules, such as zoning regulations, that can reduce damage from floods by lessening vulnerability to them; this work shows the direct influence of Dewey. White and his student Robert Kates widely applied this notion of adaptation to the study of natural hazards.

As a result of such work, the term adaptation was already widely used by the time that global concern about climate change began to grow in the late 1970s and early 1980s. It is interesting to trace the language and concepts that were used in the formation of the Intergovernmental Panel on Climate Change (IPCC) and the United Nations Framework Convention on Climate Change (UNFCCC), the two key institutions that have been central to global discussions of climate change since the late 1980s. Though the word adaptation is now closely associated with these institutions, it did not come to occupy its central role until several years after their founding; even in this new setting, the word carries associations from its earlier meanings.

The history of these institutions is well known (Bodansky, 1995; Agrawala, 1998; Schipper, 2006), and does not require extensive summary. Reviewing the development of the vocabulary used in the documents that these organizations produced is a bit like tracing the early history of the solar system, billions of years in the past, when gasses and scattered dust coalesced into larger particles that eventually formed the planets. The proto-planet Science was the first to appear; what are now the planets of Mitigation and Adaptation took somewhat longer to develop. It seems possible that, if external gravitational perturbations or random processes had operated differently, other planets might have taken form instead.

To touch on only the broadest outlines of this history, the first World Climate Conference was sponsored by the World Meteorological Organization (WMO) in 1979. It set up four working groups, three associated directly with scientific research and one with the study of impacts (see Table 9.1). It laid the groundwork for a series of workshops organized under the WMO, with the United Nations Environmental Programme (UNEP) and the International Council for Science (ICSU), held between 1980 and 1985. The last of these conferences spoke strongly about the threat of global warming. Its final conference statement not only reviewed the state of scientific knowledge and called for further study, but also pressed for concrete actions to address actual and potential impacts. The statement spoke about reduction of emissions, without using the word mitigation, and supported active programmes to accomplish such reductions. One of its major policy recommendations stated 'scientists and policy-makers should begin an active collaboration to explore the effectiveness of alternative policies and adjustments' (World Meteorological Organization, 1986). The documents employed a variety of terms to provide detail about these policies and adjustments, and used the verb 'adapt' once, without according it special emphasis: 'Support for the analysis of policy and economic options should be increased by governments and funding agencies. In these assessments the widest possible range of social responses aimed at preventing or adapting to climate change should be identified, analyzed and evaluated' (World Meteorological Organization, 1986). The conference proposed that a task

Table 9.1. *Names and tasks of working groups within the IPCC and its predecessors, 1979–2007*

1st World Climate Conference (1979)	Joint UNEP/WMO/ICSU conference (1985)	Draft resolution to UN General Assembly (1988)	1st Assessment Report (1990)	2nd Assessment Report (1995)	3rd and 4th Assessment Reports (2001, 2007)
WG I: 'climate data' WG II: 'identification of climate topics' WG IV: 'research on climate variability and change'		'science'	WG I: Science	WG I: Science	WG I: Science
		'social and economic impacts'	WG II: Impacts	WG II: Impacts, adaptation and mitigation	WG II: Impacts, adaptation and vulnerability
	WG III: 'integrated impact studies'	'possible policy responses to delay, limit or mitigate impacts'	WG III: Responses	WG III: Economic and social dimensions	WG III: Mitigation
	'encourage ... developing countries to improve energy efficiency and conservation'				
	'provide advice ... mechanisms and actions ... at the national or international levels'	'relevant treaties and other legal instruments'			
	'initiate ... consideration of a global convention'	'elements for possible future international conventions'			

force be set up to help ensure that appropriate agencies and bodies followed up on its recommendations. It also provided a list of four concrete activities that this task force should carry out (see Table 9.1); these very general activities speak only of 'mechanisms and actions', or the even more general terms such as adjustment or response, rather than adaptation. They seemed more concerned to address emissions and to set up an international organization or treaty than to cope with impacts of climate change.

In the years 1985–1988, the WMO and the UNEP moved towards establishing the IPCC. Consulting with a number of other actors, they developed a draft resolution to set up the IPCC, which was presented to the United Nations General Assembly by Malta. It proposed five major activities, the third of which, 'possible policy responses to delay, limit or mitigate impacts', contains the kernel of what is now called adaptation, but does not use the term. When one compares this list of activities to the three Working Groups of the IPCC, several points are clear. Firstly, the two activities associated with the formation of international bodies drop out, because by 1990 one could already see the momentum towards the establishment of the UNFCCC, formalized at the United Nations Conference on Environment and Development (UNCED) or Rio Summit in 1992, and the associated Kyoto Protocol, signed in 1997. Secondly, the emphasis of the first Working Group, on science, has remained unchanged. Thirdly, the attention to impacts in the second Working Group has also been a constant. Finally, it took a while for adaptation (with its associated term, vulnerability) to become defined as a major task, to be associated with impacts, and to be separated from mitigation. In other words, the term 'adaptation' does not appear in the charges to the Working Groups until the mid-1990s, relatively late in the process.

The word adaptation received its official definition, a lengthy one, in 2001 in the glossary of the Third Assessment Report of the IPCC:

Adjustment in natural or *human systems* to a new or changing environment. Adaptation to *climate change* refers to adjustment in natural or human systems in response to actual or expected climatic *stimuli* or their effects, which moderates harm or exploits beneficial opportunities. Various types of adaptation can be distinguished, including anticipatory and reactive adaptation, private and public adaptation, and autonomous and planned adaptation.

This definition was accompanied with a raft of associated concepts: adaptive capacity, adaptation benefits, adaptation costs and adaptation assessment. Detailed quantitative analyses (Janssen et al., 2006; Janssen, 2007) of academic publications show similar trends to this pattern of growing emphasis in international organizations. Fewer than 10 papers on the subject of adaptation were published per year in the 1990s. This figure increased to about 20 per year by the mid-1990s and grew steadily after that, reaching nearly 200 per year by 2005.

This term had advantages over its alternatives: a greater suggestion of positive action than 'the limiting of impacts', a sense of longer-term shifts than adjustment, a more precise focus than response, an implication of greater levels of well-being than coping. However, it carries some conceptual baggage. Firstly, it retains an association with natural hazards (Smit et al., 1999; Adger, 2000) that dates back to earlier decades, even though specialists in the area of natural hazards had begun to use the term mitigation more extensively since the late 1990s. However, natural hazards differ from climate change in their temporal patterns. Unlike climate change, natural hazards often come as brief, rare events, whose frequency and severity can be established. Moreover, the rapid change of climate change hazards is very different from the lower rates of change of most natural hazards such as floods, hurricanes, earthquakes and volcanic eruptions. This difference reflects the much greater degree of human influence in climate change than in natural hazards. Secondly, the term adaptation is coupled with the direct measurement of consequences. As the IPCC definition suggests, the notion of adaptation is associated with quantifiable effects, described in economic terms as costs and benefits or as harms and opportunities. However, some consequences may be easily measured in monetary terms, but others, linked to social identity and well-being, cannot be evaluated as readily in quantitative terms. As a result, it is difficult to tally up the total costs and benefits of alternative responses, even though the term adaptation implies such summing, and the comparisons between courses of action that such summing permits. Finally and most seriously, it offers the promise that problems are manageable. It suggests that social groups – communities, nations, all of humanity – can avoid the worst consequences of climate change by thoughtful preparation. It tends to exclude the possibility of non-adaptation from consideration. Much as the word development places all nations on a single scale, offering the suggestion that the very poorest nations of the world are developing and are moving towards the prosperity of the richer ones, so too the word adaptation places all outcomes on a single scale, offering the suggestion that the world can shift up from the less satisfactory outcomes to the better ones.

After its position as a key term in climate change debates was consolidated in the early 1990s, the term went through certain changes. For some years, a debate raged about its propriety: some argued that proposing adaptation as an important policy option only served to draw political will away from the urgent goal of mitigation, while others claimed that it promoted an urgent and necessary task. By the late 1990s, the growing recognition of the inevitability of climate change impacts, reflecting in part the persistent and effective lobbying by groups such as the Alliance of Small Island States, led to recognition of the centrality of adaptation. Moreover, the word has spread to many institutions and contexts. New mechanisms to finance adaptation have developed. At the seventh Conference of Parties (COP 7) to the UNFCCC in Marrakesh in 2001, a resolution was adopted that provided guidelines, and suggested funding, so that least developed countries could prepare national adaptation programs of action (NAPAs), which would 'address urgent and immediate needs and concerns related to adaptation to the adverse effects of climate change'. These NAPAs were a part of the diffusion of the word around the world. To take only one example, the development of the NAPA for Tanzania led to discussions in that country's parliament, which are conducted in Swahili. As a result, an existing word in that language, *kukabiliana,* took on the additional meaning of adaptation. It is derived from the verb *kabili* 'to face' and its reciprocal form *kabiliana* 'to face one another'.

A community impacted by climate change

These international discussions of adaptation, carried out on a global scale, direct attention to the groups that experience the impacts of climate change most directly. To assess the usefulness of the term adaptation in describing the possible responses of such groups, we can consider the case of a community, Phinaya, in the Peruvian Andes, where glaciers have been retreating at a rapid pace. This community is located in the department of Cusco (Canchis Province, Pitumarca District). Its lands range from about 4500 to 5300 metres above sea level. Roughly 400 km^2 in area, this community has a population of about 1100 individuals dispersed in small clusters of houses. A small nucleus of 50 or 60 households contains the school, the mayor's office and a few stores (Sendón, 2006).

For centuries the major economic activity in this region has been the raising of livestock, because the high elevation does not permit agriculture. The most common domesticated animal is the alpaca, raised for its wool and, to a lesser extent, its meat; there are also many llamas, used to transport goods in and out of the region, and smaller numbers of sheep, cattle and horses.

The raising of livestock is closely tied to the seasons of the year. Precipitation falls principally between November and March. Much of it comes as rain, though snow also falls, creating problems when it remains for several days, since alpacas and llamas do not dig through the snow to find fodder, and the herders must scrape it away. The dry season runs between May and September, during the southern hemisphere winter. This period is marked by frosts, sometimes quite severe. During the rainy season, pasture is abundant, but during the dry season it is confined to small areas along streams. The herders dig canals to bring water from streams to extend the area of pasture, but these canals often do not last for many years, since they are eroded at times of heavy flow, and also are weakened as water freezes and expands on cold nights. Moreover, the low herbaceous plants and grasses in the area cannot be harvested and stored as hay or fodder; the areas that remain moist during the dry season are therefore the key to the survival of the herds and of the herders.

The history of alpaca herding stretches back thousand of years. The wool of the alpaca was important in the major pre-Columbian civilizations of the Andes, including the Incas. The Spaniards, who conquered the Incas in 1532, introduced sheep to the Andes, but the alpacas remained important, especially at higher elevations, where they were raised by indigenous communities. Large private family-owned estates, known as haciendas, began to develop in alpaca-herding regions in the late nineteenth century, as the demand for wool grew following the Industrial Revolution, and the expansion of railways increased commerce. In these haciendas, the herders continued to live in their scattered house clusters, and were required to raise animals that belonged to the hacienda-owners in exchange for the right to graze their own animals on hacienda land. They also owed labour service to the hacienda-owners.

These haciendas continued through much of the twentieth century until the Agrarian Reform, which reached this region in the early 1970s. The hacienda-owners sold a large proportion of their herds before the reform took effect. The title to the hacienda lands passed to the indigenous communities, where the herders continued to raise their own animals. This region felt the impact of the Shining Path guerrilla movement, especially in the 1980s, and the herders were often caught between the guerrillas and the military. Drawing on their own community institutions and on the support of progressive elements of the Catholic Church, the herders organized community patrols (*rondas campesinas*) which helped protect them against the incursions of the guerrillas, and also lessened the local presence of the military, who directed their efforts elsewhere.

The 1990s and first years of the twenty-first century brought a number of changes. Among the least expected, and most warmly received, was the rebound of populations of vicuñas, a wild relative of the alpaca and llama. These animals, whose wool receives very high price on the world market, had been decimated by illegal poaching, but the community patrols blocked the poachers from entering the area. A government organization, CONACS, markets the vicuña wool on the world market. The herders organize round-ups of vicuñas several times a year. Long lines of herders, waving colourful banners, drive the animals into stone corrals. After a series of rituals that acknowledge the links between the vicuñas and the *apus* or mountain spirits, the animals are shorn and released; representatives of CONACS attend, receive the wool and pay the herders for it.

In the relatively peaceful years after the collapse of Shining Path in the early 1990s, other changes affected the area as well. Dirt roads were extended into the area, formerly only accessible on foot or horseback. These roads allow trucks to reach Phinaya, in about 8 hours from a nearby paved road. In 1997, the regional hydroelectric company EGEMSA completed a dam at Lake Sibinacocha near Phinaya, in order to increase the lake's volume and regulate its outflow, augmenting the supply of water during the dry season to the large hydroelectric plant downstream at Machu Picchu. This dam flooded a few hundred hectares of pasture, for which the local inhabitants have not received any compensation. The dam also generates power for the small station located near it, but it does not provide any power to Phinaya, located 10 km away.

Several other possible changes may develop, linked to the expansion of roads. Peruvian governments since 1990 have had strong policies favouring mines by offering owners generous concessions and imposing weak environmental regulations. Prospectors have visited the region, causing great concern among the herders who know that mines have polluted streams in nearby provinces, but no mining projects are currently under way. Some local NGOs have visited Phinaya, though they concentrate their efforts in more accessible communities at lower elevations, where populations are denser and environmental conditions are more favourable for agriculture and capital-intensive livestock raising. A few tourists have come to Phinaya, though the remoteness and lack of infrastructure discourage all but the hardiest backpackers; even that population congregates in other areas, with more established trails and more famous sights.

Settled for millennia, this area has experienced fluctuations in the glaciers, which have expanded in some periods and retreated in others. The most recent glacial advance, a rather small one, ran from about 1780 to 1880, and glaciers have been retreating since then. The pace of melting has picked up since 1940 and even more so since 1970. Though the establishment of pasture and of house clusters on newly exposed soil had been able to keep up with the slow retreat of the glaciers in the first half of the twentieth century and perhaps some years after, the more rapid retreat of glaciers at present leaves exposed till, a mix of rock, gravel and dust, on which vegetation returns only slowly.

More seriously, the retreat of the glaciers is projected to bring dry-season stream flow to very low levels by the middle of the present century. A recent study (Hüggel et al., 2003) developed models to project glacier area and volume, and runoff, in the two sections of the Andes nearest to Phinaya. The authors calibrated the model for 1962 and 1999, and projected it into the future, based on rather conservative estimates of temperature increase from 1999. This study paints an extremely bleak picture for the herders, since the absence of stream flow for several months each year would mean the end of dry-season pasture, and thus of herding in this region as well. Figures 9.1 and 9.2 below, derived from this study, show these declines. It is difficult to imagine alternatives to outmigration once the streams have dried up; at most, herders might retain seasonal camps in the area for the rainy season and travel elsewhere in the dry season. To do so, however, they would compete with other established communities, who will also face pressure on their water resources, for dry-season pasture.

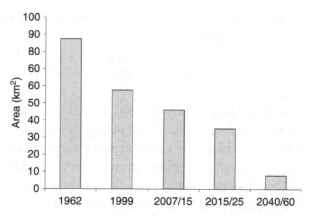

Figure 9.1 Past, current and projected areas of glaciers in the mountains near Phinaya, Peru. (Source: Hüggel et al., 2003.)

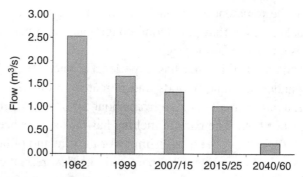

Figure 9.2 Past, current and projected stream flow for dry season in the mountains near Phinaya, Peru. (Source: Hüggel et al., 2003.)

Local perceptions of glacier retreat

In 2007, I travelled to Phinaya, and met a number of indigenous herders who lived there. This trip was a preliminary step to conducting more extensive fieldwork in Peru. During this visit, I also met a number of government officials, NGO staff members and residents of other towns and villages in the Department of Cusco, many of them with extensive experience in herding areas. I participated in meetings at an NGO, attended a major regional livestock fair, interviewed employees of the hydroelectric company EGEMSA, visited a parish priest, and spent time at an applied social science research institute. The following discussion of local perception draws on this visit and conversations.

When I arrived in Phinaya, I first presented myself to village officials, and asked their permission to meet local herders; knowing of my connection to a local

NGO, and intrigued by my ability to speak Quechua, they agreed to this request. I travelled with a minor village official, who introduced me to herders. I then explained to each individual that I had heard that the snow-mountains were changing, and that I had come to speak to local people about that topic. I asked each person if he or she would agree to talk to me about glacier retreat, and to let me take notes, and received consent in all cases. All the people offered extensive comments on glaciers and related topics. I count ten of the conversations with herders as interviews.

In Phinaya, the residents could all indicate the former extent of glaciers, and describe how they had become smaller. The most common Quechua expressions to describe this retreat are *rit'i chhullukun*, 'The ice and snow are melting', and *rit'i pisiyamun*, 'The ice and snow are diminishing.' In addition, they had words for specific features of the glaciers, indicating that they are topics of conversation. These include *yana rit'i*, 'black ice and snow', which refers to the dust-covered ice in the lower portions of glaciers; *toq'o*, 'hole', which refers to moulins, vertical shafts within glaciers formed by surface meltwater; *aqoqhata*, 'sandy hillside', which refers to glacial till, and *wayq'o*, 'gully', which refers to crevasses in the body or snout of a glacier. The local people say that these processes and features are 'in plain sight', (Quechua: *sut'i*; Spanish: *a la vista*), contrasting them with other processes and features that might require specialized knowledge or apparatuses to detect.

They link the glacier retreat to other environmental processes, particularly greater heat and more wind, which blows dust onto the glaciers, accelerating their melting. When I asked about the causes of the retreat, people offered a variety of explanations. They mentioned different kinds of pollution, including mines, factories and cities, and supernatural causes, such as punishment by God or by *apus* or mountain spirits (Ricard Lanata, 2007). They tended to speak with less certainty about the causes than about the processes themselves. They were concerned that the processes would continue, with greater scarcity of water and pasture and declines in the alpacas. A number spoke explicitly about the end of the world: *tukurapunqa vida*, 'life will come to an end', is a phrase that I heard several times. The root *tuku* can mean 'end' or 'all' or 'completion', and the suffixes *-ra-* and *-pu-*, taken together, mean 'abrupt' or 'final'. Others spoke of specific details such as the death of all the alpacas, or the arrival of a great wind that would blow all vegetation off the earth.

The interviews that I conducted represent a sample of only ten. Because of this small size, only tentative conclusions can be drawn, and future research will give firmer results. Nonetheless, the distribution of responses, even in a group this small, is striking.

The set included seven men and three women, with ages ranging from about 25 to about 55. Two of the people, both women, spoke only Quechua; the others were bilingual, and could converse in both languages. All ten were from close by: five were from the local village, two from neighbouring villages, two from adjacent provinces, and one from a more distant province in the Department of Cusco. This mobility has long characterized the herders; individuals, often women, move to other alpaca-herding communities to marry, and many travel for some weeks, months or years to earn money, as supplementary income or as a source of capital to allow them to purchase animals.

Figure 9.3 indicates the different social and spatial scale of impacts of glacier retreat that individuals mentioned, and the number of individuals who mentioned them. It is striking that no individual concentrated on the impacts on his or her household. All spoke at least of the local community, and most included other communities, whether nearby, in the region, or elsewhere in highland Peru. Interestingly, these units do not correspond to administrative units within Peru, such as the district or province; these administrative units include other areas at lower elevation, where other forms of pastoralism and agriculture are carried out, glaciers are more distant, and water issues are different. In contrast, the herders spoke of other herding communities, in other districts and provinces, with whom they share a common livelihood, a similar position in relation to glacier retreat and a variety of ties as well, including kinship, since individuals often find spouses in these communities. One man spoke at length about the way that the relations of these communities parallel the relations of the mountains below which they were located and how the spirits resident in these mountains are related to each other and speak to each other on certain occasions each year.

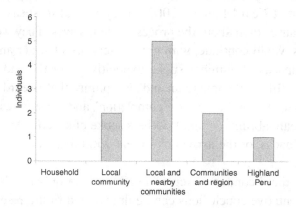

Figure 9.3 Social and spatial scale of concerns mentioned by herders in highland Peru.

Figure 9.4 shows the temporal scale that individuals used to describe the period during which glacier retreat had taken place and in which they projected future glacier retreat. They used three different kinds of temporal units. Some spoke of years, mentioning that they had seen changes since they moved in 8 years ago, or anticipated serious retreat in another 10 or 20 years. Others spoke of generations, recounting stories that they had heard from their grandparents, or speaking of concerns for their children. And still others spoke of epochs, referring back to the time of creation, or suggesting that glacier retreat might bring the end of the world, or using the Quechua word *timpu*, from the Spanish *tiempo*, 'time', to refer to an entire historical era that would end with the disappearance of the glaciers. Some anticipate a political shift that is epochal in nature. One woman asked me whether it was true that the snow peaks had been privatized, and that new owners might buy them and take them away; she did not seem reassured when I told her that this was not a possibility. (It is possible that she had heard of the mobilization in Chile to block the plan, proposed by Barrick Gold, a Canadian mining corporation, to remove and relocate three glaciers in order to extract gold ore that lay near them at the site of the Pascua Lama mine; it is also possible that she was responding to recent reforms in Peru that liberalized land and water titles, making the sale of these resources much easier.) People combined these scales in different ways. Some spoke only of one time unit, while others combined two. It is striking that just over half mention longer time units, generations and epochs, rather than simply measuring time in years. These units also suggest a collective orientation, and a concern with continuity rather than with economic growth.

The specific concerns that arose in the interviews appear in Figure 9.5. Some of these are directly linked to glacier retreat, while others have a looser connection with glacier retreat but came to mind nonetheless in the course of discussion of

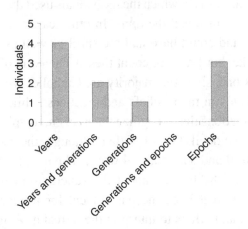

Figure 9.4 Temporal scale of concerns mentioned by herders in highland Peru.

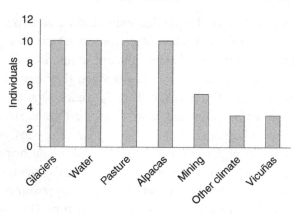

Figure 9.5 Specific issues mentioned by herders in highland Peru.

glacier retreat. The high degree of unanimity about the environmental and live-lihood issues associated with glacier retreat is very striking; all ten individuals mentioned glaciers, water, pasture and alpacas. The concerns about water centred on the availability of water, especially in the dry season, but several individuals mentioned water quality as well. One woman pointed to a large pool of water in a streambed, and said that until recently the stream had flowed throughout the year. The water in the stagnant pools that remain becomes warm, allowing parasites that infect alpacas to breed. The concern about alpacas also ranged over several topics; all were concerned about their ability to earn a living, and often voiced their worries that the alpacas were becoming thin. Several mentioned as well that they were troubled to see the alpacas suffer.

It is striking that six of the individuals directly mentioned the mountain spirits, or *apus* as they are known in Quechua. (This number is a conservative estimate, since I only counted the cases in which the individuals used the word *apu* or referred to a ritual or myth that involved the *apus*. In other cases, individuals mentioned mountains by name, and could have had the spirit as well as the mass of rock in mind when they spoke, but I did not count these.) The issue of religion among the herders is a complex one, since the majority are Catholics who combine Christian and indigenous elements in their beliefs and practices, while others are members of Protestant denominations that recognize the indigenous spirits but think of them as manifestations of the devil. For all of them, though, the mountain spirits are a real force in the natural and supernatural world. The unprecedented retreat of the ice raises questions for the herders about the supernatural world. It leads some to speculate that the retreat is divine punishment for immoral behaviour or for neglecting the spirits, and others to interpret the retreat as a sign of the sadness of the spirits about the lack of human respect for the Earth and nature.

Half of the individuals also mentioned mining, a much-feared source of pollution of streams and a matter of concern for those who see it as a violation of the physical and supernatural integrity of mountains. A smaller number mentioned other forms of climate change, particularly increasing heat and wind. Some expressed concern about possible changes for the vicuñas, animals whose increase in numbers has brought income and has also suggested that the mountain spirits or *apus* are favourably disposed towards the people, since a good season of vicuña hunts is taken as a sign that the *apus* are pleased with the offerings that the herders have made (Ricard Lanata, 2007, p. 64).

In sum, the local people are well aware of glacier retreat. They speak of a concern for themselves as a community, for neighbouring communities and other communities in the region. Some look to the near future, but most think about the long-term continuity of the collectivities through future generations and beyond. They all link glacier retreat with issues of water supply, pasture and the well-being of their herds, and many recognize other issues as well.

The question arises of whether the herders conceive of specific adaptations. Though the conversations about the future centred on difficulties rather than solutions, some people mentioned projects, understood broadly as group efforts or activities, whether locally organized or sponsored from outside, that could bring benefits or improve well-being; these projects arose spontaneously in conversation. Figure 9.6 shows the number of projects that were mentioned in interviews; the specific projects themselves, placed into groups, appear in Table 9.2. It is striking to see that just over half of the people mentioned no projects at all – a particularly high proportion, especially since the presence of foreigners predisposes rural dwellers to think of projects. It is also noteworthy that the individuals who did mention projects had an uneven distribution: two mentioned two projects, and two

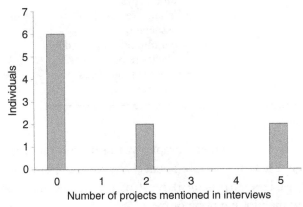

Figure 9.6 Number of projects mentioned by herders in highland Peru.

Table 9.2 *Specific projects mentioned by herders in highland Peru*

Project area	Number of interviews in which this area was mentioned	Specific projects within this area	Effectiveness as adaptations
Alpaca breeding	4	• Bringing male alpacas with long, fine white wool to inseminate females (all 4)	• Negative
Alpaca infrastructure	3	• Protecting the health of alpacas	• medium, short-term
		• Bringing medications for ill alpacas	• medium, short-term
		• Supporting the export of alpacas from Peru to other countries to increase the price of animals	• low
Water	4	• Improved irrigation for pasture	• medium, short-term
		• Construction of reservoirs to supplement stream flow	• low
		• improve supply of drinking water to the central settlement	• medium, short-term
		• creation of a bottling plant to sell water to cities	• low
Organizations	2	• Promote the reclassification of the central settlement as a higher administrative unit (district)	• Low
		• Create a regional association of communities affected by glacier retreat	• high
Regional infrastructure	1	• Improve road access to the area	• Medium

mentioned five. Though the small sample size precludes any firm conclusions, it is noteworthy that a few people talk about many projects, and many talk of none. It might have been less surprising if most of the people mentioned one or two projects, with few mentioning none or many.

I have created an admittedly arbitrary and impressionistic scale to rank these projects as possible adaptations. The only one that I ranked high, as being likely to bring significant benefits, is the proposal to create a regional association of communities affected by glacier retreat; such an organization might bring greater attention to this remote area and, with that attention, more aid. Though even a large amount of aid could not resolve the core issues of the reductions in water supply and pasture, this organization might support planned relocation efforts. The efforts to address the health of the alpacas have some medium probability of improving conditions, and at best would work in the short run; they might be compared to supplying a starving person with better housing rather than food. Similarly, the suggestions to improve roads and potable water, one of the most common requests of rural villages in Peru and throughout the developing world, might be offered some medium rank. Investments in irrigation for pasture might work for a short time, though they would not halt the overall decline in water supply. (With enough funding, a project might be able to supplement surface water with groundwater, perhaps with some system of solar energy to operate pumps; however, such a project would still face obstacles, among them reaching a dispersed population and assuring the maintenance of pumps in a remote setting.)

The one project mentioned by all four individuals – bringing male alpacas with long, fine white wool to inseminate females – would be a maladaptation, since these are the most delicate animals and hence the ones who are more vulnerable to cold and lack of adequate diet. The other projects seem to have little possibility of providing benefits, and some are completely far-fetched: this remote village, far from any paved road, is a particularly unlikely spot for a bottled water plant, especially since water supplies are decreasing. The bottled water plant only makes sense as a kind of white elephant, an impractical project that might provide some construction jobs for a few years if anyone were to fund it.

One might also envisage other projects, not mentioned by the herders. One example is tourism, which has grown rapidly in the Cusco region. This remote setting might fit a niche for backpackers or horseback riders, though other areas, closer to roads, are more likely to provide the infrastructure that tourism usually requires. And water supplies are becoming increasingly scanty precisely during the dry season, the period of peak tourism. Interestingly, none of the individuals mentioned vicuñas, even though their numbers have expanded and their wool brings considerable income. It might well be that they recognize that a government agency has a monopoly on the purchase and sale of vicuña wool, so that no new projects could come through; the herders might also identify the vicuñas so closely with the mountain spirits or *apus* that they would rather avoid involving outsiders.

Table 9.3 shows the distribution of temporal scale and number of projects. Recognizing once again the small sample size, one can note that the people who mention projects speak of a shorter temporal scale, while the people who speak of the longer temporal scales do not mention projects. As with the other association,

Table 9.3   *Number of projects and temporal scale mentioned by herders in high-land Peru*

		Temporal scale					
		Years	Years and generations	Generations	Generations and epochs	Epochs	Total
Number of	0	1	1	1	0	3	6
projects	2	2	0	0	0	0	2
men-	5	1	1	0	0	0	2
tioned	total	4	2	1	0	3	10

no causal inferences can be drawn. It is possible that people who look to the near future are more predisposed to think of projects, or that those who actively contemplate projects come to focus on shorter timescales.

Intermediary organizations and adaptation

Though one often hears about linking the global and the local (represented in this chapter by international organizations and a specific community, with their corresponding frameworks for discussing climate change), it is important to remember that specific organizations, intermediate in scale, often are the entities that carry out such linking. This section discusses four organizations that are among the most important for promoting such links in the case of Cusco, the department of Peru in which Phinaya is located. It presents them in chronological order of their foundation, and considers the ways in which they have become involved with the question of adaptation to climate change.

Oxfam

Oxfam International is a major international NGO that is involved with climate change in Peru and in Cusco. Its origins date to 1942, when it was founded in England as the Oxford Committee for Famine Relief. At this time, it was one of the local committees of the National Famine Relief Committee that lobbied the British government to allow the delivery of food to relieve famine in Nazi-occupied Greece, even though the Allies had blockaded the country. Drawing much of its early membership from social and religious activists and academics, it continued after the war. Like many other relief agencies, Oxfam gradually evolved into an organization that addressed the causes, rather than the symptoms, of famine. There are now 12 additional national offices of Oxfam, in Western Europe, North

America, Australia, New Zealand and Hong Kong as well as in the UK, which run independent programmes; despite occasional rivalries among these branches, they have co-operated, since 1995, under Oxfam International. Their work focuses on development assistance, with an emphasis on poverty alleviation and sustainable development; humanitarian work, often linked to disaster relief and risk reduction; and advocacy campaigns to address national and international policy issues. Since the 1970s, Oxfam has worked in a variety of sectors and regions in Peru, usually operating programmes through national and local partner organizations. Recent activities include reconstruction after earthquakes, support of fair trade coffee and other agrarian issues, and promotion of CONACAMI, or the National Coordinator of Communities Affected by Mining, an organization that supports peasant and indigenous communities which face conflicts with mining corporations over land and water resources.

Oxfam's core issues of sustainable development, poverty alleviation, humanitarian work and disaster relief have led it to identify adaptation to climate change as a key priority (Oxfam, 2008), since climate change is likely affect the poor, reducing their margin of survival through increased exposure to extreme events. Their plans for action include the use of new technologies, the diversification of livelihoods, disaster risk reduction and community organization. Oxfam International also promotes advocacy campaigns around this issue; it has made efforts to persuade the G8 Summit to address climate issues and to support adaptation. This emphasis on climate change is a recent shift for Oxfam, as can be seen by the organization's statements of priorities, called Briefing Papers. Briefing Paper 89 'Making the case: a national drought contingency fund for Kenya' (2006a) contains no discussion at all of climate, despite the relevancy to its theme, and Briefing Paper 91 (2006b) 'Causing hunger: an overview of the food crisis in Africa' mentions climate change as one of many reasons that hunger is endemic in Africa. In 2007, climate appears as a central issue in Briefing Paper 104 'Adapting to climate change: what's needed in poor countries, and who should pay' (2007a), where it is connected to equity, poverty and justice, and in Briefing Paper 108 'Climate alarm: disasters increase as climate change bites' (2007b), which links it to disaster relief.

Though not yet a major of activity for Oxfam in Cusco or indeed in much of Peru, Oxfam's support of adaptation to climate change is likely to expand in coming years. Oxfam participated actively in side events at COP 13 in Bali in 2007. A report published in 2008, 'Adaptation 101: how climate change hurts poor communities – and how we can help' emphasizes the point that pre-existing Oxfam projects serve the goals of adaptation to climate change, and offers as examples two cases from regions near Cusco, a project that ran from the 1990s to 2001 in the adjacent department of Puno to restore of ancient canal-building techniques in order to protect from floods and drought and a project in 2004 in the adjacent

Figure 9.7 Peruvian alpaca herder and his animals. (Source: Oxfam, 2008.)

department of Arequipa to assist an alpaca-herding community affected by frost by promoting the planting of barley and rye and the construction of sheds to protect animals against the extreme cold. A member of that community and some of his animals appear in a photograph (Figure 9.7) in that report.

Swiss Agency for Development and Cooperation

The Swiss Agency for Development and Cooperation (SDC) was founded in 1961 as the Swiss Service for Technical Cooperation. The early 1960s were a period when Western European countries and the USA expanded their foreign aid programmes and separated them administratively from diplomatic and foreign affairs offices as a result of several global forces, particularly the completion of post-war reconstruction, the deepening of the Cold War, and the emergence of newly independent nations in the former colonies. The year 1961 was marked by shifts in foreign aid in other countries as well: West Germany opened its Ministry for Economic Cooperation and Development (BMZ), the USA reorganized a number of programmes under the umbrella of the Agency for International Development and Britain established a Department of Technical Co-operation to deal with the technical co-operation side of the aid programme under the Foreign, Commonwealth Relations and Colonial Offices, a move that led to the creation of the cabinet-level Ministry of Overseas Development in 1964.

The Swiss Service for Technical Cooperation added humanitarian aid as one of its major goals in 1977. It further expanded in 1995 as a result of the transfer to it of the Office for Cooperation with Eastern Europe from the Swiss Department of Foreign Affairs. In 1996 the organization adopted the name Swiss Agency for Development and Cooperation.

Like other Swiss federal agencies, the SDC is conscious of Switzerland's small size and limited budget. It also reflects the characteristic Swiss concern to balance a history of neutrality and autonomy (Switzerland is not a member of the European Union, and only joined the United Nations in 2002) with a long-standing commitment to international co-operation and diplomacy (it was the home of the Red Cross and the League of Nations, and houses many international agencies). It therefore focuses on certain activities in which it can coordinate with the development programmes of other member nations of the OECD (Organization for Economic Co-operation and Development), particularly sustainable development, poverty alleviation, economic integration, peace and democratization. It places an emphasis on nations such as Nepal and Rwanda, which, like Switzerland, are landlocked and mountainous; within Peru, one of its priority countries, it focuses on three poor landlocked highland departments, including Cusco.

The SDC first discussed climate change as a challenge to development assistance in 2004, and emphasized it as a priority topic in 2006, offering as a justification Switzerland's active participation in the UNFCCC. Early in 2008, the SDC listed responses to climate change as one of its three environmental priorities, along with the conservation of biodiversity and efforts to limit desertification, and stressed the twin paths of mitigation and adaptation.

The SDC participates with the World Bank and the Global Environmental Facility in a multilateral programme for adaptation to climate change in Peru, Bolivia and Ecuador, particularly by providing scientific expertise. It has also begun to develop an adaptation project of its own in Peru, the Climate Change Adaptation Programme, to be carried out in Cusco and the neighbouring department of Apurimac. This programme began in 2008 and is scheduled to extend through 2011, with the possibility of renewal. Though the budget for this initial period has not been firmly established, it is likely to be about €4.3 million. The Climate Change Adaptation Programme seeks to support adaptation to climate change by reducing vulnerability in the areas of water resources, disaster relief and food security. It mentions briefly the increased risk of a specific kind of hazard, cold spells; these are discussed more fully later in this section.

This programme is coordinated by a large Swiss NGO, Intercooperation, founded in 1982 by the SDC. For a number of years, Intercooperation linked several Swiss development NGOs that ran projects for SDC, including some in Peru; it shifted to a more competitive business model in the mid-1990s, incorporating more partners and bidding for projects against other consulting firms, and reorganized as a foundation in 1997. The implementing organizations in Peru include government agencies, particularly CONAM (discussed later in this section) and the Ministry of the Environment, which is now absorbing CONAM, along with the regional governments of the departments of Cusco and Apurimac. Two NGOs are also active.

The older of these is PREDES, the Center for the Study and Prevention of Disasters. Founded in 1983, after a large El Niño event, it provides research and support services, largely to international donors, for support programmes after earthquakes, epidemics and floods and for capacity-building to reduce vulnerability. PREDES works in different regions of Peru, but has no recent or current projects in Cusco. The newer one, Libélula, was founded in 2007. It is a small consulting firm that works in the climate change area, assessing inventories of greenhouse gases and preparing sustainability reports, as well as supporting corporate social responsibility programmes and public relations. Its director is a former employee of CONAM, who was the director of PROCLIM and who represented the organization at international meetings of the UNFCCC, including in Switzerland. Concrete activities of the Climate Change Adaptation Programme in the field are scheduled to be under way late in 2008.

Practical Action

Practical Action is a sizeable international NGO focused on development and appropriate technology (see Ensor and Berger, Chapter 14). It was founded in 1966 by the economist E. F. Schumacher. His extensive experience in developing countries in the 1950s and 1960s led him to reject the then-dominant view of transferring large-scale technologies as a means of promoting economic development; he found that poor countries were unable to support these technologies, and also noted that these technologies undermined regional and national self-sufficiency and replaced traditional humanistic and religious values with narrow materialist values. He stressed the importance of locally based technologies on smaller scales, an idea that he popularized in his 1973 book *Small is Beautiful: Economics as if People Mattered*.

Founded as Intermediate Technology Development Group in 1966, it came under pressure in Latin America because its initials, ITDG, when pronounced in Spanish, sound precisely like the sentence '*y te dejé*' – 'and I left you'. It changed its name to the less awkward Practical Action in 2005. It has focused particularly on Peru, Nepal, Bangladesh, Lanka, Zimbabwe, Kenya and Sudan, with some work in other countries in Latin America, South Asia, and Southern and East Africa. It has expanded from its initial concentration on small-scale technology alone to move into training and education and then into support of community organizations, small-scale entrepreneurship and inter-community networks and consulting. Focused in its first years on agriculture, it also moved into other areas linked to livelihoods and well-being, such as health, energy, housing, domestic water supply and disaster mitigation. It has carried out this expansion while remaining centrally identified with issues of technology and the rural poor; its tagline is 'Technology Challenging Poverty'.

While still under the name of ITDG, Practical Action opened its Peru office in 1985. In that country it has concentrated on the Andean highlands, with some activities in the desert coast and upper reaches of the Amazon. It has a variety of projects in agriculture, livestock raising, irrigation, forestry, renewable energy resources, water and sanitation, housing and disaster prevention, along with support of local community organizations. The medium-scale 1986–87 El Niño event and the much larger one in 1997–98 made ITDG aware of climate variability and its connection to disasters, since these events bring flooding to northern coastal Peru and drought to the southern highlands.

In the southern part of the department of Cusco, Practical Action works on irrigation and on the promotion of native varieties of potatoes, which have grown in popularity in markets in recent decades. Practical Action also has a large project for improving alpaca production in the broad zone from 3900 to 5200 metres altitude where these animals are raised. They press herders to replace their traditional herds, which contains animals whose rather short, coarse wool includes shades of white, beige, brown, grey and black, with animals whose wool, because of its uniform white colour, length and fineness, commands a high price. They build corrals to control the mating of the animals, and construct sheds to protect the animals from the cold climate, especially during the time when the females give birth. They also stress animal health, using both modern medicines and traditional remedies that include ingredients such as herbs, soot and animal fat. They encourage alpaca herders to plant improved pasture plants, a change which is possible only in the lower and relatively milder section of the herding communities, up to 4400 metres.

In part because of their dense networks with other NGOs, including Oxfam, with whom it collaborates on disaster relief and training projects, Practical Action moved relatively early into the new field of responses to climate change. Late in 2003, it joined several organizations to oppose large dams (in their eyes, an example of inappropriate technology) as a source of hydroelectric energy and renewable energy. In 2004 they participated in the preparation of a report 'Up in smoke: the threat of global warming', which linked climate change to issues of development, poverty, equity and sustainability. They presented a small side event at the COP 11 in 2005 and a larger one at COP 12, held in Kenya, one of the countries in which they have a national office, at which they emphasized community-based adaptations, technological solutions in areas of energy, and the exchange of lessons among communities.

The Peru office began working on adaptation to climate change in 2003, coordinating with CONAM, discussed later in this paper, in the northern coastal department of Piura, one of the regions of Peru most affected by El Niño events. They drew on their expertise in disaster prevention to anticipate higher risks of disasters in this area and to propose adaptations in a wide gamut of areas, including water and soil management, agriculture, livestock raising, forestry, fisheries, health and housing. Starting in 2005, they added sites in six additional departments, including Cusco, where Practical Action uses European funding to address climate change. In the lower agricultural zones, they indicate that their long-standing programmes in maintaining native varieties of potatoes are an adaptation to climate change, since the greater diversity of potato varieties assures farmers of at least some yield in all years. For the higher herding zones, they address a kind of disaster, *friajes* or cold spells, associated with the movement of frigid air masses from near Antarctica across southern and central South America. The highest zones are most vulnerable to these periods of very low temperatures; when they occur, many animals die and people, especially infants and the elderly, suffer from increased rates of pulmonary illnesses. Practical Action states that these cold spells are becoming increasingly serious as a result of climate change. Their response is to support animal health through improved pasture, increased use of medicine and the construction of sheds; these steps support animal populations directly and, through their effects on livelihood, human health and well-being as well. Practical Action's most recent project with alpaca herders, under the rubrics of poverty reduction and climate change adaptation, has been funded by the Spanish NGO IPADE in 2005–2007, with a budget of €320 000 or US$450 000. The appealing alpacas, and the herders, often in ethnic dress, appeared as the image on the banner of the Practical Action website (www.itdg.org.pe) for much of 2008 and in their literature.

Though this poor region certainly benefits from funding, there are some questions that can be raised about this project. There is no question that herders and their flocks fall ill when particularly cold weather strikes, nor that the herders welcome the blankets and food relief that in recent years have been provided to them at such times, but there is little evidence to support the claim that these cold spells are increasing as a result of climate change. It seems likely that improved transportation networks and a greater media presence throughout Peru allow cold spells to be reported more effectively, and that the increased forecasting of cold spells by SENAMHI, the national meteorological service, has made them more visible as well. It is well established that average temperatures in the tropical Andes have been increasing (Vuille et al., 2003), raising questions why cold spells would increase. Two articles report directly on extreme, rather than average, temperatures around Cusco for the second half of the twentieth century; neither offers any confirmation of this trend. Alexander et al. (2006) present the results of coordinated research efforts by major international meteorological organizations to detect changes in extreme temperatures and precipitation around the world. Drawing on daily temperature records for the period 1951–2003 from over 2000 stations, their study provides no support for the increase in cold spells. For the region around Cusco, they report no significant change in any of the four variables associated with cold spells: the frequency of cold nights (measured as low daily minimum temperatures), the frequency of cold days (measured as low daily maximum temperatures), the duration of cold spells (measured as the number of consecutive cold days) and frost days (measured as days with temperatures below freezing). They did find significant decreases of cold nights in nearby portions of Bolivia and Chile, regions that would also affected by these large-scale *friajes*. Vincent et al. (2005) focus exclusively on South America for the period 1960–2000. Prepared with the active participation of leading meteorologists from eight South American countries, including Peru, and from North America and Europe, they draw on daily weather data from 68 stations, eight of which were over 2500 metres above sea level in elevation. Like the previous study, they find no significant trends in this area in the frequency of cold nights and cold days. They report on an additional variable, the lowest recorded temperature each year; this variable, very closely associated with *friajes*, shows no significant trends near Cusco, though in other parts of Peru and Chile, it has increased significantly.

Moreover, the all-white alpacas with long fibres, like the one in the image (Figure 9.8) from a recent Practical Action report, are more sensitive to cold, and require greater investment in medications and infrastructure; they are better suited for the lower, milder reaches of the alpaca-herding zone, rather than the higher, colder areas where the impacts of glacial retreat are felt. Moreover, there is a strong tension between their efforts to maintain traditional varieties of native crops (potatoes) and their efforts to reduce traditional breeds of native animals (alpacas). In sum, these particular efforts of Practical Action – an organization that has accomplished a great deal to relieve suffering and address issues of poverty and sustainability – seem rather ill-founded.

Figure 9.8 NGO technicians and employees. (Source: Soluciones Prácticas, 2008.)

Peruvian National Council for the Environment

CONAM is the Peruvian National Council for the Environment (Concejo Nacional del Ambiente). Established in 1994 as a decentralized national agency, it reports to the Council of Ministers rather than to any specific ministry. Its board has representatives of national, regional and local governments, as well as representatives from NGOs, academia, professional associations and other groups in civil society. The head of the board is appointed by the president of Peru. Charged loosely with supporting sustainable development, CONAM has responsibilities that include protecting biodiversity, water quality and air quality, overseeing solid waste management, supporting sustainable tourism, and representing Peru in many international environmental agreements. It is in the process of being absorbed into Peru's Ministry of the Environment (Ministerio del Medio Ambiente), established in 2008 to meet the requirements of the 2006 United States–Peru Trade Promotion Agreement. This ministry faces unusual limits for such an organization because decisions involving mining and extraction of petroleum and natural gas are managed by other agencies, closer to the industry-dominated Ministry of Energy and Mines.

CONAM is the agency that represents Peru to the UNFCCC, which Peru signed in 1992; in UNFCCC terminology, it is Peru's national focal point. It is the institutional home of a multi-agency entity in Peru, PROCLIM, The Program for Strengthening National Capacities to Manage the Impacts of Climate Change and Air Pollution, which develops national response strategies to climate change. CONAM has the responsibility of preparing National Communications, a requirement set forth in Articles 4 and 12 of the UNFCCC, for Peru. In the second and most recent national communication of 2006, glaciers feature prominently. This communication discusses the impact of glacier retreat on the supply of water and (via hydropower) of energy to cities, and suggests, with lesser emphasis, that the scenic beauty of glaciers can contribute to tourism. It also reports on a grant from the Global Environmental Facility to design and implement programmes of adaptation to climate change in the Andean region, which includes Ecuador and Bolivia as well as Peru. This grant follows on discussions among Peru, Bolivia and Ecuador that took place at the COP 10 in Buenos Aires in 2004 to develop a project to support a glacier-linked climate change adaptation in those countries, following on the experiences of the three countries in joint projects through the Andean Community. The first concrete activities in the grant are ones that CONAM will support with the National Program for Watershed Management and Soil Conservation (a branch of the Ministry of Agriculture). The first two projects are Shullcas sub-drainage in the Mantaro watershed, in the department of Junín in central highland Peru, and the Santa Teresa sub-drainage in the Urubamba watershed, in Cusco in southern highland Peru. Both of these watersheds have glacierized areas at high elevations. (Phinaya is part of the latter, since the meltwater from the glaciers around Phinaya flow into the Río Sallca which joins the Río Vilcanota, which becomes the Río Urubamba.) The first concrete action is a reforestation programme to cover 3500 hectares in the Shullcas basin, an agricultural region and an important source of water to Huancayo, the largest city in the Mantaro watershed; projected activities include monitoring the glacier at the head of the Shullcas valley. CONAM states that the water supply in the Mantaro watershed is of national importance because of its contributions to irrigated agriculture, urban water supply and hydropower generation. It lists other possible projects that could support agricultural exports, encouraged under the United States–Peru Trade Promotion Agreement: the development of new seed varieties that tolerate weather extremes, the expanded use of agrochemicals and new technologies.

Building on this start, a larger project was announced in 2008. Entitled the Adaptation to the Impact of Rapid Glacier Retreat in the Tropical Andes Project, and sponsored by the Special Climate Change Fund of the Global Environmental Facility, its budget of US$33 million (€23.3 million) supports more than four years of activities. The World Bank is the largest single international donor, contributing US$7.5 million (€5.3 million). CONAM is present in the project through its support of other climate organizations in Peru, including PROCLIM. The two pilot projects in Peru are in the Santa Teresa sub-basin in Cusco and the Shullcas sub-basin in Junín, the same ones as in the earlier grant. The stated concerns include food security, particularly in areas faced with drought, urban water supply and hydropower. In the case of Santa Teresa, the hydropower concerns are surely the largest. A large hydroelectric plant in the valley, supplied in part by meltwater from a large glacierized peak, is projected for development by EGEMSA, the same enterprise that operates the reservoir near Phinaya. Serious debris flow occurred in the Santa Teresa valley in 1998, and also destroyed another hydroelectric plant nearby. These events brought to national attention the need to stabilize slopes above such plants. By contrast, the valley has a relatively small rural population. Located at a relatively low elevation, its agriculture centres on tropical fruit and root crops, quite different from the irrigated agriculture discussed in the project. It thus seems likely that the great concern to protect the hydroelectric supply in the region from interruptions by landslides related to glacier retreat has been the major reason for raising this one small valley above all other sites – including the herders' home areas, close to glaciers – to receive a pilot project for adaptation to the impact of rapid glacier retreat.

Reflections on the term 'adaptation'

The discussion in the previous sections allows a concrete consideration of the term 'adaptation' in climate change policy, programmes and discourse in the light of the experience of a specific case, a community in the Peruvian Andes affected by the rapid melting of glaciers. It allows one to ask firstly whether the term favours certain kinds of organizations and activities, and opposes or excludes others, and, secondly, whether these organizations and activities address the concerns of the people and communities who are most directly affected by climate change.

These sections have considered three groups, all strongly engaged, though in different ways, with climate change. The first is the set of international environmental agencies. Climate change has grown as a global issue since the late 1970s. The concept of adaptation has been present since the 1980s. It consolidated its importance in the UNFCCC and IPCC framework in the early 1990s, and now, along with its twin term 'mitigation', forms one of the main pillars of climate change action. The second context is the set of development and environmental organizations active in Peru and, more specifically, the department of Cusco where the glaciers under discussion are located. Four organizations have been considered. Though founded at different times between 1942 and 1994, with different orientations and forms of support and action, they all began to emphasize climate change and adaptation in a short period, 2003 to 2007. Because of its loose, multifaceted quality, the term 'adaptation' allows the organizations to continue working in areas in which they already have expertise: small-scale technical assistance in one case, disaster relief or water development in others. It also lets them function in a familiar world of projects, in which they submit and receive proposals, manage budgets and personnel, run and evaluate the projects themselves, and produce reports and other briefings. Their use of the term also facilitates their links with international agencies, and demonstrates that they remain abreast of current trends.

The third context is the set of communities at high elevations. The herders are well aware of glacier retreat, and see its impacts on the ecosystems and livelihoods of their own communities and of neighbouring ones. They are concerned as well about aspects of glacier retreat that could be called extra-economic, or cultural, or religious, particularly its implications for the mountain spirits. Their general sense of vulnerability about glacier retreat is quite strong, as reflected in the question that a woman asked about whether the glaciers would be privatized and removed. Though some fear that further retreat will lead to an apocalypse, many consider possible lines of action; they also actively manage their herds and, through the construction of irrigation canals, their landscapes. A number of them discuss projects that they would like to see in their community. In addition to the concern about climate change, they recognize other threats, particularly mining.

Does the term 'adaptation' articulate the herders' concerns and hopes, and serve as a way to frame their actions? And can the term facilitate the support of such actions by the international, national and regional organizations that have adopted it with such alacrity? There are grounds for answering these questions both in the positive and the negative, but the bulk of this material supports the negative view.

On the one hand, the term might serve the herders well. It brings a number of organizations in Peru and in Cusco to draw support from international agencies, such as the World Bank, to serve the herding communities and to draw attention to glacier retreat. More specifically, the four organizations that were discussed have all begun to develop projects on related topics in the region. Oxfam has collaborated with organizations that support alpaca herders, though in a framework of disaster prevention rather than climate change, and Practical Action has worked steadily to promote the introduction of what they consider to be improved stock among alpaca herders. CONAM has declared glacier-linked water issues as a priority, though their emphasis is on assuring urban water supply and hydroelectric generation capacity, and the herder's negative relation with the reservoir and hydroelectric plant at Sibinacocha does not augur well for this relationship. The SDC's integrated project is likely to include alpaca herders as one of the populations whose needs it will seek to address. Though these projects, taken as a set, do not confront the critical issue of water supply as directly as they might, one might hope that the activities, once begun, will bring the herders' needs to wider attention. Additional reservations could be raised about these organizations: at times they seem to place their own interests above those of the populations in whose name they run programmes, and they develop close interrelations that at times seem overly cosy. Moreover, they have shifted to speaking of adaptation with a deftness and simultaneity that shows their long-established habit of following shifts in fashion. But their record does not suggest that they deserve sharp criticism. Their projects over the years have made many positive contributions, showing poverty that has been alleviated, hazards that have been addressed and steps towards sustainability that have been taken. They are far from the most smug or self-serving of development organizations.

One might view this case in the context of the early history of the term adaptation, itself only a part of the early history of humanity's efforts to come to terms with climate change. Faced with the problems that the term presents, one might counsel optimism and patience. Perhaps the term will grow, much as the word development – itself a product of a historical moment – moved from a narrow framing of economic growth to include issues of equity, quality of life and sustainability. Perhaps the explicit attention to cost–benefit analysis within current discussions of adaptation will lead organizations to look more broadly at the consequences of climate change and the responses to it, and that as a result they will take more seriously the long-term, collective and cultural concerns of the herders.

Already one can see within the adaptation literature some signs of such expansion. The notion of migration as an adaptation was once taboo because it infringed on the principle of national sovereignty; to suggest that people might need to leave their country was to suggest that other countries should receive them. But now it is broached more often, in both internal and international terms. Many people are familiar with the relocation of indigenous Alaskan villages, which have been moved away from bluffs that are eroding as a consequence of sea level rise, loss of sea ice and melting of permafrost. In similar fashion, the inhabitants of the Carteret Islands, low-lying atolls in Papua New Guinea which are subject to flooding exacerbated by sea level rise, are being moved to a larger, higher island, Bougainville, a migration that has also received significant attention in the press. The fate of the herders, already a mobile population, is likely to consist of movement to other parts of Peru. With their extensive social networks in neighbouring communities, they may seek to attach themselves to other households of herders, though these, too, face increasing pressures on their water supplies. They might shift to herding sheep, but even if they continue to herd alpacas, they will lose the important link to vicuñas. Attuned to the possibility of migration as an adaptation, the international, national and regional agencies might seek to find ways to assist the herders in such a relocation, to allow them to return on visits to their former home, and to avoid joining the masses of other migrants who leave behind their property, skills and communities to move to the already-crowded cities.

On the other hand, much evidence weighs against this sanguine outlook. A series of obstacles suggest that the term adaptation may not serve well to frame herders' actions and to build links with other organizations. Though one can note the logistical obstacles to current projects, particularly the remoteness of the communities, the harsh weather that is unappealing to many staff members of organizations, and the dispersed populations, other problems are greater. Indeed, this case illustrates the three conceptual obstacles associated with the term adaptation that were highlighted in an earlier section of this chapter. Firstly, the link between adaptation and hazards has led to a kind of obfuscation of climate issues. Organizations that specialize in hazards issues have presented the occasional spells of extreme cold at high elevations as a major problem associated with climate change, despite the absence of firm evidence that these spells have increased or are likely to increase. This emphasis directs attention towards short-term acute problems of moderate importance, and away from long-term chronic problems of greater importance, particularly water availability. Secondly, the emphasis within the adaptation framework on a particular form of valuation, cost–benefit analysis, directs attention towards the problems – often genuine ones – that can be easily measured, such as economic well-being, and away from the ones – often equally genuine ones – that cannot be easily measured, such as cultural and religious well-being. The long time horizons of the herders are hard to incorporate into such valuation as well. The herders' concern for non-humans also disappears from view within this framework; the animals, whose suffering is of concern to the herders, simply become an income source, and the mountain spirits vanish altogether, even though they matter a great deal to the herders. Even if their incomes were maintained in a new area, they would still experience wrenching dislocation. Finally and most seriously, the term 'adaptation' contains the promise that problems can be addressed, following the view, firmly established in the IPCC definition, that the harms associated with climate change can be modified and beneficial opportunities exploited. This perspective cannot match up with the view of a number of herders that glacier retreat will bring an epochal shift, or with the concern of all of them that their basic way of life is under serious threat.

The case of the alpaca herders makes it difficult to sustain optimism and patience. Major changes have already taken place, with streams and pastures drying up. These changes will continue and accelerate, and the alternatives are few. The meliorist language of adaptation seems unsuited to this great challenge. At best the herders are wary counterparts of the international, national and regional organizations that have adopted this new term, as they have adopted earlier ones. Only time will tell whether these early years of the term will have been a brief opportunity that was well used, one that allowed a broadening and deepening of concern, or whether they merely became a time of successful marketing. The herders' voices may genuinely come to form part of the vast global conversation about climate change, or the herders' faces may only serve to provide images that are included in publications by organizations that sought to obtain funds. It would be a sorry outcome if those organizations were to neglect the herders, one of the groups that have felt earliest and most profoundly the impacts of climate change, the process from which no one on this Earth can escape.

Acknowledgements

I recognize the support of the National Science Foundation in funding this research through grant SES-0345840 to the Center for Research on Environmental Decisions at Columbia University. Mourik Bueno de Esquita, Xavier Ricard and Gustavo Valdivia provided valuable assistance during my stay in Cusco. Neil Adger, Walter Baethgen, Christian Hüggel, Irene Lorenzoni, Carolyn Mutter, Karen O'Brien, Martin Scurrah, Kevin Welch and Steve Zebiak gave helpful comments on earlier drafts, and offered useful suggestions during the writing of this chapter.

References

Adger, W. N. 2000. 'Institutional adaptation to environmental risk under the transition in Vietnam', *Annals of the Association of American Geographers* **90**: 738–758.

Agrawala, S. 1998. 'Context and early origins of the Intergovernmental Panel on Climate Change', *Climate Change* **39**: 605–620.

Alexander, L. V., Zhang, X., Peterson, T. C., Caesar, J., Gleason, B., Klein Tank, A., Haylock, M., Collins, D., Trewin, B., Rahimzadeh, F., Tagipour, A., Ambenje, P., Rupa Kumar, K., Revadekar, J., Griffiths, G., Vincent, L., Stephenson, D., Burn, J., Aguilar, E., Brunet, M., Taylor, M., New, M., Zhai, P., Rusticucci, M. and Vazquez-Aguirre, J. L. 2006. 'Global observed changes in daily climate extremes of temperature and precipitation', *Journal of Geophysical Research* **111** (D5): D05109.

Bodansky, D. M. 1995. 'The emerging climate-change regime', *Annual Review of Energy and the Environment* **20**: 425–461.

Darwin, C. 1859. *On the Origin of the Species by Natural Selection*. London: John Murray.

Hüggel, C., Haeberli, W., Kääb, A., Ayros, E. and Portocarrero, C. 2003. 'Assessment of glacier hazards and glacier runoff for different climate scenarios based on remote sensing data: a case study for a hydropower plant in the Peruvian Andes', EARSeL Workshop, *Observing Our Cryosphere from Space*, Bern, Switzerland.

Janssen, M. A. 2007. 'An update on the scholarly networks on resilience, vulnerability, and adaptation within the human dimensions of global environmental change', *Ecology and Society* **12**: 9.

Janssen, M. A., Schoon, M. L., Ke, W. and Borner, K. 2006. 'Scholarly networks on resilience, vulnerability and adaptation within the human dimensions of global environmental change', *Global Environmental Change* **16**: 240–252.

Oxfam 2006a. 'Making the case: a national drought contingency fund for Kenya', Oxfam Briefing Paper No. 89. Oxford: Oxfam.

Oxfam 2006b. 'Causing hunger: an overview of the food crisis in Africa', Oxfam Briefing Paper No. 91. Oxford: Oxfam.

Oxfam 2007a. 'Adapting to climate change: what's needed in poor countries, and who should pay', Oxfam Briefing Paper No. 104 . Oxford: Oxfam.

Oxfam 2007b. 'Climate alarm: disasters increase as climate change bites', Oxfam Briefing Paper No. 108. Oxford: Oxfam.

Oxfam 2008. *Adaptation 101: How Climate Change Hurts Poor Communities – And How We Can Help*. Boston: Oxfam America.

Ricard Lanata, X. 2007. *Ladrones de sombra: El universo religioso de los pastores del Ausangate*. Lima: Instituto Francés de Estudios Andinos.

Schipper, E. L. 2006. 'Conceptual history of adaptation in the UNFCCC process', *Review of European Community and International Environmental Law* **16**: 82–92.

Schumacher, E. 1973. *Small Is Beautiful: A Study of Economics as if People Mattered*. London: Blond and Briggs.

Sendón, P. F. 2006. 'Los términos de parentesco quechua qatay y qhachun según los registros entohistóricos y etnográficos: una interpretación', *Revista Andina* **43**: 9–58.

Smit, B., Burton, I., Klein, R. J. T. and Street, R. 1999. 'The science of adaptation: a framework for assessment', *Mitigation and Adaptation Strategies for Global Change* **4**: 199–213.

Soluciones Prácticas 2008. *Memoria annual 06–07*. Lima: Soluciones Prácticas.

Vincent, L. A., Peterson, T. C., Barros, V. R., Marino, M. B., Rusticucci, M., Carrasco, G., Ramírez, E., Alves, L. M., Ambrizzi, T., Berlato, M. A., Grimm, A. M., Marengo, J. A., Molion, L., Moncunill, D. F., Rebello, E., Anunciacão, Y. M. T., Quintana, J., Santos, J. L., Baez, J., Coronel, G., García, J., Trebejo, I., Bidegain, M., Haylock, M. R. and Karoly, D. 2005. 'Observed trends in indices of daily temperature extremes in South America 1960–2000', *Journal of Climate* **18**: 5011–5023.

Vuille, M., Bradley, R. S., Werner, M. and Keimig, F. 2003. '20th century climate change in the tropical Andes: observations and model results', *Climatic Change* **59**: 75–99.

White, G. F. 1945. *Human Adjustment to Floods*, Geography Research Papers No. 29. Chicago: University of Chicago Department of Geography.

World Meteorological Organization 1986. *Report of the International Conference on the Assessment of the Role of Carbon Dioxide and of Other Greenhouse Gases in Climate Variations and Associated Impact*, WMO Report No. 661, 9–15 October 1985, Villach, Austria.

10

Do values subjectively define the limits to climate change adaptation?

Karen L. O'Brien

Introduction

Climate change adaptation is increasingly seen as both a necessary and urgent response to a changing climate, and much research is being undertaken to identify barriers and constraints to successful adaptation. Most discussions focus on limited adaptive capacity as a constraint to adaptation to climate change, and emphasise technological, financial and institutional barriers (Grothmann and Patt, 2005; Yohe and Tol, 2002). It is presumed that once these external barriers are removed or overcome, society will be able to successfully adapt to a changing climate. It has, however, also been suggested that adaptation to climate change may be limited by the irreversible loss of places and identities that people value (Adger et al., 2009a, 2009b). Adger et al. (2009b) argue that social and individual characteristics may likewise act as deep-seated barriers to adaptation. Such perspectives raise important questions about the role that individual and societal values play in adapting to climate change: is adaptation a successful strategy for maintaining what is valued? How do adaptation measures taken by some affect the values of others? In the case of value conflicts, whose values count?

Values are, in effect, an interior and subjective dimension of adaptation. In contrast to systems and behaviours that can be objectively measured and observed, values subjectively influence the adaptations that are considered desirable and thus prioritised. There has, however, been very little analysis in the climate change literature of the relationship between values and climate change adaptation, or more generally of the psychological dimensions of adaptation (Grothmann and Patt, 2005). This research gap can be considered important for three reasons. First, the interior or subjective ability of human actors to adapt can be very different from the objective ability, and these differences can contribute to the underestimation

Adapting to Climate Change: Thresholds, Values, Governance, eds. W. Neil Adger, Irene Lorenzoni and Karen L. O'Brien. Published by Cambridge University Press. © Cambridge University Press 2009.

or overestimation of adaptive capacity (Grothmann and Patt, 2005). Second, adaptations to climate change may affect what individuals or groups value, particularly in cases where adaptation measures are imposed by others (for example, by government institutions or private actors) and create their own ancillary or secondary impacts. For example, the construction of barriers and sea walls may limit access to coastal areas and influence coastal processes, affecting what local residents and fishermen value. This draws attention to the importance of recognising how adaptation measures are enacted from – and impact upon – differing prioritised values. Third, prioritised values change as individuals and societies change, thus any outcome of climate change adaptation that is considered acceptable today may be evaluated differently in the future. The relationship between adaptation and changing values thus needs to be assessed. Research on values places a greater focus on the interior dimensions of adaptation, and can provide new insights on the limits to adaptation as a response to climate change.

This chapter discusses the relationship between climate change adaptation and values. I first discuss values and the diverse ways that they are studied, both within and across cultures. I then consider how values are related to human needs, motivations, and worldviews, and discuss how these may change over time. Next, I present specific examples of key values that are evident in Norway, and reflect on how different values may influence adaptation priorities, particularly in relation to changes in snow cover associated with climate change. This preliminary exploration suggests that the limits to adaptation may be subjectively defined, rather than defined solely by objective criteria. Consequently, values that are compromised by climate change and not addressed through response measures may represent limits to adaptation for some individuals, communities and groups in Norway – a country that, as a whole, is considered to have a high capacity to adapt to changing climate conditions. Understanding the relationship between the subjective and objective dimensions of climate change adaptation may provide important insights on the limits to adaptation as a response to climate change, both for present and future generations.

The analysis here acknowledges the diverse understandings of the role that values play in global change processes. The literature on values is diffuse and there is a lack of agreement as to what influences values, and how and why they change (Rohan, 2000). Sociological perspectives tend to emphasise social structural explanations of cultural values and psychological variables, whereas anthropological approaches emphasise values as core elements of culture that are integral to a culture's worldview and that provide purpose and meaning in people's lives (Gecas, 2008). Political science perspectives emphasise the links between economic development, democratisation and changes in values (Inglehart and Welzel, 2005). As Williams (1979, p. 17) notes, '[i]n the enormously complex universe of value phenomena, values are simultaneously components of psychological processes, of social interaction, and of cultural patterning and storage'. While there is little agreement across disciplines about what is meant by values and how they are formed, there seems to be a consensus that they can be considered as important predictors of behavior and attitudes, that they are contextually conditioned but somewhat resistant to change, and that they are intergenerationally transmitted and cherished across cultures (Pakizeh et al., 2007).

Values and worldviews

Values can be defined in many ways: the term has been used to refer to a wide variety of concepts, including interests, pleasures, likes, preferences, moral obligations, desires, wants, goals, needs, aversions and attractions (Williams, 1979). Values are generally considered to be core conceptions of 'the desirable' within every individual and society. Rokeach (2000, p. 2) argues that '[t]hey serve as standards or criteria to guide not only action but also judgment, choice, attitude, evaluation, argument, exhortation, rationalization, and, one might add, attribution of causality'. It is widely recognised that values differ between individuals, groups, institutions, societies, cultures and other supra-individual entities. Yet it is also acknowledged that values are not unlimited or random. Despite great cultural diversity across the globe, 'the number of human values [is] small, the same the world over, and capable of different structural arrangements ...' (Rokeach, 2000, p. 2). Although essential features of values may be shared, they are nonetheless expressed uniquely, depending on culture and context: 'Values always have a cultural content, represent a psychological investment, and are shaped by the constraints and opportunities of a social system and of a biophysical environment' (Williams, 1979, p. 21).

Both individuals and groups have associated value systems, which are described by Rohan (2000, p. 270) as meaning-producing cognitive structures, or 'integrated structures within which there are stable and predictable relations among priorities on each value type'. Personal value systems, or 'judgments of the capacity of entities to enable best possible living', are distinguished by Rohan (2000, p. 265) from social value systems, which reflect people's perceptions of other's judgements about value priorities. Personal or social value systems can be used to select objects and actions, resolve conflicts, invoke social sanctions, and cope with needs or claims for social and psychological defences of choices that are either made or proposed (Williams, 1979). Value systems thus can be considered to play an important role in responding to climate change, both in terms of mitigation of greenhouse gas emissions and adaptation to changing climate conditions.

Value priorities have been measured using the Rokeach Value Survey (Rokeach, 1973). This method, based on a ranking of words representing terminal (i.e. goals)

Figure 10.1 Theoretical model of relations among ten motivational types of values. (Source: Schwartz, 2006).

or instrumental (i.e. modes of conduct) values, is based on the understanding that individuals organise their beliefs and behaviours in ways that will serve to maintain and enhance their self-conceptions as moral and competent human beings (Rokeach, 1973). However, as Rohan (2000) notes, the survey offers no theory about the underlying value system structure. Such a structure was proposed by Schwartz (1994), who considers values as integrated, coherent structures that may be influenced by factors such as age, life stage, gender and education. Schwartz identifies ten types of universal values that are found in all cultures and societies: security, tradition, conformity, power, achievement, hedonism, stimulation, self-direction, universalism and benevolence (Schwartz, 1994).

Schwartz's (1994) 'Values Theory' holds that the distinguishing feature among values is the type of motivational goal that they express. Motivationally distinct personal value orientations are, according to Schwartz, derived from three universal requirements of the human condition: 'needs of individuals as biological organisms, requisites of coordinated social interaction, and survival and welfare needs of groups' (Schwartz, 2006, p. 2). Schwartz recognises that there are dynamic relations among values, and argues that a single motivational structure organises the relations among sets of values and behaviour (Bardi and Schwartz, 2003). Schwartz's structure is represented as a circle that captures conflicts and congruities among the ten basic values, with an emphasis on values that focus on organisation, individual outcomes, opportunity and social context (see Figure 10.1). The motivations and needs described by Schwartz are structured such that priorities on adjacent value types in the value system will be similar, while those that are opposite each other represent maximum differences. Schwartz (1996) argues that values are most likely to be activated, entered into awareness, and used as guiding principles in the presence of value conflicts. Importantly, he points out that '[t]his integrated motivational structure of relations among values makes it possible to study how whole systems of values, rather than single values, relate to other variables' (Schwartz, 2006, p. 4).

Seligman and Katz (1996) challenge the traditional view of a value system as a single ordered set of values that is important to self-concept and helps guide thought and action, and argue instead that values systems are dynamic and creatively applied to situations, rather than rule bound. Their research shows that value systems are only stable in a particular domain, and are very much dependent upon context. Using a study about environmental values as an example, they found that 'value reordering takes place depending on whether individuals are asked to rank order values as they are important to them as general guiding principles or as they are important to them with regard to a specific issue' (Seligman and Katz, 1996, p. 63). Their view is compatible with Schwartz's value structure, but suggests that different value types may be reordered in different contexts and for different purposes.

Worldviews

Worldviews describe the basic assumptions and beliefs that influence much of an individual or group's perceptions of the world, their behaviour, and their decision-making criteria (Kearney, 1984). The concept of worldview, or *Weltanschauung*, has developed along various religious and philosophical trajectories, leading Sire (2004) to conclude that how one conceives of a worldview is dependent on one's worldview. From the postmodern perspective of Foucault, a worldview can be neither true nor false in any objective sense, and is instead linked to relationships between knowledge and power (Naugle, 2002; Sire, 2004). Rohan (2000) notes that worldviews and ideologies are often erroneously labelled as values, but argues that there is nonetheless an inescapable link between people's personal value priorities and the way they view the world, and that value system structure can be used to guide investigations of people's worldviews. At the personal level, worldviews have been linked to cognitive structures, which have been shown to change as individuals develop (Kegan, 1982, 1994). Inglehart (1997, 2000) describes how values are linked to traditional, modern and postmodern worldviews, and shows through a series of World Values Surveys, that there are links between the values identified by Schwartz and traditional, modern and postmodern worldviews. Traditional worldviews may, for example, place a greater emphasis on the set of values associated with conservation, which include tradition, conformity and security. Modern worldviews may place emphasis on values associated with self-enhancement, such as power, achievement and hedonism. Values linked to openness to change, such as stimulation and self-direction, may bridge both modern and postmodern worldviews. Finally a postmodern worldview may emphasise values that focus on self-transcendence, such as universalism and benevolence. The conflicts between opposing values in Schwartz's Value Theory may potentially be associated with differing worldviews, with consequences for social change and democratisation (Inglehart and Welzel, 2005).

Although there has generally been a greater emphasis on value differences than value change, the theoretical and empirical links between values, needs, cognition and worldviews suggest that values do change over time. Rokeach (1979) identifies two factors that influence value changes and related changes in attitudes and behaviour: (1) changes in self-conceptions or definitions of the self; and (2) increases in self-awareness about hypocrisies, incongruities, inconsistencies or contradictions between self-conceptions or self-ideals and one's values, related attitudes and behaviours. At the personal level, value changes can be linked to changes in social status or age, which are generally accompanied by changes in self-conceptions and consequently, by changes in value systems and in value-related attitudes and behaviour (Rokeach, 1979). At the social level, '[a]ny society must change in its value constitution to cope with changing adaptive problems, yet it must retain some coherence in its appreciative system (based on some minimal consensus) or the social order will break down' (Williams, 1979 p. 21). Values thus result from both psychological needs and societal demands, both of which may change as a result of changes in society, life situation, experiences, self-conception and self-awareness (Rokeach, 1979).

Maslow's holistic–dynamic theory of a 'hierarchy of needs' holds that an individual's dominating goal at any stage is a strong determinant of their worldview and philosophy of the future, as well as of their values (Maslow, 1970). A hierarchy of needs suggests that values change as needs become satisfied and new motivations emerge. This has been confirmed through longitudinal studies of values carried out through the World Values Survey, which shows that socio-economic development tends to produce intergenerational value differences and a shift from survival values to self-expression values (Inglehart and Welzel, 2005). Indeed, the human development and developmental psychology literatures show that individual and societal value structures change over time, and may in fact be evolving to new structures and worldviews in the future (Maslow, 1970; Williams, 1979; Kegan, 1994; Inglehart, 1997; Wilber, 2000).

It is important to point out that although value priorities may shift with changing worldviews, values associated with earlier worldviews do not necessarily disappear – they simply decrease in priority. Traditional values and modern values remain within postmodern worldviews, but they may be considered to be a lower priority and visible only in some contexts and situations. Economic stagnation and political collapse may lead to a re-prioritisation of these values (Inglehart, 1997).

Rokeach (1979, p. 3) emphasises that 'changes in values represent central rather than peripheral changes, thus having important consequences for other cognitions and social behaviour'. In other words, values can change, but such changes are neither trivial nor arbitrary.

Different and dynamic values have significance for climate change adaptation. The values associated with traditional, modern and postmodern worldviews are hypothesised to correspond to different priorities for climate change adaptations. Traditional worldviews may prioritise adaptation strategies that emphasise needs for belongingness and group identity, that recognise local knowledge, and that support traditional sectors and livelihoods and preserve cultural icons and identities (including, for example, strong connections to nature). Modern worldviews may prioritise adaptations that reduce climatic threats to economic modernisation and growth through, for example, rational, scientifically based technological adaptations based on cost–benefit analyses and quantified scenarios of future climate change. They may also emphasise responses that promote freedom and achievement, particularly market-based strategies for responding to climate change. Postmodern worldviews may prioritise adaptations that promote well-being, equity and justice, with attention to the poor and marginalised, future generations and the role of ecosystem services.

The potential for value conflicts in adaptation to climate change must be recognised. Adaptations that are imposed or enacted by a modern state may, for example, influence the values of individuals or communities with a more traditional worldview. As mentioned in the introduction, a 'modern' adaptation response to storm surges and sea level rise might involve the construction of sea walls and floodgates to prevent damage to property, infrastructure and individual lives. Such coastal defences may be effective in reducing loss of income and lives, yet they may have a negative impact on local knowledge, traditional livelihoods, a sense of belonging or cultural identity. They may also negatively influence postmodern values such as ecosystem integrity and social equity. The following section considers some of the factors that will be explored in future empirical research on the relationship between values and adaptation to climate change in Norway. The key assertion is that values do matter in adaptation decisions and strategies, and that value conflicts may result if values are not overtly acknowledged. Ignoring values can lead to misleading conclusions about the limits to adaptation.

Climate change adaptation and values in Norway

Different and dynamic values mean that climate change adaptations prioritised by some actors may not be considered as successful responses by others. In fact, some adaptation measures may directly affect the values of others, both in the present and future. In theory, the inability to respond to different value priorities may represent a limit to adaptation. For some individuals, communities and cultures, climate change may lead to the irreversible loss of objects, places, species or ecosystem functions that are valued by current generations, not to mention a loss of experiences and perceived rights that are valued. In this section, I present general examples of traditional, modern and postmodern values in Norway and discuss how they are changing. I then consider how adaptations to changing snow cover may correspond to different values in Norway, which may lead to conflicts within present generations, or with future generations.

'Norwegian values' are frequently associated with nature, rural livelihoods, simplicity, honesty and humility (Eriksen, 1993). However, Norway's national identity and culture are continually being constructed and created, and they embody many of the contradictions that exist between traditional, modern and postmodern values (Eriksen, 1993). Although Norwegian identity is closely linked to traditional values (for example, an emphasis on rural areas, nature and the family), there is at the same time an increasing emphasis on modern values (for example, individualism, economic development, material wealth, technology and scientific progress) (Slagsvold and Strand, 2005). Yet there is also evidence of the emergence of postmodern, pluralistic values in Norway. Norwegian identity has been characterised as egalitarian individualism, which includes a pluralistic rejection of social hierarchies and the promotion of equity across gender and classes, and between rural and urban areas (Eriksen, 1993). An emphasis on social democracy, equality and individual integrity includes traditional and modern values, but also transcends these values to embrace a broader notion of 'Norway'. As articulated by the Norwegian Foreign Minister Jonas Gahr Støre (2006), Norway is in the process of shaping a new and bigger 'we' that is valid for all and that can be adjusted as Norway changes. Norway increasingly sees its role in the international community as one of responsibility, and its commitment to international peace processes and high levels of development assistance might be interpreted as part of its postmodern identity. Norway – and the Nordic countries in general – is one of the countries described by Inglehart (2000) as having shifted over the past decades towards a post-materialist, postmodern worldview, which is reflected in the current government's world-centric social discourse and ethics (for example, democracy, equality and social responsibility).

While the predominant discourse appears to be moving from modern to postmodern, as evidenced through the World Values Survey (Inglehart, 1997), a full spectrum of values coexists in Norway. Distinctions between traditional, modern and postmodern structures can be clearly observed at the individual and community levels in Norway, where there are likely to be value distinctions between rural and urban areas, and between generations and social classes. Below, I draw attention to some very general values that are associated with these three worldviews in Norway, and then discuss how they may be changing.

Traditional worldviews

In Norway, values associated with traditional worldviews include an emphasis on family, equality, belonging to the local community, identity and security. Traditional values favour recollectivisation over individualism and cultural homogeneity over diversity (Aukrust and Snow, 1998). The agricultural landscape in particular provides a sense of stability, historical connection, identity and a sense of belonging (Lindland, 1998). Norwegian social welfare policy in recent decades has emphasised family values and economic and social security, as evidenced by increases in old-age pensions, the extension of parental leave and the introduction of a Family Cash Benefit scheme (Botten et al., 2003). The Lutheran state church dominates religious life in Norway, and an estimated 88% of Norway's population of 4.3 million were members in 1999 (Leirvik, 1999). In some parts of the country, a strong Protestant influence actively tries to prevent the moral decay of the simple Norwegian identity.

Modern worldviews

The rise of modernity first appeared in Norwegian cities, where it culminated as 'classic modernity' in the 1950s and 1960s (Gullestad, 1996). Individuals that valued progress, technology and development transformed Norway into an oil nation with enormous economic power. Modernity combined with wealth placed increasing emphasis on individualism, materialism and the role of the private sector. The modern social welfare system has placed a greater focus on private welfare sources, such as the family, the market and voluntary organisations, and on the idea of 'mutual obligations' and 'personal responsibility' (Botten et al., 2003). Even outdoor recreation is increasingly being carried out in a modern context which, according to Riese and Vorkinn (2002) can be expected to influence the process of meaningful construction. The Norwegian notion of *friluftsliv* ('outdoor life') is constructed as a traditional Norwegian value, yet it has been transformed and adapted to modern values, and indeed can be considered 'both a consequence of and a reaction against the industrialized and urbanized society' (Sandell, 1993, p. 2).

Postmodern worldviews

Many individuals and groups in Norway exhibit postmodern worldviews and associated values, which emphasise self-expression and self-realisation, pluralism and integration. The transmission of values in families has gone from the notion of 'obedience' to the notion of 'being oneself' (Gullestad, 1996). Gullestad (1996, p. 37) argues that '[t]hese new tendencies resonate with the kinds of flexibility and creativity needed in the present stage of capitalism'. In recent years there has been a call for a new architecture for social welfare, which challenges universalism and instead focuses on improving the welfare of the poorest (Botten et al., 2003). Since the 1970s, religious pluralism has increased in Norway, mainly as a result of Muslim immigration (Leirvik, 1999).

The different values associated with traditional, modern and postmodern worldviews are not static among individuals, communities or social groups. Rather, they are changing in response to a constellation of factors, including economic changes (neo-liberal economic policies, increased material wealth and consumption), demographic changes (urbanisation and an aging population), cultural changes (an increase in immigrants and changing youth cultures) and geopolitical changes (consideration of European Union membership, increased competition for natural resources in the Arctic). There is evidence that traditional values in Norway have become more liberal (Statistics Norway, 1996). Although differences between traditional and modern values have been closely linked to differences between rural and urban areas, Bæck (2004) found that many young people in rural areas express values and preferences that are closely associated with urban settings, or what he refers to as an urban ethos, which is closely linked to modern values. The difference in values between rural and urban areas is decreasing as rural areas gain better access to communication, media, and the spread of lifestyles and modes of living. Furthermore, Inglehart and Baker (2000, p. 49) found that '[i]ndustrialization promotes a shift from traditional to secular–rational values, while the rise of postindustrial society brings a shift toward more trust, tolerance, well-being, and postmaterialist values. Economic collapse tends to propel societies in the opposite direction.' However, their research also shows that the influence of traditional values is likely to persist, as belief systems can exhibit both durability and resilience. In any case, modern values are not unproblematic in Norway, and there is a concern that increased materialism may erode support for the social welfare system, particularly among younger generations (Edlund, 1999).

Adaptations to changes in snow cover

How might these different and dynamic values in Norway influence adaptation to changes in snow cover associated with climate change, and how might values be affected by adaptation measures? It is well recognised that climate change will result in differential impacts within Norway (RegClim, 2005). Vulnerability to these impacts is, however, considered to be a function of exposure, sensitivity and adaptive capacity (McCarthy et al., 2001). The capacity to adapt to climate change is frequently considered to be a function of wealth, technology, education, information, skills, infrastructure, access to resources, and stability and management capabilities (McCarthy et al., 2001). Norway ranks high in all of these areas, thus in theory has a high capacity to adapt to a changing climate (O'Brien et al., 2004). However, empirical research shows that this capacity is not always translated into successful adaptations (Naess et al., 2005), and this has contributed to a growing recognition that there are barriers to adaptation, both in countries with developing and developed economies (Adger et al., 2007).

Values are seldom considered as an important factor within the wider discourse on adaptation. They represent an interior and subjective dimension of adaptation that is not easily observed and measured. Nonetheless, the relationship between values and climate change adaptation can be studied and analysed by looking at how the impacts and adaptations associated with a decreasing snow cover affect traditional, modern and postmodern values in Norway. It is projected that snow cover will decrease in many areas of Norway as temperatures rise over the next century. Climate models project that winter temperatures will increase by 2.5–4 °C by 2100, and that the number of mild days (with temperatures above freezing) will increase at lower elevations and in the Arctic. Precipitation is expected to increase in many parts of Norway, including during winter in the eastern part of the country (RegClim, 2005). In terms of skiing conditions, it is projected that there will be an average of 60 days with conditions suitable for skiing by 2050, which represents a 40% decrease compared to the period 1981–1999 (RegClim, 2005).

These changes will translate into different impacts for individuals and communities in Norway, depending not only on where they are located, but also on what they value. Traditional values associate snow cover and winter sports with local or national identity, and many communities are dependent upon winter tourism for income and livelihoods. The link between traditional values, identity and national heritage was particularly visible during the planning of the Winter Olympics in Lillehammer in 1994 (Eriksen, 1993). Traditional modes of winter transportation, including cross-country skis, the *spark* and the *pulk* (two types of sleds) are likely to become less viable and visible as snow cover decreases. While these changes are often considered trivial in comparison to the impacts of climate change on the basic needs for food, water and shelter in many parts of the world, the point is that they will directly affect what many people in Norway value. Adaptations to climate change directed towards traditional values might therefore emphasise the preservation of heritage, tradition and identity, which often occurs through the preservation of traditional landscapes and cultural icons, such as the Holmenkollen ski jump in Oslo (Antrop, 2005). Acknowledging the decrease in snow cover as a loss, preserved through museums and festivals, may be one way of adapting to change, but transforming livelihoods and maintaining a sense of community and belonging could represent a greater challenge to adaptation under climate change.

Modern values emphasise snow as a medium for winter sports, particularly skiing, which is considered an important economic sector in Norway because of the links to tourism, winter cabins, producers of equipment, and local businesses. Adaptations to decreased snow cover that are directed at modern values may include advanced snow-making technologies, indoor snow domes, artificially cooled cross-country ski tracks, and other technological responses. In terms of identity, modern societies are capable of reconstructing identities fairly easily, whether it is through roller-skis (i.e. cross-country skis on wheels) or skating on synthetic ice. These adaptations are unlikely to appeal to the values associated with traditional worldviews. In other words, from the perspective of traditional values, artificial snow on green mountains may not be a satisfactory replacement for snow-covered mountains, and roller skis may not be an acceptable substitute for traditional winter sports. Furthermore, reduced access to snow may turn cross-country skiing into an elite sport for those with access to resources, rather than a sport available to all Norwegians. Alpine ski centres at higher elevations may benefit from the loss of competition from other ski areas in Europe, while those at lower elevations may reinvent themselves as centres for recreation and relaxation. However, as Lund (1996) notes, 'The striking tendency of alpine skiing to reinvent itself every decade may be invisible to those relatively new to the sport but it is certainly not lost on longtime skiers who can all remember, very clearly, just how skiing used to be.'

Postmodern values are likely to view changes in snow cover from a larger, systems perspective. The role of snow in biological, physical and social systems may be emphasised, with the integrity of social–ecological systems considered a priority. Adaptations to climate change may address not only human needs, but the needs of different species, as well as ecosystem functions and services. Such values are not unique to postmodern worldviews, and instead may have a strong basis in some traditional worldviews. For example, snow cover is important to reindeer, thus snow is likely to be valued by Saami reindeer herders in Northern Norway. As Reinert et al. (2008, p. 5) point out, a loss of nature quality cannot be compensated by a gain in other values: 'The cultural values of Saami reindeer herding, in the past and the present, are intertwined with the nature values of the tundra landscape, and the values that need to be preserved must be understood in terms of the spatio-temporal particularity they represent.'

Postmodern values may emphasise the relationship between snow cover and hydrological regimes, including the implications of melting snow for sea level rise.

The relationship between less snow cover, decreases in the planetary albedo and the global energy balance may be a concern, as this could accelerate warming (Holland et al., 2006). The distant impacts of climate change on other populations and groups are also likely to be of relevance to postmodern values, as they raise issues of equity, justice and rights. Adaptations that take into account postmodern values may very well focus on creating dramatic changes in energy systems in order to reduce greenhouse gas emissions. Such changes are often discussed separately as examples of climate change mitigation, but they nonetheless represent an important adaptive response to a changing climate.

The potential for value conflicts in relation to climate change adaptation has not been widely discussed in the literature on climate change. To successfully address different and dynamic values, climate change adaptations may have to both recognise and address a wide spectrum of values, including threats to physiological needs and safety needs (both in Norway and elsewhere), as well as values that influence modern and postmodern values such as individual identity, achievement, universalism and ecosystem integrity. Human development research has shown, however, that the values that emerge as priorities from a postmodern perspective (for example, equity, justice and ecosystem integrity) may not be prioritised by those holding traditional or modern worldviews (Maslow, 1970; Kegan, 1998; Wilber, 2007). Similarly, modern values such as those related to growth, technological advances and scientific rationalism may not be recognised or prioritised by individuals and communities with traditional worldviews. Furthermore, those with postmodern worldviews may not recognise or prioritise the values associated with 'post postmodern' worldviews, which might, for example, include a greater emphasis on aesthetic and spiritual values, such as the experience of snow, a sense of place, or non-dual relationships with other living organisms. Some of these 'post postmodern' values are, however, dominant values in some traditional societies, a fact that may be captured by the circular structure of Schwartz's 'Values Theory'. Nonetheless, the fact that many of these values may not be recognised or addressed through adaptations potentially represents a limit to adaptation as a response to climate change.

Conclusion

What do different and dynamic values and worldviews imply for adaptation to climate change? On the one hand, one could argue that climate change adaptations should first and foremost satisfy security and survival values that are linked to physiological needs, safety needs and social order. Such adaptations can be considered as a foundation for human development and human security. On the other hand, one could argue that climate change adaptations should aim to preserve values that are associated with postmodern and other worldviews, such as universalism, benevolence, altruism and biospheric values. These values may dominate in future generations, if material needs and survival values are satisfied (Inglehart, 1997). Surprisingly, there is an implicit assumption in most current discussions of climate change adaptation that what is valued by individuals and societies today is likely to be equally valued by future generations. An exception is future economic values, which are often addressed through discounting (Toman, 2006). However, as Adger et al. (2009a, p. 15) point out, '[t]he loss of place and its psychosocial and cultural elements (the loss of a "world") can arguably never be compensated for with money'.

The challenge then is to identify adaptation strategies that acknowledge and address a spectrum of values. If this is not feasible, it is important to identify value conflicts and consider whose values count. The capacity to respond to different and dynamic values may be closely linked to the perspectives of those holding power, those making adaptation decisions, and those carrying out the adaptations. The values and worldviews of so-called stakeholders who are directly involved in climate change adaptation thus matter, both to present and future generations. As Williams (1979, p. 23) emphasises, '[v]alues make a difference; they are not epiphenomenal'.

If values subjectively define the limits to adaptation as a response to climate change, as much or more so than objective factors, then the positive and negative outcomes of climate change cannot be assessed without considering what different individuals and communities value, both in the present and future. Successful adaptation will depend on the capacity of individuals and societies to perceive and respond to a spectrum of legitimate values that extend beyond those that are relevant to oneself or one's group. One clear challenge of climate change adaptation is to take into account values that correspond to diverse human needs and multiple perspectives and worldviews. This includes values that many individuals and groups do not currently prioritise, yet which are likely to become important as humans further develop. As values change, the outcomes of climate change are likely to be reassessed and re-evaluated. The emergence of more pluralistic, integral and holistic worldviews would suggest that aggressive reductions in greenhouse gas emissions may turn out to be the adaptation that is most valued by future generations.

Acknowledgements

I thank Michael van Niekerk and Marianne Bruusgaard for research assistance, and Gail Hochachka, Lise Kjølsrød, Svein Jarle Horn, Jonathan Reams, Johanna Wolf, Irene Lorenzoni and Neil Adger for valuable comments on earlier drafts. This research is part of the PLAN project on 'The Potentials of and Limits to Adaptation in Norway', funded by the Research Council of Norway.

References

Adger, W.N., Agrawala, S., Mirza, M.M.Q., Conde, C., O'Brien, K., Pulhin, J., Pulwarty, R., Smit B. and Takahashi, K. 2007. 'Assessment of adaptation practices, options, constraints and capacity', in Parry, M.L., Canziani, O.F., Palutikof, J.P., Van der Linden, P.J. and Hanson C.E. (eds.) *Climate Change 2007: Impacts, Adaptation and Vulnerability. Contribution of Working Group II to the Fourth Assessment Report of the Intergovernmental Panel on Climate Change.* Cambridge: Cambridge University Press, pp. 717–743.

Adger, W.N., Barnett, J. and Ellemor, H. 2009a. 'Unique and valued places at risk', in Schneider, S.H., Rosencranz, A. and Mastrandrea, M. (eds.) *Climate Change Science and Policy.* Washington, DC: Island Press, in press.

Adger, W.N., Dessai, S., Goulden, M., Hulme, M., Lorenzoni, I., Nelson, D., Naess, L.-O., Wolf, J. and Wreford, A. 2009b. 'Are there social limits to adaptation?', *Climatic Change,* **93**: 335–354.

Antrop, M. 2005. 'Why landscapes of the past are important for the future', *Landscape and Urban Planning* **70**: 21–34.

Aukrust, V.G. and Snow, C.E. 1998. 'Narratives and explanations during mealtime conversations in Norway and the US', *Language in Society* **27**: 221–246.

Bardi, A. and Schwartz, S.H. 2003. 'Values and behavior: strength and structure of relations', *Personality and Social Psychology Bulletin* **29**: 1207–1220.

Bæck, U.N. 2004. 'The urban ethos', *Nordic Journal of Youth Research* **12**: 99–115.

Botten, G., Elvbakken, K.T. and Kildal, N. 2003. 'The Norwegian welfare state on the threshold of a new century', *Scandinavian Journal of Public Health* **31**: 81–84.

Edlund, J. 1999. 'Trust in government and welfare: attitudes to redistribution and financial cheating in the USA and Norway', *European Journal of Political Research* **35**: 341–370.

Eriksen, T.H. 1993. 'Being Norwegian in a shrinking world: reflections on Norwegian identity', in Kiel, A.C. (ed.) *Continuity and Change: Aspects of Modern Norway.* Oslo: Scandinavian University Press, pp. 11–37.

Gecas, V. 2008. 'The ebb and flow of sociological interest in values', *Sociological Forum* **23**: 344–350.

Grothmann, T. and Patt, A. 2005. 'Adaptive capacity and human cognition: the process of individual adaptation to climate change', *Global Environmental Change* **15**: 199–213.

Gullestad, M. 1996. 'From obedience to negotiation: dilemmas in the transmission of values between generations in Norway', *Journal of the Royal Anthropological Institute* **2**: 25–42.

Holland, M.M., Bitz, C.M. and Tremblay, B. 2006. 'Future abrupt reductions in the summer Arctic sea ice', *Geophysical Research Letters* **33**: L23504.

Inglehart, R. 1997. *Modernization and Postmodernization: Cultural, Economic, and Political Change in Forty-Three Societies.* Princeton: Princeton University Press.

Inglehart, R. 2000. 'Globalization and post-modern values', *Washington Quarterly* **23**: 215–228.

Inglehart, R. and Baker, W.E. 2000. 'Modernization, cultural change, and the persistence of traditional values', *American Sociological Review* **65**: 19–51.

Inglehart, R. and Welzel, C. 2005. *Modernization, Cultural Change, and Democracy: The Human Development Sequence.* Cambridge: Cambridge University Press.

Kearney, M. 1984. *World View.* Novato: Chandler and Sharp.

Kegan, R. 1982. *The Evolving Self: Problem and Process in Human Development.* Cambridge: Harvard University Press.

Kegan, R. 1994. *In Over Our Heads: The Mental Demands of Modern Life*. Cambridge: Harvard University Press.

Leirvik, O. 1999. 'State, church and Muslim minority in Norway', paper presented at the *Dialogue of Cultures Conference*, 21–23 April 1999, Berlin.

Lindland, J. 1998. 'Non-trade concerns in a multifunctional agriculture: implications for agricultural policy and multilateral trading system', paper presented at *OECD Workshop on Emerging Trade Issues in Agriculture*, 26–27 October 1998, Paris.

Lund, M. 1996. 'A short history of Alpine skiing', *Skiing Heritage* **8**: 1. Available at www.skiinghistory.org/history.html

Maslow, A. H. 1970. *Motivation and Personality*. London: Harper and Row.

McCarthy, J. J., Canziani, O. F., Leary, N. A., Dokken, D. J. and White, K. S. (eds.) 2001. *Climate Change 2001: Impacts, Adaptation and Vulnerability. Contribution of Working Group II to the Third Assessment Report of the Intergovernmental Panel on Climate Change*. Cambridge: Cambridge University Press.

Næss, L. O., Bang, G., Eriksen, S. and Vevatne, J. 2005. 'Institutional adaptation to climate change: flood responses at the municipal level in Norway', *Global Environmental Change* **15**: 125–138.

Naugle, D. K. 2002. *Worldview: The History of a Concept*. Cambridge: Eerdmans.

O'Brien, K. L., Sygna, L. and Haugen, J. E. 2004. 'Resilient or vulnerable? a multi-scale assessment of climate impacts and vulnerability in Norway', *Climatic Change* **64**: 193–225.

Pakizeh, A., Gebauer, J. E. and Maio, G. R. 2007. 'Basic human values: inter-value structure in memory', *Journal of Experimental Social Psychology* **43**: 458–465.

RegClim 2005. *RegClim: Norges klima om 100 år: Usikkerheter og risiko*. Oslo, Norway. Available at http://reglim.met.no

Reinert, E. S., Aslaksen, I., Eira, I. M. G., Mathiesen, S., Reinert, H. and Turi, E. I. 2008. *Adapting to Climate Change in Reindeer Herding: The Nation–State as Problem and Solution*, Working Papers in Technology Governance and Economic Dynamics No. 16. Norway: The Other Canon Foundation and Tallinn: Tallinn University of Technology.

Riese, H. and Vorkinn, M. 2002. 'The production of meaning in outdoor recreation: a study of Norwegian practice', *Norwegian Journal of Geography* **56**: 199–206.

Rohan, M. J. 2000. 'A rose by any name? The values construct', *Personal and Social Psychology Review* **4**: 255–277.

Rokeach, M. 1973. *The Nature of Human Values*. New York: Free Press.

Rokeach, M. (ed.) 1979. *Understanding Human Values: Individual and Societal*. New York: Free Press.

Rokeach, M. 2000. *Understanding Human Values*, 2nd edn. New York: Simon and Schuster.

Sandell, K. 1993. 'Outdoor recreation and the Nordic tradition of "friluftsliv": a source of inspiration for a sustainable society', *Trumpeter* **10**. Available at: www.icaap.org/iuicode?6.10.1.10

Schwartz, S. H. 1994. 'Are there universal aspects in the structure and contents of human values?', *Journal of Social Issues* **50**: 19–45.

Schwartz, S. H. 1996. 'Value priorities and behaviour', in Seligman, C., Olson, J. M. and Zanna, M. P. (eds.) *The Psychology of Values, Ontario Symposium, vol. 8*. Mahwah: Lawrence Erlbaum, pp. 1–24.

Schwartz, S. H. 2006. 'Les valeurs de base de la personne: théorie, mesures et applications' ('Basic human values: theory, measurement and applications'), *Revue française de sociologie* **4**. (English version available at www.fmag.unict.it/Allegati/convegno%207-8-10-05/Schwartzpaper.pdf)

Seligman, C. and Katz, A. N. 1996. 'The dynamics of value systems', in Seligman, C., Olson, J. M. and Zanna, M. P. (eds.) *The Psychology of Values, Ontario Symposium, vol. 8*. Mahwah: Lawrence Erlbaum, pp. 53–75.

Sire, J. W. 2004. *Naming the Elephant: Worldview as a Concept*. Downers Grove: IVP Academic.

Slagsvold, B. and Strand, N. P. 2005. 'Morgendagens eldre: blir de mer kravstore og mindre beskjedne?' in Slagsvold, B. and Solem, P. E. (eds.) *Morgendagens Eldre: En Sammenligning av Verdier, Holdninger og Atferd Blant Dagens Middelaldrende og Eldre*, Nova Report No. 11/05. Oslo: NOVA, pp. 23–50.

Statistics Norway 1996. *Values in Norway 1996*, Notes 97/19. Oslo: Statistics Norway.

Støre, J. G. 2006. 'Å skape et nytt og større "vi"', speech delivered by Norwegian Foreign Minister Jonas Gahr Støre on 8 December 2006. Available at www.regjeringen.no/nb/dep/ud/dep/utenriksminister_jonas_gahr_store/taler_artikler/2006/A-skape-et-nytt-og-storre-vi.html?id=437985

Toman, M. 2006. 'Values in the economics of climate change', *Environmental Values* **15**: 365–379.

Wilber, K. 2000. *Integral Psychology: Consciousness, Spirit, Psychology, Therapy*. Boston: Shambhala.

Wilber, K. 2007. *Integral Spirituality: A Startling New Role for Religion in the Modern and Postmodern World*. Boston: Integral Books.

Williams Jr, R. M. 1979. 'Change and stability in values and value systems: a sociological perspective', in Rokeach, M. (ed.) *Understanding Human Values: Individual and Societal*. New York: Free Press, pp. 15–46.

Yohe, G. and Tol, R. S. J. 2002. 'Indicators for social and economic coping capacity: moving toward a working definition of adaptive capacity', *Global Environmental Change* **12**: 25–40.

11

Conceptual and practical barriers to adaptation: vulnerability and responses to heat waves in the UK

Johanna Wolf, Irene Lorenzoni, Roger Few, Vanessa Abrahamson and Rosalind Raine

Introduction

The health impacts of global climate change have long been a focus of discussion for researchers and policy-makers. In recent years the number of studies and reports on the theme has risen significantly, as reflected in the extended list of citations in the human health chapter of the latest Intergovernmental Panel on Climate Change (IPCC) assessment report (Confalonieri et al., 2007). Most of the analysis to date has concentrated on the epidemiological dimensions of disease and climate, investigating how climatic trends may alter the distribution, prevalence and health burden of diseases, and assessing how changes in extreme weather events and associated hazards may impact on health (for recent overviews see McMichael et al., 2003; Epstein and Mills, 2005; Watson et al., 2005). Increasingly, however, this work has begun to broaden and stimulate debate in the public health arena, with investigation extending to responses by individuals, communities and health system institutions. Within this mounting body of work, there has been an increasing movement towards a public health agenda for adaptation (for example Grambsch and Menne, 2003; Füssel and Klein, 2004; Ebi et al., 2005; Menne and Ebi, 2006). Yet, despite this progress, there remain surprisingly few empirical studies that concentrate on the intersection between climatic hazards, health, vulnerability and behaviour (Matthies et al., 2003; Few, 2007).

This chapter discusses a novel interdisciplinary approach to understanding the vulnerability of individuals to the effects of climate change and variability. It draws on and links to the relevant, but distinct, literatures identified above, in the context of early results from an empirical study in the UK investigating how elderly people perceive their own vulnerability to the effects of heat waves. Our research findings challenge some of the accepted theoretical perspectives on social capital and argue

Adapting to Climate Change: Thresholds, Values, Governance, eds. W. Neil Adger, Irene Lorenzoni and Karen L. O'Brien. Published by Cambridge University Press. © Cambridge University Press 2009.

that cognitive and behavioural barriers to adaptation can be better understood and addressed by drawing upon research literatures hitherto distinct.

Background

Stress arising from extreme heat is an important example of the health impacts of a changing climate. The latest projections for Europe suggest that periods of intense and prolonged heat are likely to become more frequent, more intense and longer (Christensen et al., 2007). We can distinguish indirect impacts, such as climatic shifts leading to changing patterns of infectious diseases (which are mediated by an intermediary, for example the malaria-transmitting mosquitoes), from direct impacts such as heat waves. For example, during the summer of 2003 between 27 000 and 40 000 excess deaths are thought to have been caused by the heat wave across Europe (Kovats and Jendritzky, 2006). Epidemiological studies suggest that specific groups in society are more vulnerable to the impacts of heat stress than others. Vulnerability to temperature-related impacts of climate change is determined by physiological and socio-economic factors (Kovats and Jendritzky, 2006), including age and disease profile, housing conditions, prevalence of air conditioning and behaviour (McMichael et al., 2003). Predisposing factors, for example, physical conditions (particularly chronic obstructive pulmonary disease (COPD) and diabetes), age and drug intake lead to increased heat-related morbidity and mortality (Koppe et al., 2004). The literature points to the elderly as a vulnerable group due to its age and illness profile. This conception of vulnerability is based on risk factors which increase the likelihood of morbidity and mortality and is very prominent in research exploring health impacts of climate change. However, it is largely uninformed by individuals' own perception of vulnerability and how these may affect adaptation behaviour.

Two of the main areas of research that have informed climate change adaptation are first, studies of human exposure to and coping with natural hazards and extreme events and hence the role of vulnerability (for example Burton et al., 1993; Adger et al., 2001; Wisner et al., 2004; Adger, 2006), and second, risk perception research (Slovic et al., 1981; Bord et al., 1998; O'Connor et al., 1999; Stamm et al., 2000; Zwick and Renn, 2002; Leiserowitz, 2005.) The main contribution of hazard research to adaptation is what Wisner et al. (2004, p. 49) call 'the social production of vulnerability'; a key consideration to understand how, where and why the impacts of climate variability and change manifest. This has led to an explicit account of how social, and indeed individual, processes shape vulnerability of people and places, and is now key to adaptation research. On the other hand, risk perception research in developed countries shows that while climate change is perceived a real threat, people fail to relate this to their personal lives (cf. for example Bord et al., 1998).

Study outline, framing and empirical findings:
four types of response strategies

Interviews were conducted with independently living elderly people aged between 72 and 94 and their carers[1] (including family members and friends, aged between 24 and 87) during the summer 2007,[2] a total of 105 participants. Interviewees are referred to by their anonymous synonyms that indicate their location, age and gender.[3] The interview questions focused on perception of and coping strategies for hot weather, daily routine and changes to it in hot weather, social activity level and contact with other people, outlook on life and health status. Interviews were analysed using qualitative coding techniques that follow Grounded Theory (Glaser and Strauss, 1967). After open coding (Strauss and Corbin, 1990), the emerging categories were condensed and refined using iterative axial coding (Glaser and Strauss, 1967; Ezzy, 2002), merging and expanding categories where necessary.

We use Few's 'health impact pathway' as a framework to analyse responses by the elderly. Few (2007) proposed tracing a 'health impact pathway' as a simple tool with which to map out how extreme weather events translate into health impacts, and thereby to highlight how the different factors that contribute to vulnerability and response to risk come into effect. In this way the health impact pathway also serves to locate opportunities for intervention to reduce risk – opportunities that might alternatively be articulated as entry points for adaptation.

Figure 11.1 indicates how heat wave events can generate the macro-environmental and micro-environmental conditions that may lead to heat stress in individuals and ultimately to health outcomes such as heatstroke and cardiovascular disease. At each step in this chain there are potential interventions that can be made to reduce risk and interrupt the health impact pathway. In the context of this research we show here examples of personal behaviours that might act (or might be perceived to act) in this way: these are indicated in the shaded boxes. Most are reactive forms of response, although anticipatory action can be implicit in technological investments and pre-arranged behaviours such as travel. Considering these response options, it is possible to envisage more clearly how 'personal' factors such as

---

[1] The term carer commonly refers to 'those who provide (without pay) care or assistance to people who are ill or need help with personal activities of daily living' (O'Reilly et al., 2008). In this research, however, the majority of elderly are fully independent, and the term carer is used broadly to refer to a person identified by the elderly participant as taking an important role in their lives and/or being available to provide care, advice or assistance (including reciprocal between spouses). Most often, this person would be a spouse, a brother or sister, a child or grandchild, or a friend of the person. The carers were identified in part in response to a question about to who the participant would turn in case he/she really needed help with something.

[2] One consequence of interviewing during a summer with no heat wave may be that participants did not as easily relate to the effects of heat. The timing, however, also provided an opportunity to explore anticipatory adaptation, (adjustments made in preparation for an event, rather than during it), and recall of strategies used previously in the absence of cues.

[3] For example, NC-07-F51 is carer for elderly participant number 7 in Norwich, female, aged 51. NE-07-M76 is Norwich elderly participant number 7, male, aged 76.

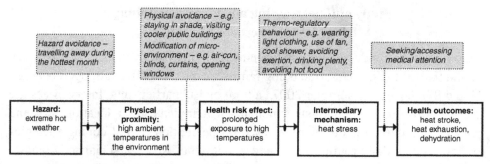

Figures 11.1 Simplified health impact pathway for heat waves in the UK, with examples of response mechanisms.

knowledge, self-efficacy and social networks of individuals help shape specific health risk behaviours and hence vulnerability. Although beyond the scope of this study, it also suggests how 'external' factors such as care services provision and urban planning policy come into play, and highlights the role of 'internal' (health status) factors such as impaired mobility, impaired thermoregulation or chronic disease in constraining response options or heightening susceptibility to health outcomes (Few, 2007).

Drawing on this framework, the following sections discuss four main categories of action in light of elderly responses to this study: anticipatory strategies (such as provision of air conditioning, and seasonal travel during the hottest time of the year); modification of the micro-environment (such as use of blinds, curtains, opening windows and fitting air conditioning); avoidance (such as staying in the shade, and visiting cooler public buildings); and thermoregulatory behaviour (such as relaxing, drinking plenty, avoiding hot food, showering). The role of healthcare provision was outside the scope of this particular study.

Anticipatory behaviour

Anticipatory behaviour aims to prepare for heat before it occurs and is intended to be used again in future events. Because of these characteristics it requires forethought and a certain investment. Behaviour of this type includes fitting a house or room with air conditioning and leaving home for the hottest month of the year. The type of behaviour that anticipates a heat event is least evident in this research. This type of response is employed only by those who identify heat as a threat to them personally and is explicitly linked to preparing for heat in the future. As a result, this type of behaviour is the consequence of anticipation of risk and would qualify as individual anticipatory adaptation (see Smit and Pilifosova, 2001). In practice, the link between anticipating heat and perceiving it as a threat is rare. This type of response fails to be enacted because of this mismatch. Here, the failure to perceive the threat, combined with lack of anticipation, constitute barriers to adaptation to heat events.

Modification of environment

Attempts to modify the environment of people's homes by using fans, blinds and opening or closing windows (depending on the time of day and how hot it is outside compared to inside) were found to be somewhat common responses. When tools such as fans or blinds are bought specifically for the purpose of staving off heat these adjustments bear an anticipatory dimension; the use of these tools in future heat events seems likely. At the time of acquisition however, these measures often occur in response to rather than in preparation for a heat event. Behaviour of this type is initially reactive but can take on the character of long-term adaptation if adopted again in the future (see Smit and Pilifosova, 2001). Potential changes people could employ to the home that can keep indoor temperatures low include adding insulation and double-glazed windows, both of which are labour- and cost-intensive adjustments usually only available to home-owners, and unlikely to be initiated during a heat event. Such changes could constitute anticipatory adaptation if they were implemented to avoid future exposure to heat. The evidence here suggests that adjustments may not be driven by heat alone. For blinds or curtains, and also opening or closing windows, participants may act because of the Sun's rays (for example damaging furniture) or to get fresh air (or in case of closing windows to avoid traffic noise or air pollution) rather than heat. Heat was not a key concern for most participants; thus it is questionable whether this type of response would occur at the right time to avoid future heat stress.

Avoidance behaviour

Many participants report avoiding the heat. In most cases this means staying indoors where it is cooler than outside, staying in the shade and avoiding the midday heat. This type of behaviour is obviously adopted when it is already hot and little forethought and no investments are required. As a result, this type of behaviour constitutes reactive adaptation (see Smit and Pilifosova, 2001). Some participants suggested that nothing can be done about heat and that avoidance is the best, and possibly the only, strategy to prevent its ill effects. This suggestion of having done all they can may in itself act as a barrier to further preventative activities. This is particularly the case for those participants who feel they have little agency to act in the face of heat.

Thermoregulatory behaviour

When it is already very hot, participants report adopting thermoregulatory behaviour to try to cool themselves off. These measures include taking a shower or splashing water on face and arms. This type of behaviour constitutes a completely reactive, short-term adjustment meant to abate immediate risk and would classify as coping rather than adaptation (see Adger, 2000; Berkes and Jolly, 2001). Similarly to the avoidance behaviour above, this type of behaviour may act as a barrier to further action as participants believe it is all they can do. As a result, other adjustments are not considered and participants who act in this way, purely reactively, may be at heightened risk, depending on their circumstances.

We note that the further along the pathway a response occurs, the less it constitutes an anticipatory adaptive response (Figure 11.1).

Perceptions and responses by the elderly

The above exploration of potential interventions and action on heat waves using Few's framework show how participants of this research reported that they respond to heat. Three salient aspects shape participants' behavioural responses and these are detailed below. Drawing on literatures on vulnerability, risk perception and related psychology, and social networks, this section explores how participants' responses can be explained in relation to adaptation to health impacts of climate change.

Perceptions of vulnerability

Epidemiological literature suggests that elderly people with particular pre-existing medical conditions, such as chronic obstructive pulmonary disease (COPD) (Rydman et al., 1999), and those living alone and not leaving the house each day (Semenza et al., 1996) are at greatest risk from heat illness, as outlined above. However, the interviews with independent elderly people in this research consistently reveal that most participants do not perceive themselves as at risk of suffering from the effects of heat waves – despite belonging to a group of the population that is known to be among the most vulnerable. Most participants do not perceive heat as a threat to them personally and as a result generally do not prepare for heat events. If participants adjust their routine, or employ any coping strategies, these are largely reactive and are undertaken when it is already hot. Those few participants who say heat can be a problem for them elaborate that they have always been susceptible to hot weather, since childhood.

Numerous participants were unable to think of groups that are more vulnerable to heat than the general population. Those who named groups seen as susceptible to extreme heat referred to people who are obese, of ill health (high blood pressure, heart conditions, asthma and other diseases contributing to COPD), fair skinned and poorly acclimatised (immigrants from cooler countries). Participants also pointed to groups by age, for example, babies and young children, and those with mobility impairments. There is evidence for some confusion between effects of heat (i.e. temperature) and Sun (i.e. UV radiation), a finding that has been well documented in other research (see for example Ungar, 2000).

An important finding is that even those individuals who identified the elderly as a group relatively more vulnerable to the effects of heat on health, and are over 75 years of age, do not attribute themselves to this group. Further, participants who named conditions from which they suffer do not identify with the vulnerability that arises from these conditions. Some literature suggests that potential hazards to society are usually evaluated to be higher than individual threats; individuals tend to underestimate their personal probability of experiencing negative events (Sjöberg, 2000). Similarly, research on elderly people's perceptions of falls suggests that the elderly themselves, while identifying falls as a significant risk to older people, do not generally perceive falls as a risk to them personally (Braun, 1998; Hughes et al., 2008). The results presented here support such results in a different context, and are also congruent with findings of a recent North American survey with elderly people aged 65 and above suggesting that 60% of those interviewed do not think that heat poses a significant threat to them personally (Sheridan, 2007).

Cognitive dimensions of responding to heat

The range of influences on individual attitudes and behaviours to health and environmental issues is well documented in the risk and psychology literatures. Attempts at modelling these (for example Ajzen, 1991; Stamm et al., 2000) highlight the complex relationships between different individual and contextual characteristics. Amongst these, both factual knowledge and cognition of personal exposure to risks (the rational base of assessing risk), and feelings of risk (the affective base) bear a significant influence on individuals' behavioural responses. Risk perceptions and judgements of risk acceptability are driven by the balance of perceived risks with tangible benefits, assessed on both rational and affective bases (for example Slovic et al., 2004).

Other important influences on behaviour are perceived controllability (the belief that one has volition control over the performance of a behaviour, or internal locus of control) and self-efficacy (i.e. the belief that one is capable of attaining a certain outcome or effect through one's actions) (for example Hines et al., 1987; Armitage and Conner, 2001). According to Bandura's (1997) social cognitive theory, individuals regulate the effort they dedicate to a particular task based on the outcome they expect from their actions and will be more inclined to undertake a series of actions if they believe they can succeed. People with low self-efficacy tend to avoid engaging in tasks; on the contrary, those with high self-efficacy will act. It has been postulated, however, that high self-efficacy can result in negative outcomes, when people overestimate their ability to complete a particular task or series of actions. While this study did not employ the standard quantitative measures to estimate self-efficacy, the interview data give insights into attitudes that arguably are related to the general self-efficacy (Bosscher and Smit, 1998; Luszczynska et al., 2005) of these participants.

Our analysis of interviewees' responses suggests a distinction between elderly who viewed themselves as capable of coping with the effects of a heat wave, and those elderly who, on the contrary, imply that little if anything can be done about heat or they are unsure what to do when discussing their responses to a possible heat wave. The former imply they are eager to maintain their own independence, some expressing this also by refusing help from others. This could be interpreted as both an indication that they actively overestimate their own ability to respond to periods of intense heat and concurrently the desire to not impose (or be a 'burden') on others. Those elderly who feel there is nothing they could do about the heat feel they have little or no agency in (a) changing the situation (when it is hot, the heat cannot be turned down), and (b) improving their own ability to cope with the heat. These attitudes could constrain the response of these individuals and possibly contribute to their vulnerability.

The role of social networks

It has been suggested that belonging to a strong social network, including having friends locally and participating in group activities, can have a protective effect against heat illness (Semenza et al., 1996) and that social isolation is a further risk factor (Naughton et al., 2002). Critical reviews of social capital indicate that the presence of networks (links between groups) or bonding (relationships between individuals) social capital may, albeit not necessarily, lead to an increase in resilience in societies and are associated with survival and recovery from natural disasters (Adger, 2003; Pelling, 2003; Pelling and High, 2005). This literature raises an expectation that high social capital, in the form of bonding support networks and bridging capital, could decrease vulnerability (see for example Woolcock and Narayan, 2000) but leaves unclear under which conditions this might occur.

With this in mind, the present study examined the level and type of social involvement of participants, and to what extent these influence adjustments to heat stress along the health impact pathway. We found the support networks of independent elderly participants vary widely. Some participants draw on family members and neighbours for social contact and advice while some others are isolated and effectively alone. The level of contact with others ranges between once a day to about once every two weeks, but most participants fall in between, having some social contact two or three times a week. The extent of the social network also varies significantly and is dependent on active involvement in social activities. Some participants put significant effort into socialising in clubs or groups while others have little interest in such activities.

What seems at least as important as the frequency of the contact and extent of the social network in this research is what type of relationship participants entertain and what sort of advice the might obtain from their social networks. In some cases, the friend/relative identified by the elderly interviewed as carer in this study was not the member of family living closest to them, and this shows that complex relationships have to be considered in an analysis of the social networks. Numerous friends and relatives interviewed as carers were unable to identify correctly the risks heat poses for health. Further, even those who name the elderly as a group vulnerable to the effects of heat fail to think of their cared-for as belonging to this group. For example, when asked whether she could do anything for her father to keep him comfortable in the heat, one respondent replies 'No, he just stays indoors', despite having identified being elderly as a factor of risk and listing strategies to keep herself comfortable in the heat (NC-07-F51). Some carers identify coping strategies for themselves but often do not translate these into advice for elderly Taking those interviewed as carers as an aspect of the bonding social networks of participants, this suggests that the type of advice these elderly are likely to receive from their social networks, if any, may be less than helpful to cope in the event of extreme heat. It could also imply that their cared-for have previously turned down offers of help and the carer therefore does not think it useful to offer help again. These results point to more complex interactions between social networks and maintaining low vulnerability than the above literature suggests.

Discussion: five key factors in shaping vulnerability

Emerging from the results above are five key factors that shape vulnerability to the effects of heat waves among the participants of this study. This section explores these factors in more detail.

Many elderly do not recognise their vulnerability

Even those participants who can identify groups that are vulnerable, and can point to factors they think contribute to vulnerability, fail to link these to themselves and their own condition. This is found also among people who live alone, are house-bound, suffer from COPD or diabetes, and other conditions that contribute to their vulnerability. In particular, there seems to be very little association between climate change and heat-related morbidity and mortality. Together, these two findings suggest that the public fails to associate both climate with local impacts on health, and health impacts with themselves.

This finding also relates to the risk perception literature on climate change, which suggests that people are concerned about climate change and perceive it as a reality, yet one that occurs in geographically removed places (Bord et al., 1998). Public perception of risk has recently been found to be influenced significantly by affective imagery and underlying values (Slovic, 2000). This means that 'sights, sounds, smells, ideas, and words to which positive and negative affect or feeling states have become attached through learning and experience' play a crucial role in making an issue salient to individuals (Slovic et al., 1998, p. 3). In this context, it seems at least plausible that the lack of association between health impacts and the self, combined with a lack of affective imagery of such impacts, helps reproduce 'not here – not me' perceptions.

'There's nothing you can do'

Reflecting findings in the literature, our analysis indicates that those individuals who feel they are unable to take action successfully to resolve a particular issue or problem tend to reinforce this belief by distancing themselves from the issue in question. Such individuals tend to externalise the locus of control and thus feel less able to take action (see for example Kollmuss and Agyeman, 2002). In the context of our findings, it seems possible that the vulnerability of those elderly who felt nothing can be done about hot spells may be increased and this could thus constrain proactive adaptive behaviour that would aim to prevent exposure. This situation could be compounded if they are removed from any social support networks. We argue that, based on our findings, the attitude of elderly who do not believe they have agency in preventing heat stress contributes to their vulnerability and reduces their resilience and thus ability to adapt.

'I don't like asking for help'

Those individuals who perceived themselves as able to cope on their own or with minimal help from others with situations to which they would objectively be at high risk, underline the importance of considering the relationship between objective and subjective perceptions of the self. We had no means in this study of assessing whether the elderly we interviewed would have successfully coped with a heat wave.

However, their responses beg the question of whether perceived self-sufficiency among some individuals may indeed represent a hindrance and in fact may perpetuate and further their vulnerability if these views were sustained, for example, for the duration of a heat wave. If so, this could arguably occur as a result of individuals overestimating their capacity to react while underestimating the risk. In the cases of some participants external support is, if accepted at all, considered with significant reluctance and this could further exacerbate their vulnerability to such an extreme event. It may also be reasonable to consider, however, that in extreme situations some of the elderly who feel able to cope independently may recognise their own limitations, thus accepting help and support from others. The study presented in this chapter, however, raises questions about the role of networks in increasing elderly resilience to extreme heat, the next key factor below.

Social capital does not necessarily reduce vulnerability

Among the people interviewed as carers, many perceive their family member or friend not to be vulnerable to the effects of heat – often despite identifying the elderly as more affected by heat than other groups in society. Numerous of them therefore fail to see the need for preparation. Knowledge of how heat affects health and what can be done to prevent effects is often poor. Carers were also often unaware of how the medical conditions of their family member/friend mediate how the person can be affected by heat. The combination of these factors implies that the vulnerability of the elderly could be amplified because they rely either on poor advice or cope without effective help. This result challenges conceptions of social capital that suggest that access to and involvement in social networks brings about adaptive capacity and helps prevent negative outcomes of extreme events (see for example Woolcock and Narayan, 2000). This research suggests that the relationship between social capital, understood as social contact and participation in networks, and adaptation is neither unidirectional nor necessarily positive.

Responses constitute reactive adaptation,
not anticipatory adaptation

Among those who do adjust behaviour during hot spells, the majority of responses occur when it is already very hot and are therefore purely reactive as a means to counteract prolonged exposure to heat and prevent heat stress. There is little evidence of longer-term, proactive adaptation in terms of hazard avoidance or modification of the home environment. The reason for this appears to relate to the point made above, because the failure to perceive oneself as vulnerable, or to regard heat as a major issue, does not warrant any longer-term strategic preparations to adapt.

Another reason may be that possible responses to heat tend to be seen as 'common-sense' behavioural adjustments that respondents report having used primarily to cope when it was already very hot. These results directly inform the adaptation literature, highlighting that proactive adaptation that looks to adapt long term does not readily happen, and must hurdle a series of cognitive and motivational barriers. The left-hand side of Figure 11.1 in particular might be seen to offer entry points for anticipatory adaptation, but, as Few (2007) underlines, the success with which such opportunities are taken up is determined by a much wider set of enabling and constraining factors. Technical and financial capacities to adopt adaptive measures are inevitably key, but human behaviour in the face of extreme weather hazards is also fundamentally shaped by perceptions and attitudes, including self-efficacy. The research raises questions about the adaptive capacity of specific vulnerable groups of people in developed countries and therefore challenges the conception that these countries necessarily have high adaptive capacity.

Conclusion

The results discussed in this chapter contribute four key insights to the discussion on barriers to adaptation. First, we demonstrate that the way in which heat, as an example of a climate-related event, is perceived in relation to oneself directly affects whether or not an individual is motivated or inclined to adapt. When heat is not considered a major issue, few adaptive adjustments are made. Second, the discussion highlights that individuals' perceptions of their own vulnerability are crucial in shaping whether, and if so how, they respond to heat events. We show that elderly people, a group relatively vulnerable to heat stress, do not perceive this vulnerability and therefore respond reactively without adapting in the longer term. Third, personal characteristics, such as self-efficacy, are shown to influence whether individuals act in response to or preparation for a heat event, or whether they act at all. Both perceiving oneself as self-sufficient and believing there is nothing one can do about heat can act as barriers to adaptation when they support disempowered attitudes on one hand, and cause individuals to refuse help on the other. Fourth, we challenge the ways in which social networks and social capital relate to adaptive capacity at the individual level. Involvement in social networks and support from them can mean poor advice and reliance on poorly informed individuals, and therefore could exacerbate vulnerability to heat stress. This suggests that social capital per se may not necessarily a positive asset and rather that its value depends on the nature of the interaction and the characteristics of the networks involved.

Our results support O'Brien et al. (2004) in concluding that there are highly vulnerable groups within affluent developed countries, here in the UK. Beyond this, however, this research highlights that health-related vulnerability is not simply a function of certain measurable characteristics, such as age or disease profile, but that it is also an outcome of individuals' perceptions and their traits which shape how responses to extreme events such as heat waves are enacted. In fact, all these factors interact to produce vulnerability. Because of the direct nature of heat impacts, individuals may have relatively more agency to respond to these stresses than to many indirect effects of climate change. Accordingly, perceptions matter more here and understanding their implications is paramount in overcoming the barriers to long-term proactive adaptation. This research also indicates that results from the risk perception literature are as important to climate change adaptation as they are to questions of mitigation. To answer important questions of cognitive and behavioural barriers to adaptation, research must bring further together not only the resilience and vulnerability literatures, but also that on risk perception, cognitive and behavioural psychology, and individual resilience. We conclude that to overcome these barriers, both deeper social inquiry – to better understand their social and individual origins – and accordingly informed policy to protect the most vulnerable are required.

Acknowledgements

This research was undertaken as part of the adaptation programme of the Tyndall Centre for Climate Change Research, funded by NERC, ESRC and EPSRC, and of a University College London research project titled 'Heat waves in the UK: impacts and public health responses', funded by a Medical Research Council grant (id. 76585). The authors thank Neil Adger, Bridget Fenn, Sari Kovats and Paul Wilkinson for a fruitful collaboration. This research drew on helpful support from the North Central London Research Consortium (NoCLoR), the Norfolk Primary Care Trust and the SPHERE Primary Care Research Network. This research was approved by the Charing Cross Research Ethics Committee, National Research Ethics Service (Ref. no. 07/Q0411/37), and by institutional research ethics committees of University College London and the London School of Hygiene and Tropical Medicine. The authors are indebted to the participants without whom this research would not have been possible.

References

Adger, W.N. 2000. 'Social and ecological resilience: are they related?', *Progress in Human Geography* **24**: 347–364.

Adger, W.N. 2003. 'Social capital, collective action and adaptation to climate change', *Economic Geography* **79**: 387–404.

Adger, W.N. 2006. 'Vulnerability', *Global Environmental Change* **16**: 268–281.

Adger, W. N., Kelly, P. M. and Huu Ninh, N. (eds.) 2001. *Living with Environmental Change: Social Vulnerability, Adaptation and Resilience in Vietnam*. London: Routledge.

Ajzen, I. 1991. 'The theory of planned behaviour', *Organizational Behavior and Human Decision Processes* **50**: 179–211.

Armitage, C. J. and Conner, M. 2001. 'Efficacy of the theory of planned behaviour: a meta analytic review', *British Journal of Social Psychology* **40**: 471–499.

Bandura, A. 1997. *Self-Efficacy: The Exercise of Control*. New York: W. H. Freeman.

Berkes, F. and Jolly, D. 2001. 'Adapting to climate change: social–ecological resilience in a Canadian western Arctic community', *Conservation Ecology* **5**: 18.

Bord, R. J., Fisher, A. and O'Connor, R. E. 1998. 'Public perceptions of global warming: United States and international perspectives', *Climate Research* **11**: 75–84.

Bosscher, R. J. and Smit, J. H. 1998. 'Confirmatory factor analysis of the General Self-Efficacy Scale', *Behaviour Research and Therapy* **36**: 339–343.

Braun B. L. 1998. 'Knowledge and perception of fall-related risk factors and fall-reduction techniques among community-dwelling elderly individuals', *Physical Therapy* **78**: 1262–1276.

Burton, I., Kates, R. W. and White, G. F. 1993. *Environment as Hazard*. London: Guilford.

Christensen, J. H., Hewitson, B., Busuioc, A., Chen, A., Gao, X., Held, I., Jones, R., Kolli, R. K., Kwon, W.-T., Laprise, R., Magaña Rueda, V., Mearns, L., Menéndez, C. G., Räisänen, J., Rinke, A. Sarr, A. and Whetton, P. 2007. 'Regional climate projections', in Solomon, S., Qin, D., Manning, M., Chen, Z., Marquis, M., Averyt, K. B., Tignor, M. and Miller, H. L. (eds.) *Climate Change 2007: The Physical Science Basis. Contribution of Working Group I to the Fourth Assessment Report of the Intergovernmental Panel on Climate Change*. Cambridge: Cambridge University Press, pp. 847–940.

Confalonieri, U., Menne, B., Akhtar, R., Ebi, K. L., Hauengue, M., Kovats, R. S., Revich, B. and Woodward, A. 2007. 'Human health', in Parry, M. L., Canziani, O. F., Palutikof, J. P., Van der Linden, P. J. and Hanson, C. E. (eds.) *Climate Change 2007: Impacts, Adaptation and Vulnerability. Contribution of Working Group II to the Fourth Assessment Report of the Intergovernmental Panel on Climate Change*. Cambridge: Cambridge University Press, pp. 391–431.

Ebi, K. L., Smith, J. B. and Burton, I. (eds.) 2005 *Integration of Public Health with Adaptation to Climate Change: Lessons Learned and New Directions*. London: Taylor and Francis.

Epstein, P. R. and Mills, E. (eds.) 2005. *Climate Change Futures: Health, Ecological and Economic Dimensions*. Boston: Center for Health and the Global Environment, Harvard Medical School.

Ezzy, D. 2002. *Qualitative Analysis: Practice and Innovation*. London: Routledge.

Few, R. 2007. 'Health and climatic hazards: framing social research on vulnerability, response and adaptation', *Global Environmental Change* **17**: 281–295.

Füssel, H.-M. and Klein, R. J. T. 2004. *Conceptual Frameworks of Adaptation to Climate Change and their Applicability to Human Health*, PIK Report No. 91. Potsdam: Institute for Climate Impact Research.

Glaser, B. G. and Strauss, A. 1967. *The Discovery of Grounded Theory: Strategies for Qualitative Research*. London: Weidenfeld and Nicholson.

Grambsch, A. and Menne, B. 2003. 'Adaptation and adaptive capacity in the public health context', in McMichael, A., Campbell-Lendrum, D., Corvalan, C., Ebi, K., Githeko, A., Scheraga, J. and Woodward, A. (eds.) *Climate Change and Human Health: Risks and Responses*. Geneva: World Health Organization, pp. 220–236.

Hines, J.M., Hungerford, H.R. and Tomera, A.N. 1987. 'Analysis and synthesis of research on responsible environmental behaviour: a meta-analysis', *Journal of Environmental Education* **18**: 1–18.

Hughes, K., Van Beurden, E., Eakin, E.G., Barnett, L.M., Patterson, E., Backhouse, J., Jones, S., Hauser, D., Beard, J.R. and Newman, B. 2008. 'Older persons' perception of risk of falling: implications for fall-prevention campaigns', *American Journal of Public Health* **98**: 351–357.

Kollmuss, A. and Agyeman, J. 2002. 'Mind the gap: why do people act environmentally and what are the barriers to pro-environmental behaviour?', *Environmental Education Research* **8**: 239–260.

Koppe, C., Kovats, S., Jendritzky, G. and Menne, B. 2004. *Heat-Waves: Risks and Responses, Health and Global Environmental Change Series*. Copenhagen: World Health Organization.

Kovats, S. and Jendritzky, G. 2006. 'Heat-waves and human health', in Menne, B. and Ebi, K.L. (eds.) *Climate Change and Adaptation Strategies for Human Health*. Darmstadt: Steinkopff, pp. 63–97.

Leiserowitz, A. 2005. 'American risk perceptions: is climate change dangerous?', *Risk Analysis* **25**: 1433–1442.

Luszczynska, A., Scholz, U. and Schwarzer, R. 2005. 'The general self-efficacy scale: multicultural validation studies', *Journal of Psychology: Interdisciplinary and Applied* **139**: 439–457.

Matthies, F., Few, R. and Kovats, S. 2003. 'Social science and adaptation to climate change', *IHDP Update* issue 03/2003: 15.

McMichael, A.J., Campbell-Lendrum, D.H., Corvalan, C.F., Ebi, K.L., Githeko, A., Scheraga, J.D. and Woodward, A. (eds.) 2003. *Climate Change and Human Health: Risks and Responses*. Geneva: World Health Organization.

Menne, B. and Ebi, K. (eds.) 2006. *Climate Change and Adaptation Strategies for Human Health*. Darmstadt: Steinkopff.

Naughton, M.P., Henderson, A., Mirabelli, M.C., Kaiser, R., Wilhelm, J.L., Kieszak, S.M., Rubin, C.H. and McGeehin, M.A. 2002. 'Heat-related mortality during a 1999 heat wave in Chicago', *American Journal of Preventive Medicine* **22**: 221–227.

O'Brien, K., Sygna, L. and Haugen, J.E. 2004. 'Vulnerable or resilient? A multi-scale assessment of climate impacts and vulnerability in Norway', *Climatic Change* **64**: 193–225.

O'Connor, R.E., Bord, R.J. and Fisher, A. 1999. 'Risk perceptions, general environmental beliefs, and willingness to address climate change', *Risk Analysis* **19**: 461–471.

O'Reilly, D., Connolly, S., Rosato, M. and Patterson, C. 2008. 'Is caring associated with an increased risk of mortality? A longitudinal study', *Social Science and Medicine* **67**: 1282–1290.

Pelling, M. 2003. *The Vulnerability of Cities: Natural Disasters and Social Resilience*. London: Earthscan.

Pelling, M. and High, C. 2005. 'Understanding adaptation: what can social capital offer assessments of adaptive capacity?', *Global Environmental Change* **15**: 308–319.

Rydman, R.J., Rumoro, D.P., Silva, J.C., Hogan, T.M. and Kampe, L.M. 1999. 'The rate and risk of heat-related illness in hospital emergency departments during the 1995 Chicago heat disaster', *Journal of Medical Systems* **23**: 41–56.

Semenza, J.C., Rubin, C.H., Falter, K.H., Selanikio, J.D., Flanders, W.D., Howe, H.L. and Wilhelm, J.L. 1996 'Heat-related deaths during the July 1995 heat wave in Chicago', *New England Journal of Medicine* **335**: 84–90.

Sheridan, S. C. 2007. 'A survey of public perception and response to heat warnings across four North American cities: an evaluation of municipal effectiveness', *Journal of Biometeorology* **52**: 3–15.

Sjöberg, L. 2000. 'Factors in risk perception', *Risk Analysis* **20**: 1–11.

Slovic, P. 2000. *The Perception of Risk.* London: Earthscan.

Slovic, P., Fischhoff, C. and Lichtenstein, S. 1981. 'Perceived risk: psychological factors and social implications', *Proceedings of the Royal Society of London A* **376**: 17–34.

Slovic, P., MacGregor, D. G. and Peters, E. 1998. *Imagery, Affect, and Decision-Making.* Eugene: Decision Research.

Slovic, P., Finucane, M. L., Peters, E. and MacGregor, D. G. 2004. 'Risk as analysis and risk as feelings: some thoughts about affect, reason, risk and rationality', *Risk Analysis* **24**: 311–322.

Smit, B. and Pilifosova, O. 2001. 'Adaptation to climate change in the context of sustainable development and equity', in McCarthy, J. J., Canziani, O. F., Leary, N. A., Dokken, D. J. and White, K. S. (eds.) *Climate Change 2001: Impacts, Adaptation and Vulnerability. Contribution of Working Group II to the Third Assessment Report of the Intergovernmental Panel on Climate Change.* Cambridge: Cambridge University Press, pp. 879–912.

Stamm, K. R., Clark, F. and Eblacas, P. R. 2000. 'Mass communication and public understanding of environmental problems: the case of global warming', *Public Understanding of Science* **9**: 219–237.

Strauss, A. and Corbin, J. 1990. *Basics of Qualitative Research: Grounded Theory Procedures and Techniques.* London: Sage.

Ungar, S. 2000. 'Knowledge, ignorance and the popular culture: climate change versus the ozone hole', *Public Understanding of Science* **9**: 297–312.

Watson, R. T., Patz, J., Gubler, D. J., Parson, E. A. and Vincent, J. H. 2005. 'Environmental health implications of global climate change', *Journal of Environmental Monitoring* **7**: 834–843.

Wisner, B., Blaikie, P., Cannon, T. and Davis, I. 2004. *At Risk: Natural Hazards, People, Vulnerability and Disasters.* London: Routledge.

Woolcock, M. and Narayan, D. 2000. 'Social capital: implications for development theory, research and policy', *World Bank Research Observer* **15**: 225–249.

Zwick, M. M. and Renn, O. (eds.) 2002. *Perception and Evaluation of Risk: Findings of the Baden-Württemberg Risk Survey 2001.* Stuttgart: Centre of Technology Assessment in Baden-Württemberg and the University of Stuttgart, Sociology of Technologies and Environment.

12

Values and cost–benefit analysis: economic efficiency criteria in adaptation

Alistair Hunt and Tim Taylor

Introduction

In this chapter we explore the extent to which the measures of value incorporated in cost–benefit analysis (CBA) can be utilised to guide decision-making in adapting to climate change. Our motivation derives from the fact that whilst CBA is now a key element in the project and policy appraisal process in a number of European sectoral contexts (for example air quality in Europe: Holland et al., 2005), the timescales over which climate change adaptation considerations range are beyond those normally considered in such appraisals. As a result, the assumption normally made that unit monetary values utilised in CBA should be based on current preferences and resource scarcity patterns is questionable. Using stated preference techniques Layton and Brown (2000) begin to explore this issue in the context of greenhouse gas mitigation. This chapter pursues this further in the context of adaptation to climate change. Adaptation is understood here to include the spectrum from specific actions, or options, designed to mitigate specific climate risks, to the socio-economic and cultural conditions (i.e. adaptive capacity), that facilitate adaptation to the full range of identified climate change risks. Decisions relating to the adaptation to climate change risks can then be seen to include both sectoral-specific responses and those that shape social and economic development more generally.

The chapter addresses three aspects of CBA related to preference revelation that are applicable to the long time horizons relevant to decisions related to adaptation but which have not been discussed in this context to date. The discussion outlines the theoretical and conceptual issues before providing illustrations based on recent empirical elicitation research. The second section considers time preference

Adapting to Climate Change: Thresholds, Values, Governance, eds. W. Neil Adger, Irene Lorenzoni and Karen L. O'Brien. Published by Cambridge University Press. © Cambridge University Press 2009.

discounting and risk aversion – closely associated with each other through their relationships with uncertainty – taking the perspective developed in the Stern Review of the Economics of Climate Change (Stern, 2006) as a starting point. The third section discusses the role of the future socio-economic context in determination of preferences and values for use in CBA of adaptation options. The fourth section then briefly demonstrates their use in the context of CBA of heat early-warning systems and cultural heritage maintenance where non-market valuation techniques are adopted in order to derive measures of adaptation benefits. The final section presents some conclusions.

Time preference discounting and risk aversion

The practice of discounting in social CBA reflects the observed time trade-offs in money markets where the interest rate is determined by equating, at the margin, individuals' savings behaviour (reflecting what is known as consumption discounting) and producers' borrowing for investment purposes (reflecting what is known as the opportunity cost of capital). However, since in reality there are multiple market rates of interest that do not correct for market failures such as externalities, rates used in social CBA are often constructed from observations and judgements relating to the behaviour of the two sides of the market separately. Thus, annual rates used in the UK in recent years have ranged from 6% (1991–2003), based on the opportunity cost of capital, to 3.5%, declining after 30 years (2003 to date), based on consumption discounting. The treatment of discounting in the recent Stern Review on the Economics of Climate Change (Stern, 2006), was controversial (see for example Maddison, 2006; Tol and Yohe, 2006; Weitzman, 2007), for adopting a new, lower, rate of 1.4% to discount the monetised impacts of climate change and costs of greenhouse gas reduction. Since adaptation is recognised as being a likely alternative form of response to expected climate change impacts, and is likely – to some degree at least – to be subject to social CBA in its determination, it is worth considering the appropriateness of the Stern rate, and discounting more generally, in this context.

The Stern Review modelled the costs of climate change impacts to the year 2200 using the PAGE2002 Integrated Assessment Model (Hope, 2003). The discount rate parameter used in this modelling was derived from the social time preference rate (STPR), originally defined by Ramsey (1928) in its traditional form as:

$$i = z + n \times g \tag{12.1}$$

where:

z is the rate of pure time preference (impatience – utility today is perceived as being better than utility tomorrow) plus catastrophe risk, *g* is the rate of growth of real consumption per capita, and

n is the percentage change (fall) in the additional utility derived from each percentage change (increase) in consumption (*n* is referred to as the 'elasticity of the marginal utility of consumption').

The values of *n* and *g* used by Stern are 1% and 1.3%, respectively. Note that *n* also captures a measure of risk aversion to differing levels of *g*; it also measures inequality aversion. The value of *z* used is 0.1% and represents the annual risk of catastrophe eliminating society, only. The overall social discount rate is therefore 1.4%. Controversy has centred on the legitimacy of adopting a pure rate of time preference of zero. To use the terminology from the IPCC (for example Arrow et al., 1996), the discussion has centred on the relative merits of using prescriptive or descriptive values. Stern uses the prescriptive argument in the intergenerational context to justify a zero rate of pure time preference rate when asserting that there is no a priori reason to weight the utility of one person at one point in time different than that of another person at another point in time (as argued by Broome, 1992). Critics such as Weitzman (2007) argue that individual preferences have a sovereign role in welfare economics and so should be represented in the discount rate used in social CBA; positive market interest rates therefore suggest adoption of a positive pure time preference rate. In fact, the positions need not be incompatible; investments with current-generation consequences could be discounted at (positive) pure time preferences derived from current-generation preferences whilst intergenerational consequences could be discounted at a zero rate. This possibility is recognised by Stern (2006, p. 54) but not utilised.

As highlighted above, adaptation actions have consequences in both the near future and the distant future. Potential inconsistencies therefore arise in a number of contexts. First, there is a discrepancy between emission mitigation decisions utilising the Stern discount rate of 1.4% and adaptation decisions for example, in public healthcare, using the Treasury rate of 3.5%. (Indeed, many investments serve both objectives and so there is a question as to which to use.) Second, public sector actions are likely to be discounted at a different (lower) rate from decisions made by other agents (for example, using a market rate of interest). Third, the value of *n* may imply a different treatment of distributional issues in adaptation decision-making than in comparable public sector projects.

The previous arguments suggest that for short-term futures considered in adaptation decision-making there may be both philosophical and practical reasons for retaining positive discount rates, consistent with those used in other public and private decisions. In the longer term, however, decisions about adaptation necessarily turn from consideration of specific options towards the provision of adaptive capacity which – like sustainability more generally – can be seen as the availability of capital of all forms. Toman (2006) notes that positive discount rates imply that capital resources are substitutable. If capital substitutability cannot be assumed, however, it suggests that discounting in adaptation assessment may be best utilised in conjunction with the use of capital constraints similar to the notion of environmental stewardship suggested by for example Howarth (1995). This approach reflects the pluralistic ethical framework suggested by Norton and Toman (1997) to encompass alternative value systems concerning the future, combining utilitarian-based CBA decision rules with the rights-based rules outlined by for example Sen (1982). Whilst this combination approach appears to offer a less restrictive solution to that of adopting a prescriptive approach to discounting, its practical operation is still likely, however, to rely on some form of public prescription. What form this should take is left unresolved.

Alternative grounds for adopting decision rules other than CBA arise from the reconsideration of the role of risk-determined preferences. As identified above, the determination of the social time preference rate is to some extent determined by the degree of risk aversion that an individual has to alternative possible future income and wealth levels. However, it is quite possible – as originally recognised by Savage (1951) – that when uncertainty is too great, the parameter is not amenable to quantification. In this case, risk aversion can only manifest itself in an alternative decision rule such as employment of the minimax criterion. This decision rule finds the loss associated with each alternative future and then selects the strategy that minimises the worst loss (Arrow et al., 1996). The possibility of catastrophe or some form of irreversibility is likely to exacerbate the difficulty of risk aversion parameterisation; as Adger et al. (2009) illustrate, more fundamental losses associated with climate change may include the loss of cultural identity which is not likely to be reducible to such parameterisation. Whilst irreversibility does not imply that CBA is necessarily inappropriate there are clearly issues of representation that restrict the application of CBA when spiritual and cultural identities and assets are involved. In this case, alternative decision rules are likely to provide more appropriate means with which to assess the merits of adaptation options.

Socio-economic change and individuals' preferences

Discounting and risk aversion address issues related to resource-related preferences and values contingent on time and uncertainty; Stern, for example, ties time-contingent values to absolute levels of consumption (proxied by GDP per capita) and probabilities of society's extinction. However, for consistency, other factors determining individuals' preferences should also be represented in adaptation decision-making. These factors may themselves be dependent on socio-economic conditions and so be expected to change over time as society develops.

These preferences may be estimated by using identified historical relationships between socio-economic variables and monetary values to extrapolate across future time periods under socio-economic scenarios. A more sophisticated variant of the extrapolation model would be to develop simulation value functions constructed from a combination of observed relationships and understandings of value determination elicited from for example household-based interviews (see for example Kilbourne et al., 2005). Alternatively, survey-based approaches may be used that require respondents to hypothesise what their values might be under alternative socio-economic scenarios, but removing pure time preferences since they are represented in the discount rate.

These techniques rely on the utilisation of socio-economic scenarios, themselves principally derived through extrapolation of historic trends. Whilst some applicable scenario elements such as GDP projections are available and established (for example in the IPCC emission scenarios: Nakicenovic and Swart, 2000), other aspects such as design technologies are less easy to define or depend on definition at a sub-national scale to be useful to adaptation decision-making. A first step to meeting these difficulties has been taken by Berkhout et al. (2002) who construct scenarios for the UK that are framed by the change elements, 'governance' and 'values'. Though their interdependence is recognised, combining these elements along a 'governance' axis from (for example, regional) *autonomy* to (for example, global) *interdependence*, and a 'values' axis from *individual* to *community*, serves to allow the creation, in outline, of four distinct socio-economic scenarios. These scenarios, aspects of which have been developed in stakeholder consultations, may then be used to formulate time-dependent simulations of specific willingness to pay (WTP) values.

Such socio-economic scenarios, through their demographic projections, provide data that can be used to define the scale of vulnerability and in their descriptions of governance and values allow us to speculate about their possible influence on susceptibility. They also can provide, or inform, projections of resource costs – primary energy costs from the IPCC emission scenarios being one example. Reflecting the issue of intra versus intergenerational values raised in the discussion of discounting, there is, however, likely to be a temporal limit to the validity of using such approaches to project future changes in preferences and resource costs. Whilst short-term changes (say <30 years) may be modelled, using current preferences and resource allocation patterns as a starting point, this approach appears to be less defensible in the longer term when technological change and societal development renders resource mixes and households consumption baskets unknown, or at least highly uncertain. This being the case, there seems to be a point in time – which may differ according to the sectoral context – at which the capital-based decision rules identified above become more practical to use than economic efficiency-based rules such as CBA. Thus, the expression of values may become reflected in the manner in which social decisions relating to adaptation are made in response to indicators of adaptive capacity, projected for future time periods under alternative policy scenarios.

Though perhaps more appropriate to a country at a lower level of material development than the UK – the geographical focus of this paper – a further constraint to the applicability of modelling WTP to socio-economic change is that the types of socio-economic change brought about by climate change impacts may be such that a climate change-induced socio-economic future is structurally different from the baseline socio-economic future used to model economic values. This may most obviously be the case in the event of non-marginal, socially contingent, climate impacts such as the economic multiplier effects associated with migration, the loss of an economic sector and socio-political stress and conflict. Downing et al. (2005), for instance, present the example of the Sahel which is a semi-arid region that corresponds to the current physical limit of agricultural and pastoral production systems and where climatic episodes, primarily of drought, already create stresses and may, with more frequent or intense episodes under climate change futures, tip the system into an increasingly instable state. Furthermore, as Yohe et al. (2007) make clear, this type of non-marginal socio-economic change is likely to constrain some forms of adaptation currently being considered.

Illustrations

The previous discussion reflects on the conceptual limits to the use of market prices and resource costs in adaptation decision-making involving long time horizons. This section provides evidence from two stated preference studies that shed light on two of the issues discussed above: time preferences and socio-economic determinacy of WTP. In so doing, they help to illustrate the potential for such study methods to identify the current generation's preferences over the future, as well as to highlight the challenges that they face in generating robust, defensible values for use in CBA. Consequently, they also serve to demonstrate potential limits to the validity of CBA as a decision-rule to inform climate change adaptation.

Heat warning systems

Heat waves can have significant impacts on human health. These include increased mortality rates and increased morbidity including heat strokes, dehydration, neurological conditions, renal disease and mental illness (Kovats and Ebi, 2006).

Following the events of summer 2003, the British Government established planning systems for responses to extreme summer conditions. The Heat Wave

Plan for England sets out measures to reduce the health impacts of heatwaves (NHS, 2007). This establishes both short-term and longer-term measures to adapt to health consequences of climate change. Longer-term actions include the greening of the built environment with 'trees, plants and green space act[ing] as natural air conditioners'; shading and insulating housing and buildings; and increasing energy efficiency through energy-neutral cooling systems. Responsive, short-term actions include staff training in care homes and in the NHS, and enhancing preparedness, for example, the provision of cool rooms. The establishment of a 'Heat-Health Watch' system provides a framework for the management and triggering of emergency responses.

Kovats and Ebi (2006) highlight the difficulties in assessing heat wave preparedness plans, but call for regular monitoring of the effectiveness of the systems. This monitoring would include assessment of the costs of setting up the system, the costs of maintenance and the costs of warnings (both direct and indirect costs). Kovats and Ebi suggest that benefits should be defined in terms of the number of (years of) lives saved. This represents a cost-effectiveness approach; the monetisation of health benefits would allow a CBA to be undertaken. To date, little work has been done in this area in applying such metrics to assess the Heat Wave Plan, though the NHS acknowledge that this is an issue on which further research will be needed (Nurse, pers comm.).

Our recent work in this area, reported in Alberini et al. (2006), explores the potential for the monetisation of health benefits; specifically the reduction in risk of premature deaths brought about by an investment into a Heat Wave Plan. WTP valuation of this impact relies on non-market valuation techniques. We utilised a survey-based, contingent valuation, technique to value such risk reductions in the UK, as well as France and Italy. In this survey, we ask respondents for their WTP to reduce their annual risk of death by a 5 in 1000 probability – a probability change broadly consistent with the effectiveness of such an investment. Two questions ask the respondent for their WTP for the risk change spread over the next ten years, and for the ten years from age 70, respectively. The sample size in the UK was 330; the sample was restricted to those aged 40 and over – pre-testing having identified that younger age groups struggle to comprehend such risk changes. The WTP results for the 5 in 1000 WTP question, multiplied by 200, give central values of a prevented fatality (VPF) of approximately £450 000 – a value consistent with the range currently suggested by Defra for project appraisal in the UK – and an implicit value of a life year of £24 000. The VPF, derived using the WTP to value the risk change over the ten years from age 70, is approximately £150 000.

How can these results be interpreted in the current context of projecting mortality benefits to, say, 2050? Comparison between the WTP values for current and future risk changes shows that individual respondents discount future health risks – statistical analysis reveals an implicit 6% annual discount rate. This rate is in the range of current market interest rates and so superficially at least, this result appears to support the use of a positive pure time preference rate for the present generation. Two caveats should, however, be borne in mind: first, the WTP results do not tell us whether the respondent was considering identical 'goods' differentiated only by time; for example, it is possible that the respondent considered it likely that her health condition would be poorer in the future, and that the risk reduction would restore her to a lower health status than currently. Second, it has been argued for example, by Broome (1992) that, in contrast with material goods, the benefits of health improvements generally are independent of when they occur – a life of a 20-year-old saved this year is as valuable as that of a 20-year-old saved next year. Clearly, the descriptive versus prescriptive debate over which discount rate to use is relevant here, symbolised by the contrast between intra and interpersonal preference ordering, respectively.

Regressing respondents' WTP for the current risk reduction against their socio-economic characteristics reveals a positive relationship between WTP and income. Temporal transfer of this WTP to future years may then be made on the basis of this positive income coefficient, which is here found to be significant. However, the validity of using the derived income elasticity may be reduced as income levels increase from current levels to a multiple of for example, 1.013^{50} of these levels, in 50 years' time, as in the A2 socio-economic scenario adopted by Stern, since the implied income level is significantly higher than the high side variance in current income levels. In any case, income may be regarded as a minimum requirement in explaining temporal variation in WTP, and future studies undertaken with temporal transfer in mind would profit from identifying a wider variety of WTP determinants that could be interpreted under socio-economic scenarios.

If this analysis suggests that there may be limits to the use of CBA in evaluating adaptation strategies with regard to human health, it is imperative to explore the potential for use of alternative decision rules. Cost-effectiveness criteria initially appears more appealing, though similar difficulties in quantification of the costs are likely to arise. Capital- or rights-based rules then become attractive though their practical operation is, at present, not defined clearly.

Cultural heritage in Sussex

Whilst CBA is routinely used in flood management appraisal in the UK, it does not generally attempt to include monetised values for cultural heritage impacts – partly because of the dearth of literature in this area (see Navrud and Ready, 2002 for a review). In the heritage context in the UK, work on the valuation of cultural heritage is gaining increasing importance in the establishment of programmes (see Eftec, 2005).

Lewes has a number of buildings regarded as culturally important. CBA of adaptation to future flood risk should therefore consider incorporating these values. In order to explore the possibility for doing so, we undertook a stated preference study in the town – the first to our knowledge in the climate change context – see Hunt and Taylor (2006) for a full description.

Two buildings were selected for specific attention based on focus group discussions – the St Thomas à Becket church and Harvey's Brewery, both in Lewes, Sussex (see Figures 12.1, 12.2 and 12.3). Both buildings were impacted by the October 2000 floods, characterised as a 1-in-100-year event. The impacts of the October 2000 floods were presented as described in the three Figures along with a scenario for increased risk of flooding outlined in Table 12.1. A dichotomous choice question was posed to elicit WTP to avoid such damages in the future.

A postal survey of 1000 properties in the Lewes District Council area elicited 132 replies (excluding blank returns), of which 76 gave non-protest responses to both the WTP questions and the income variables. This sample can hence not be considered 'representative' per se, but the results here show some of the potential for the use of this methodology in this context.

The mean WTP was estimated to be £23.48 (£19.22 to £31.22 with a 95% confidence interval). The median WTP was estimated to be £17.52 (£14.71 to £20.79 with a 95% confidence interval). These are lower than values for air pollution

Figure 12.1 Harveys Brewery, Lewes, Sussex. Founded in 1790, Harvey's Brewery is situated on the River Ouse. It was rebuilt in 1880 to a traditional design. It contains an old fermenting room and is a popular local landmark. In the 2000 floods, there occurred extensive flooding to the brewery site, with water above 1 metre in depth. In particular, there was flooding to the fermenting room and damage to the brickwork of the building where flood waters reached. This was repaired, though some modern materials were used in place of traditional materials that were not available.

Figure 12.2 St Thomas à Becket at Cliffe, Lewes, Sussex (front view): a small Anglican parish church steeped in history, dating back to the twelfth century. The body and tower of the church date from the fourteenth and fifteenth centuries, while a vestry was added in the nineteenth century. It is a Grade 2 listed building.

Figure 12.3 St Thomas à Becket (interior view). In the 2000 floods, contaminated water flooded the church to a depth of 1 metre. As the water subsided, this left thick deposits of silt on the walls and floor of the church. A 16-stop nineteenth-century organ was extensively damaged, but was able to be restored. Virtually all of the contents of the church were affected. A thick layer of silt covered the pews and these were restored. The library of the church was ruined. Vestments on the altar and in the vestry were extensively damaged and were not repairable.

Table 12.1 *Increased probability of flooding in Lewes based on UKCIP02 Medium–High Climate Scenario*

Now	2011–2040	2041–2070	2071–2100
1 in 100	1 in 100	3 in 100	7 in 100

Source: Derived from Hulme et al. (2002).

related welfare damages from Lincoln Cathedral, for which a WTP of £49.77 per household per year was found in Lincoln, with distance decay leading to an average for Lincolnshire of £26.77 (Pollicino and Maddison, 2002). However, it is comparable to another study for Newcastle which showed a WTP of £10 for remediating Grainger Town (Garrod and Willis, 2002).

In the regression analysis, distance is significant, with a negative sign, highlighting the fact that those further away from Lewes are willing to pay less to preserve heritage buildings in Lewes. The buildings in question are not 'mega-buildings' like many of those valued in previous studies. Thus, their appeal is more local/region-specific. The Lewes identity was felt important by the focus group, and this is perhaps shown in that a larger proportion of zero bids in the final survey came from locations in the southern part of the Lewes District Council region, areas such as Peacehaven which identify more closely with Brighton than Lewes. The distance decay evidenced in this study is more significant than that of buildings of significant cultural interest surveyed previously, supporting the hypothesis that local identity is important.

An interesting demographic trend that may affect the values placed on cultural heritage is that of increased labour mobility. This may affect different aspects of value placed on sites of interest, as interactions with the buildings in question change values placed on them. This may affect future survey designs in this field.

The focus group revealed some important issues that may influence the values attributable to climate change risks to heritage. A first point is that some of the respondents had difficulty comprehending the relatively small risks associated with the flooding event and processing these in such a way as to express preferences in a consistent way. The problem of cognition of low probability–large impact events has previously been noted in other contexts, such as the valuation of changes in mortality risks above (see also Fischhoff et al., 1978) and represents a challenge to the WTP valuation of extreme events associated with climate change.

Associated with this point, focus group respondents made it clear that the strength of preferences was strongly influenced by the fact that an event of this magnitude – the October 2000 floods – had recently been experienced.

Coincidentally, the perception of objective risks in the minds of the respondents appeared to be further distorted by the fact that the focus group took place during a heat wave – another extreme weather event. Whilst it may be that the recent experience helped respondents to understand the impact and so anchor preferences better, it also appeared to be the case that the proximity of the recent event – exacerbated by the contemporaneous heat wave – helped to exaggerate the objective size of the risks and so upwardly bias the WTP preferences. Again, this serves to highlight a generic problem with the elicitation of WTP preferences in the context of climate change related extreme events.

A further point relates to the valuation of cultural heritage under socio-economic scenarios for the purpose of undertaking CBA of climate change adaptation. Focus group respondents argued that the value of a cultural asset was to some degree influenced by the use of the building. Churches are increasingly being converted to commercial purposes. Similarly, the local character and significance of a place to the community (and the values attributed to it) may be affected by issues such as the increasing degree of mobility in the labour force, changes in the social structure of society and, for buildings with religious significance, changing demographics in terms of religious observance. Thus, it might be expected that future socio-economic development paths would lead to changes in use and hence individual preferences. It is possible to envisage constructing plausible patterns of ownership and use under alternative development scenarios. However, the modelling of preferences under such scenarios is less straightforward; one method to explore is the presentation of alternative uses to respondents who would subsequently be asked to rank or weight such alternatives.

Application of standard CBA approaches is, hence, complicated in terms of the impacts of climate change and adaptation strategies and should be undertaken with care. In the absence of robust benefit measurement, cost-effectiveness analysis and multi-criteria analysis perhaps provide the most appropriate techniques to assess potential policies to adapt to climate change in the heritage context. In the more general context, multi-hazard risk assessment of historic structures including information gap decision theory (Ben-Haim, 2001) offers a more holistic consideration of climate risk to heritage structures. Again, however, where the risks cannot be quantified sufficiently, it may be more appropriate to treat the stock of cultural heritage as an asset and so include within a capital-based decision rule.

Conclusion

This chapter briefly reviews the issues involved in defining monetary values in decision rules applied to adaptation. With respect to discounting, it finds that whilst there appear to be both practical and philosophical reasons relating to consistency for positive discounting over relatively short time periods, the long term requires prescription either through the discount rate or by using measures of capital maintenance. The latter is favoured conceptually by the authors for its greater flexibility, though methods for its implementation are not yet developed.

There is the potential to explore changes in preferences due to socio-economic change through the use of socio-economic scenarios. A number of techniques are identified. However, as illustrated by the use of the stated preference case studies, the modelling of preferences, perhaps particularly those relating to non-market goods for which people are not used to express WTP preferences in any case, may only be viable within certain temporal bounds. Thus, whilst future research would be useful in investigating the potential of stated preference techniques for exploring how preferences relating to time and socio-economic context are determined and utilised in scenario analysis, it should be complemented by more bottom–up, local-scale, exercises to measure and monitor capital stocks that constitute adaptive capacity.

References

Adger, W. N., Dessai, S., Goulden, M., Hulme, M., Lorenzoni, I., Nelson, D., Naess, L.-O., Wolf, J. and Wreford, A. 2009. 'Limits and barriers to adaptation', *Climatic Change* **93**: 335–354.

Alberini, A., Hunt, A. and Markandya, A. 2006. 'Willingness to pay to reduce mortality risks: evidence from a three-country contingent valuation study', *Environmental and Resource Economics* **33**: 251–264.

Arrow, K. J., Cline, W. R., Maler, K. G., Munasinghe, M., Squitieri, R. and Stiglitz, J. E. 1996. 'Intertemporal equity, discounting and economic efficiency', in Bruce, J. P., Lee, H. and Haites, E. F. (eds.) *Climate Change 1995: Economic and Social Dimensions of Climate Change.* Cambridge: Cambridge University Press, pp. 129–144.

Ben-Haim, Y. 2001. *Information Gap Decision Theory.* London: Academic Press.

Berkhout, F., Hertin, J. and Jordan, A. 2002. 'Socio-economic futures in climate change impact assessment: using scenarios as learning machines', *Global Environmental Change* **12**: 83–95.

Broome J. 1992. *Counting the Cost of Global Warming.* Cambridge: White Horse Press.

Downing, A. D., Butterfield, R., Ceronsky, M., Grubb, M., Guo, J., Hepburn, C., Hope, C., Hunt, A., Li, A., Markandya, A., Nyong, A., Tol, R. S. J. and Watkiss, P. 2005. *Scoping Uncertainty in the Social Cost of Carbon,* Final project report: *Social Cost of Carbon: A Closer Look at Uncertainty.* London: Department of Environment, Food and Rural Affairs.

Eftec. 2005. *Valuation of the Historic Environment.* London: English Heritage, the Heritage Lottery Fund and Department of Transport.

Fischhoff, B., Slovic, P., Lichtenstein, S., Read, S. and Coombs, B. 1978. 'How safe is safe enough? A psychometric study of attitudes towards technological risks and benefits', *Policy Sciences* **9**: 127–152.

Garrod, G. and Willis, K. 2002. 'Northumbria castles, cathedrals and towns' in Navrud, S. and Ready, R. (eds.) *Valuing Cultural Heritage: Applying Environmental Valuation Techniques to Historic Buildings.* Cheltenham: Elgar, pp. 40–52.

Holland, M., Hunt, A., Hurley, F., Navrud, S. and Watkiss, P. 2005. *Methodology for the Cost–Benefit Analysis for CAFE*, Vol. 1, *Overview of Methodology*. Available at http://ec.europa.eu/environment/air/cafe/pdf/cba_methodology_vol1.pdf

Hope, C. 2003. *The Marginal Impacts of CO_2, CH_4 and SF_6 Emissions*, Research Paper No. 2003/10. Cambridge: Judge Institute of Management.

Howarth, R.B. 1995. 'Sustainability under uncertainty: a deontologist approach', *Land Economics* **71**: 417–427.

Hulme, M., Jenkins, G.J., Lu, X., Turnpenny, J.R., Mitchell, T.D., Jones, R.G., Lowe., J., Murphy, J.M., Hassell, D., Boorman, P., McDonald, R. and Hill, S. 2002. *Climate Change Scenarios for the United Kingdom: The UKCIP02 Scientific Report*. Norwich: Tyndall Centre for Climate Change Research, School of Environmental Sciences, University of East Anglia.

Hunt, A. and Taylor, T. 2006. 'Buildings', in Metroeconomica (ed.) *Project E: Quantify the Cost of Future Impacts*. London: Defra.

Kilbourne, W., Grünhagen, M. and Foley, J. 2005. 'A cross-cultural examination of the relationship between materialism and individual values', *Journal of Economic Psychology* **26**: 624–641.

Kovats, R.S. and Ebi, K. 2006. 'Heatwaves and public health in Europe', *European Journal of Public Health* **16**: 592–599.

Layton, D. and Brown, G. 2000. 'Heterogeneous preferences regarding global climate change', *Review of Economics and Statistics* **82**: 616–624.

Maddison, D. 2006. *Further Comments on the Stern Review*. Birmingham: Department of Economics, University of Birmingham. Available at www.economics.bham.ac.uk/maddison/Stern%20Comments.pdf

Nakicenovic, N. and Swart, R. (eds.) 2000. *Emissions Scenarios 2000: Special Report of the Intergovernmental Panel on Climate Change*. Cambridge: Cambridge University Press.

Navrud, S. and Ready, R. (eds.) 2002. *Valuing Cultural Heritage: Applying Environmental Valuation Techniques to Historic Buildings*. Cheltenham: Elgar.

NHS. 2007. *Heat Wave Plan for England*. London: Department of Health.

Norton, B.G. and Toman, M. 1997. 'Sustainability: ecological and economic perspectives', *Land Economics* **73**: 553–568.

Pollicino, M. and Maddison, D. 2002. 'Valuing the impacts of air pollution on Lincoln Cathedral' in Navrud, S. and Ready, R. (eds.) *Valuing Cultural Heritage: Applying Environmental Valuation Techniques to Historic Buildings*. Cheltenham: Elgar.

Ramsey, F.P. 1928. 'A mathematical theory of saving', *Economic Journal* **38**: 543–59.

Savage, L. 1951. 'The theory of statistical decisions', *Journal of the American Statistical Association* **46**: 55–67.

Sen, A.K. 1982. 'Approaches to the choice of discount for social benefit–cost analysis', in Lind, R.C. (ed.) *Discounting for Time and Risk in Energy Policy*. Washington, DC: Resources for the Future, pp. 325–353.

Stern, N. 2006. *Stern Review on the Economics of Climate Change*. London: HM Treasury and Cabinet Office.

Tol, R. and Yohe, G. 2006. 'A review of the Stern Review', *World Economics* **7**: 233–250.

Toman, M. 2006. 'Values in the economics of climate change', *Environmental Values* **15**: 365–379.

Yohe, G.W., Lasco, R.D., Ahmad, Q.K., Arnell, N.W., Cohen, S.J., Hope, C, Janetos, A.C. and Perez, R.T. 2007. 'Perspectives on climate change and sustainability', in Parry, M. L., Canziani, O. F., Palutikof, J. P., van der Linden P. J. and Hanson C. E. (eds.) 2007. *Climate Change 2007: Impacts, Adaptation and Vulnerability. Contribution of Working Group II to the Fourth Assessment Report of the Intergovernmental Panel on Climate Change*. Cambridge: Cambridge University Press, pp. 811–841.

Weitzman, M. L. 2007. 'A review of the Stern Review on the Economics of Climate Change', *Journal of Economic Literature* **45**: 703–724.

13

Hidden costs and disparate uncertainties: trade-offs in approaches to climate policy

Hallie Eakin, Emma L. Tompkins,
Donald R. Nelson and John M. Anderies

Introduction

As policy-makers struggle to define the policy agenda to address the challenge of climate change, three distinct influential approaches to climate policy are emerging in the climate change literature: implementing climate change adaptation; reducing social vulnerability; and managing ecosystem resilience. Each of these approaches has been developed in specific policy contexts associated respectively with natural hazard mitigation, poverty and social welfare investment, and natural resource management. In these contexts each approach has met with varying levels of past success. The fact that climate change is characterized by a high probability of surprise events; significant scientific uncertainty; and a need for long-term planning horizons only makes policy development more difficult.

In this chapter we argue that each of the three approaches involves implicit trade-offs in both the process of policy formation and in policy outcomes. These trade-offs are rarely considered in the evaluation of policy options, yet may have important implications for social welfare and sustainability. Through the analysis of case studies of adaptation to climate variability and change, we illustrate how the different ways of approaching the process of adjusting to future change can inadvertently lead to, for example, the privileging of efficiency over equitable distribution of resources (for example, risk-based adaptation approach), equity at the expense of cost (for example, social vulnerability approach), or intergenerational equity over political legitimacy (resilience approach).

In the next section, we first create caricatures of the three approaches to climate policy: risk-based adaptation, vulnerability and resilience. Most policy decisions can be identified primarily with the core tenets of one of these approaches. Each approach is associated with parallel and occasionally overlapping intellectual

Adapting to Climate Change: Thresholds, Values, Governance, eds. W. Neil Adger, Irene Lorenzoni and Karen L. O'Brien. Published by Cambridge University Press. © Cambridge University Press 2009.

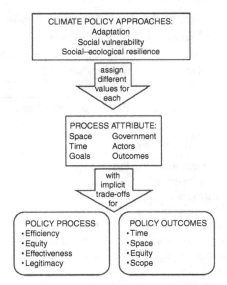

Figure 13.1 Trade-offs implicit in the adoption of different climate policy approaches.

trajectories in the theoretical literature on vulnerability, resilience and adaptation to climate change (for example O'Brien et al., 2004; Tompkins and Adger, 2004; Adger, 2006; Eakin and Lemos, 2006; Smit and Wandel, 2006). These intellectual trajectories each implicitly prioritize different aspects of policy development and policy outcomes (see Figure 13.1).

Second, using three empirical examples, we explore the key differences and the implicit trade-offs inherent in each of these approaches. Making trade-offs explicit is essential to avoid or address potentially important and costly externalities that can emerge as part of the policy process. As the three examples illustrate, while trade-offs may not be so stark in the literature, in practice, they must be clarified and explored to ensure that policy-makers are aware of the potentially significant costs associated with each approach. We base our evaluation of trade-offs on the economic efficiency, equity, effectiveness and legitimacy of the decisions or policy choices (Adger et al., 2003). This typology requires a consideration of the relative importance of these four elements in decision-making and has been applied only at the conceptual level in relation to climate change (see Adger et al., 2005).

Finally, we analyse the implications of these policy trade-offs in light of the specific challenges of a changing climate. Our analysis makes clear that some of the trade-offs associated with particular approaches are accentuated in light of the large uncertainties, potential for surprise and long-term planning horizons demanded in climate policy. In this context, a combination of approaches may be appropriate.

Table 13.1 *Implicit trade-offs in the development of climate policy using the adaptation, resilience and vulnerability approaches*

Process criteria	Adaptation Approach	Vulnerability Approach	Resilience Approach
Spatial scale of implementation	Sector focus	Places, communities, groups	Large-scale coupled social–ecological systems, for example, populated watersheds
Temporal emphasis of implementation	Short-term and medium-term future risks	Past and present vulnerabilities	Long-term future
Actors	Public–private partnerships	Public sector, vulnerable groups	Civil society, public sector
Policy goal	Address known and evolving risks	Protect populations most likely to experience harm	Enhance systems capacity for recovery and renewal
Desired outcome	Maximum loss reduction at lowest cost	Minimize social inequity and maximize capacities of disadvantaged groups	Minimize probability of rapid, undesirable and irreversible change

We thus conclude that trade-offs and assumptions must be made explicit during the policy process in order for stakeholders to adequately evaluate policy options.

Intellectual heritage of vulnerability, adaptation and resilience approaches

The adaptation, vulnerability and resilience frameworks are currently being applied in climate change decision and policy-making by NGOs (Simms, 2005), international institutions (Sperling, 2003) and national governments (Defra, 2000). However, climate change presents unique challenges to policy-makers, including issues of uncertainty and extensive planning horizons, and each of the frameworks addresses these in different and unequal ways. To assess the trade-offs implicit in the use and application of these approaches, we focus on five areas: the spatial scale of implementation, the temporal emphasis; the actors involved, the policy goals; and the desired outcomes (see Table 13.1).

What we are calling 'risk-based adaptation' has evolved from the need to manage the technological and environmental hazards facing society (White, 1986; Burton et al., 1993). In relation to climate change, this approach has typically involved identifying the most significant climatic hazards affecting a place or economic sector, estimating the probability of exposure and likelihood of damage (*risk*) and assessing the most cost-effective and expedient means of reducing that risk to a level perceived as tolerable to the society exposed (see overview in Eakin and Luers, 2006). Policy is thus driven by known risks, or risks to which a certain degree of certainty can be calculated, and thus has an advantage of political expediency and the appearance of economic efficiency.

Increasingly the discourse has moved from a focus on specific risk-reducing investments (for example investments in a dyke or sea wall) to 'building adaptive capacity' and creating an enabling environment in which specific actions can be defined (West and Gawith, 2005; Mortimore and Manvell, 2006; Smit and Wandel, 2006). In both national policy and international efforts to promote adaptation, much of the discourse focuses on national investment in the creation of appropriate incentive structures, enabling environments and decision-support tools to facilitate the adaptations of other actors (for example, private sector, civil society, local government) (Eakin and Lemos, 2006).

Vulnerability reduction approaches emerged as a moral response to observed social inequities in the impacts of social and environmental change (Hewitt, 1983; Wisner et al., 2004). Policy interventions in this approach target specific population groups (or species) that are considered to be at disproportionate risk to loss and harm (O'Brien and Leichenko, 2003; Vogel and O'Brien, 2004; Paavola and Adger, 2006). Vulnerability thus focuses on reducing exposure or sensitivity of these disadvantaged populations, and/or enhancing their capacities to manage risks and cope with loss via transfers of resources and targeted investments. Given the moral motivation for vulnerability-focused policy, issues of economic cost and efficiency are of secondary importance to the primary goal of reducing harm to the target population. Climate policy formed using this approach thus may raise politically charged issues relating to both the distribution of power in existing governance and injustice in exposure to environmental risk (Liverman, 2001). Vulnerability-based climate policy may thus focus attention on deep-rooted inequities in acquiring basic needs and entitlements, such as land, education, political enfranchisement, employment, housing and market opportunities.

Resilience policy interventions are designed to enhance system attributes to avoid an abrupt 'flip' of a coupled social–ecological system (SES) into a less-desired state. Resilience, the capacity of a coupled system to continue to function in the face of change or to transform into an improved system, is observed through its ability to experiment and learn (and thus make mistakes and experience loss), buffer disturbance and self-organize in response to uncertain stressors (Folke et al., 2002). The resilience approach to natural resource management emerged from ecological theory in the 1970s (Holling, 1973) to better address the dynamic, highly uncertain and non-linear dynamics of ecosystems (Gunderson, 2001; Gunderson and Holling, 2002; Carpenter and Brock, 2004). The focus of resilience-based policy is on the aggregated capacity of the system to adjust and respond rather than the differential (and potentially unequal) capabilities for and barriers to organization, learning and action for different elements within the system. This approach is characterized by the premise that there is no long-run trade-off between ecological integrity and human welfare; sustainability is only achieved through conceiving the social–ecological system as integrated and interdependent over time. Policy informed by resilience implies a high tolerance of variability, volatility and localized loss for the greater benefit of the system's integrity.

Examples of adaptation, vulnerability and resilience approaches

The differences between the three approaches are not trivial: they entail implicit trade-offs with potentially important implications for those affected by policy. Thus the selection of a particular framework for policy development also implies an implicit acceptance of a set of economic, political and social costs. In this section, we discuss the implicit trade-offs in the process of climate change policy development, entailing trade-offs among efficiency, equity, effectiveness and political legitimacy.

Adaptation to climate change in the UK

Adaptation as an approach to addressing the challenge of climate change in its most simplistic sense is based on an evaluation of a specific threat, followed by an analysis – for example, cost–benefit – of the most expeditious means of addressing the threat (Sarewitz et al., 2003). Defined this way, adaptation may tend to privilege the 'here and now' over the 'there and then' in order to achieve a reduction in outcome risk as efficiently as possible. Implicitly or explicitly, such responding to the immediate threats of a system may entail applying a high discount rate to future losses, and less attention to enhancing capacities to address unknown risks and surprise events in the future (Markandya and Halsnaes, 2001). The adaptation approach is thus responsive to place-based risks posed to specific activities and populations, but at the potential expense of intergenerational equity and cross-spatial externalities of adaptive actions.

One of the more widely applauded responses to the challenge of climate change has been the United Kingdom's Climate Impact Programme (UKCIP), which was initiated in 1995. The main strategy of the British government was to create a semi-autonomous agency, stakeholder-led organization, dedicated to raising the capacity of the UK to adapt to climate change through a better understanding of the impacts of climate change (McKenzie-Hedger et al., 2000; McKenzie-Hedger et al., 2006). During its 12 years of operation, the programme has increasingly focused its attention on applying a risk-based approach to adaptation (Willows and Connell, 2003; West and Gawith, 2005). This shift in focus has produced an implicit prioritization of two elements in the delivery of adaptation support: efficiency and effectiveness. Nevertheless, there are some indications that this prioritization has come at the expense of providing an equitable delivery of services and support to all parts of UK society.

First, UKCIP has operated principally on the basis of voluntary interactions with diverse stakeholders, with the assumption that demand will identify sectors of greatest need. However socially marginalized communities, institutions and sectors of lesser economic importance may have little capacity for organization or the initiative to seek UKCIP advice – regardless of their vulnerability. While UKCIP has made significant efforts to engage those who are vulnerable, for example the elderly and those prone to the ill-effects of heat, the institutional structure of the UKCIP and its mission to be stakeholder-led has paradoxically created a barrier to widespread public engagement. UKCIP however is aware of this and is working to overcome this (West and Gawith, 2005).

Second, because the outputs from the collaborations are not funded by the UKCIP or the British government but by the voluntary partner, the costs to taxpayers of providing climate change adaptation support are relatively low, but perhaps less effective in terms of addressing the underlying and persistent vulnerabilities associated with less vocal and economically capable sectors. Thus there is no guarantee that UKCIP will deliver support equitably, or that vulnerable populations will be assisted.

Managing vulnerability in north-east Brazil

Vulnerability reduction is designed to benefit those populations that are least able to respond to particular risks, thereby saving lives and minimizing harm. An approach to climate policy that is based in vulnerability assessment and analysis thus prioritizes improving equity in *outcomes* over economic efficiency and equity in *process,* by ensuring that the impacts of climate change do not disproportionately affect particular sectors or populations. For this reason of distributive justice, the public sector's development role is particularly important in vulnerability. Nevertheless, the equity of the process and the long-term effectiveness of policy interventions based on vulnerability approaches may be questioned by other stakeholders who are not immediate beneficiaries and/ or who are called upon to underwrite the transfer of resources to the most vulnerable. As with any policy that transfers benefits to a subsector of a population, there is also an incentive for rent-seeking behaviour and political manipulation in the allocation of scarce resources and public benefits.

These issues are apparent in the case of drought management policy in north-east Brazil. In response to the deaths of over 500000 individuals during a late-nineteenth-century drought, the Brazilian government moved from a drought-management approach based on water infrastructure to a vulnerability approach. The latter involved two policy goals: (1) to reduce social inequalities and develop capacities in general, and (2) to reduce exposure to the negative outcomes of individual drought events through humanitarian relief. Arguably, in relation to the first goal, the programme with the largest impact on reducing vulnerability to drought is the federal social security scheme set up in the late 1980s as a poverty reduction measure for the entire country. The social security scheme provides income that does not co-vary with the agricultural cycle, and estimates show that there are nearly four additional beneficiaries for each social security recipient (Ministério da Previdência Social, 2005). More recently, the federal government launched a safety net programme (*Bolsa Família*) which targets the poorest households to encourage school attendance and assist with basic household expenses. As with social security this source of income is critical during drought events.

The second policy goal is addressed through emergency relief efforts targeted at the most vulnerable populations. In response to a drought emergency, the public sector funds the delivery of food and water and cash-for-work programmes. While the goals of the programme have remained constant, the delivery mechanism has been continually refined over the last century partially in an effort to reduce corruption. Because the goal is not equity in *process* but rather addressing inequities in *outcome*, policy-makers are able to trade access to resources during drought events for political support (D. R. Nelson and T. J. Finan, unpublished data). The potential for corruption can lead to less than effective outcomes and problems in efficiency – although, as in the case of Brazil – when the entire population is relatively poor, meeting the needs of a few (the vulnerable via more equitable income distribution), can also be politically expedient for those in power. Nevertheless, it is unclear how much longer an increasing urban population – with its own significant needs – will continue to support the vast outpouring of money to support a dwindling and economically marginal population.

Ignoring system resilience in the Goulburn Broken Valley

The resilience approach focuses not on specific human populations or particular risks, but on developing the capacity of the whole coupled system to cope with change and uncertainty. The system focus suggests an implicit trade-off between the economic efficiency and equity of *process,* and the prioritization of the long-term functioning of the social–ecological system. This may well lead to systematic underestimation of the cost of the process, inadequate assessment of the social distribution of this cost, and a trade-off between future well-being and immediate risk reduction. In contrast to the vulnerability approach, resilience may also imply valuing the needs of the social–ecological system as a whole over the vulnerabilities of specific individuals or groups – a perspective that trades off effectiveness in long-term outcomes against legitimacy and political expediency in the adaptation process, and public over private benefits. These points have several implications for operationalizing resilience-based policies, as can be seen in the case of the enhancement in the robustness (resilience, narrowly defined) of agricultural production to short-term fluctuations in rainfall of the Goulburn Broken Valley in south-eastern Australia.

The large-scale irrigation system in the Goulburn Broken Valley was developed over the twentieth century in a process that led to an expansion of intensive agriculture and the replacement of the native vegetation that once covered the catchment with cropping systems. The irrigation system was put in place to reduce the variation in water supply for farming by increasing the catchment's robustness in face of short-term rainfall fluctuations. Nevertheless, the policy choice – essentially an adaptation-based approach to a single stressor – over time produced negative environmental outcomes. By replacing the native vegetation with a relatively high evapotranspirative capacity with crops and grasses with relatively low evapotranspirative capacity, water tables rose and brought to the surface salt mobilized from palaeo-salt deposits. The result has been waterlogging, increased soil salinity and increased salinity in rivers. During heavy rainfall the system is now susceptible to flood crisis when water tables rise to the surface, as occurred in 1973. Ironically, in becoming more robust to dry periods, the system has become less robust to wet periods.

The state government responded to the crisis by introducing new institutions and devolving responsibility for management to regional communities. Yet the communities and government continue to focus on investment in the generation of more complex, large-scale institutional structures to manage an increasingly sensitive system through enhanced technical efficiency and engineering solutions. Re-establishing the natural control of the water tables through revegetation was not part of the new approach, although this is what would be required to enhance resilience (Anderies; 2005, 2006).

In retrospect, the Goulburn Broken Valley provides an example of how focusing on technical adaptation options for specific threats and for particular groups has compromised the *system's* capacity to cope with change (Anderies, 2006). In contrast, a resilience approach would probably entail high costs. A shift away from heavy reliance on irrigated agriculture would likely generate economic and social hardship well beyond farmers in the valley, although the local agriculturalists would probably bear the brunt of such a policy. Implementing such a policy would obviously meet with political opposition, and, in terms of the policy process, would likely be challenged by stakeholders in the region.

Implications for climate policy

Addressing climate change presents unprecedented policy challenges and among them, three characteristics of the climate change problem are particularly difficult for decision-makers: (1) high scientific uncertainty, (2) the potential for surprise events, and (3) the need for long-term planning horizons. As the examples of the three policy approaches described above illustrates, each approach entails important trade-offs, and these trade-offs circumscribe the efforts of policy-makers to address the unique challenges of climate change.

Policy trade-offs and addressing scientific uncertainty

Scientific uncertainty in the climate change debate leads to divergent opinions regarding contemporary actions. For some, certainty must precede action. For others, the magnitude of *possible* change justifies taking immediate action based on the current state of knowledge (for example United Nations, 1992). Social and biophysical data and our modelling capabilities contribute to the level of uncertainty. Uncertainty pervades estimates of future emissions scenarios as well as our understanding of the range of possible political, social, economic and physical impacts (Moss and Schneider, 2000).

Currently, the high uncertainty about how the climate will change, how we can stop it and how we should respond has stymied the policy process, as different interest groups use the existence of uncertainty as a rationale to argue for policies that favour their immediate and anticipated future needs. In the face of uncertainty in climate change outcomes, addressing the needs of populations who are currently economically marginalized, and thus can be argued to be most vulnerable, is a politically expedient and, arguably, an economically efficient approach (assuming that uncertainty generates a high future discount rate). Similarly, addressing *known* risks and outcomes – for example, risks that can be reliably extrapolated from past trends and experience – through appropriately targeted adaptation measures is also economically efficient, if future uncertainty is ignored. Neither of these policy approaches specifically responds to future uncertainty. A vulnerability approach is concerned with the needs of today and often assumes that decreasing social vulnerability will automatically lead to a corresponding increased capacity to deal with future disturbances. Adaptation approaches cut future uncertainty out of the debate by focusing on risks that we are aware of and can predict with some certainty. The Goulburn Broken Valley example, however, demonstrates the shortcomings of these approaches in the face of high uncertainty and complexity in social–ecological interactions. Of all the approaches, only resilience specifically considers future uncertainty. The resilience approach is designed to build system capacity to cope with change through maintaining flexibility. However, because of the trade-offs between efficiency and flexibility (Redman, 1999) a resilience framework acknowledges that vulnerabilities are inherent to any system (Nelson et al., 2007).

Trade-offs and surprises

Surprises are a function of uncertainty. They refer to any discontinuity between actual process or outcomes and the expected processes and outcomes (Kates and Clark, 1996; Gunderson, 2003). They include local, discrete events and cross-scale interactions, as well as novel situations that are unique to a particular area or population (Gunderson, 2003). In relation to climate change, surprise relates not only to the occurrence of an event, but also its timing and intensity. Climate change is predicted to increase the variability of climate events (IPCC, 2007), which will increase the possibility of surprises.

Climate change offers no guarantee that the past will be a good predictor of future risk, or that technological adaptations will be adequate or appropriate. Adaptation activities directed towards specific risks reduce the range of choice available in the future to deal with surprises (Walker et al., 2006). For example, significant investments in infrastructural improvements designed to address a specific threat (such as sea level rise) are not easily reversed or adjusted in response to new emergent risks, such as coastal ecosystem collapse. In the absence of information about feedbacks and potential new risks involved in the interaction of changing social, economic and ecological conditions the vulnerability approach can lead to a failure to anticipate new, emergent vulnerabilities affecting distinct populations. The north-east Brazil case demonstrates that business as usual in relation to climate variability – humanitarian relief – may no longer suffice under increasing variability and surprise events. The social and economic strain may become incompatible with future climate regimens. In contrast, a resilience approach is designed to address surprise. Timmerman (1986) distinguishes between negative and positive surprises, what he terms 'catastrophes' and 'epiphanies' and argues that a resilient system is one that avoids catastrophes and fosters epiphanies. Therefore, recognition of the potential for unanticipated feedbacks and non-linear responses of a system becomes part of the process of adaptation. Although some actions may result in a system less adapted to current conditions

(due to trade-offs between adaptation and resilience), a resilience-based policy framework is intended to enhance the capacity of society to learn from such
outcomes and adjust goals and interventions accordingly (Nelson et al., 2007).
Nevertheless, 'preparing for surprise' is not a policy panacea (Anderies et al.,
2007). Similar to dealing with uncertainty, under a resilience framework, preparing for surprise requires tolerance of short-term losses.

Trade-offs and long-term planning horizons

Weather is a phenomenon that occurs at very discrete intervals, from minutes to
days, at the most and a weather forecast is used for planning short-term activities. This is familiar to most of us. Unlike weather, climate is defined by averages
of weather over a period of years. Seasonal climate forecasts – with a lead of
several months to a year – are used for disaster management, agricultural decisions
and energy allocation decisions. Although users are faced with a steep learning
curve, these forecasts are being integrated in decision-making structures in various
sectors. Climate change forecasts, on the other hand, can provide time horizons
of 100 years or more. Planning over this time horizon is outside the experience of
most individuals and businesses, and certainly outside the normal planning horizon for public policy decisions.

The resilience approach prioritizes long-term system functioning. Whereas
adaptation and vulnerability approaches are biased towards the sensitivities and
exposures of the present generation, resilience demands an intergenerational viewpoint. Perrings (2007; p. 15 180), for example, argues that probabilistic predictions
of potential policy outcomes are needed in order to take [public] 'action now to
protect the interests of future generations everywhere'. An adaptation approach,
though based on cost–benefit assessments by private agents, and thus apparently
cost efficient, provides no guarantee of longer-term sustainable outcomes for the
affected system. In the vulnerability approach, the public sector attends to the
immediate needs of small sectors of a population as those needs arise, while potentially ignoring signals that the sustainability of the entire system that supports
that population and others may be at risk. As illustrated in the Goulburn Broken
Valley case, past practice may be politically expedient on a year-to-year basis but
not particularly adaptive to longer-term systemic change. In the UK, for example,
the national government has for decades protected the relatively small settlements
along the Norfolk coast from sea level rise and coastal erosion at considerable
national expense. Only recently has this vulnerability approach been revaluated as
too expensive and ultimately ineffective. Nevertheless, a resilience approach faces
significant obstacles in terms of political legitimacy if it privileges the long-term
public good over immediate human needs.

While it is undeniable that climate policy will require attention to existing populations that face high exposure and sensitivity to climate-induced harm, it is also possible that attention to the position of these populations within the broader social system may enhance policy flexibility, without necessarily loss of political expediency, equity or legitimacy. For example, a recent global attitude survey suggests that even very impoverished populations do see the big picture and understand the potential negative implications of a trade-off between their short-term gains in welfare and longer-term exposure to risk and harm through environmental degradation (Pew Research Foundation, 2007). Governments can help vulnerable populations by partially buffering short-term harm while investing to reduce uncertainties associated with future broader social welfare gains. Current work in north-east Brazil does just this with a planning approach that distinguishes between preparation for specific drought events and long-term development needs (Nelson and Finan, 2009). The challenge in a vulnerability approach is creating a buffer over the short term while making gains in addressing the underlying inequalities.

Conclusions

From this comparison of the three approaches it has become clear that the approach used in climate policy formulation can have subtle and not-so-subtle impacts. Whether a vulnerability, adaptation or resilience approach is used brings with it a host of hidden assumptions and implicit trade-offs. Only through the open discussion of the problem and the explicit clarification of priorities associated with different solutions can a policy be developed that addresses the myriad needs that are exposed by climate change. These needs include those of the already vulnerable to economic and social change; the newly vulnerable to weather-related impacts; and those who will be required by the policies developed to assist those who are affected. These debates need to be played out in the public arena to ensure that there is negotiated consensus on the means to move forward. Without such consensus the legitimacy of any climate change policy is debateable. A starting point for this debate is to clarify policy goals and explicitly identify winners and losers to ensure that adequate buffering systems can be put in place during transformation, for example in the case of Brazil from an area supported by drought relief to one which is not, or in the case of Goulburn Broken Valley from an agriculturally intensive area to a less intensively farmed area.

Trade-offs will be associated with the adoption of each policy approach to deal with climate change or variability. On the one hand, in climate policy it is clear that the moral/ethical concerns of vulnerability are central to decision-making – and that is the strength of vulnerability approach – nevertheless, vulnerability approaches can result in counterproductive policy, if these approaches prohibit investment in future systemic resilience. On the other hand, while resilience approaches appear efficient and appropriate for many challenges in ecological system management, enhancing the resilience of social–ecological systems requires engaging with the reality of policy-makers who are morally obligated to protect their most sensitive constituencies – and face a general political intolerance for disturbance and loss of any scale. Adaptation approaches may be the most palatable and expedient in the context of Western democratic decision systems, yet run very high risks of ignoring the most vulnerable and enhancing the sensitivity of the social–ecological system to multiple, interacting stressors at different spatial and temporal scales. Ultimately all approaches may well need to be combined to ensure that climate change policy addresses the many needs of the multiple stakeholders.

To better understand the trade-offs inherent in these approaches further research is required to identify where each approach is best utilized. Are there conditions wherein only one approach or a combination of two or three will generate the most supported and the most effective solution? More work on the identification and the relative importance of decision-makers' values in policy-making is also needed, both to assist with the decision-making process, but also to ensure that the hidden assumptions and disparate uncertainties in climate change policy are clearly and openly debated.

References

Adger, W. N. 2006. 'Vulnerability', *Global Environmental Change* **16**: 268–281.

Adger, W. N., Brown, K., Fairbrass, J., Jordan, A., Paavola, J., Rosendo, S. and Seyfang, G. 2003. 'Governance for sustainability: towards a "Thick" understanding of environmental decision-making', *Environment and Planning A* **35**: 1095–1110.

Adger, W. N., Arnell, N. W. and Tompkins, E. L. 2005. 'Successful adaptation to climate change across scales', *Global Environmental Change* **15**: 77–86.

Anderies, J. M. 2005. 'Minimal models and agroecological policy at the regional scale: an application to salinity problems in southeastern Australia', *Regional Environmental Change* **5**: 1–17.

Anderies, J. M. 2006. 'Robustness, institutions and large-scale change in social–ecological systems: the Hohokam of the Phoenix Basin', *Journal of Institutional Economics* **2**: 133–155.

Anderies, J. M., Rodriguez, A. A., Janssen, M. A. and Cifdaloz, O. 2007. 'Panaceas, uncertainty and the robust control framework in sustainability science', *Proceedings of the National Academy of Sciences of the USA* **104**: 15 194–15 199.

Burton, I., Kates, R. W. and White, G. F. 1993. *The Environment as Hazard*. New York: Guilford.

Carpenter, S. R., and Brock, W. A. 2004. 'Spatial complexity, resilience and policy diversity: fishing on lake-rich landscapes', *Ecology and Society* **9**: 8.

Defra 2000. *Climate Change: The UK Programme*. London: Defra.

Eakin, H. and Lemos, M. C. 2006. 'Adaptation and the state: Latin America and the challenge of capacity-building under globalization', *Global Environmental Change* **16**: 7–18.

Eakin, H. and Luers, A. 2006. 'Assessing the vulnerability of social–environmental systems', *Annual Review of Environment and Resources* **31**: 365–394.

Folke, C., Carpenter, S., Elmqvist, T., Gunderson, L., Holling, C. S. and Walker, B. 2002. 'Resilience and sustainable development: building adaptive capacity in a world of transformations', *Ambio* **31**: 437–440.

Gunderson, L. H. 2001. 'Managing surprising ecosystems in southern Florida', *Ecological Economics* **37**: 371–378.

Gunderson, L. H. 2003. 'Adaptive dancing: interactions between social resilience and ecological crises', in Berkes, F., Colding, J. and Folke, C. (eds.) *Navigating Social–Ecological Systems: Building Resilience for Complexity and Change*. Cambridge: Cambridge University Press, pp. 33–52.

Gunderson, L. H. and Holling, C. S. (eds.) 2002. *Panarchy: Understanding Transformations in Human and Natural Systems*. Washington, DC: Island Press.

Hewitt, K. 1983. 'The idea of calamity in a technocratic age', in Hewitt, K. (ed.) *Interpretations of Calamity (Risks and Hazards)*. Winchester: Allen and Unwin, pp. 3–32.

Holling, C. 1973. 'Resilience and stability of ecological systems', *Annual Review of Ecology and Systematics* **4**: 2–23.

IPCC 2007. Solomon, S., Qin, D., Manning, M., Chen, Z., Marquis, M., Averyt, K. B., Tignor, M. and Miller, H. L. (eds.) *Climate Change 2007: The Physical Science Basis. Contribution of Working Group I to the Fourth Assessment Report of the Intergovernmental Panel on Climate Change*. Cambridge: Cambridge University Press.

Kates, R. W. and Clark, W. C. 1996. 'Expecting the unexpected', *Environment* **38**: 6–18.

Liverman, D. M. 2001. 'Vulnerability to global environmental change', in Kasperson, J. X. and Kasperson, R. E. (eds.) *Global Environmental Risk*. Tokyo: United Nations University, pp. 201–216.

Markandya, A. and Halsnaes, K. 2001. 'Costing methodologies', in *Climate Change 2001: Mitigation. Contribution of Working Group III to the Third Assessment Report of the Intergovernmental Panel on Climate Change*. Cambridge: Cambridge University Press, pp. 451–498.

McKenzie-Hedger, M., Gawith, M., Brown, I., Connell, R. and Downing, T. E. (eds.) 2000. *Climate Change: Assessing the Impacts – Identifying Responses. The First Three Years of the UK Climate Impacts Programme*. Oxford: UKCIP and DETR.

McKenzie-Hedger, M., Connell, R. and Bramwell, P. 2006. 'Bridging the gap: empowering decision-making for adaptation through the UK Climate Impacts Programme', *Climate Policy* **6**: 201–215.

Ministério da Previdência Social. 2005. 'Previdência e estabilidade social: curso formador em previdência social', in *Coleção Previdência Social*. Brasília: Ministério da Previdência Social/Secretaria de Previdência Social, p. 125.

Mortimore, M. and Manvell, A. 2006. 'Climate change: enhancing adaptive capacity', in NRSP (ed.) *NRSP Briefs*. Hemel Hempstead: Natural Resources Systems Programme, p. 8.

Moss, R. H. and Schneider, S. H. 2000. 'Uncertainties in the IPCC TAR: recommendations to lead authors for more consistent assessment and reporting', in Pachauri, R., Taniguchi, T. and Tanaka, K. (eds.) *Guidance Papers on the Cross-Cutting Issues of the Third Assessment Report of the IPCC*. Geneva: World Meteorological Organization, pp. 33–51

Nelson, D. R. and Finan, T. J. 2009. 'Weak winters: dynamic decision making and extended drought in Ceará, Northeast Brazil', in Jones, E. C. and Murphy, A. D.

(eds.) *The Political Economy of Hazards and Disasters*. Walnut Creek: AltaMira Press.

Nelson, D. R., Adger, W. N. and Brown, K. 2007. 'Adaptation to environmental change: contributions of a resilience framework', *Annual Review of Environment and Resources* **32**: 395–420.

O'Brien, K. L. and Leichenko, R. M. 2003. 'Winners and losers in the context of global change', *Annals of the Association of American Geographers* **93**: 89–103.

O'Brien, K., Eriksen, S., Schjolden, A. and Nygaard L. P. 2004. *What's in a Word? Conflicting Interpretations of Vulnerability in Climate Change Research*, Working Paper No. 2004:04. Oslo: CICERO.

Paavola, J. and Adger, W. N. 2006. 'Fair adaptation to climate change', *Ecological Economics* **56**: 594–609.

Perrings, C. 2007. 'Future challenges', *Proceedings of the National Academy of Sciences of the USA* **104**: 15 179–15 180.

Pew Research Foundation. 2007. *World Publics Welcome Global Trade – But not Immigration*, Global Attitudes Project Report. Washington, DC: Pew Research Foundation.

Redman, C. L. 1999. *Human Impact on Ancient Environments*. Tucson: University of Arizona Press.

Sarewitz, D., Pielke Jr, R. and Keykhah, M. 2003. 'Vulnerability and risk: some thoughts from a political and policy perspective', *Risk Analysis* **23**: 805–810.

Simms, A. 2005. *Africa: Up in Smoke?* The 2nd report from the Working Group on Climate Change and Development. London: New Economics Foundation and International Institute for Environment and Development.

Smit, B. and Wandel, J. 2006. 'Adaptation, adaptive capacity and vulnerability', *Global Environmental Change* **16**: 282–292.

Sperling, F. (ed.) 2003. *Poverty and Climate Change: Reducing the Vulnerability of the Poor through Adaptation*. London: World Bank and Department for International Development.

Timmerman, P. 1986. 'Nature myths and how systems cope (or fail to cope) with surprise', in Clark, W. C. and Munn, R. E. (eds.) *Sustainable Development of the Biosphere*. Cambridge: Cambridge University Press, pp. 445–449.

Tompkins, E. L. and Adger, W. N. 2004. 'Does adaptive management of natural resources enhance resilience to climate change?', *Ecology and Society* **9**: 10.

United Nations 1992. *United Nations Framework Convention on Climate Change*. Rio de Janeiro: United Nations.

Vogel, C. and O'Brien, K. L. 2004. 'Vulnerability and global environmental change: Rhetoric and reality', *Aviso: An Information Bulletin on Global Environmental Change and Human Security* **1**: 1–8.

Walker, B., Gunderson, L. H., Kinzig, A., Folke, C. and Schultz L. 2006. 'A handful of heuristics and some propositions for understanding resilience in social–ecological systems', *Ecology and Society* **11**: 13.

West, C. C., and Gawith, M. J. (eds.) 2005. *Measuring Progress: Preparing for Climate Change through the UK Climate Impacts Programme*. Oxford: UKCIP.

White, G. F. 1986. 'Human adjustment to floods', in Kates, R. and Burton, I. (eds.) *Geography Resources and Environment*. Chicago: University of Chicago Press, pp. 11–25.

Willows, R. and Connell, R. (eds.) 2003. *Climate Adaptation: Risk, Uncertainty and Decision-Making*. Oxford: UKCIP.

Wisner, B., Blaikie, P., Cannon, T. and Davis, I. 2004. *At Risk: Natural Hazards, People's Vulnerability and Disasters*. London: Routledge.

14

Community-based adaptation and culture in theory and practice

Jonathan Ensor and Rachel Berger

Introduction

This chapter explores the relationship between culture and adaptation in theory and practice. Our aim is to make clear the important role that culture plays in enabling adaptation, and show how community-based adaptation is well placed to promote, rather than challenge, individual and shared concepts of well-being.

Our intention is to step back from the typical jumping-off point for 'good development', which emphasises community participation, and to unpack the relationships between individual well-being, culture, community and adaptation. To this end, we commence with a brief review of the principles of community-based adaptation, before exploring its relationship to notions of the individual, community and culture taken from political philosophy. In particular, we rely on thinking that has emerged from debates over the role of culture and community in the life of the individual that have emerged from attempts to resolve liberal (predominantly individualist) and communitarian (community focused) views of society.

This body of work is instructive as it reveals, first, the importance of culture to individual well-being, and second, the limits that culture places on the freedom of individuals and communities to embrace change. In the final section, the implications of this understanding for community-based adaptation are drawn out through examples of Practical Action's experiences, demonstrating how the 'community based' approach is able to recognise and respond to the different roles that culture assumes in adaptation.

It should be recognised that this chapter reflects Practical Action's experiences of adaptation and as such is a first step in exploring the complex relationship between cultures, communities and climate change. However, it is hoped that by

Adapting to Climate Change: Thresholds, Values, Governance, eds. W. Neil Adger, Irene Lorenzoni and Karen L. O'Brien. Published by Cambridge University Press. © Cambridge University Press 2009.

explicitly identifying the importance of culture, this paper will help NGOs work more effectively in support of communities that are affected by climate change.

Cultures, communities and adaptation

It is the premise of this paper that culture has an important role to play in the process of adaptation. Stavenhagen (1998) suggests three definitions of culture: as capital, as creativity, or as a total way of life. Here, the third view is assumed in reference to culture, meaning 'the sum total of the material and spiritual activities and products of a given social group...a coherent and self-contained system of values and symbols as well as a set of practices that a specific group reproduces over time and provides individuals with the signposts and meanings for behaviour' (Stavenhagen, 1998, p. 5).

However, the need to adapt to climate change may pressure individuals and communities into changing livelihoods, lifestyles or patterns of behaviour, potentially challenging existing notions of culture. A series of questions arise: does, and if so how does, a shared culture provide, alter or limit the options for adaptation? How and why do individuals within communities respond to the prospect of changes to their lives and livelihoods? And importantly, what lessons emerge for those working to secure lives and livelihoods in the face of climate change?

Stavenhagen's definition of culture suggests an important interdependence between groups and individuals. The prospect of the autonomy of individuals being limited by the communally held cultural environment has long been of particular interest to political philosophers, as it forms a focus of the disagreement between the liberal and communitarian schools. Will Kymlicka and Joseph Raz in particular have sought to clarify the nature of the relationship between the individual and their cultural community. Raz and Kymlicka are by no means the only philosophers to approach this issue, which is central to the communitarian school, nor are they by any means the first. See, for example, Van Dyke's earlier work on the liberal approach to group rights (Van Dyke, 1977), or, earlier still, Hegel's critique of liberalism and the interdependency of the individual and community (Kymlicka, 2002). Their work offers valuable insights for those attempting to understand the role of culture in the life of individuals and how it relates to the challenge of adaptation.

Well-being and cultural context

Raz (1988) considers the nature of communal living in terms of personal well-being, which he deconstructs via *goals* and *social forms*. A person's goals are important to Raz's analysis because well-being is considered from the point of view of the individual. Goals are different from the biological needs of shelter, food and so forth, and instead incorporate plans, relationships and ambitions, are consciously held, and play an implicit role in the actions and reactions of the individual. Thus, 'improving the well-being of a person can normally only be done through his goals. If they are bad for him the way to help him is to get him to change them, and not to frustrate their realization' (Raz, 1988, p. 291).

Social forms are defined similarly to Stavenhagen's culture, as shared beliefs, folklore, collectively shared metaphors and the like. Social forms pervade an individual's decisions, such that 'a person's well-being depends to a large extent on success in socially defined and determined pursuits and activities' (Raz, 1988, p. 309). Thus, 'a person can have a comprehensive goal only if it is based on existing social forms, i.e. on forms of behaviour which are widely practiced in his society'. Which is to say that goals cannot be selected by an individual in a purely objective manner; the important aspects of one's life are deemed so with reference to social forms. Thus 'engaging in the same activities will...have a different significance in the life of the individual depending on the social practices and attitudes to such activities' (Raz, 1988, p. 311). This latter point has particular significance for those working internationally on adaptation, as it draws attention not only to the potential inappropriateness of her or his own views on the best adaptation options, but also to the fact that a successful adaptation approach in one location will not necessarily translate to a different cultural context.

Raz's model allows analysis of the mechanisms of social interaction and personal well-being. Importantly, the structure of goals (important life plans) and social forms (culture) helps to place individual autonomy and well-being in a cultural context. Kymlicka (1989) offers broadly similar analysis, but refers to the need to 'see value' in the activities that make up our goals. However, the range of options in which we may see value is limited by the cultural 'context of choice'. Analogous to social forms, the context of choice determines the importance of our actions, as those actions 'only have meaning to us because they are identified as having significance by our *culture*, because they fit into some pattern of activities which can be culturally recognised as a way of leading one's life' (Kymlicka 1989, p. 189, emphasis in original). The phrasing here is evocative of Raz. Interestingly, Kymlicka also offers empirical evidence to demonstrate the interdependence of the individual and society. For example, the practice by oppressive regimes of attacking identity and culture provides evidence of the importance of cultural heritage in providing individuals with 'emotional security and personal strength' (Kymlicka, 1989, p. 193). This points to the power of identity politics, and is the complement of the subjective sense of group membership that is heightened when communities are placed under threat (Cohen, 1999).

The views of Kymlicka and Raz invite tentative answers to the questions raised by Stavenhagen's definition of culture at the start of this section. Kymlicka summarises his overall conception as 'how freedom of choice is dependent on social practices, cultural meanings, and a shared language'. Accordingly, '[o]ur capacity to form and revise a conception of the good is intimately tied to our membership in a societal culture, since the context of individual choice is the range of options passed down to us by our culture' (1995, p. 126). Similarly, for Raz, well-being is defined in reference to social forms, as it depends 'to a large extent on success in socially defined and determined pursuits and activities' (1988, p. 309). The importance of maintaining well-being or self-respect suggests that adaptation options are limited: not every available option will resonate with local social forms (a truism in even the most laissez-faire cultures). For example, to suggest the use of human waste in composting would be met with abhorrence in some cultures. Yet this approach is commonplace in some communities, reinforcing Raz's view that the same activities have a different significance in different places. However, a more positive reading of Raz and Kymlicka's analyses suggests a productive role for culture, in which social forms and the context of choice define opportunities for adaptation, and suggests that for success, proposed changes should be rooted in or build on local culture.

The above analysis also suggests that individuals and communities will respond to the prospect of change differently depending on how and why change emerges. Kymlicka points out that the importance of community and identity can alter in different circumstances, a fact that becomes shockingly evident when identity politics spills over into violent conflict. The subjective sense of group identity may be reinforced amongst communities that are disenfranchised, face competition for resources or are threatened by existential danger. However, this observation is not the same as suggesting that culture is necessarily resistant to change, and it is wrong to paint cultures as fixed unchanging entities across space or time. Rather, it suggests that changes that are perceived as a threat to culture are likely to be resisted, closing down opportunities for developing the local context of choice. For example, proposing petty trading as a livelihood opportunity for women in some areas of Pakistan, where women's freedom to work outside the home is strongly constrained, is likely to meet with resistance. Moreover, the proposal itself, showing lack of sensitivity to social forms, could constrain future dialogue. Changes imposed from outside should therefore be avoided, and instead the full involvement of communities in the process of adaptation should be promoted: in short, change should be developed from within cultures rather than from without.

Culture and community-based adaptation

The community-based adaptation approach implicitly acknowledges these principles. However, it is our belief that explicit reference to the role of culture is necessary to ensure that this inherent strength of community-based adaptation is delivered in practice. Community-based adaptation has been defined as a process focused on those communities that are most vulnerable to climate change, based on the premise of understanding how climate change will affect the local environment and a community's assets and capacities (Huq and Reid, 2007). It is rooted in the local context and requires those working with communities to engage with indigenous capacities, knowledge and practices of coping with past and present climate-related hazards. The difference between a community-based adaptation project and a standard development project is not principally in the intervention, but in the way the intervention is developed: not what the community is doing, but why and with what knowledge. The aim is to enable the community to understand and integrate the concept of climate risk into their livelihood activities in order to increase their resilience to immediate climate variability and long-term climate change. Community-based adaptation is essentially an action research approach to the problem of climate change impacts on livelihoods.

By seeking to work with communities to identify local problems and locally appropriate solutions, community-based adaptation can naturally build on social forms (working with cultures) and provides an opportunity to extend the local context of choice (effecting change from within, through dialogue and developing a local understanding of the challenges of climate change). Rooting the process of adaptation in communities allows important communal practices and collectively held metaphors or sayings to be identified and used to facilitate change from within, rather than attempting to force change from without. In particular, cultures that lack a tradition or history of adaptation (to climate or other environmental challenges) require an approach that builds from existing social forms and is sympathetic to local notions of well-being. This may be more or less of a challenge depending on how radical a transformation is required and whether the existing cultural context of choice is narrowly defined or deeply entrenched.

The role of culture outlined here indicates the need for a nuanced approach to adaptation that is grounded in a highly developed appreciation of local social dynamics. It may be that situations are encountered 'where "local culture" is oppressive to certain people' and may rob the most vulnerable within a group of a voice (Cleaver, 2001, p. 47). However, an understanding of culture can also help to transcend a simplistic view of power relations as oppressive, and instead point to the complex role that power holders play in well-being and the entry points that they offer into communities (Ensor, 2005, p. 266). Community-based adaptation demands that a line is walked between the 'we know best' and 'they know best' positions (Cleaver, 2001, p. 47) by seeking to work within cultures to build and develop dialogue on the challenges of climate change. As Twigg notes in the context of disaster-resilient communities, success will see the approach to adaptation become a shared community value or attitude. However, reaching this point also

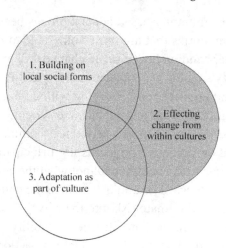

Figure 14.1 Culture can play multiple roles in community-based adaptation.

requires an enabling environment – a 'political, social and cultural environment that encourages freedom of thought and expression, and stimulates inquiry and debate' (Twigg, 2007, p. 26). Engendering such an environment may be a key challenge in the process of moving community-based adaptation beyond isolated development projects and into institutionalised policy frameworks.

Experiences of culture and adaptation

This section reflects on the framework presented above, using examples from Practical Action's experience to illustrate how community-based adaptation can be successful when it builds on existing social forms and effects change from within cultures, and where communities are well versed in the need for adaptation to harsh environments. These three elements are the central implications of the analysis of culture, well-being and adaptation, but do not operate in isolation from each other. As Figure 14.1 illustrates, these dimensions can overlap when the process of adaptation is played out.

Figure 14.1 highlights the need to understand the cultural context when considering approaches to adaptation. In particular, the overlaps between the circles are instructive: cultures without deeply entrenched cultural norms and that have experience of adaptation may still be more responsive to approaches that recognise rather than challenge important social forms (overlap between circles 3 and 1). Similarly, adaptive cultures may be best served by approaches that effect change through seeking to extend the local context of choice, which (however flexible) will still have limits (overlap between circles 3 and 2). The identification of important social forms may be sufficient on its own to suggest adaptation options, or can be used as a starting point to build change from within cultures (overlap between circles 1 and 2). The following sections explore these approaches with reference to Figure 14.1.

Building on local social forms

Key concerns around climate change in the upland regions of Nepal are soil erosion, landslides and flooding, which have increased as poor forest management and the clearance of trees on sloping land have combined with increasingly severe rainfall. In affected communities, Hinduism and Buddhism provide strong and long-standing social forms, framing and informing individual and communal life. Both religions have close links to nature, suggesting an entry point for developing natural resource management responses to climate change. In Hinduism it is regarded as a sin to destroy the banyan and the closely related peepul or bodhi trees (*Ficus religiosa*), which are sacred and associated with temples. Rivers are also sacred and importance is attached to keeping water sources clean through the protection of trees and plants at the spring or source. In Buddhism avoiding harming nature is a central concept, suggesting synergies with biodiversity conservation and building environmental resilience.

Whilst religious beliefs and practices are a core part of the local culture and value system, they have not translated into a wider demand for conservation or natural resource management. In working with the local communities, Practical Action has been able to frame soil and tree management in terms of their religious value, generating an environment in which the community is receptive to tree planting and new land management techniques. By adopting an approach that resonates with local social forms, the intervention has been readily adopted, with the effect of empowering people to protect their assets and resources and reinforce their livelihood security in the face of increasing rainfall variability.

Similar lessons are found in the very different context of the pastoralist communities of north-eastern Kenya. Living in areas where rainfall is below 250 mm in a normal year and crop cultivation is not feasible, a tribal culture has evolved over centuries built around livestock breeds that are capable of surviving through grazing and browsing the sparse vegetation. Social status, activities and roles by gender and age are all based on the finding of water and pasture for livestock, through patterns of seasonally based nomadism. Community and family resilience to disasters (droughts, El Niño-related floods and tribal conflict) is increased through diversity of livestock and herd management, based on the differing preferences for grazing and needs for water of goats, sheep, cattle and camels. Social identity for pastoralists is strongly linked to livestock, even when individuals move out of their local social groups through education. It is common for such people, even though they may be city-dwellers, to own large herds that are managed by family members.

Several factors have impacted over decades to threaten the viability of pastoralist livelihoods. However, experiences of climate variability, manifested in increased length and frequency of droughts, floods and disease, have increased the vulnerability of these communities to the point of threatening the viability of a culture and livelihood that is in tune with a very harsh environment. In recent years, a succession of crises precipitated by severe droughts has led to high death rates in livestock, and has forced large numbers of pastoralists into settlements as displaced persons. Traditional coping strategies based on systems of livestock gifts and loans have been eroded through the massive loss of herds. With small remaining herds, there is a real problem in terms of livelihoods, particularly for young unmarried men, who would have had the role of warrior–herders in remote camps. The Kenyan education system, which values agriculture and settled existence, further undermines pastoralist values. So, as education becomes available to more families as they settle close to schools, this further erodes the tribal society's coping mechanisms: as children attend school instead of learning their role as herders, vital environmental knowledge is not passed on. This in turn erodes the status of elders as keepers of traditional knowledge because traditional knowledge is no longer seen as relevant.

Community-based adaptation efforts have differentiated between the newly settled communities and those who remain as livestock keepers. The newly settled communities include the most vulnerable members of society – female-headed households and the elderly. A central adaptation challenge here is to identify new livelihoods that can fit into a strongly value-laden society. Moreover, alternatives must be appropriate for groups that possess very few resources, limited in the best case to a donkey and a few goats. Discussion with the whole community, including elders, revealed options including donkey-based small businesses development, such as carrying water and other goods using donkey and carts and cultivation and marketing of aloe for export.

Women are the traditional donkey handlers within pastoralist society, making donkey-based haulage an attractive option as it is both livestock-based and open to women. Wild aloe harvesting was an activity carried out by women in the nomadic camps but is now prohibited as aloe is a protected species. Aloe cultivation, however, enables those in settlements to continue this work. Both of these interventions have therefore sought to capitalise on existing social forms and have been well received by the whole community as a result.

For those still able to be livestock keepers, Practical Action's work has explicitly focused on securing livelihoods that are the key element of a deeply entrenched local culture and value system. During a severe drought in 2006 community elders strongly expressed a need for support to enable their livestock to survive. As it was clear that development work could not continue if livestock mortality was very high, fodder (rather than food aid) was provided as a response to the extreme conditions. The success of this intervention has been reinforced through the provision of basic animal healthcare training for community-selected individuals. As a result, survival rates of livestock during drought and floods have improved, herds have been maintained, and a few people have been able to earn a new livelihood as animal healthcare workers.

Each of the strategies outlined above are versions of those illustrated by circle 1 in Figure 14.1, in that they not only build on existing social forms but support and respond to the needs of the local culture. As a result they have been welcomed by the communities involved. However, a key challenge that remains in the Kenyan context is finding new livelihood options for young men in the warrior age group that build on social forms and resonate with their need to demonstrate courage and strength. Options for adaptation will need to identify and retain those aspects of their traditional activity that are key markers of social status – even if confrontational displays of strength are no longer an option. Failure to recognise and work with the traditional needs of this group risks disenfranchising and alienating a key section of society who may, in the worst case, seek to reinforce their identity through recourse to violence. Strategies to address this issue lie in the intersection between circles 1 and 2 in Figure 14.1, building on important existing cultural markers but in such a way as to extend the context choice and effect a change that is acceptable within the community. Ensor (2005) discusses a successful example of this strategy, in which female genital mutilation is recognised as a key signifier of social status for all generations of a girl's family. In this example, an alternative to the traditional rites of passage was developed, known as 'circumcision by words'. The approach retains all the important cultural elements of a traditional coming of age ceremony, but without damaging and dangerous cutting.

The strength and depth of the ties between the pastoralist livelihood and culture mean that the broader question of livelihood viability in the face of climate change is also tied to cultural survival. Pastoralism will depend on the government of Kenya developing an institutional and policy framework that recognises and seeks to support, rather than undermine, pastoralist peoples' culture and values.

Effecting change from within

In Pakistan, Practical Action has been working with a partner, the Rural Development Policy Institute (RDPI), with a formerly pastoralist society in the Thal desert region in Punjab between the Chenab and Sindh/Indus rivers. Until 50 years ago, the area's population lived in scattered hamlets of six to ten households comprising one extended family, owning large herds of camels (up to 600 animals) grazing a wide area of desert scrub, with each hamlet lying close to a well-based water source. In the 1960s–70s, huge irrigation channels were developed by the government, drawing water from the rivers to open up the land to agriculture. Grazing lands which had been a common property resource were taken and people were encouraged to grow wheat and grain: as the availability of grazing land diminished, there was little choice. Perceptions of climate change in this region are that rains are less regular, and that seasons are becoming harsher. However, it is hard to separate climatic impacts from those due to the ecosystem disruption caused by the transition to agriculture. While agriculture has brought increased wealth for some communities, rising input prices and increasingly erratic rainfall patterns are leading to high levels of indebtedness for many farmers. In the last ten years many community members have migrated to towns in search of employment, leading to the desertion of hamlets by all except the elderly.

Traditionally, livelihoods were based around sound natural resource management and many customs and sayings linked to the weather were used to regulate behaviour. For example, in a study of local knowledge on climate, members of several villages talked about the emergence of the Sohal star as an indication that the severity of summer was coming to an end, and how sighting it would determine the change of seasons and a change of activities. This knowledge does not apply to a crop-based livelihood, and is no longer seen as relevant by younger generations, who prefer to listen to the radio for their weather forecasts.

Although in some years agriculture yields a good income, it is becoming less viable. Alternative rural livelihood options are not obvious. Through a process of identifying cultural values, community-based adaptation has focused on raising awareness of traditional knowledge of sound environmental management. Changes in lifestyles and livelihoods have been brought about through a process of discussing the value of planting trees that were once employed for multiple uses: for fruit, fodder and fuel. Similarly, by reawakening an appreciation of the value of the camel in withstanding drought through the revival of traditional festivals of camel dancing, the status of the camel has been revived. In the process, those with land that has not yielded well have been encouraged to allow their land to revert to natural desert vegetation that is suitable for grazing. These adaptation strategies are predominantly circle 2 approaches (Figure 14.1), in which the focus is on extending the context of choice beyond existing options (migration). By reviving traditional knowledge and the festivals that celebrate and communicate it, acceptable and appropriate alternative livelihood strategies have re-entered the communal consciousness and thus the context of choice.

However, as in northern Kenya, climate change adaptation in this region will require government policies that support local culture and traditional land uses that are environmentally appropriate and work with the climate.

Adaptation as part of culture

It is important to recognise that just as some cultures may be deeply and narrowly defined and thereby resistant to change (or certain types of change), in others adaptation and flexibility are or have become part of life (circle 3 in Figure 14.1). This can be seen to an extent in the Kenyan pastoralist system, in which flexibility has emerged as a necessary strategy to exploit a harsh environment, but is constrained within strong cultural norms. The precarious and marginal livelihoods of sandbank dwellers in Bangladesh have engendered a more fundamental adaptability. Bangladesh is the floodplain for several large rivers flowing from the Himalayas. The rivers change their course annually during the monsoon season when vast areas of the country are routinely submerged. Whilst the flow of these rivers is projected to increase as glaciers melt under the influence of climate change, even at present it is not unusual for large areas of land to disappear on one side of the river, while new sandbanks emerge on the other. Despite this, the banks support a very marginalised population who, due to population growth, are forced to live on the erosion-prone banks of large rivers. Erosion results in community members losing most of their assets (and of course their land) several times in a decade, forcing them to relocate. For these sandbank-dwellers, resilience and adaptability are key survival skills, and have become, perhaps, part of their defining culture.

Practical Action has worked with these communities to identify key vulnerabilities and develop technologies that build on practices already in use in similarly affected communities so that resilience is increased and livelihoods strengthened. For example, fish is a key ingredient in local diet, but during the monsoon season, the river's flow is too strong for local fishing boats. Flood water however creates additional temporary water bodies, giving the opportunity for fish cultivation. By training people to construct cages from bamboo and netting, families are enabled to breed fish for food and income generation. As floods worsen and longer periods of inundation are experienced, the planting of crops is delayed. By developing floating vegetable gardens – a practice prevalent in coastal regions – using locally available materials, seedlings can be reared ready for planting as soon as flood waters recede. Practical Action's experience of working with these communities has been that their lack of resources and limited government support, information and infrastructure has left them with a cultural context of choice that is neither deeply entrenched nor narrowly defined. Rather, they have embraced adaptability as a part of their cultural response to their harsh environment. The people have shown themselves to be open to developing new practices and livelihood options that strengthen their coping strategies. Note, however, that whilst this suggests a context that is receptive to changes in livelihood strategies, strong cultural forms still need to be recognised: reflections on Practical Action's experiences in Bangladesh highlight how 'pre-existing [formal and informal institutions and patterns of behaviour] should not only be acknowledged, but incorporated into policy or project design and approach, rather than bypassed or challenged.' (Lewins et al., 2007, p. 33). Adaptation interventions must acknowledge that even in the most flexible societies, the mechanisms of change will inevitably be framed by a cultural context that may be the entry point for interventions (intersection of circles 3 and 1 in Figure 14.1) and offer opportunities for change (intersection of circles 3 and 2 and the intersection of circles 3 and 2 and 1).

Conclusion

Community-based adaptation is well placed to work with a fuller understanding of the relationship between culture and adaptation. The analysis presented here provides insights into the interplay between culture, well-being and adaptation, and offers potentially overlapping options for action: through building on social forms, through extending the context of choice and effecting change from within, and by recognising and capitalising on the adaptability inherent within culture. The examples illustrate that by seeking to work with communities to identify local problems and locally appropriate solutions, community-based adaptation can naturally build on local social forms. Rooting the process of adaptation in communities allows important communal practices to be identified and used to facilitate change from within, rather than attempting to force change from without. Whilst some cultures exhibit a readiness to embrace change, others lack a tradition or history of adaptation and require an approach that builds from the existing cultural context and is sympathetic to local notions of well-being. It is our hope that by better understanding the underlying and ever-present relationship between culture and adaptation, those working with communities affected by climate change will be in a position to capitalise on the inherent ability of community-based adaptation to support a process of change that both addresses the climate challenge and maintains individual and collective well-being.

References

Cleaver, F. 2001. 'Institutions, agency and the limitations of participatory approaches to development', in Cooke, B. and Kothari, U. (eds.) *Participation: The New Tyranny?* London: Zed Books, pp. 36–55.

Cohen, R. 1999. 'The making of ethnicity: a modest defence of primordialism', in Mortimer, E. (ed.) *People, Nation and State: The Meaning of Ethnicity and Nationalism*. London: I. B. Tauris, pp. 3–11.

Ensor, J. 2005. 'Linking rights and culture: implications for rights-based approaches', in Gready, P. and Ensor, J. (eds.) *Reinventing Development? Translating Rights-Based Approaches from Theory into Practice*. London: Zed Books, pp. 254–277.

Kymlicka, W. 1989. 'Liberalism, individualism and minority rights', in Hutchinson, A.C. and Green, L.J.M. (eds.) *Law and the Community.* Toronto: Carswell, pp. 181–204.

Kymlicka, W. 1995. *Multicultural Citizenship.* Oxford: Clarendon Press.

Kymlicka, W. 2002. *Contemporary Political Philosophy.* Oxford: Oxford University Press.

Lewins, R., Coupe, C. and Murray, F. 2007. *Voices from the Margins: Consensus Building and Planning with the Poor in Bangladesh.* Rugby: Practical Action Publishing.

Raz, J. 1988. *The Morality of Freedom.* Oxford: Clarendon Press.

Reid, H. and Huq, S. 2007. *Community-Based Adaptation: A Vital Approach to the Threat Climate Change Poses to the Poor.* London: International Institute for Environment and Development.

Stavenhagen R. 1998. 'Cultural rights: a social science perspective', in Niec, H. (ed.) *Cultural Rights and Wrongs.* Paris: UNESCO, pp. 1–20.

Twigg, J. 2007. *Characteristics of a Disaster-Resilient Community.* London: Benfield Hazard Research Centre, University College London.

Van Dyke, V. 1977. 'The individual, the state and ethnic communities in political theory', *World Politics* **29**: 343–369.

15

Exploring the invisibility of local knowledge in decision-making: the Boscastle Harbour flood disaster

Tori L. Jennings

Introduction

On 16 August 2004, the Boscastle Harbour flood occurred in north Cornwall, England. One hundred tons of water per second funnelled through Boscastle that day, destroying shops, uprooting trees, and sweeping cars and caravans out to sea. I use the case of Boscastle as an illustration of the subordination of local knowledge regarding adaptation to climate, and the implications these actions might have for climate change policy. Flooding is a regular event in the region, and indigenous Cornish have a long history of adapting to it. Their knowledge about local hydrology contrasts sharply with that of recent non-Cornish incomers to the area and of government officials, who viewed the flood as an unusual event possibly caused by global warming.

In this chapter, I illustrate how various policy decisions implemented since Boscastle ceased to be a manor in 1946, and particularly due to the influence of tourism, have reversed the effectiveness of local adaptive behaviours. Although the focus of this chapter is on Boscastle (as shown in Figure 15.1), the observations shed light on related fields of enquiry, most significantly the importance of using historical knowledge to inform decisions about development and management of local environments, as well as the problem of public participation in formulating adaptive responses to climate change. This research is based on 20 months of ethnographic fieldwork carried out from September 2003 to May 2005 in north Cornwall, with two bouts of follow-up research in December 2005 and January 2007.

The notion that climate is a social construct and not an intrinsic feature of nature has for some time received the attention of human geographers and historians of

Adapting to Climate Change: Thresholds, Values, Governance, eds. W. Neil Adger, Irene Lorenzoni and Karen L. O'Brien. Published by Cambridge University Press. © Cambridge University Press 2009.

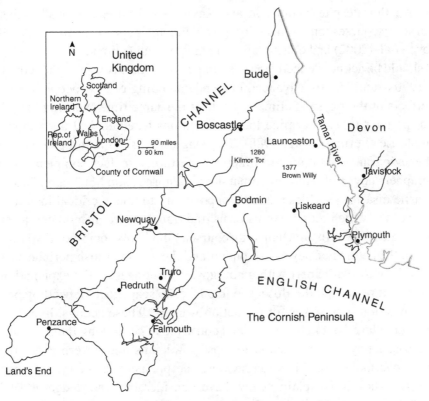

Figure 15.1 Location of Boscastle, Cornwall, south-west England.

science (for example Glacken, 1967; Stehr, 1997; Fleming, 1998; Thornes, 1999; Jankovic, 2000). Hulme (2008) for example, points out that the physical dimensions of climate are imbued with cultural meaning, and to detach climate from culture diminishes both its significance and our ability to understand it. Anthropologists by contrast have taken a relatively modest interest in the way climate is socially constructed, and have generally ignored the socio-political and cultural dimensions of climate change, which have largely been obscured by debates over its causes and effects (Strauss and Orlove, 2003). This oversight is rooted in the way various scholars implicitly view the concept of climate (Outhwaite, 1983). Social theory suggests that concepts like 'landscape' and 'environment' are socially and culturally constructed, yet we have generally not applied these insights to 'climate'. We tend to view climate less as a concept than an objective constraint on human action, which partly explains why human adaptation to climate variability has long dominated the interest of anthropologists (for example Moran, 2000). In this way, anthropologists remain surprisingly unaware of cultural elaborations that produce what we label as 'climate', or how climate might be implicated in unequal power relations, this despite Thompson and Rayner's (1998) exceptional analysis of cultural discourses on climate, or given the insights of Cruikshank (2001), Orlove et al. (2002) and others into oral narratives and climate.

It should be acknowledged, however, that anthropologists throughout the 1980s and 1990s significantly advanced our understanding of how people ameliorate the effects of the physical climate, while at the same time exposing inequitable social policies that undermine local adaptations to ecological uncertainty (for example Leatherman et al., 1988). The long-established anthropological focus on behavioural adaptation to climatic uncertainty in fact complements what geographers now call 'cultural climatology' (Thornes and McGregor, 2003). The Boscastle disaster provides a unique opportunity to take a critical look not only at the ways in which tourism transforms traditional environmental coping strategies, but also how climate change discourse(s) mediates socio-political relations between the centres of power in London and the rural Cornish periphery. Put in the words of a river expert with a non-government agency, the explanation that climate change caused the Boscastle flood was a 'scapegoat' depending on 'how people position themselves for the action wanted'. These actions, like the flood itself, were played out in the media and consisted of competing interests between the tourist lobby, environmental groups, politicians and government officials. Largely overlooked in this struggle were the perspectives of local Cornish in Boscastle who, as I explain below, have considerable knowledge about local hydrology but are at a distinct disadvantage when it comes to shaping policies in their own district.

The lack of knowledge about Boscastle on behalf of non-Cornish incomers, the national press, and government officials led to misunderstandings about the history of flooding in Boscastle and its coexisting cultural communities, which in this chapter I refer to broadly as 'locals' and 'incomers'.[1] Many locals contend that the 2004 flood was the result of inept government land management practices, as much as of natural causes or extreme weather associated with anthropogenic climate change. Their local knowledge about flooding contrasts sharply with that of recent non-Cornish incomers to the area and government agents who viewed the flood as an unusual event possibly caused by global warming. Numerous scholars agree that the social and economic divide separating working-class Cornish from the class of Cornish-born but English-educated middle class, and non-Cornish incomers or 'emmets' as they are pejoratively known, has contributed to a distinctive Cornish cultural identity (for example Deacon and

---

[1] No formal mechanism (e.g. census) exists to track which individuals in the village are 'local'. The principal sources for this information include postal letter carriers and other locals. Throughout my research, I consistently found the term 'local' applied to individuals whose surname reflects kinship relations in a particular locality. Locals are 'related to almost everyone in the village' (personal communication).

Payton, 1993; Westland, 1997; Hale, 2001). As we shall see, misunderstandings about the different communities of Boscastle and history of flooding in the region positioned the needs of tourism above a well-informed flood defence strategy (Clark, 2006).

Climate and society scholars have taken a nuanced approach to the complex role climate plays in mediating notions of class and social policy. Klinenberg (2003), for example, established that uneven social policies, more than climatic conditions, disproportionately impacted the elderly and poor during the 1995 Chicago heat wave that killed 700 people (see also Jankovic, 2006). I suggest that in the case of Boscastle rhetoric about climate change in both the media and government publications concerning the recovery effort of Boscastle, redirected scrutiny away from questions about government land management practices. It also produced a socially homogeneous and coherent view of the village where the opinions of Cornish locals, many of whom already feel marginalised in their own village by the business interests of non-Cornish incomers, were largely ignored. That is to say, climate change discourse unintentionally obscured the social complexity and history of Boscastle. Rural Cornish villages such as Boscastle, in the words of parish priest, Derek Carrivick, must be understood as 'communities of communities'.

The purpose of this chapter however is neither to detract from the very real threat climate change poses to people and the environment, nor to argue one way or the other about the role climate change played in the Boscastle flood. Instead, I focus attention on the subordination of local knowledge in instances of decision-making despite widespread efforts to enhance participation and public input into decision-making (O'Riordan and Ward, 1997; Birch and ODPM, 2002; Few et al., 2007). I argue that climate change adaptation requires not only an appreciation of local knowledge, it requires a fundamental shift in the way bureaucracies manage their authority. Roger Pielke has long argued that political views about anthropogenic climate change produce a 'built-in bias against adaptation' which encourages policy-makers to implement costly and ineffective mitigation policies to cut carbon emissions, while prejudicing them against policies geared toward climate change adaptation (2005, p. 546). Although Pielke analyses the implications of political bias against adaptation, his purpose is not to discern the pre-existing motivations behind the politics. Yet these motivations become critical when formulating adaptive responses to climate change that necessarily involve public participation, because as the literature on participation and indigenous knowledge suggests, the incorporation of local people's knowledge into bureaucratic decision-making often has unexpected consequences (Nadasdy, 2005).

Critics of participation discourse contend that participation is deeply embedded in relations of power (for example Agrawal, 1995; Cooke and Kothari,

2001; Nadasdy, 2003; Few et al., 2007). They conclude that community-based participatory models, which endeavour to integrate so-called 'indigenous knowledge', 'local knowledge', 'traditional ecological knowledge' or 'experiential knowledge' into existing institutional structures, merely reproduce the views of government officials and extend the power of the state (Nadasdy, 2003, 2005). Participation arose during the 1970s and 1980s as a well intentioned alternative to centralised 'top–down' models of development. Believing that 'bottom–up' participatory approaches would both empower local people and be more sympathetic to local needs, proponents of participation have settled on the view that when participation fails, the problem is a technical rather than political one (Agrawal, 1995; Nadasdy, 2005). But these assumptions have been widely questioned in the literature for several important reasons. For one, anthropologists have long regarded 'knowledge' as a process that develops within cultural, political, economic and institutional contexts (Bulmer, 1967; Evans-Pritchard, 1976; Cruikshank, 1998; Nadasdy, 2003). Colin Scott, Michel Foucault, Peter Brosius, Paul Nadasdy and numerous Science and Technology Studies scholars such as Bruno Latour and Sheila Jasanoff have deftly explored the relationship between knowledge construction and power. Nadasdy for instance has demonstrated that 'environmental knowledge is produced and used in the context of state power and bureaucratic management' (2003, p. 264). Most significant for the purposes of this chapter is how indigenous environmental knowledge fares in the face of bureaucracy, despite overtures to public participation in decision-making noted by Few et al. (2007), O'Riordan and Ward (1997) and others.

Nadasdy's critical work on participation and co-management reveals that participation is a 'culturally charged political undertaking' that obscures the cultural nature of the relationships between local people and bureaucrats who formulate and implement policy (2005, p. 220). He argues that in order for local knowledge to be integrated into decision-making, it must first conform to those criteria sanctioned by state power, a process that profoundly distorts or 'constrains' the way local people think about events. (Nadasdy, 2003, 2005; see also Cruikshank, 1998). Rather than inviting an 'alternative view' into decision-making, participation distills the 'lived-experience' of local people into what is relevant to bureaucrats (Nadasdy, 2005). This was illustrated in Boscastle on several occasions. In 2004, for instance, a household survey was commissioned by the North Cornwall District Council in partnership with the Environment Agency. The survey asked Boscastle residents about the flood and what changes they wanted to see in the village (for survey see Atlantic, 2005). The survey did not differentiate locals and incomers. The Environment Agency also conducted formal exhibitions to inform residents about the proposed Valency Flood Defence Scheme, and supposedly to solicit public input into a replacement bridge design.

While formal exhibitions and household surveys appeared to engage Boscastle residents, the reality from the perspective of locals proved otherwise. In the words of one local, the Environment Agency 'say they consulted with locals... They use technical language that locals do not understand... and then they say they found agreement.' By contrast, the Environment Agency in its final public report on the Boscastle flood entitled, *Living with the Risk* states: 'Rebuilding Boscastle has been a team effort involving residents, businesses and organisations with an interest in the village' (2005b, p. 25). A local woman explained to me that she and other locals were in fact 'disillusioned' by the questions asked in the household survey and during exhibitions. In her words, they 'didn't ask the right questions... they asked questions that villagers would not have wanted asked.' By way of illustration, she explained: 'They would ask, "Do you want a pink bridge or a blue bridge?" They didn't ask, "Do you want a bridge?"' She argued that the questions posed by officials merely gave the appearance of choice. Not only did some local people feel that government officials failed to consider their views regarding the 2004 flood, they had little expectation that government officials would. 'It's a bit of a waste of time filling it [the questionnaire] out because we've seen it all before. If they don't get the answer they want, they find reason to ignore it.'

As Few et al. (2007) and others have shown, illusions of participation (whether intentional or not) 'can potentially do more harm than good' (Few et al., 2007, p. 11). With these significant critiques of participation in mind, participation remains an often repeated objective in UK environmental management (O'Riordan and Ward, 1997; Few et al., 2007). In this chapter I counter the prevailing discourses about climate change that subvert the local, and propose instead, that from the local, we can understand how discourses on climate change obscure historically embedded and implicit power relations – relations that undermine policies intended to tackle global warming.

Background

On 16 August 2004, an unusually powerful thunderstorm dumped 196 mm of rain in 4 hours at the head of the Valency catchment which drains into the medieval fishing village cum heritage tourist site of Boscastle Harbour in north Cornwall, England, resulting in major damage in the village (Golding, 2005) (Figure 15.2). On that day, I was in the rural moorland parish of Blisland some ten miles south of Boscastle where I had been living and conducting research into cultural constructions of weather and climate since September 2003. The most useful part of my research was participating in and observing the daily activities of farmers, fishermen and villagers in north Cornwall. When the Boscastle flood occurred, I was already established in the community and it was largely through informal

Figure 15.2 Debris dam, at Boscastle, Cornwall. (Photo by David Flower.)

conversations that I tried to understand how local people thought about the flood. In the months following the flood, I also conducted interviews with climate scientists, Ministers of Parliament, BBC weather forecasters, Environment Agency officials and non-government agents.

The British media played a significant role in perpetuating misunderstandings about Boscastle following the August 2004 flood. National press coverage was dramatic, often inconsistent in its details, and frequently attributed the disaster to global climate change. For instance, 24 hours after the flood the *Independent* asked: 'A flash flood in the pan or a rainstorm caused by global warming?' And 48 hours after the flood the *Guardian* warned: 'Britain should expect more dangerous flash floods, catastrophic rain and hail storms, droughts and heatwaves from the rapid changes in rainfall patterns brought by global warming.' Such headlines characterised the barrage of tabloid, broadsheet and BBC responses to the Boscastle flood. Meteorologists and climate scientists by contrast were more cautious. David Griggs, Met Office Director of Climate Research Division, was loath to openly associate this isolated event with global climate change: 'All we can say is we would expect to see more events like Boscastle with climate change', he told me in February 2005. 'We would never attribute the event to global warming.' Boscastle locals were themselves circumspect about implicating climate change in the 2004 flood because, as local historian and author Anne Knight told me, 'this [flooding] has happened many times'. Incredulous about the media focus on climate change, she articulated the view of many local Cornish adding that climate change merely made the event 'relevant to people up country'. It would be incorrect to suppose, however, that local people deny climate change or the problems it will likely bring to the region. Rather, their experiences reveal how our assumptions about climate change obscure other things.

The assumption that Boscastle residents were victims of climate change is problematic because it ignores the host of historical factors and institutional arrangements that contributed to the flood. Questions about the extent to which tourism might have played a role in the 2004 flood, for instance, were eclipsed by concerns about how the flood would affect tourism in future. Tourism in Cornwall has increasingly become an object of social analysis (for example Urry, 1990; Thornton, 1994; Andrew, 1997; Westland, 1997; Hale, 2001), but tourism in the context of climate change has received little attention despite the British government's concern over the economic impacts of climate change. The needs and interests of the tourist industry in Boscastle went unquestioned because tourism is regarded by many government officials and county councillors as integral to the overall economic improvement of Cornwall today. The Cornwall Tourist Board points out that because Cornwall is one of the poorest counties in the UK, which has qualified it for EU Objective One funding since 1999, the tourist industry provides a vital source of revenue and jobs. Tourism makes up more than a quarter of the Cornish economy and is heavily subsidised through indirect government taxation.

Even though economic alternatives to tourism are difficult to imagine in Boscastle, some locals are sceptical about the benefits tourism brings to their community. 'We are told we cannot live without the tourist industry,' a local woman said to me, 'but we are paying for the bloody thing [through taxes, higher housing prices, car park charges, fuel, and food costs]!' Critics of tourism in Cornwall suggest that tourism might have negative impacts on indigenous industries (Young, 1973; Andrew, 1997), that tourism provides only low-skill, low-wage jobs (Deacon et al., 1988; Thornton, 1994; Payne et al., 1996), and that the majority of tourist-related enterprises are owned by people outside Cornwall – mostly coming from the south-east of England where institutions of centralisation are gathered (Urry, 1990). Detractors further argue that local taxation income to resource services for tourism is especially burdensome for the Cornish rural working poor, who subsidise an industry that unduly benefits holidaymakers and part-time residents from outside Cornwall. The 2004 flood reinforced the negative view about tourism held by many local people. For instance, when I expressed my bewilderment to a Cornish couple about the apparent influence of non-Cornish business people in the community, they bluntly explained that the tourist industry 'has a loud voice', and then added, 'They [incomers] think they are locals after living here only five years... [but] they have no experience of flooding.' In other words, the notion of 'local' is appropriated by incomers who produce new narratives about the community and environment based upon experiences developed outside Boscastle.

Geological evidence indicates that flash flooding has occurred in the Valency Valley where Boscastle is located since the Pleistocene (Anderson, 2007). The catchment above Boscastle is approximately 20 km²; it rises to 300 m AOD (Above Ordnance Datum) and is approximately 7 km long. Not only does the River Valency descend abruptly into Boscastle, it converges with the similarly steep River Jordan in the lower portion of the village known locally as 'Bridge'. In recent times, Boscastle experienced flooding in 1882, 1894, 1903, 1926, 1940s, 1950, 1958, 1963, 2004 and 2007 (Clark, 2006). Based on modelling data by the Institute of Hydrology at Wallingford, flood consultants to the Environment Agency, the Agency concluded that the 2004 flood was an 'extreme event' not likely to occur more than once in 400 years, thereby authorising harbour regeneration and a new flood protection scheme (Environment Agency, 2005b). The magnitude of the 2004 flood compared with previous events is not in dispute. Even so, along with Clark (2006), this chapter suggests that both the Environment Agency and Wallingford group failed to accurately record and interpret historical flood data, or adequately employ local knowledge during their investigation of the 2004 flood. According to local people with whom I spoke, government agencies were indifferent to the skills and knowledge of locals 'whose practical experience', explained one villager, 'was disregarded in preference for academic qualifications'.

On the surface, Boscastle today is a village of second homes, holiday lets, tourist shops and restaurants, and 'Boscastle's own families can no longer afford to live in the homes that had been theirs for over 800 years' (Knight and Knight, 2004, p. 38). Of the approximately 1000 full and part-time residents living in Boscastle, fewer than 200 are local Cornish. Yet as I have already suggested, these figures obscure the social complexity of Boscastle. Following the 2004 flood, June Runnalls, a Cornish woman in her eighties, wryly remarked to me, 'Proper Cornish would know better than to put their cottages at the bottom of that narrow valley.' Her comment possessed dual meaning. On the one hand, it was an observation about the conversion of ancient harbour fish cellars and warehouses into holiday homes and tourist shops in the valley bottom where flooding is historically known to occur. On the other hand, it was an implicit statement about the meaning of 'local' and its relation to knowledge of place. For example, the original medieval and Elizabethan manor houses of Boscastle were both located in 'Top Town' near the High Street (Main Street). Fore Street and High Street developed together as the original areas of market and trade in Boscastle. Most holidaymakers in Boscastle, however, do not that know Top Town exists because, as a local couple explained to me, the 'tourist part of the village' is located in 'Bridge' and 'Quaytown' in the valley bottom. The strenuous climb from Bridge to Top Town suggests that people made a conscious decision many centuries ago to build Fore Street and High Street, the true heart of the village, in a protected area far from the valley bottom. To the annoyance of locals, both national and local press coverage of the 2004 flood described the valley bottom as the High Street and 'heart of the village'.

From fishing village to tourist destination: a brief history of Boscastle

To explore further how Cornish locals understand the environmental history of Boscastle, and how they historically adapted to the vagaries of weather and climate, requires at least a cursory examination of Boscastle's history. To begin, Boscastle survived as a manor for almost 900 years; all the inhabitants, shops and ventures of the village and surrounding farms existed in relation to the manor, the manor Lord, and his agents. In the Manor of Boscastle, manor workmen carried out the maintenance of the Rivers Valency and Jordan and other watercourses built to control runoff and mitigate flooding. One of the descendants of these manor workmen, Roy Pickard, who is now in his seventies and still lives in Boscastle, remembers how his grandfather managed the historic watercourses of Boscastle. Manor workmen cleared and maintained meticulously built slate storm drains, slipways and culverts, they managed banks and hedges built to direct runoff or strand debris in the floodplain that might otherwise cause debris dams, and each spring and autumn they cleared saplings and brush from the river banks and throughout the valley bottom. The subsistence strategies of local tenants likewise contributed to river and valley maintenance. Marshes planted with fruit trees, for example, attenuated water runoff. Tenants routinely collected faggot bundles and the manor gave tenants permission to gather fallen timber for their cooking and heating needs as well as permitting them to cut hazel sticks for bean rows. These activities reduced the number of trees, dead limbs and brush that might otherwise wash downstream and block the upper and lower bridges of the harbour during heavy rainstorms, which proved to be a serious problem in the 2004 flood.

Profound economic and social change in Britain and in much of the world anticipated the sale of Boscastle Manor in 1946; its properties were dispersed a decade later. In turn, the maintenance of historical watercourses, as well as the economic, social and environmental relations of the village gradually transformed. According to Roy Pickard, maintenance of the Rivers Valency and Jordan began to decline during the war years and the subsequent decade under the Rivers Authority. Initially, the Cornwall River Board (1950–1965) and Cornwall River Authority (1965–1973) controlled land drainages, streams and rivers in and around Boscastle. Internal Drainage Boards composed of district authorities and farmers also took part, all vying for limited funds divided between schools, roads, water drainages and flood controls. Ultimately, the National Rivers Authority (1989–1996), the forerunner of the Environment Agency, and then the Agency itself, took responsibility from the former ten regional water authorities of England and Wales in 1996. Today the Environment Agency has responsibility for flood controls and land drainages throughout Cornwall. Environmental management, in other words, shifted from generations of tenants and workmen trained as apprentices and knowledgeable about the local environment to centralised decision-makers and increasingly distant bureaucratic agencies.

Development and change of occupancies during the post-war years gradually altered or destroyed many of the historical watercourses of Boscastle including culverts, slipways, meadows and hedges. Incomers renovated cottages and converted defunct shipping warehouses and fish cellars located in the valley bottom near the harbour into holiday accommodations and tourist shops (the area most damaged by the 2004 flood). Roy Pickard despairs at the changes: 'These culverts, after the manor was sold in 1946, gran'fer said to me, he said, "boy, nobody will know where they [culverts] are and I don't suppose they will be touched and they will grow in" ... .' Roy Pickard explained further that knowledge about the location of these culverts was lost with the deaths of the manor workmen who once maintained them. Cornish locals continue to prize this sort of experiential knowledge. 'The old men knew far better,' reflected Roy, 'now they do everything on computers.'

As further evidence of changes wrought by tourism, another villager explained that the North Cornwall District Council converted the Manor Meadows (Valency Fields), a natural floodplain traditionally left undeveloped and used for village fetes, into a tourist car park between 1970 and 1973, which they extended onto National Trust land in 1981 (see Figure 15.3). The car park destroyed the system of hedges and banks that slowed runoff and collected debris that washed down the River Valency. As recently as 2006, the Environment Agency approved an additional extension of the tourist car park, even though the car park and cars themselves received considerable attention as having contributed to the destructiveness of the 2004 flood. Debris dams remain at the centre of ongoing disputes between the Environment Agency and local people. Locals maintain that debris dams were a major factor in the 2004 flood, thus implicating the Environment Agency in incompetent river management. The Agency in its defence cites the conclusions of the Wallingford Report, which states that flooding would have occurred 'even if the bridge had not been blocked by [debris] during the flood' (2005b, p. 131).

Government expertise and local knowledge: can participation work?

So far I have attempted to show the various ways different interest groups perceived the 2004 Boscastle flood. I have also shown that giving voice to local people in

Figure 15.3 Bridge looking toward Manor Meadows and Valency Valley, Boscastle, circa 1920s. (Boscastle Archive photo.)

Boscastle reveals the prevalence of adaptive strategies to extreme environmental conditions that existed long before climate change became a dominant discourse of the media, the British government, and more recently, the public. Yet tourism, development and changes in government regulation have eroded most, if not all, of those pre-existing adaptive strategies. 'The aim of anthropological theory,' says Alfred Gell, 'is to make sense of behavior in the context of social relations.' The Boscastle flood provides a window into the dissimilar agendas and long-established asymmetrical socio-political relations between centralised institutions of power and local people. It also shows that government interest in tourism for economic growth and rhetoric about combating climate change work at cross-purposes because tourism often displaces the adaptive strategies and experiential knowledge necessary to ameliorate the effects of extreme weather events. In order for governments to give more than lip service to climate change adaptation, as Pielke (2005) indicates they must, then policy-makers will need to alter not merely their policies, but how they exert authority.

Despite the fact that climate change was never formally cited by the Environment Agency as the chief cause of the 2004 flood, it became a scientific narrative that allowed the 'experts' and media to make claims to objectivity about Boscastle denied to indigenous Cornish. Climate change also framed the debate about the future of Boscastle and the Environment Agency's proposed Valency Flood Defence Scheme, which got fully under way in January 2007 at a cost of £4.6 million. In what might be viewed as 'stealth issue advocacy' (Pielke, 2007), the Environment

Agency included statements about the threat of climate change in its public exhibitions and publications about the flood scheme, implicitly suggesting that Boscastle was a prelude of things to come unless proper action is taken to protect Britain from climate change (Environment Agency, 2005a). The Agency's approach to ameliorating flooding in Boscastle ultimately disappointed many locals, who viewed the flood protection scheme with scepticism and even derision. Their concerns were not without merit. Contrary to the 1-in-75-year design standard the Environment Agency claims their flood scheme provides and insurance companies expect, Colin Clark, a historical hydrologist from the Charldon Hill Research Station, Somerset, concludes that the proposed flood scheme will leave Boscastle vulnerable to floods at more than 1 in 35 years. When Anne Knight took the unusual step of writing to the Chief Executive of the Association of British Insurors about this discrepancy, the ABI replied in writing that it 'follows the directions of the Environment Agency' and 'does not set policy'.

Given the concerns about participation raised in this chapter and elsewhere, how can participation duties be achieved in climate change adaptation? First, negative outcomes such as the distrust and frustration described in this study along with case studies by O'Riordan and Ward (1997) or Few et al. (2007), should not be treated as technical problems that can be overcome by introducing some new approach (Nadasdy, 2005). As Nadasdy has shown, new approaches merely replicate the root problem. Second, participation must be viewed in the context of social relations which are fundamentally about power. Until government agencies acknowledge the inequality of these relationships, little will change. Finally, the Cornish locals of Boscastle were largely invisible to government officials and the media in part because of historical relations, but also because officials (perhaps unconsciously) found it easier to engage with incomers whose interests and concerns about tourism approximated those of the government. According to Adger et al. (2009), cultural meaning and values are an 'important resource for guiding adaptation to climate change'. Perhaps one of the best ways to access this meaning is by analysing the social relationships behind the participatory process.

Acknowledgements

This chapter would not have been possible without the critical feedback and assistance of Anne and Rodney Knight of Boscastle. I am also indebted to the inspiration and useful insights of Paul Nadasdy, Samuel Randalls, James Porter and the editors of this volume whose comments have significantly strengthened this paper. This research was assisted by a fellowship from the International Dissertation Field Research Program of the Social Science Research Council with funds provided by the Andrew W. Mellon Foundation.

References

Adger, W. N., Barnett, J. and Ellemor, H. 2009. 'Unique and valued places at risk', in Schneider, S. H., Rosencranz, A. and Mastrandrea, M. (eds.) *Climate Change Science and Policy*. Washington, DC: Island Press, in press.

Agrawal, A. 1995. 'Dismantling the divide between indigenous and scientific knowledge', *Development and Change* **26**: 413–439.

Anderson, J. 2007. Regional Coordinator for Regional Important Geological and Geomorphological Sites. Interview with author, 7 June 2007. Crackington Haven.

Andrew, B. P. 1997. 'Tourism and the economic development of Cornwall', *Annals of Tourism Research* **24**: 721–735.

Atlantic, C. 2005. *Boscastle Framework Plan Final Report*. Truro: Roger Tym and Partners.

Birch, D. and Office of the Deputy Prime Minister 2002. *Public Participation in Local Government: A Survey of Local Authorities*. London: Office of the Deputy Prime Minister.

Bulmer, R. 1967. 'Why is the cassowary not a bird', *Man* **2**: 5–25.

Clark, C. 2006. 'Planning for floods: will we ever learn?', *International Water Power and Dam Construction* 58: 20–27.

Cooke, B. and Kothari, U. (eds.) 2001. *Participation: The New Tyranny*. London: Zed Books.

Cruikshank, J. 1998. *The Social Life of Stories: Narratives and Knowledge in the Yukon Territory*. Lincoln: University of Nebraska Press.

Cruikshank, J. 2001. 'Glaciers and climate change: perspectives from oral tradition', *Arctic* **54**: 377–393.

Deacon, B. and Payton, P. 1993. 'Re-inventing Cornwall: culture change on the European periphery', in Payton, P. (ed.), *Cornish Studies*, vol. 1. Exeter: University of Exeter Press, pp. 62–79.

Deacon, B., George, A. and Perry, R. 1988. *Cornwall at the Crossroads?* Redruth: The Cornish Social and Economic Research Group.

Environment Agency 2005a. *The Climate Is Changing: Time to Get Ready*. Bristol: Environment Agency.

Environment Agency 2005b. *Living with the Risk: The Floods in Boscastle and North Cornwall 16 August 2004*. Bristol: Environment Agency.

Evans-Pritchard, E. E. 1976. *Witchcraft, Oracles and Magic among the Azande*. Oxford: Clarendon Press.

Few, R., Brown, K. and Tompkins, E., 2007. 'Climate change and coastal management decisions: insights from Christchurch Bay, UK', *Coastal Management* **35**: 255–270.

Fleming, J. R. (1998). *Historical Perspectives on Climate Change*. Oxford: Oxford University Press.

Glacken, C. J. 1967. *Traces of the Rhodian Shore: Nature and Culture in Western Thought from Ancient Times to the End of the Eighteenth Century*. Berkeley: University of California Press.

Golding, B. (ed.) 2005. *Boscastle and North Cornwall post Flood Event Study: Meteorological Analysis of the Conditions Leading to Flooding on 16 August 2004*. Exeter: Met Office.

Hale, A. 2001. 'Representing the Cornish: contesting heritage interpretation of Cornwall', *Tourist Studies* 1: 185–196.

Hulme, M. 2008. 'Geographical work at the boundaries of climate change', *Transactions of the Institute of British Geographers* 33: 5–11.

Jankovic, V. 2000. *Reading the Skies: A Cultural History of English Weather, 1650–1820*. Chicago: University of Chicago Press.

Jankovic, V. 2006. 'Change in the weather', *Bookforum* February/March: 39–40.

Klinenberg, E. 2003. *Heat Wave: A Social Autopsy of Disaster in Chicago*. Chicago: University of Chicago Press.

Knight, R. and Knight, A. 2004. *The Book of Boscastle: The Parishes of Forrabury and Minster*. Tiverton: Halsgrove.

Leatherman, T. L., Thomas, R. B. and Luerssen, S. 1988. 'Challenges to seasonal strategies of rural producers: uncertainty and conflict in the adaptive process', *Coping with Seasonal Constraints, Masca Research Papers in Science and Archaeology* **5**: 9–20.

Moran, E. F. 2000. *Human Adaptability*. Boulder: Westview Press.

Nadasdy, P. 2003. *Hunters and Bureaucrats: Power, Knowledge, and Aboriginal–State Relations in the Southwest Yukon*. Vancouver: University of British Columbia Press.

Nadasdy, P. 2005. 'The anti-politics of traditional ecological knowledge: the institutionalization of co-management discourse and practice', *Anthropologica* **47**: 215–232.

O'Riordan, T. and Ward, R. 1997. 'Building trust in shoreline management: creating participatory consultation in Shoreline Management Plans', *Land Use Policy* **14**: 257–276.

Orlove, B. S., Chiang, J. C. H. and Cane, M. 2002. 'Ethnoclimatology in the Andes: a cross-disciplinary study uncovers a scientific basis for the scheme Andean potato farmers traditionally use to predict the coming rains', *American Scientist* **90**: 428–435.

Outhwaite, W. 1983. *Concept Formation in Social Science*. London: Routledge and Kegan Paul.

Payne, S., Henson, B., Gordon, D. and Forrest, R. 1996. *Poverty and Deprivation in West Cornwall in the 1990s*. Bristol: School of Policy Studies, University of Bristol.

Pielke, R. A. 2005. 'Misdefining climate change: consequences for science and action. *Environmental Science and Policy* **8**: 548–561.

Pielke, R. A. 2007. *The Honest Broker: Making Sense of Science Policy and Politics*. Cambridge: Cambridge University Press.

Stehr, N. 1997. 'Trust and climate', *Climate Research* **8**: 163–169.

Strauss, S. and Orlove, B. S. (eds.) 2003. *Weather, Climate, Culture*. Oxford: Berg.

Thompson, M. and Rayner, S. 1998. 'Cultural discourses: expert and lay perceptions of climate change' in M. Thompson and S. Rayner (eds.) *Cultural Choice and Climate Change*, vol. 1. Columbus: Batelle Press, pp. 265–343.

Thornes, J. E. 1999. *John Constable's Skies*. Birmingham: University of Birmingham Press.

Thornes, J. E. and McGregor, G. R. 2003. 'Cultural climatology', in Trudgill, S. T. and Roy, A. (eds.) *Contemporary Meanings in Physical Geography*. London: Arnold, pp. 173–197.

Thornton, P. 1994. 'Tourism in Cornwall: recent research and current trends', in Payton, P. (ed.), *Cornish Studies*, Vol. 2. Exeter: University of Exeter Press, pp. 108–127.

Urry, J. 1990. *The Tourist Gaze*. London: Sage.

Wallingford 2005. *Flooding in Boscastle and North Cornwall, August 2004: Phase 2 Studies Report* (No. EX5160). Wallingford: Institute of Hydrology.

Westland, E. (ed.) 1997. *Cornwall: The Cultural Construction of Place*. Penzance: Patten Press.

Young, G. 1973. *Tourism: Blessing or Blight?* Harmondsworth: Penguin.

16

Adaptation and conflict within fisheries: insights for living with climate change

Sarah Coulthard

Introduction

Continuous over-fishing and declining fish stocks have caused many scholars to proclaim a global crisis in fisheries (Pauly et al., 1998, 2002; Worm et al., 2006; Clark, 2007). The extent and cause of this crisis are complex and heavily contested. Despite this uncertainty, there is growing consensus that the future of fisheries is severely threatened and urgent action is required if marine eco-systems are to function in the future (FAO, 2000; Pauly et al., 2002; WCSD, 2002). It is within this era of alarm that fisheries must now contend with an additional threat – that of human-induced climate change. Most research on fish-eries and climate change focuses on the predicted impacts of climate change on the fish – fish ecosystems (Walther et al., 2002), fish abundance (Stenevika and Sundbya, 2007) and fish habitats (Pittock, 1999). More recently, the impact of cli-mate change on the economic future of fisheries has received growing attention (Eide, 2008). This chapter expands on this debate through an exploration into how the fisher and fishing society is adapting (or not) to the many changes and challenges now faced.

Fishing is an inherently risky and unpredictable business. It remains one of the most dangerous of all human livelihoods (Perez-Labajos, 2008) and fishers often live with high levels of income fluctuation and uncertainty. Despite this, the 'thrill of the hunt' and 'finding of fish' has been interpreted as a valued aspect of fish-ing life – the adventure and challenge of fisheries playing an important role in job satisfaction (Pollnac and Poggie, 2006 p. 333). Pollnac et al. (1998) argue that this type of psychology serves to adapt fishers to the dangers and risks of their occupation. Fishers also frequently display occupational diversity and innovation in their income-earning activities (McCay, 1978; Allison and Ellis, 2001). Despite

Adapting to Climate Change: Thresholds, Values, Governance, eds. W. Neil Adger, Irene Lorenzoni and Karen L. O'Brien. Published by Cambridge University Press. © Cambridge University Press 2009.

the inherent risks in fisheries, fishing societies have lived, and lived well, for generations with the unpredictability of the sea and its bounty. Fishers, worldwide, have perhaps already earned the right to be considered as expert adapters.

Whilst the livelihood of fishing may embed a degree of adaptability in its very nature, the continued pressures from the fisheries crisis has led many fishers to the limits of their experienced coping range. It is this aspect that deserves greater attention from wider debates on adaptation to climate change. Many fishers world-wide are faced with tough decisions – to remain within an increasingly uncertain livelihood or to change their livelihood, heritage and way of life in order to sur-vive. Adaptation responses emerge from a range of fishing countries and different cultures and are able to inform research about what to expect from adaptation and the challenges raised in supporting adaptation successfully. As Adger et al. (2003) argue, the present challenge in adaptation research is to recognize the varied sen-sitivities to climate change exhibited by different sectors, as well as the need for enhanced adaptation to face future climate change *outside* of the 'experienced cop-ing range' (emphasis is author's own).

The aims of this chapter are 2-fold: to provide insight into how adaptation is cur-rently being negotiated by fishing societies, and to contribute towards conceptual thinking on adaptation in the context of the fisheries crisis. In doing so, it combines literature and theory from the arenas of fisheries management, rural livelihoods and climate change. A recurrent focus of the chapter is the role of culture in the process of negotiating adaptation in fisheries. In this sense it contributes to wider debates on the centrality of culture and values in facilitating adaptation to climate change.

The chapter starts with an overview of the fisheries crisis, its many drivers and contestations and the recent additional threat from human-induced climate change. Adaptation by fishing society is clearly necessary within this crisis. However, pres-sures to adapt stem not only from falling fish stocks but also from fisheries manage-ment and policy approaches. A popular solution to the fisheries crisis is the removal of fishers, as a means of reducing fishing pressure on threatened stocks. The chapter explores some of the implications of this policy, in particular the growing conflict between livelihood needs and conservation measures. The choices fishers face can be hypothesized as two extremes: a temporary 'coping' with the problems in fisheries in the hope for better days ahead, or 'adaptation' to pressure by leaving the fishery entirely and finding an alternative livelihood. The role of culture is pivotal to this decision-making process and can act as either a facilitator or barrier to adaptation.

The crisis in fisheries

Under present trends, some scholars (Worm et al., 2006) predict the collapse of all commercial fish stocks within the next 50 years, and yet millions of people are directly dependent upon fishing for a livelihood and as a source of protein rich food security (FAO, 2000; Allison and Ellis, 2001). The consequences of such a prediction are clearly catastrophic at a global scale, affecting both mankind and ocean functioning and diversity. However, it is important to note that, comparable to debates on climate change, the extent of the crisis in fisheries remains scientifically uncertain and politically contested. Assessing the viability of fish stocks is a complex science (Hilborn and Walters, 2003), which often produces recommendations that are unpalatable and unpopular in the political and public spheres. As the nature of the crisis is complex, so too are the drivers. Assignment of blame for the crisis varies across different scientific disciplines, each with its own method of explanation and understanding.

A clear contributor to the fisheries crisis has been our historical approach to fisheries development and a maximization of fishing capacity and pressure. Until quite recently, the world's oceans had been assumed as an inexhaustible resource, which could improve the lives of fishers worldwide through the upgrading of technology and properly marketed produce. During the first half of the twentieth century Western countries therefore industrialized their fisheries. Developing countries, aiming to build up their economies and improve lives for millions of poor, followed suit in the period after World War II. International institutions such as the World Bank and the Food and Agriculture Organization of the United Nations (FAO) assisted in modernizing fishing fleets and improving access to global markets. One outcome of such development measures is that in the present day, a substantial proportion of the world population – particularly in the South – is directly or indirectly dependent upon the existence of fisheries. This number of dependents is growing. During the past three decades, the number of fishers has grown faster than the world's population: in 2006, an estimated 30 million people worked as fishers, the great majority of these in Asia, Africa and Latin America (FAO, 2006). In very basic terms, there are 'too many people chasing too few fish', a phrase commonly used to express the large population of fishers and their growing technological capacity to increase the size of catch.

The crisis in fisheries also reflects a crisis in management (Kooiman et al., 2005). The open access nature of fisheries is commonly cited as an underlying driver of over-exploitation. This was characterized in Hardin's (1968) 'Tragedy of the commons', which predicts that all open-access regimes, such as oceans, will succumb to over-exploitation and degradation due to a lack of defined property rights. Open access denotes that access to the resource is unregulated, free and open to everyone and as such, the resource is susceptible to misuse and overuse. However, a growing body of scholars has countered the assumptions of this tragedy, illustrating the capabilities of many societies to organize themselves and govern natural resources is a sustainable way (Feeny et al., 1990; Dietz et al., 2003). Increasingly, it is argued that the erosion of *traditional* management laws and institutions contribute to the crisis in fisheries management, particularly in small-scale, or artisanal fisheries. Subsequently, there are growing calls for better recognition of traditional property rights and greater incorporation of local knowledge in fisheries management (McCay and Acheson, 1987; Dyer and McGoodwin, 1994; Finlayson and McCay, 1998; Berkes, 2004).

Despite multiple interpretations of the fisheries crisis and its causes, the world's scientists and policy-makers have reached a consensus that, variation accepted, fisheries worldwide are threatened by over-utilization at an unsustainable rate (FAO, 2000; Pauly et al., 2002; WCSD, 2002). It is within this atmosphere of urgency and alarm that fears develop over yet another threat to fisheries – that of human-induced climate change. We have known for some time that climate has a strong influence over fish abundance. A good example is the El Niño–Southern Oscillation (ENSO) phenomenon, an ocean–atmosphere event influencing sea surface temperatures, oceanic circulation and nutrient availability (Wang and Schimel, 2003). Located in the tropical Pacific, El Niño and La Niña events have a worldwide impact on weather and climate (IPCC, 2001) and cause serious oscillations in many fish stocks especially in Latin America (Ibarra et al., 2000). This type of global climatic variation makes the assessment of fish stocks and fishing trends particularly difficult (Watson and Pauly, 2001). Nevertheless, the last decade has seen a flurry of research attempting to assess and predict the impacts of human-induced climate change on marine ecosystems and fisheries (Walther et al., 2002). The latest IPCC (2007) reports on climate change state, with high confidence, that observed changes in marine and freshwater systems are associated with rising water temperatures. These include shifts in ranges and changes in algal, plankton and fish abundance in high-latitude oceans, and range changes in fish migration rivers. Emphasis is placed on expected *regional changes* and *shifts* in the distribution and productivities of fisheries. Local species extinctions will occur at the edge of ranges, whilst in some cases productivity can increase (Easterling et al., 2007, p. 275).

The vulnerability of fisheries from human-induced climate change therefore seems 2-fold. First, given the multiple facets of the fisheries crisis it is difficult to segregate impacts from climate change from all other pressures which marine ecosystems currently face. It will therefore be difficult to know where and when regional shifts and changes occur, making prediction and response to impacts tricky. Vulnerability becomes more pronounced when we consider who might be at risk from regional changes in fisheries. Large commercial fishing fleets arguably have some capacity to follow fish abundance wherever it occurs in the world's oceans. But the small-scale, artisanal fisher has a much poorer migratory choice, at least at a multi-regional scale. The vulnerability of small-scale fishers has been recognized by the IPCC, who state (with high confidence) that complex and localized impacts will be felt by 'small holders, subsistence farmers and fishers worldwide' (IPCC, 2007, p. 29)

Adaptation in fisheries: necessities and pressures

At a global scale, fisheries are clearly changing and their future is undecided. At a regional and local scale, declining fish catches are a serious problem for many coastal communities (Marshall, 2001; Atta-Mills et al., 2004; Gutberlet et al., 2007). Fishers are well aware of the problems in fisheries and frequently live with diminished returns and fragile profits. However, it is not only unpredictable catches that spur the need for adaptation. The management *response* to the fisheries crisis is, in many ways, creating a crisis of its own amongst fishing peoples.

Increasingly, the dominant discourse of fisheries science and management interventions focus on preservation and conservation of fish stocks. This is understandable given the urgency with which over fishing must be addressed, and with recognition that conservationists have spent years trying to prioritize the fisheries crisis on the world's political agenda. A popular solution is to reduce the number of fishers and boats, as a necessary means of conserving fish stocks, biodiversity and habitats. This 'removal of fishers' from fisheries is often enacted through providing alternative livelihoods, a now frequently recommended management approach (Pomeroy et al., 1997; Kühlmann, 2002; Pauly et al., 2005). Another, related, approach is the establishment of Marine Parks with no-take zones, which again require fishers to stop fishing to allow ecosystem recovery and replenishment of stocks (Roberts, 1997; Agardy, 2000; Pauly and Watson et al., 2005). Such mechanisms often come hand in hand; Marine Park initiatives are now advised to aim for some level of livelihood security for fishers (Pomeroy et al., 2004). This is often in the form of compensation, involvement in management and reserve protection, buy-back schemes and alternative employment in tourism (Roberts and Hawkins, 2000). However, as is discussed extensively in development and conservation literature marrying the aims of conservation, sustainable livelihoods and economic development is extremely challenging (Salafsky et al., 2001; Majanen, 2007). Initiatives for reducing fishing pressure frequently marginalize important actors whom, as a result, withhold support (Brown et al., 2001). Marine reserves in particular have been an arena for conflict between 'protecting' and 'utilizing' fisheries (Faasen and Watts, 2007). As Christie (2004) argues, a marine protected area (MPA) may be a biological 'success' – resulting in increased fish abundance and diversity – but a social 'failure' – lacking participation in management and conflict resolution mechanisms. Such conflicts have spurred calls for a pro-poor approach to conservation (Brockington et al., 2006; Kaimowitz and Sheil, 2007) and better linkages between conservation and development research (Brown, 2002; Brown et al., 2002; Balint, 2006; Baral et al., 2007).

Conservation of fish stocks and finding alternatives to fishing livelihoods can be viewed as an adaptive strategy in response to the fisheries crisis. However, fisheries biologists and managers have, it seems, underestimated the ease with which people can be moved out of fisheries – and this now has repercussions on the workability and acceptability of 'adaptation' through fisheries management. Fishing societies have lived with the dynamic sea for generations; their cultures are thus inextricably bound to the marine environment and its many changes. What is missing from the biology-dominated discourse on fisheries is the strong cultural (McCay, 1978; McGoodwin, 2001), traditional (Dyer and McGoodwin, 1994) and even religious (Kraan, 2007) associations, which are so evident in fisheries – and their centrality to making management work.

Adaptation is often discussed in an international setting as governments debate how to support adaptation and who is responsible for doing so (Paavola and Adger, 2006). However, adaptation is enacted at the individual level where it is not autonomous but framed by social structures and process (Adger, 2003; Paavola and Adger, 2006). Fishers who face a declining catch or/and pressure from fisheries management have a degree of choice in how they respond. They can either hang up their nets, and leave the fisheries sector, or continue to fish in the hope of future improvements. The latter may be chosen in spite of diminishing returns, growing household poverty and increased risk associated with non-compliance with fisheries restrictions. Illegal and unregulated fishing remains widespread (Sumaila et al., 2006) and non-compliance to fishing law is common (Nielsen and Mathiesen, 2003). Binkley (2000) describes in detail the coping strategies employed by struggling Nova Scotia fishing-dependent households who are 'getting by in tough times'. She states: 'by restructuring their work fishing-dependent households hope that they can get by until the fishery bounces back' (Binkley, 2000, p. 323). These days, many fishers are caught 'between a rock and a hard place' – there is growing doubt that the fish will come back; it is even more doubtful that the fisheries legislators will leave.

Adaptation is thus necessary – however, it is important to conceptualize and differentiate the degree of adaptation that is chosen and possible. Amid hanging up one's net or continuing to fish in defiance of growing hardship, there are in-between options to be mediated, namely *livelihood diversification*.

From diversifying livelihoods to diversifying out of fisheries: coping or adapting

One way of organizing these concepts of choice, diversification and adaptation within fisheries is by better distinguishing the terminology of coping and adaptation.

Within climate change discourse the terms coping strategies and adaptation are often used interchangeably and sometimes even intermixed, such as adaptive coping mechanisms as described by Brouwer et al. (2007). Whilst coping with and adapting to climate change imply different aspects of time and process, the time element and permanence which are associated with adaptation deserve greater attention. Within fisheries it seems particularly important to distinguish between the two. Many fishers, worldwide, exhibit multiple livelihood activities as a means of *coping with* lean fishing seasons and low income. As Allison and Ellis (2001) argue, fisherfolk diversify their livelihood for very good reasons; the high risk of the occupation, seasonal fluctuation in the resource, and the need to reduce the risk of livelihood failure by spreading it across more than one income source. Marschke and Berkes (2006) go as far as equating livelihood diversity with the well-being of fishers in rural Cambodia. In their study, Cambodian fishers emphasized that the ability of household members to access multiple types of fishing gear and techniques or to combine livelihood skills such as raising animals contributed to the adaptability of the household and successful well-being (Marschke and Berkes, 2006).

As the fisheries crisis deepens, however, fishers are increasingly under pressure to permanently leave the fishery – to hang up their nets, sell their boats and find a new way of living. This movement out of fisheries is a longer-term adaptation and a more permanent response to the fisheries crisis. Marschke and Berke's article clearly shows an association between multiple livelihoods and well-being as a recognized source of resilience to risk and threat – this is, however, quite different from the act of leaving the fishery and choosing a different livelihood as a permanent change. Marschke and Berkes (2006) emphasize the benefits of multiple livelihoods, which are pursued largely in addition to their fishing way of life. Would fishers describe the same sense of well-being if they were asked to remove themselves from fishing entirely?

Livelihood studies give a wealth of theoretical debate on the relationships between coping and adaptation. Davies (1996) argues that adaptation occurs when coping strategies become permanently incorporated in the normal cycle of activities. She argues that coping strategies are too often seen as an inherently good thing (Davies, 1996, p. 61) and that adaptation can be either positive or negative.

Positive adaptation is of choice and can be reversed if fortunes change and usually leads to increase security and sometimes wealth. It is concerned with risk reduction and is likely to involve an intensification of existing livelihood strategies or a diversification into neighboring livelihood systems… Negative adaptation is of necessity, tends to be irreversible and frequently fails to contribute to a lasting reduction in vulnerability. It occurs when the poor are forced to adapt their livelihoods because they can no longer cope with short-term shocks and need to alter fundamentally the ways in which they subsist. (Davies and Hossain, 1997, p. 5; Van der Geest and Dietz, 2004).

In advocating the necessity for adaptation, do we sufficiently consider the different types of adaptation which may emerge: positive and negative; chosen or enforced adaptation and all the shades in between?

Adaptation in fisheries could be envisaged as a clear decision to move out of fisheries – to 'hang up one's net' and substantially rely on non-fisheries means to provide a household income. Adaptation in this sense is closely related to the concept of transformability in socio-ecological systems thinking. Transformability has been defined as 'the capacity to create a fundamentally new system when ecological, economic, or social (including political) conditions make the existing system untenable' (Walker et al., 2004, p. 7). According to fisheries conservationists, this may be the only viable option – a transformation of the system, through substantial adaptation by humankind, to secure sustainability for future fisheries. Walker et al. (2004) argue that transformability is closely related to adaptability – it is only when adaptability fails that transformability takes place. With this in mind, it seems that fisheries management has gone straight to advocating a phase of transformability without sufficiently considering the opportunities for adaptation and resilience that may exist within fishing society itself. Transformability requires capacity 'to create untried beginnings from which to evolve a new way of living' (Walker et al., 2004, p. 8). This is a tall expectation from fishing societies – not only are many fishers locked into their livelihoods through heavy investment in gears and knowledge but this also means leaving one's history, heritage, culture and traditions behind. The cultural appropriateness of transformability through change will be fundamental in any attempts to bring about a sustained reduction in fishing pressure.

Conclusion

This chapter posits that in the midst of a fisheries crisis, fishers worldwide are reacting to change and new pressures. Responses by fishers can be visualized in terms of coping, non-temporary responses such as livelihood diversification, or adaptation, a permanent departure from the fishery into a non-fishing related occupation. Choices and decisions about whether to cope with, or adapt to, the fisheries crisis are socially and culturally driven (Adger, 2003). In the context of livelihood diversification, Frank Ellis reminds us that 'Choice, or the lack of it, does not obey some sort of definable break point between two mutually exclusive states. There are many instances where individual choice may be socially circumscribed at standards of living well above the survival minimum... households and individuals can also move back and forth between choice and necessity seasonally and across years' (Ellis, 2000, p. 56). If we consider those fishers who choose to make ends meet in the hope of a revival in the fishery, as illustrated by Binkley's (2000) description of fishers getting by in Nova Scotia, these points become particularly pertinent.

The decision of one fisher to leave the fishery is likely to be influenced by his or her fellow fishers and the social institutions in which fisheries are embedded (Coulthard, 2006, 2008). Necessity to adapt within the fishery will also change seasonally and between years. Fish catches oscillate and this may serve to relight hopes of a return to better incomes – a reason to remain a fisher, and disagree with the perceived crisis.

Perceived necessities to adapt thus wax and wane over any given period in the fishery. Understanding how social relationships function in such an uncertain environment, both within and between individual, household and societal levels, requires long term observation of the social aspects of a fishery. Adger (2003) illustrates how collective action and social capital can be the foundation of coping with extremes in weather and for community-based management of natural resources. However, it can also be a barrier to adaptation – people may rely on each other to stand firm against the pressures to adapt, where this involves perceived unacceptable levels of change. Strike action may hold similar parallels in so much as it depends on identification of a common cause, institutional organization and fights for a collective identity (Beckwith, 1998).

In conclusion, if adaptation interventions are to be effective and sustainable, they must be built on an understanding of the values people have and the cultural settings by which these are structured and influenced. However, the challenges of learning from local cultures in propagating adaptation are steep. In particular, it will be difficult to link nationwide polices with local realities and cultural specifics. Adaptive co-management may be one way to progress, especially given its claims to bridge different levels of organization and to enable system governance to be more supportive of self-learning at the local level (Folke et al., 2005). Adaptive co-management has been defined as a process by which institutional arrangements and ecological knowledge are tested and revised in a dynamic, ongoing, self-organizing process of learning by doing (Folke et al., 2002; Olson et al., 2004). It combines the dynamics of adaptive management (Holling, 1978) with the collaborative management and sharing of power and responsibility between multiple actors (Olsson et al., 2004).

In fisheries, many fishers know their systems well and there are certainly opportunities for effective management through collaboration, as well as evidence of its success (Pomeroy et al., 2001; Nielsen et al., 2004). It is important to recognize, however, that the nature of fishers' responses to change is shaped by a negotiation of, and trade-off with, other factors. Co-management of resources often prioritizes the sustainability of the resource (the fish). The well-being of the fisher may be stimulated by other factors, a preserved identity, culture and social cohesion being amongst them. Olsson et al. (2004) argue that successful adaptive approaches depend upon flexible institutions, which are able to respond to feedback and support resilience.

However, many cultures, religions and beliefs are inflexible and non-negotiable, which may make adaptive co-management problematic. Furthermore, adaptive co-management recognizes the importance of working with key leaders. They provide leadership, trust, vision, meaning, and they help transform management organizations toward a learning environment: 'Lack of leaders can lead to inertia in social–ecological systems' (Olsson et al., 2004, p. 451). In the context of fisheries, this creates another potential challenge. Key persons in a fishing society, those with position, trust and influence, may well be those least able to leave the fishery: boat owners, heads of fisherman associations, exporters. Those with most influence to bring about co-management may also be those with the most to lose from a change in the fishery, especially where that change instigates a decline in fishing.

Culture, it seems, can act as either a facilitator or barrier to adaptation and its role must be better understood, and worked with, in both fisheries and climate change management. It is therefore important to consider the substantial potential for learning between the realities of adaptation in fisheries and wider academic and policy debates that encircle adaptation to climate change. There may be learning opportunities from the successes and failures of adaptation and its propagation in fisheries, which can inform the shape of current adaptation strategies in climate change. Certainly within fisheries, the social and cultural aspects that motivate fisher behaviour need as much attention as the ecological changes which fuel the current crisis. Culture evolved alongside society's ability to live with change in the past, and thus must be central in living with change in the future.

References

Adger, W. N. 2003. 'Social capital, collective action, and adaptation to climate change', *Economic Geography* **79**: 387–404.

Adger, W. N., Huq, S., Brown, K., Conway, D. and Hulme, M. 2003. 'Adaptation to climate change in the developing world', *Progress in Development Studies* **3**: 179–195.

Agardy, T. 2000. 'Effects of fisheries on marine ecosystems: a conservationist's perspective', *ICES Journal of Marine Science* **57**: 761–765.

Allison E. H., and Ellis, F. 2001. 'The livelihoods approach and management of small-scale fisheries', *Marine Policy* **25**: 377–388.

Atta-Mills, J., Alder, J. and Sumaila, U. R. 2004. 'The decline of a regional fishing nation: the case of Ghana and West Africa', *Natural Resources Forum* **28**: 13–21.

Balint, P. J. 2006. 'Improving community-based conservation near protected areas: the importance of development variables', *Environmental Management* **38**: 137–148.

Baral, N., Stern, M. J. and Heinen, J. T. 2007. 'Integrated conservation and development project life cycles in the Annapurna Conservation Area, Nepal: is development overpowering conservation?', *Biodiversity and Conservation* **16**: doi 10.1007/s10531.

Beckwith, K. 1998. 'Collective identities of class and gender: working-class women in the Pittston coal strike', *Political Psychology* **19**: 147–167.

Berkes, F. 2004. 'Rethinking community-based conservation', *Conservation Biology* **18**: 621–630.

Binkley, M. 2000. 'Getting by in tough times: coping with the fisheries crisis', *Women's Studies International Forum* **23**: 323–332.

Brockington, D., Igoe, J. and Schmidt-Soltau, K. 2006. 'Conservation, human rights, and poverty reduction', *Conservation Biology* **20**: 250–252.

Brown, K. 2002. 'Innovations for conservation and development', *Geographical Journal* **168**: 6–17.

Brown, K., Adger, W. N., Tompkins, E., Bacon, P., Shim, D. and Young, K. 2001. 'Trade-off analysis for marine protected area management', *Ecological Economics* **37**: 417–434.

Brown, K., Tompkins, E. and Adger W. N. 2002. *Making Waves: Integrating Coastal Conservation and Development*. London: Earthscan.

Brouwer, R., Akter, S., Brander, L. and Haque, E. 2007. 'Socioeconomic vulnerability and adaptation to environmental risk: a case study of climate change and flooding in Bangladesh', *Risk Analysis* **27**: 313–326.

Christie, P. 2004. 'Marine protected areas as biological successes and social failures in Southeast Asia', in Shipley, J. B. (ed.) *Aquatic Protected Areas as Fisheries Management Tools: Design, Use, and Evaluation of These Fully Protected Areas*. Bethesda: American Fisheries Society, pp. 155–164.

Clark, C. W. 2007. *The Worldwide Crisis in Fisheries: Economic Models and Human Behaviour*. Cambridge: Cambridge University Press.

Coulthard, S. 2006. 'Developing a people-centred approach to the coastal management of Pulicat lake: a threatened coastal wetland in South India', Ph.D. thesis, Department of Economics and International Development, University of Bath.

Coulthard, S. 2008. 'Adapting to environmental change in artisanal fisheries: insights from a South Indian lagoon', *Global Environmental Change* **18**: 479–489.

Davies, S. 1996. *Adaptable Livelihoods: Coping with Food Insecurity in the Malian Sahel*. Basingstoke: Macmillan.

Davies, S. and Hossain, N. 1997. *Livelihood Adaptation, Public Action and Civil Society: A Review of the Literature*, Working Paper No. 57. Falmer: Institute of Development Studies, University of Sussex.

Dietz, T., Ostrom, E. and Stern, P. C. 2003. 'The struggle to govern the commons', *Science* **302**: 1907–1912.

Dyer, C. L. and McGoodwin, J. R. (eds.) 1994. *Folk Management in the World's Fisheries: Lessons for Modern Fisheries Management*. Boulder: University Press of Colorado.

Easterling, W. E., Aggarwal, P. K. and Batima, P. 2007. 'Food, fibre and forest products', in Parry, M. L., Canziani, O. F., Palutikof, J. P., Van der Linden, P. J. and Hanson, C. E. (eds.) 2007. *Climate Change 2007: Impacts, Adaptation and Vulnerability. Contribution of Working Group II to the Fourth Assessment Report of the Intergovernmental Panel on Climate Change*. Cambridge: Cambridge University Press, pp. 273–313.

Eide, A. 2008. 'An integrated study of economic effects of and vulnerabilities to global warming on the Barents Sea cod fisheries', *Climatic Change* **87**: 251–262.

Ellis, F. 2000. *Rural Livelihoods and Diversity in Developing Countries*. Oxford: Oxford University Press.

FAO 2000. *The State of World Fisheries and Aquaculture*. Rome: Food and Agriculture Organization of the United Nations.

FAO 2006. *The State of the World Fisheries and Aquaculture 2006*. Rome: Food and Agriculture Organization of the United Nations.

Faasen, H. and Watts, S. 2007. 'Local community reaction to the no-take policy on fish-ing in the Tsitsikamma National Park, South Africa', *Ecological Economics* **64**: 36–46.

Feeny, D., Berkes, F., McCay, B. J. and Acheson, J. 1990. 'The tragedy of the commons: twenty-two years later', *Human Ecology* **18**: 1–19.

Folke, C., Carpenter, S., Elmqvist, T., Gunderson, L., Holling, C. S. and Walker, B. 2002. 'Resilience and sustainable development: building adaptive capacity in a world of transformations', *Ambio* **31**: 437–440.

Folke, C., Hahn, T., Olsson, P. and Norberg, J. 2005. 'Adaptive governance of social-eco-logical systems', *Annual Review of Environment and Resources* **30**: 441–473.

Gutberlet, J., Seixas, C. S., Thé, A. P. G. and Carolsfeld, J. 2007. 'Resource conflicts: chal-lenges to fisheries management at the São Francisco River, Brazil', *Human Ecology* **35**: 623–638.

Hardin, G. 1968. 'The tragedy of the commons', *Science* **162**: 1243–1248.

Hilborn, R. and Walters C. J. (eds.) 2003. *Quantitative Fisheries Stock Assessment: Choice, Dynamics and Uncertainty.* Berlin: Springer-Verlag.

Holling, C. S. 1978. *Adaptive Environmental Assessment and Management.* London: Wiley.

Ibarra, A. A., Reid, C. and Thorpe, A. 2000. 'Neo-liberalism and its impact on overfish-ing and overcapitalisation in the marine fisheries of Chile, Mexico and Peru', *Food Policy* **25**: 599–622.

IPCC 2001. *Climate Change 2001: The Scientific Basis. Prepared by Working Group I of the Intergovernmental Panel on Climate Change.* Cambridge: Cambridge University Press.

IPCC 2007. *Climate Change 2007: Synthesis Report. Contribution of Working Groups I, II and III to the Fourth Assessment Report of the Intergovernmental Panel on Climate Change.* Geneva: IPCC.

Kaimowitz, D. and Sheil, D. 2007. 'Conserving what and for whom? Why conservation should help meet basic human needs in the tropics' *Biotropica* **39**: 567–574.

Kooiman, J., Bavinck, M., Jentoft, S. and Pullin, R. 2005. *Fish for Life: Interactive Governance for Fisheries.* Amsterdam: Amsterdam University Press.

Kraan, M. 2007. ' "God's time is the best": the role of religion in fisheries management of Anlo Ewe Beach seine fishermen in Ghana', paper presented at *People and the Sea IV*, 5–7 July 2007, Centre for Maritime Research, Amsterdam.

Kühlmann, K. J. 2002. 'Evaluations of marine reserves as basis to develop alterna-tive livelihoods in coastal areas of the Philippines', *Aquaculture International* **10**: 527–549.

Majanen, T. 2007. 'Resource use conflicts in Mabini and Tingloy, the Philippines', *Marine Policy* **31**: 480–487.

Marshall, J. 2001. 'Connectivity and restructuring: identity and gender relations in a fish-ing community', *Gender, Place and Culture* **8**: 391–409.

Marschke, M. J. and Berkes, F. 2006. 'Exploring strategies that build livelihood resil-ience: a case from Cambodia', *Ecology and Society* **11**: 42. Available at www.ecologyandsociety.org/vol11/iss1/art42

McCay, B. J. 1978. 'Systems ecology, people ecology, and the anthropology of fishing communities', *Human Ecology* **6**: 397–422.

McCay, B. J. and Acheson, J. M. (eds.) 1987. *The Question of the Commons: The Culture and Ecology of Communal Resources.* Tucson: University of Arizona Press.

McGoodwin, J. R. 2001. *Understanding the Cultures of Fishing Communities: A Key to Fisheries Management and Food Security,* Fisheries Technical Paper No. 401. Rome: Food and Agriculture Organization of the United Nations.

Nielsen, J.R. and Mathiesen, C. 2003. 'Important factors influencing rule compliance in fisheries: lessons from Denmark', *Marine Policy* **27**: 409–416.

Nielsen, J.R., Degnbol, P., Viswanathan, K.K., Ahmed, M., Hara, M. and Abdullah, N.M.R. 2004. 'Fisheries co-management: an institutional innovation? Lessons from South East Asia and Southern Africa', *Marine Policy* **28**: 151–160.

Olsson, P., Folke, C. and Berkes, F. 2004. 'Adaptive co-management for building resilience in social–ecological systems', *Environmental Management* **34**: 75–90.

Paavola, J. and Adger, W.N. 2006. 'Fair adaptation to climate change', *Ecological Economics* **56**: 594–609.

Pauly, D., Christensen, V., Dalsgaard, J., Froese, R. and Torres, F. 1998. 'Fishing down marine food webs', *Science* **279**: 860–863.

Pauly, D., Christensen, W., Guénette, S., Pitcher, T.J., Sumaila, R., Walters, C.J., Watson, R. and Zeller, D. 2002. 'Towards sustainability in world fisheries', *Nature* **418**: 689–695.

Pauly, D., Watson, R. and Alder, J. 2005. 'Global trends in the world fisheries: impacts on marine ecosystems and food security', *Philosophical Transactions of the Royal Society of London B* **360**: 5–12.

Perez-Labajos, C. 2008. 'Fishing safety policy and research', *Marine Policy* **32**: 40–45.

Pittock, A.B. 1999. 'Coral reefs and environmental change: adaptation to what?', *American Zoologist* **39**: 10–29.

Pollnac, R.B. and Poggie, J.J. 2006. 'Job satisfaction in the fishery in two southeast Alaskan towns', *Human Organization* **65**: 332–342.

Pollnac, R.B., Poggie, J.J. and Cabral, S.L. 1998. 'Thresholds of danger: perceived risk in a New England fishery', *Human Organization* **57**: 53–59.

Pomeroy, R.S., Pollnac, R.B., Katon, B.M. and Predo, C.D. 1997. 'Evaluating factors contributing to the success of community-based coastal resource management: the Central Visayas Regional Project-l, Philippines', *Ocean and Coastal Management* **36**: 97–120.

Pomeroy, R.S., Katon, B.M. and Harkes, I. 2001. 'Conditions affecting the success of fisheries co-management: lessons from Asia', *Marine Policy* **25**: 197–208.

Pomeroy, R.S., Parks, J.E. and Watson, L.M. 2004. *How Is Your MPA Doing? A Guidebook of Natural and Social Indicators for Evaluating MPA Management Effectiveness*. Gland: IUCN.

Roberts, C.M. 1997. 'Ecological advice for the global fisheries crisis', *Trends in Ecology and Evolution* **12**: 35–38.

Roberts, C.M. and Hawkins. J.P. 2000. *Fully Protected Marine Reserves: A Guide*. Washington, DC: World Wide Fund for Nature.

Salafsky, N., Cauley, H., Balachander, G., Cordes, B., Parks, J., Margoluis, C., Bhatt, S., Encarnacion, C., Russell, D. and Margoluis, R. 2001. 'A systematic test of an enterprise strategy for community based biodiversity conservation', *Conservation Biology* **15**: 1585–1595.

Stenevika, E.K. and Sundbya, S. 2007. 'Impacts of climate change on commercial fish stocks in Norwegian waters', *Marine Policy* **31**: 19–31.

Sumaila, U.R., Alder, J., and Keitha, H. 2006. 'Global scope and economics of illegal fishing', *Marine Policy* **30**: 696–703.

Van der Geest, K. and Dietz, T. 2004. 'A literature survey about risk and vulnerability in drylands, with a focus on the Sahel', in Dietz A.J., Ruben, R. and Verhagen, A. (eds.) *The Impact of Climate Change on Drylands, with a Focus on West Africa*. Dordrecht: Kluwer, pp. 117–146.

Walker, B., Holling, C. S., Carpenter, S. R. and Kinzig, A. 2004. 'Resilience, adaptability and transformability in social–ecological systems', *Ecology and Society* **9**: 15. Available at www.ecologyandsociety.org/vol9/iss2/art15

Walther, G. R., Post, E., Convey, P., Menzel, A., Parmesan, C., Beebee, T. J. C., Fromentin, J., Hoegh-Guldberg, O. and Bairlein, F. 2002. 'Ecological responses to recent climate change', *Nature* **416**: 389–395.

Watson, R. and Pauly, D. 2001. 'Systematic distortions in world fisheries catch trends', *Nature* **414**: 534–536.

WCSD 2002. *Report of the World Summit on Sustainable Development*, 26 August–4 September 2002, Johannesburg, South Africa. New York: United Nations.

Worm, B., Barbier, E. B., Beaumont, N., Duffy, J. E., Folke, C., Halpern, B. S., Jackson, J. B. C., Lotze, H. K., Micheli, F., Palumbi, S. R., Sala, E., Selkoe, K. A., Stachowicz, J. J. and Watson, R. 2006. 'Impacts of biodiversity loss on ocean ecosystem services', *Science* **314**: 787–790.

17

Exploring cultural dimensions of
adaptation to climate change

Thomas Heyd and Nick Brooks

Introduction

The latest report of the IPCC states that 'Warming of the climate system is unequivocal' and that most of the warming over the past half-century is '*very likely* due to the observed increase in anthropogenic [greenhouse gas] concentrations' (IPCC, 2007a, pp. 1, 4). A range of potentially damaging impacts of climate change is anticipated, some of which may be abrupt and irreversible, with potentially severe impacts on human and natural systems (IPCC, 2007b). It is a reasonable proposition that, in light of these conclusions, ethically responsible decision-makers ought to take appropriate action, be it in terms of prevention, mitigation or adaptation (see Jamieson, 2001; Gardiner, 2004).

Though anthropogenic climate change may be new, significant local and regional variations in climate have occurred throughout the historical period, and prehistoric modern humans lived through repeated periods of abrupt and severe climate change that was often global in nature, responding and adapting to environmental change and variation with varying degrees of success and a variety of different outcomes (for example Roberts, 1998; Brooks, 2006).

In this chapter, we propose that culture plays an important role in mediating human responses to environmental change. In particular, we argue that these responses depend heavily on the extent to which societies see themselves as separate from or part of the wider physical or 'natural' environment. A detailed discussion of the social construction of nature is beyond the scope of this chapter (but see Heyd, 2007). For the purposes of this chapter, the term nature is used here to refer to the suite of biogeophysical and biogeochemical systems and processes that serve to regulate the physical environment over a wide range of spatial and temporal scales. These systems are not isolated from human influence, but may be viewed

Adapting to Climate Change: Thresholds, Values, Governance, eds. W. Neil Adger, Irene Lorenzoni and Karen L. O'Brien. Published by Cambridge University Press. © Cambridge University Press 2009.

as autonomous in relation to human beings (i.e. have a tendency to maintain their structure even if subject to human influence). In fact, many 'natural' systems are manipulated by human beings to a greater or lesser extent, and even heavily managed or 'artificial' (for example, agricultural) 'human' systems depend on natural processes (for example, hydrological and geochemical cycles).

The difficulty in defining what is 'natural' merely serves to highlight the problems with discourses based on rigid distinctions between the realms of humanity and nature, discourses which are the subject of this chapter. Here we discuss the influence of culture on conceptions of, and behaviour towards, natural systems and processes in a non-Western context, and compare this example with the mainstream of Western societies. Next, we illustrate how certain conceptions of the relation of human beings to the natural environment may lead to serious policy errors, with disastrous effects for human populations. We follow this up with a discussion of the role of culture both as a source of 'maladaptation' and as a generator of useful coping strategies, in the context of environmental change and variability. We sum up by noting that culture may serve as a resource in two ways, in relation to the 'management' of the non-human sphere and in relation to the development of governance processes, and conclude that a deeper understanding of the cultural mediation of responses to environmental dynamism may be of significant value in the development of resilience to accelerating climate change.

The role of culture

In addition to prevention and mitigation, adaptation increasingly is becoming the focus of discussions of human responses to climate change. Most of these discussions, however, focus on physical, economic or managerial options (for example IPCC, 2007b). We propose that the effectiveness of these ways of addressing adaptation to climate change may be crucially dependent on the underlying cultural fabric of the human groups involved for their successful implementation.

The notion of culture, and how to understand it insofar as it contributes to our adaptation to the environments in which we live, is itself a debated topic (see for example Ingold, 1994). For the limited purposes of this chapter we will speak of culture as comprising the ways of living involving values, beliefs, practices and material artefacts that condition the production of tangible as well as intangible goods and services needed for the satisfaction of a human group's needs and wants. Certainly, we should not think of cultures as neat, homogeneous, isolatable units that can be apportioned to discrete human groups. The culture of any group has to be conceived of as dynamic, subject to constant transformation and in regular interaction with that of other groups, especially given the interrelationship of human populations in today's increasingly globalising context. Moreover, any set of values, beliefs or practices common to a human group is mediated by power relations, and is not simply the result of adaptation to objective conditions of the natural environment. Nevertheless, particular cultural patterns are among the factors that distinguish human groups, and may play a crucial role in the ability of these groups to cope with (sometimes severe) environmental changes (for example, driven by changes in climate).

In the following section we introduce one account of responses to natural phenomena that illustrate an alternative cultural pattern to those prevalent in contemporary Western societies. This account is then contrasted with certain Western cultural approaches to the natural environment, and followed by an example of how the importation of certain Western cultural values has led to maladaptation in a globalised context.

Sentient landscapes

The noted Canadian anthropologist Julie Cruikshank writes about cultural responses to natural phenomena during a period of important changes in climate of pre-colonial north-western North America, recounting some of the oral traditions of the coastal Alaska Tlingit and the Yukon First Nations (Cruikshank, 2001, 2002). She retells stories about glaciers that, in response to actions of some individuals, swallow up whole villages, while their icy surfaces also served as virtual 'highways' that the intrepid travelled between the interior of the continent and the coastal areas, which enjoyed a more moderate climate and are located beyond the glaciated mountain ranges. These oral traditions show that glaciers were not merely conceived as inert masses of ice but as entities that pay attention and respond to human behaviour patterns, such as speaking carelessly, spilling blood, making noise and cooking with grease in their vicinity (Cruikshank, 2001, pp. 385, 387, 388).

Cruikshank describes these peoples' way of conceiving the ensemble of animate and inanimate beings by using the term 'sentient landscapes'. This term takes note of the assumption that, from the perspective of the Alaska Tlingit and Yukon First Nations, the land is not merely inert matter but alive, capable of something akin to perception and action. It also means that the diverse animate and inanimate components of land are not treated as mere resources or obstacles, but as *active counterparts* to human beings (also cf. Ingold's (2000) notion of 'sentient ecology'). Cruikshank describes the type of relationship between people and land exhibited in these oral traditions as involving 'social responsibility', which for these people arises from 'the social nature of all relations between humans and nonhumans, that is, animals and landscape features, including glaciers' (2001, p. 382). She argues that this 'local knowledge embedded in oral traditions' displays 'commitment to an active, thoroughly positioned human subject whose behavior is understood to have consequences' (2001, p. 391). In her analysis, the type of relationship displayed in these approaches to landscape underscores 'the social content of the world and the importance of taking personal and collective responsibility for changes in that world' (2001, p. 391). The idea that is expressed in the relationship with natural processes discussed here is that these components of the natural landscape have their own kind of *agency*.[1]

To people who have not been raised in the cultural *milieux* where these stories originate, the notion of sentient landscapes, and the accounts on which it is based, may seem irrational, belonging to the realms of myth and superstition. However, the idea that the natural environment is home to a variety of forces that somehow perceive and respond to human actions, and that impose constraints on human activity and act to shape human society, is 'wide and persistent from Plato and Aristotle to Lovelock's Gaia hypothesis' (M. Chase, pers. comm.). The animistic beliefs of hunter–gatherer groups, the naturalistic polytheism of many non-Western and pre-Christian European societies, and some contemporary environmentalist discourse, all conceive of the natural world as active and to a certain extent sentient, capable of influencing and limiting human activity.

Nature and culture in Western societies

We might contrast belief systems that treat landscapes as sentient entities that interact with human beings with the post-Enlightenment Western tradition which views humanity and nature as essentially separate (Merchant, 1980; Horigan, 1988; Peterson, 2001). The distancing of humanity from nature, and the separation of the natural and cultural spheres, has been viewed by many thinkers as a prerequisite for the smooth functioning of society, with civilisation serving to elevate us above nature. Hobbes (1651/1985) famously credited the institutions of government and the state with preventing humanity from existing in a 'state of nature' in which life would be 'solitary, poore, nasty, brutish and short'. Freud (1930/2002) similarly viewed civilisation as being in beneficial opposition to humankind's 'original, autonomous disposition' towards violence and aggression. The psychoanalyst Erich Fromm (1955/2002) wrote about human beings 'falling out of nature' as the result of an 'awakening' from an unthinking animal state driven by biological and cultural evolution. To Fromm, humanity's subsequent cultural evolution 'is based on the fact that [humanity] has lost [its] original home, nature – and … can never return to it' (2002, p. 24).

---

[1] This aspect of their approach to landscape, of course, is not unique to Alaska Tlingit and the Yukon First Nations, but common to many peoples who have deep roots in their lands, including the Inuit and the Indigenous People of the Russian North, as well as the Mapuche and Quechua, of South America's Andes mountain ranges.

In the Western, and particularly the Anglo-Saxon, world, a belief that humanity is separate from nature has played a major role in influencing attitudes towards, and interactions with, the wider physical or 'natural' environment. On the one hand, this point of view has fostered a view of nature as a resource to be exploited for the material benefit of human beings, and/or preserved as a luxury valued for its aesthetic and therapeutic qualities (for example Merchant, 1980). On the other, a lack of appreciation of the extent to which the cultural and natural spheres are linked, and the prioritisation of culture (that is, the human-produced) over nature (understood as not-human produced), has acted to diminish (i) the perceived importance of the environmental consequences of human activity, and (ii) our grasp of the power of the natural environment to exert its influence on human affairs. Even when adverse impacts of *human activity* on the environment are identified, this limited view of the relationship between culture and nature downplays the complementary role of the *agency of the non-human*, and its potential for adverse impacts on human systems via environmental pathways (see Heyd, 2005).

Separation and maladaptation

The failure to recognise that human systems are embedded within, and thoroughly dependent on, the wider physical environment (comprised of a variety of 'natural' systems that interact with each other and with human systems), and that our actions have consequences that are mediated through that environment, can have profound developmental consequences. The African Sahel provides an example of how development can result in the exacerbation of vulnerability to climatic change and variability when this development is decoupled from considerations of environmental agency and limits, and when – as a result of this – the environment is implicitly treated as if it were static.

The Sahel is the semi-arid transition zone between humid tropical Africa and the arid Sahara desert, characterised by a high degree of temporal and spatial variability in rainfall, and by alternating periods of relative humidity and aridity which may last from years to centuries (Brooks, 2004). Traditional methods of subsistence in the Sahel developed to accommodate climatic variability, with mobile pastoralism allowing nomadic groups to exploit changing patterns of resource availability over a range of timescales, and reciprocal arrangements between nomadic pastoralists and settled farmers providing mechanisms for coping with periods of hardship (Swift, 1977; Thébaud and Batterby, 2001). However, throughout the twentieth century, colonial and post-colonial governments sought to move away from traditional land and resource management systems, in favour of intensive, 'modern' food production systems aimed at delivering rapid economic growth and based on Western developmental models. The result was the commercialisation of agriculture and the marginalisation and settling of nomadic populations in the name of progress and modernisation – a recurring theme in African 'development' (Cooper, 1997).

Of particular note within the context of this paper is the fact that 'modernisation' of food production systems accelerated in the Sahel during the 1950s and 1960s, a period of unusually high rainfall (at least within the context of the period of meteorological records: Brooks, 2004). This process involved the introduction of modern infrastructure, for example boreholes and cement-lined wells, the rapid expansion of agriculture northwards into areas that previously had been unsuitable for agriculture, the adoption of sedentary agriculture by some pastoral groups, the privatisation of common rangelands, and the restriction of pastoralists' movements (Swift, 1977; Thébaud and Batterby, 2001). This process of modernisation undermined traditional mechanisms for coping with drought, and resulted in increased competition for resources, also contributing to conflict. Increased stocking levels, agricultural intensification, and particularly the northwards expansion of agriculture into areas that were productive in the anomalously wet 1950s and 1960s but which were historically marginal, resulted in a large increase in exposure and vulnerability to drought. This exacerbation of vulnerability led to disaster when rainfall declined dramatically in the early 1970s, resulting in the collapse of agricultural and pastoral systems, particularly in marginal areas. During the severe drought of 1972–73, hundreds of thousands of people, and millions of livestock, are estimated to have died across the Sahel as a result of famine (Sheets and Morris, 1976; Hill, 1989).

The initial response of the international research community to the Sahelian drought and famine of the early 1970s was to blame environmental degradation caused by overgrazing and inappropriate land use practices (for example Charney et al., 1975). It was suggested that there was a need for Western development agencies to educate the inhabitants of the Sahel about the detrimental effects of their traditional methods of resource management, and the need to adopt 'modern' techniques (Lamprey, 1975). What was lacking from the discourse in the 1970s (and in much of the subsequent literature) was an understanding of disaster risk resulting from the complex interaction of a suite of socio-economic factors with climatic and environmental change and variation, and of how traditional systems of resource management had evolved within the context of environmental change and variation in this marginal region characterised by extreme rainfall variability on multiple timescales. It is now accepted that there was no region-wide, systemic land degradation or desertification in the Sahel, and that the apparent desertification of the region in the 1970s and 1980s was simply a manifestation of the natural (and historically common) oscillation of the 'desert boundary' associated with multi-decadal scale rainfall variability (Brooks, 2004).

The Sahelian famine of the 1970s appears ultimately to have been driven by the pursuit of development policies that paid little or no attention to the environmental context in which they were implemented. In the pursuit of 'progress' and economic growth, colonial bureaucrats, development agency staff and post-colonial governments assumed that the technocratic approaches to development pursued in the 'more advanced' societies of Europe and North America were by definition superior to 'primitive' traditional African systems of resource management and could be applied universally. The demands and constraints of the very different Sahelian environment were not considered; nor was the sustainability of development policies in the event of future episodes of increased aridity, which were inevitable given the nature of the Sahelian climate. Foreign agencies and colonial and African governments alike adopted a Western European developmental paradigm which implicitly viewed the natural environment as static and ignored the *agency* inherent in natural systems, even as it saw nature as something to be overcome.

The result was an over-extension of agriculture and pastoralism into areas in which these activities were not viable in the long term. This unsustainable development was unable to accommodate natural decadal-scale variations in climate, triggering a highly disruptive regional socio-economic transition which is still being played out today. The Sahel teaches us how our perceptions of nature, and the culture that mediates those perceptions can lead to maladaptation – the pursuit of policies and practices which make people more vulnerable to changes in the natural environment in which human systems are embedded.

Cultural sources of maladaptation

While Western ideological systems that emphasise humanity's 'separateness' from nature have downplayed agency in natural systems, partly as a reaction against 'irrational' myth and superstition (Merchant, 1980), rational scientific enquiry in the Enlightenment tradition actually has supported the assumption of agency inherent in natural systems, particularly since early geological studies demonstrated that the environment has changed on long timescales. It must also be recognised that the Western cultural perspective outlined here, in which human beings are seen as separate from nature, has always been the subject of 'internal' challenges, for example from the Romantic and environmental movements. Consequently, we do not propose to revive the supposition of a radical difference between a degraded, necessarily blind-sighted, 'Western' culture, on the one hand, and pure 'non-Western' or 'traditional' cultures espousing a harmonious coexistence with the natural environment, on the other.

Nonetheless, the assumption that humanity is separate from or 'above' nature, and the dominance of ideas of social (as opposed to environmental) agency in the social sciences, have resulted in developmental models that pay little or no attention to how human systems affect and are affected by the behaviour of natural

systems, particularly on timescales longer than a few years. Linked with an ideology of progressive social evolution that legitimises the social and economic
restructuring of societies along Western lines in the name of progress, these models
have been spread across the world through the vectors of European colonialism
(Conklin, 1997), economic globalisation (Gray, 1995), international development
programmes (Cooper and Packard, 1997; Ebrahim, 2001), as well as political
ideology. For example, in her seminal account of the many adverse environmental
impacts of the Chinese Cultural Revolution, Shapiro (2001, p. 3) states that 'The
Maoist adversarial stance towards the natural world is an extreme case of the mod-
ernist conception of humans as fundamentally distinct and separate from nature.'
While ideas of progressive social evolution were discredited in academic circles in
the twentieth century, they persist in the language of politics and development (for
example Gray, 2007). While China is not a Western country, Maoist ideology was
based on Marxist–Leninist ideology, and Chinese communism was essentially a
Western import, albeit one that took on a distinctive character once coupled with
Chinese culture and Mao-era politics (Shapiro, 2001).

While awareness of the potential for changes in natural systems (for example, driven by human impacts on the environment) to affect human societies is
growing, and despite the growth of environmental and ecological economics as
academic disciplines, the dominance of economics as a framework for decision-
making means that the translation of concerns about phenomena such as climate
change into meaningful policies or actions is extremely limited.[2] The primacy of
ideas based on progressive economic development is evident in the prioritisation
of economic growth over concerns about environmental degradation and anthropogenic climate change. In the context of climate change, current development
policies act as obstacles to both mitigation (for example, the reduction of greenhouse gas emissions) and adaptation (for example, the adoption of practices that
reduce exposure to risks associated with existing or anticipated climate-related
hazards). This is the case not only in wealthy Western nations, but also in rapidly
developing countries likely to be worst affected by climate change. China, beset
by environmental problems and vulnerable to the impacts of climate change, now
matches the United States in its total greenhouse gas emissions (*Nature*, 2007;
Zeng et al., 2008). In Mexico, agricultural restructuring associated with market
liberalisation has reduced the resources at farmers' disposal, decreasing flexibility
and adaptive capacity at a time of increasing climatic variability and uncertainty
(Eakin, 2005).

---

[2] Greenhouse gas emissions are rising faster than projected even under the most pessimistic scenarios used
by the IPCC, and there is little hope of stabilising concentrations of CO_2 at or below 450 parts per million,
the figure usually associated with the 2 °C 'guardrail' value of maximum advisable increase in global mean
surface temperature.

The pursuit at the macro-scale of policies that amplify exposure and vulnerability to natural hazards (a trend we might label as unsustainable development, or maladaptation) is echoed at the local scale. Examples include: the rebuilding of settlements along low-lying coasts devastated by the 2004 tsunami in Sumatra (Steinberg, 2007); the degradation of systems that protect coastal areas from flood and storm hazards (for example Gössling, 2003); the rebuilding of houses and settlements in areas damaged by floods in 2003 in Central Europe and then devastated again in 2005 (Leroy, 2006); the expansion of settlements into low-lying flood plains in the United Kingdom (IPPR, 2005).

Explanations for these patterns of maladaptation must involve various social, political, geographical and economic factors. For example, people often end up rebuilding devastated settlements in the same places and exposing themselves to the same risks that led to earlier episodes of destruction as a result of social and economic marginalisation, the imposition of restrictions on their activities by government, the opportunity costs associated with relocation, a lack of support for adaptation measures by government, or because of the perception that risks are balanced or outweighed by benefits such as access to resources (Leroy, 2005). However, more generally, the failure to deploy more effective coping and adaptation strategies, particularly by governments and individuals in wealthy nations, suggests that there is a fundamental and broadly based *cultural inadequacy*, characterised by an inability to fully comprehend or act on certain risks associated with environmental variability and change even when information on these risks is widely available. Put another way, a lack of certain cultural resources leads to a reduction of the adaptive capacity of individuals, groups and societies which otherwise seem to possess adequate resources and mechanisms for adaptation.

Cultural mediation of adaptive strategies

It is instructive to contrast cultures that view landscapes as sentient entities that respond to human actions with those that view nature as passive and occupying a separate space to human social life. In many 'traditional' societies, the conception of nature as a sentient force which actively and constantly interacts with human beings provides a framework through which considerations of human–environment interactions are internalised within social discourse.

There is considerable research being carried out into the coping behaviours and other cultural consequences, if any, engendered by 'natural' disasters in various societies. It seems that such events typically will persist in the cultural memory of cohesive social groups for a time-span equivalent to one lifetime. This is the case with regard to awareness of the signs of impending tsunamis and volcanic eruptions among some populations living in Papua New Guinea and the Solomon Islands, for example (Davies, 2002, pp. 37–38). Other adaptive behaviours following natural disasters may be directed more toward the long term, such as the permanent relocation of villages or cities (Davies, 2002, pp. 39–40; also see Fagan (2000) on the Moche relocation of their capital after its first-time near-total destruction). Such experiences may also lead to more indirectly adaptive behaviours, such as the creation of myths and the establishment of taboos about occupying certain areas of the land (Lowe et al., 2002, p. 138). In those cases the direct, concrete cultural memory of the disastrous effects of the event may become lost but not before leading to an adaptation that exhibits respect for the natural phenomena at issue through habitual, ritual or mythical means.

In contrast, the marginalisation of nature in many 'modernised' societies mitigates against such modes of internalisation. As a result, measures to address environmental problems often end up as 'add-ons' to existing policies and plans that have been developed with little or no consideration of the wider environmental contexts within which they must be implemented. This is evident in the increasing use of the concept of 'climate-proofing' by governments and development agencies, aimed at 'protecting' existing developmental practices and infrastructure from climate change (for example Asian Development Bank, 2005). The acknowledgement that such existing developmental models and systems might be fundamentally unsustainable in the face of climate variability and change, that protection might not be possible, and that climate change might require the abandonment of existing developmental goals, strategies and policies, and also of some geographical areas, is resisted in developmental contexts. Adaptation is seen as a means of securing economic development and growth within existing developmental frameworks, rather than as a process through which the relationship between humanity and nature may be redefined in order to develop systems that can accommodate the variability inherent in a fundamentally dynamic environment, which is becoming more dynamic as a result of anthropogenic climate change.

Where Western societies, and the international institutions they dominate, do attempt to address adverse anthropogenic impacts on natural systems, existing economic frameworks tend to limit the scope for action, mitigating against remedial actions aimed at sustaining or restoring natural systems and processes where such actions incur costs in the societal sphere. Studies such as the Millennium Ecosystem Assessment Program (2005) have explicitly addressed interactions between human and non-human systems, and their mutual dependency. However, the primacy of short-term economic criteria as a basis for decision-making, and the development of concepts of ecosystem goods and services, means that the preservation of natural systems tends to be justified (at least at the policy level) only insofar as these ecosystems can be (passively or actively) exploited in order to deliver immediately identifiable developmental and economic benefits for human beings.

Culture as resource

The dominant approach in Western societies that treats 'nature' and 'culture' as separate spheres represents a formidable obstacle to adaptation and sustainable development, preventing the internalisation of relations with the non-human environment into everyday life, political discourse and policy formulation. Nonetheless, Western societies do possess two broad types of 'cultural resource' with which they might address these issues. On the one hand, scientific descriptions that increasingly characterise natural systems as having a kind of self-organisation or agency that may be controllable or predictable *only to a very limited extent* provide potential frameworks within which to interrogate our conceptions of the relations between human and non-human systems (Maturana and Varela, 1973, 1980; Prigogine and Stengers, 1984; Kaufmann, 1995, 2000). Such perspectives may lead to policies that accommodate and cope with climatic and environmental change by making adjustments to human systems, which in turn allow natural systems and forces space and time for their own expression. Such an approach is already being implemented in some instances, for example in coastal management schemes in countries such as the UK, and on river systems in Central Europe (Ledoux et al., 2005).

On the other hand, new approaches to governance may provide a means of internalising human–environment interactions into policy formulation and implementation. Western (and many non-Western) societies traditionally have been highly centralised, meaning that policies have been formulated by decision-makers who are often remote from the realities of the localised contexts within which human–environment interactions are played out. An increasing emphasis on decentralisation and participatory decision-making involving local stakeholders may provide a context in which the consideration of relations between human and non-human actors can be internalised in policy development. This 'localised' approach will be particularly relevant to issues of adaptation to environmental change, while the scientific understanding of human–environment interactions on a planetary scale will feed into systems of global environmental governance to address issues such as climate change mitigation. Insofar as governance fundamentally has to do with 'the manifold ways in which humans regulate their affairs to reach common goals and react to a changing environment' (Pattberg, 2007, p. 1), and normally involves the institutionalisation of rules and general norms (Pattberg, 2007, p. 14), it will be of key importance in the transformation of our societies in preparation for climate change. The evolution of governance institutions is dependent on cultural preconditions that either favour or undermine the institutionalisation of rules and norms (Jaeger et al., 1993; Macnaghten and Jacobs, 1997; Bulkeley and Mol 2003; Ostrom, 2005; Bromley, 2006). Coupled with a wider awareness of the interdependency of human and non-human systems (for example through science, public debate and a questioning of the supposition that humans are separate from nature), the involvement of citizens in governance may generate new institutions and influence decision-making bodies in favour of more environmentally sustainable approaches to the global environment.

Conclusion

The challenges posed by phenomena such as climate change certainly call for the scientific research and technological innovations that will enable us to develop appropriate physical and socio-economic modifications to our environment. However, true resilience may also require more fundamental changes in our modes of living (see for example Homer-Dixon, 2006). We argue that the readiness to make the necessary modifications, and to fundamentally change our common patterns of living, may be best understood as dependent on the complex that we refer to as *culture*. In a world of intensifying environmental risks, it is of fundamental importance that we determine the ways in which human values and practices are mediated by ideas about the relationship between humanity and the wider 'natural' environment, and the processes through which adaptive (or maladaptative) cultural patterns come about, are fundamentally important steps that can complement the development of technological, engineering and managerial coping and adaptation strategies. Cruikshank encapsulates this perspective in her statement that 'our human ability to come to terms with global environmental problems will depend as much on human values as on scientific expertise' (Cruikshank, 2001, p. 390).

References

Asian Development Bank 2005. *Climate Proofing: A Risk-Based Approach to Adaptation,* Pacific Studies Series. Madaluyong City: Asian Development Bank.

Bromley, D. W. 2006. *Sufficient Reason: Volitional Pragmatism and the Meaning of Economic Institutions.* Princeton: Princeton University Press.

Brooks, N. 2004. *Drought in the African Sahel: Long-Term Perspectives and Future Prospects,* Working Paper No. 61. Norwich: Tyndall Centre for Climate Change Research, University of East Anglia. Available at www.tyndall.ac.uk.

Brooks, N. 2006. 'Cultural responses to aridity in the Middle Holocene and increased social complexity', *Quaternary International* **151**: 29–49.

Bulkeley, H. and Mol, A. P. J. 2003. 'Participation and environmental governance: consensus, ambivalence and debate', *Environmental Values* **12**: 143–154.

Charney, J., Stone, P. H. and Quirk, W. J. 1975. 'Drought in the Sahara: a biogeophysical feedback mechanism', *Science* **187**: 434–435.

Conklin, A. L. 1997. *A Mission to Civilize: The Republican Idea of Empire in France and West Africa, 1895–1930.* Stanford: Stanford University Press.

Cooper, F. 1997. 'Modernizing bureaucrats, backward Africans, and the development concept', in Cooper, F. and Packard, R. (eds.) *International Development and the Social Sciences.* Berkeley: University of California Press, pp. 64–92.

Cooper, F. and Packard, R. (eds.) 1997. *International Development and the Social Sciences*. Berkeley: University of California Press.

Cruikshank, J. 2001. 'Glaciers and climate change: perspectives from oral tradition', *Arctic* **54**: 377–393.

Cruikshank, J. 2002. 'Nature and culture in the field: two centuries of stories from Lituya Bay, Alaska, Knowledge and Society', in de Laet, M. (ed.) *Research in Science and Technology Studies: Knowledge and Technology Transfer* No. 13. Amsterdam: JAI and Elsevier Science, pp. 11–44.

Davies, H. 2002. 'Tsunamis and the coastal communities of Papua New Guinea', in Torrence, R. and Grattan, J. (eds.) *Natural Disasters and Cultural Change*. London: Routledge, pp. 28–32.

Ebrahim, A. 2001. 'NGO behavior and development discourse: cases from Western India', *International Journal of Voluntary and Nonprofit Organizations* **12**: 79–101.

Fagan, F. 2000. *Floods, Famine, and Emperors: El Niño and the Fate of Civilization*. New York: Basic Books.

Freud, S. 1930/2002. *Civilisation and its Discontents*, translated D. McLintock. London: Penguin.

Fromm, E. 1955/2002. *The Sane Society*. London: Routledge.

Gardiner, S.M. 2004. 'Ethics and global climate change', *Ethics* **114**: 555–600.

Gray, J. 1995. *Enlightenment's Wake*. London: Routledge.

Gray, J. 2007. *Black Mass: Apocalyptic Religion and the Death of Utopia*. London: Allen Lane.

Gössling, S. 2003. 'Market integration and ecosystem degradation: is sustainable tourism development in rural communities a contradiction in terms?', *Environment, Development and Sustainability* **5**: 383–400.

Heyd, T. (ed.) 2005. *Recognizing the Autonomy of Nature: Theory and Practice*. New York: Columbia University Press.

Heyd, T. 2007. *Encountering Nature: Toward an Environmental Culture*. Aldershot: Ashgate.

Hill, A.G. 1989. 'Demographic responses to food shortages in the Sahel', *Population and Development Review* **15**: 168–192.

Hobbes, T. 1651/1985. *Leviathan*. London: Penguin.

Homer-Dixon, T. 2006. *The Upside of Down: Catastrophe, Creativity and the Renewal of Civilization*. Toronto: Knopf Canada.

Horigan, S. 1988. *Nature and Culture in Western Discourses*. London: Routledge.

Ingold, T. (ed.) 1994. *Companion Encyclopedia of Anthropology, Humanity, Culture and Social Life*. London: Routledge.

IPCC 2007a. *Climate Change 2007: Synthesis Report. Contribution of Working Groups I, II and III to the Fourth Assessment Report of the Intergovernmental Panel on Climate Change*. Geneva: IPCC.

IPCC 2007b. Parry, M.L., Canziani, O.F., Palutikof, J.P., Van der Linden, V.J. and Hanson, C.E. (eds.) *Climate Change 2007: Impacts, Adaptation and Vulnerability. Contribution of Working Group II to the Fourth Assessment Report of the Intergovernmental Panel on Climate Change*. Cambridge: Cambridge University Press.

IPPR 2005. *The Commission on Sustainable Development in the South East*, Final Report. London: Institute for Public Policy Research.

Jaeger, C., Dürrenberger, G., Kastenholz, H. and Truffer, B. 1993. 'Determinants of environmental action with regard to climatic change', *Climatic Change* **23**: 193–211.

Jamieson, D. 2001. 'Climate change and global environmental justice', in Edwards, P. and Miller, C. (eds.) *Changing the Atmosphere: Expert Knowledge and Global Environmental Governance*. Cambridge: MIT Press, pp. 287–307.

Kaufmann, S. A. 1995. *At Home in the Universe: The Search for the Laws of Self-Organization and Complexity*. Oxford: Oxford University Press.

Kaufmann, S. A., 2000. *Investigations*. Oxford: Oxford University Press.

Lamprey, H. F. 1975. *Report on the Desert Encroachment Reconnaissance in Northern Sudan*, 21 Oct. to 10 Nov. Paris: UNESCO/UNEP.

Ledoux, L., Cornell, S., O'Riordan, T., Harvey, R. and Banyard. L. 2005. 'Towards sustainable flood and coastal management: identifying drivers of, and obstacles to, managed realignment', *Land Use Policy* **22**: 129–144.

Leroy, S. 2005. 'Rapid environmental changes and civilisation collapse: can we learn from them?', paper presented at *Rapid Landscape Change and Human Response in the Arctic and Sub-Arctic Conference*, 15–17 June 2005, Whitehorse, Canada.

Leroy, S. 2006. 'From natural hazard to environmental catastrophe: past and present', *Quaternary International* **158**: 4–12.

Macnaghten, P. and Jacobs, M. 1997. 'Public identification with sustainable development: investigating cultural barriers to participation', *Global Environmental Change* **7**: 15–24.

Maturana, H. and Varela, F. 1973/1980. 'Autopoiesis and cognition: the realization of the living', in Cohen, R. S. and Wartofsky, M. W. (eds.) *Autopoiesis and Cognition: The Realization of the Living*, Boston Studies in the Philosophy of Science No. 42. Dordrecht: Reidel, pp. 59–138.

Merchant, C. 1980. *The Death of Nature*. San Francisco: Harper.

Millennium Ecosystem Assessment 2005. *Ecosystems and Human Well-Being: Synthesis*. Washington, DC: Island Press.

Nature (Editorial) 2007. 'The heat is on', *Nature* **450**: 319.

Ostrom, E. 2005. *Understanding Institutional Diversity*. Princeton: Princeton University Press.

Pattberg, P. H. 2007. *Private Institutions and Global Governance: The New Politics of Environmental Sustainability*. Cheltenham: Elgar.

Peterson, A. L. 2001. *Being Human: Ethics, Environmnent and Our Place in the World*. Berkeley: University of California Press.

Prigogine, I. and Stengers, I. 1984. *Order Out of Chaos: Man's New Dialogue with Nature*. Toronto: Bantam.

Roberts, N. 1998. *The Holocene: An Environmental History*, 2nd edn. Oxford: Blackwell.

Shapiro, J. 2001. *Mao's War against Nature*. Cambridge: Cambridge University Press.

Sheets, H. and Morris, R. 1976. 'Disaster in the desert', in Glantz, M. H. (ed.) *The Politics of Natural Disaster: The Case of the Sahel Drought*. New York: Praeger, pp. 25–76.

Steinberg, F. 2007. 'Housing reconstruction and rehabilitation in Aceh and Nias, Indonesia: rebuilding lives', *Habitat International* **31**: 150–166.

Swift, J. 1977. 'Sahelian pastoralists: underdevelopment, desertification, and famine', *Annual Review of Anthropology* **6**: 457–478.

Thébaud, B. and Batterby, S. 2001. 'Sahel pastoralists: opportunism, struggle, conflict and negotiation: a case study from eastern Niger', *Global Environmental Change* **22**: 69–78.

Zeng, N., Ding, Y., Pan, J., Wang, H. and Greggs, J. 2008. 'Climate change: the Chinese challenge,' *Science* **319**: 730–731.

18

Adapting to an uncertain climate on the Great Plains: testing hypotheses on historical populations

Roberta Balstad, Roly Russell, Vladimir Gil
and Sabine Marx

Theoretical context

We begin this chapter with two questions: first, how do people adapt to climate uncertainty? Second, can the historical past serve as a laboratory for testing and understanding human responses to climate uncertainty? We examine these questions in the context of a study of adaptation to climate variability in a county in the northern Great Plains of the United States during and after its initial agricultural settlement in the late nineteenth century. The purpose of the study is to understand how a newly arrived population of European immigrants and European-origin settlers from the Eastern United States adapted to the harsh and uncertain climate conditions on the Great Plains. But rather than constructing a historical narrative on how climate influenced the settlement experience, we will instead examine current theory on decision-making related to adaptation to climate uncertainty and study how the settlement experience in the Great Plains can illuminate and contribute to this body of theory.

There is a large body of research on decision-making that can be related to human adaptation to uncertainty. It deals with such topics as the need for predictability and control, overconfidence in judgements, the role of available or recent and thus easily predictable models in anticipating future events, risk communication and management, and others (Weber, 2006; Marx et al., 2007). Here, however, we will focus specifically on three aspects of the corpus of research on decision-making under uncertainty: (a) patterns of information processing; (b) the finite pool of worry; and (c) the bias toward a single action in response to uncertainty. These ideas have been the focus of research by social scientists in the Center for Research on Environmental Decisions (CRED) at Columbia University, funded by the US National Science Foundation.

Adapting to Climate Change: Thresholds, Values, Governance, eds. W. Neil Adger, Irene Lorenzoni and Karen L. O'Brien. Published by Cambridge University Press. © Cambridge University Press 2009.

Research on information processing suggests that individuals tend to process information (and respond to risky situations, including climate uncertainty and change) via two different parts of the brain. These are the affective processing system, which is the source of emotions and experiences associated with emotions, and the analytic processing system, which is capable of dealing with what is traditionally deemed scientific information – abstract, statistical and probabilistic (Marx et al., 2007). Affective or experience-based reasoning is based on one's own experience or that of others. It can involve a strong emotional response to what is often vivid presentation or description. Analytic reasoning involves the evaluation of information obtained from scientists or others who are in positions of authority or hold authoritative credentials, and frequently can be statistically represented. It often has less impact on individual decision-making than experiential or anecdotal information. These two types of information processing often interact in decisions about climate and adaptation to climate. Marx et al. point out that experiential information processing is generally dominant in decision-making under uncertainty because it produces output faster. When the output of the two systems is in conflict, the affective system is more likely to determine the behaviour. However, analytical information processing can often moderate the affective and experiential responses to risk, especially when a problem calling for a decision is discussed in a group.

Social scientists have also found that decisions related to risk management are largely driven by worry, and that individuals' perceptions of specific problems change as worry about one type of risk increases or decreases (Linville and Fischer, 1991). Linville and Fischer showed that if two negative events occur during the same time period, they must share the loss-buffering resources available during that time period. Although this line of research has been widely applied in organizational psychology, consumer psychology, behavioral decision-making and to some extent in the health sciences, the application of the concept to climate-related decisions is fairly recent. In their study of Argentine farmers facing climate, political and economic risks, Hanson et al. (2004) found that when individuals are confronted with new worries (or multiple worries simultaneously), there is a tendency to focus on a single area of concern and pay less attention to other worries, even though there has been no change in the risks previously perceived as high and worrisome. The phenomenon is described as the 'finite pool of worry' because they found that a single, significant worry has a tendency to dominate and crowd out other worries in the 'pool'. Because increases in worrying about one problem can lead to diminished concern about other problems, there is a limit to the number of problems related to adaptation to climate variability and uncertainty that individuals will choose to deal with at any point in time (Weber, 2006). It also suggests that by focusing on a single worry, individuals tend to reduce the complexity of the spectrum of problems that they face.

Another suboptimal response to risk that is related to the role of affect and which has a similar tendency to focus and simplify individual decision-making and adaptation to climate variability is the 'single action bias'. Weber (1997) found that individuals responding to an external threat tend to respond in terms of a single action, even when it is in their best interest to adopt a more diverse or multifaceted approach. This study of farmers in the Midwest found that they tended to adapt to climate variability in a single way, such as changing production practices, altering pricing practices or seeking government intervention, but never engaged in more than one of these practices. Similarly, Argentine farmers who were able to store grain on their farms did so and as a consequence were less likely to adopt multiple or additional safety measures, such as irrigation or crop insurance, than those who had little or no capacity to store their grain (Hanson et al., 2004). This too suggests that individuals may be satisfied with adapting to climate risks in a narrower and more focused way than the situation warrants. The same phenomenon has been observed in medical diagnostics where radiologists searching x-rays for lesions stopped their search after discovering one lesion, leaving additional abnormalities unnoticed (Berbaum et al., 1991). These examples illustrate the tendency in disparate settings for humans to engage in only one response to threat and then to take no further action because their feeling of worry or vulnerability has been reduced.

The rationale for historical research

The rationale for conducting historical research on climate impacts and responses to these impacts goes beyond understanding the past for its own sake (Endfield, 2008). Much of the existing social science theory on climate adaptation and decision-making under uncertainty, including research on patterns of information processing, the finite pool of worry, and the single action bias, is the product of either laboratory studies or behavioural observation, and sometimes both. What these approaches lack, however, is the perspective that can be obtained through examining adaptive behaviour and linking perceptions, experiences and behaviour over long time periods and in the context of complex and often interacting economic, technological, policy and climate systems. Climate adaptation decisions are always made in a temporal context, as well as in meteorological, political, economic and cultural contexts. The temporal context may influence adaptation decisions in regard to the sequence and timing of influences on decision-making or to variations in lag times between the perception of external problems and individual responses. Although the single action bias and the finite pool of worry are most evident in the immediate context of a specific decision, they are shaped, in part, by the decision-maker's previous experiences and assessments of the decision setting.

Studies of behaviour in contemporary laboratory settings cannot provide the temporal perspective needed to understand the cumulative role of time and experience on decision-making. However, historical studies, because they can focus on behaviour over long time periods, can provide valuable insight into these influences on decision-making.

A related but different issue is whether historical data and other records of past behaviour can be analysed in terms of theory developed in the contemporary laboratory. If the goal of research is to identify regular patterns of individual and social behaviour so as to understand, evaluate or even anticipate future behaviour, it is reasonable to test the underlying theory against behaviour at multiple time periods and in multiple places where conditions may be similar or may vary in theoretically useful ways. This type of research follow-up is usually costly, however, and consequently is rarely done. If historical research on real (historical) populations can be combined with, or even substituted for, interviews or experiments using contemporary populations, it could advance decision theory related to climate adaptation and, in Popper's words, help determine whether theory can 'stand up to the demands of practice'. In both cases, the overall cost of research may be reduced and the epistemological benefits increased (Popper, 1934).

Historical background

The rapid settlement of the trans-Mississippi West after the US Civil War in the 1860s was encouraged by the federal government, which systematically negotiated reductions in the amount of land it had earlier ceded to the Native Americans. It then initiated sweeping new land and transportation policies affecting the use of these lands. As a result, the transformation of this region from a largely uninhabited rangeland to a settled checkerboard of towns, farms and eventually cities was especially rapid and became one of the defining elements of the country's history (Turner, 1893). Because of its limited rainfall, the Great Plains were characterized by rolling expanses of treeless grasses which had the advantage of not requiring deforestation before the land could be planted in commercial crops. There was from the start some controversy as to whether the Plains were suitable for agriculture. Among the early voices was that of John Wesley Powell, who explored the area for the US Geological Survey and declared that irrigation was necessary if agriculture on the Plains were ever to be productive (Powell, 1878; Morris, 1926). Arrayed against his cautionary advice were the railroad companies, which had been granted land by the government in exchange for extending the railroads and which advertised widely about the benefits of settling in the American West. They sold both town plots and farm land to would-be settlers. Those who wished to obtain their land at lower cost registered claims with the federal land office under the Homestead Act, passed in 1862, which provided up to 160 acres of surveyed, unclaimed land to individuals in exchange for occupying the land for five years, paying minimal registration fees, and making certain improvements on the land (Gates, 1968). Consequently, the cost of entry for farming on the Great Plains was low, and would-be farmers were attracted to the territory from nearby states, from the eastern USA and Canada, and from northern Europe and the British Isles. What they did not know was that although the cost of entry was low, the cost of staying could be high.

This study focuses on Kingsbury County in the eastern part of the Dakota Territory an area opened to settlement near the end of the 1870s. This was originally the northernmost part of the Louisiana Purchase from France in 1803. It was formalized as a territory in 1861 in anticipation of white settlement, and it was subdivided into two states, North and South Dakota, in 1889. Workers on the railroads were the first to arrive in the area that became Kingsbury County, and farmers and merchants followed soon after the railroad tracks were laid. Although the popular image of the frontier settler is that of a farmer engaged in subsistence agriculture, the settlers of eastern South Dakota were often both townspeople and commercial farmers from the start. The railroads were a commercial lifeline, bringing settlers, food and supplies into the towns of the Territory and leaving with the fruits of farmers' labour.

The settlement of Kingsbury County, like most of the Dakota Territory, was effected by the combined efforts of the territorial government, the railroad companies, and thousands of land-hungry, would-be settlers who were attracted to the newly opened agricultural lands in the west. These settlers obtained much of their information about the area from word of mouth or through publications prepared by the territorial government and the railroads that were designed to attract settlers. The Dakota Territorial government had an active publication program under the supervision of the Commissioner of Immigration that sent documents describing the fertility of Dakota soils and the Territory's benign climate to readers in many countries in the language of that country. These documents obviously were meant to counter what were recognized as negative impressions of the region. In 1887, for example, the Commissioner of Immigration published a statement on the climate of Dakota. It began with the statement: 'Scarcely anything connected with Dakota is the subject of greater misconception than its climate' (Commissioner of Immigration, 1887). The climate of the Territory was described to potential immigrants as 'the pure, exhilarating, healthful climate of Dakota'; readers were told that the visitor 'who has once drunk deep draughts of this prairie oxygen, is under the charmer's spell, and can never again content himself to live without the Territory' (Commissioner of Immigration, 1887).

By 1880, there were over 1000 new residents in Kingsbury Country, about 30% of whom had been born outside the United States. Many of those who were born in the USA, however, were themselves the children of immigrants. A sample of households in Kingsbury County in 1880 shows that roughly 60% of the settlers were born abroad or were raised in a household by parents who were born abroad. This suggests that the cultural background of settler households was more often that of European immigrants than of families who came to the Dakota frontier from other parts of the United States. The previous agricultural experience of these settlers was largely in areas quite different from the Great Plains.

Whether or not settlers experienced the exhilaration of the prairie oxygen predicted by official publications selling the Territory, the first white settlers, who generally came from the humid eastern United States or the rainy countries of northern Europe, faced agricultural conditions in the Dakota Territory for which their previous experiences had not prepared them. They had little or no experience farming in areas with rainfall as low or as variable as it was in the eastern part of the Dakota Territory. On average, the annual precipitation was sufficient to support agriculture, particularly in the eastern part of the state. However the precipitation was irregular, both annually and in terms of the distribution of rainfall during the growing season, when it was most needed for agriculture. Not only was the Dakota climate drier than the areas left by the settlers who flocked to the newly opened lands, but it had highly variable precipitation patterns and strong, hot winds that caused what soil moisture there was to evaporate, leaving it even drier than suggested by the annual rainfall levels. In the nineteenth century, the variability in rainfall raised the question of whether the Great Plains could support agriculture, a question that remains controversial today (Riebsame, 1991; Cunfer, 2005; Parton et al., 2007). An example of the extreme variability in precipitation in the region is the rainfall in the James River Valley of what is now South Dakota. In 1881, when the area was first being settled by agriculturalists, there was 40 inches of rain. In 1894, there was only 14 inches (Schell, 2004). Between these two points, there were years with adequate rainfall and years of drought (and years when the rainfall arrived during critical points in the growing cycle and years in which the seasonal distribution of rainfall hurt crops), but in general, rainfall was high in the early 1880s and low in the late 1880s and early 1890s (Kepfield, 1998). In almost all cases, immigrants to the area had learned to farm in more humid European and eastern North American climates.

The winter climate on the plains was also unlike anything most of the settlers had previously experienced. Publications intended for prospective settlers emphasized that Dakota was located on the same latitude as Harrisburg, Pennsylvania, and Cleveland, Ohio, and in Europe it was on the same latitude as France and Austria. The clear implication of this comparison was that the Dakota weather was similar to that of these other places. Government publications also emphasized that there was considerably less snow in Dakota than in the eastern states and implied that summer generally began shortly after the winter broke in March (Commissioner of Immigration, 1887). Not surprisingly, there was no discussion in these publications of the disastrous impact of Dakota blizzards on cattle, humans and even railroads, stopping the settlers' access to food and farm implements. In 1880, there was a blizzard that stopped the trains for over six months, leaving the newly arrived settlers in Kingsbury County with shortages of food, firewood and cattle feed. They ground seed, intended for planting in the spring, in coffee grinders to get flour for bread. Another disastrous blizzard took place in 1888. In both years, many settlers were lost in the swirling snow and died of exposure or hunger (Wilder, 1940; Laskin, 2004).

In addition to the annual and seasonal variations in the levels of rainfall and the continuing possibility of devastating winter blizzards, the Dakota settlers faced a variety of other unexpected climate-related events in their new homes. There were fires, sparked by dry conditions and spread by high winds, that swept across the plains, destroying crops in their way; tornados or cyclones that were strong enough to destroy buildings, including a brick church in DeSmet, the largest town in Kingsbury County; floods that followed winters with heavy snows; plagues of grasshoppers in the 1870s and as late as 1880 that totally destroyed crops and other vegetation in their path; lightning strikes; and destructive hail (Crothers, no date; Robinson, 1904). As a settler in a story by Willa Cather, who grew up on the plains, describes the climate, 'He had seen it smitten by all the plagues of Egypt. He had seen it parched by drought, and sogged by rain, beaten by hail, and swept by fire, and in the grasshopper years he had seen it eaten as bare and clean as bones that the vultures have left. After the great fires he had seen it stretch for miles and miles, black and smoking as the floor of hell' (Cather, 1896). The severity of these weather events, combined with the difficulty in predicting them, provides a research setting for looking at behavioural responses to unexpected climate changes.

Climate adaptation and decision-making

We look at the impact of these weather events on the Great Plains by examining two types of behaviour in the initial settler population: first, residential persistence over time in Kingsbury County, and second, economic investment in the County through land ownership. We have taken a systematic sample of the population of heads of household in the County in 1880, using the manuscript schedules of the Federal Census for that year in order to trace evidence of these individuals in other sources over time to determine who stayed and to infer why. Information on individual land ownership in the nineteenth century is available through the Bureau of Land Management. One of the advantages of focusing on Kingsbury County is that the events of the early settlement period, including climate events and their impacts on the settler population, have been chronicled in great detail by observant residents of the county from 1879 to 1894, including Laura Ingalls Wilder, author of ten books on the Great Plains, the best known of which are the 'Little House on the Prairie' books.

The basic conditions of settlement were similar for most people, whether they came from a few hundred or many thousand miles away. The commercial foundations for a thriving agricultural area were in place. As mentioned earlier, land for commercial agriculture was available to all essentially for free through homesteading or for purchase at competitive prices, and land in the railroad towns that sprang up as the farmers arrived, was also available for purchase. The climate was a surprise, and variable and extreme weather events were to pose adaptation problems for the settlers. But during the initial years of what was called the Great Dakota Boom (late 1870s), the increase in annual rainfall gave newcomers the impression that this was a fertile land (Schell, 2004). This increase in precipitation in the late 1870s was popularly attributed to the fact that the settlers were beginning to plow the prairies, and the saying 'Rainfall follows the plow', was heard frequently to explain the rainfall. The availability of sufficient water and land for the taking, and, because of improved economic conditions in the United States, the construction of railroads that provided access to eastern markets made the area attractive to many. The population of eastern Dakota increased rapidly in the decade following 1878 and millions of acres were registered as new homesteads each year, with the number of acres reaching its height in 1883 (Schell, 2004). In short, the result of federal land and transportation policies was to encourage commercial wheat farming in the Dakota Territories. Unfortunately, many of the newly arrived farmers took advantage of these policies to pursue a familiar type of agriculture that was less well suited to the territory than the cattle-grazing culture that preceded them on the land.

The decline in precipitation after 1888 created economic hardships for local farmers who saw their yields decline with the rainfall. The impact of lower rainfall was combined with irregular weather events, like hailstorms, that destroyed crops even more rapidly than drought (*Kingsbury County News*, 1888). By this time, many of the farmers had already taken out loans or mortgages for farm machinery. This debt intensified the economic impact of the weather events, and farmers were forced to delay payments when they lost their crops. By the late 1880s, local newspapers began to print lists of delinquent tax sales of land (*Kingsbury County News*, 1888). By the 1890s, the newspapers were printing lists of foreclosures and mortgage sales as well (*Kingsbury County Independent*, 1893).

Almanzo and Laura Wilder provide an illustration of adaptation to climate impacts in Kingsbury County over a long period of time. They had both moved to the County as soon as the lands were opened to homesteading in 1879, and they married in 1885. But after a severe hailstorm destroyed their thriving crops in the summer of 1886, they were forced to rethink their economic plan for the year. Rather than selling their wheat and oats as planned, they took out a mortgage on their homestead, rented their tree claim to another farmer, and harvested and sold wild hay on the Chicago market (Wilder, 1971). The Wilders responded to the destruction caused by the hailstorm with energy, inventiveness and good humour and cited the irony that it reversed the Irish proverb, 'The rich man gets his ice in summer and the poor man gets his in the winter.' The hail, whatever it had done to the crops, provided the poor Plains farmers with ice in summer.

The Wilders survived that year to face increasing drought, high winds and cyclones, fires and other difficulties in subsequent years. They diversified their farming, adding sheep, which did better in the dry years than the crops, lowered their living standards and tried various economic remedies. But conditions were difficult. Almanzo Wilder's brother and sister each had homesteaded property for nearly a decade, and in 1888 their lands were sold for delinquent taxes (*Kingsbury County News*, 1888; Anderson, 1985). Laura Wilder's parents, who had a homestead and also owned commercial property in the town of DeSmet, kept their land, perhaps because their commercial property provided an income somewhat independent of agriculture and less immediately tied to predictable climate. In 1894, the Wilders decided to leave Kingsbury County. In her diary of the trip to Missouri, where they finally settled permanently, Laura Wilder began: 'For seven years there had been too little rain. The prairies were dust...Crop after crop failed...The agony of hope ended when there was no harvest and no more credit, no money to pay interest and taxes; the banker took the land. Then the bank failed' (Wilder, 1962). The Wilders and their neighbours faced a double problem: an inhospitable climate where they lived and a nationwide economic panic that tied their hands economically. On the back of the wagon belonging to the family that accompanied them to Missouri was a hand-lettered notice, 'Rear Guard of Coxey's Army', linking their pilgrimage to a new home in the milder climate of Missouri to the protest march on Washington of 500 unemployed workers from Ohio (Anonymous, 1999).

Despite the difficulties that farmers like the Wilders and their families faced, a large proportion of those who moved to the County when it was first opened to settlement stayed, unlike settlers in western parts of the Dakota Territory where most of the settlers moved away within a few years. The next stage of this project will be to analyse the landowning and agricultural persistence patterns of the sample of settlers in 1880 to determine what might have influenced the decision to stay or to move elsewhere.

Ties between theory and history

We found that the settlers who came to the Great Plains had little or no direct experience with the climate or familiarity with scientific assessments of the land before they moved there. Their information came from the publications of the Territorial government or the railroads, both of which had an economic motive in attracting settlers to the Territory. An even stronger message than the descriptions of the benign climate was the promise of free land, which appealed directly to potential immigrants. Once they were living in the Territory, however, they experienced a climate that belied the 'milk and honey' messages that they had received from formal governmental and corporate sources. It was at this point that the settlers began to have their initial, first-hand experiences with the climate of the Plains. For most people, the climate made life difficult, but they adjusted to its extremes – until the climate threatened their ability to make a living. Then the settlers felt they had to decide whether to stay or to leave. The limit to adaptation, then, came when climate uncertainties and disincentives reached a threshold that interfered with critical economic aspects of their lives.

In terms of processing information, there were three rather than two patterns that can be observed among the Dakota settlers. Basic analytic information came from publications, laws and advertising about the feasibility of moving to the Territory. Second, experiential information about the value of land ownership and commercial markets for grain provided a critical push to the Plains. In settlers' decisions, the emotional desire for land far outweighed both the analytical information that they received on the climate and their own early experience of the harsh and variable climate of the Plains. But we found a third type of information processing that helped the settlers cope with the conditions in which they found themselves. This involved domesticating problems by fitting them into existing frames of reference with sayings or aphorisms. A typical example can be seen in Caroline Ingalls' response to her daughter's statement that the prairie was beautiful, but it seemed they had to fight it all the time. Ingalls responded, 'This earthly life is a battle … If it isn't one thing to contend with, it's another. It always has been so, and it always will be. The sooner you make up your mind to that, the better off you are' (Wilder, 1941).

The concept of the 'finite pool of worry' was based on observations of farmers in the North American Midwest and in Argentina. At first glance, it does not seem to apply to farmers in the Dakota Territory. In part, this is because the 'pool' of climate-influenced worries, such as drought, were often inseparable from other worries (mostly socio-economic). But during blizzards, floods or fires, climate events could be directly related to physical survival. The response of the Wilders, their families and friends, to these conditions took place on several levels. We found that their expectations (and worries) were fluid over time and involved both gradual adaptation to extreme conditions and rejection of living under these conditions at different times. The change from acceptance to rejection could be related to the successive impacts of climate events or to the timing and linkage of climate problems with economic problems.

A third element we examine is the existence of a 'single action bias' in adaptation to the climate variability on the Great Plains. Looking at the experiences of specific individuals in Kingsbury County in the face of periodic, unexpected and previously unexperienced climate-related crises, we find that the impact of climate events had the power to affect multiple aspects of the settlers' lives, including their financial security, their homes, their safety and their immediate economic livelihood. Observing the experiences of specific settlers on the Plains suggests that they responded to climate-induced problems with multiple strategies. This included strategies related to improving yields through mechanization and plowing more acres and improving profitability by shifting to new crops and markets. They also rented land to others and started new businesses in the railroad towns to supplement their income. In large part, this flexibility may be the result of their responses, not to climate events alone, but to climate events in the context of a complex and expanding economic system.

Conclusions

We began with two questions: how people adapt to climate uncertainty, and whether the historical past can serve as a laboratory for testing and understanding human responses to climate uncertainty. We have found that adaptation strategies tend to be closely related to economic conditions and opportunities and to the context framed by government policies. For the settlers of eastern Dakota Territory, the physical climate they experienced was a surprise, one with which they had little or no previous experience. Their adaptation to that climate was moulded by the availability of fungible resources and opportunities in other areas – human resilience, policy support, access to technology and the state of the larger economy.

Because climate is rarely isolated from socio-economic phenomena, adaptation to climate extremes is generally mediated and interpreted through the lens of economic, policy and technological resources and perceived opportunities. In essence, adaptation to climate impacts is influenced as much by social, economic and technological policies and possibilities as by the climate events themselves. Cultural backgrounds may also influence adaptation, but because of the levelling effects of social and economic opportunities and technology, culture does not appear to be the most important influence on adaptation in this system.

The second issue in this paper is whether historical analysis can provide a useful approach to understanding adaptive behaviour. The answer, we believe, is yes, but it is most useful as one of several analytical approaches to understanding climate adaptation. We conclude that historical data can be used to *analyse* adaptive strategies under specified conditions. They can also be used to *compare* behaviour observed in laboratory studies with that in real situations, and the findings can be used to feed ideas back into laboratory experiments. However, the lack of interactive data, and the impossibility of interacting directly with historic populations, makes it impossible to *test* specific experiments at the microscale in a way that parallels laboratory experiments. In the end, it is necessary to understand both individual behaviour and broad community response patterns if we are to understand adaptation to climate change. Historical approaches to climate adaptation can help us to do so.

References

Anderson, W. 1985. *A Wilder in the West*. Brookings: Reynolds Printing Co.

Anonymous. 1999. 'The Cooleys and the Wilders: friends, neighbors, traveling companions', in *Laura Ingalls Wilder Lore* **25**: 1–5.

Berbaum, K. S., Franken Jr, E. A., Dorfman, D. D., Miller, E. M., Caldwell, R. T., Kuehn, D. M. and Berbaum, M. L. 1998. 'Role of faulty visual search in the satisfaction of search effect in chest radiography', *Academic Radiology* **5**: 9–19.

Cather, W. 1896. 'On the divide', in Faulkner, V. (rev. ed.) 1970. *Willa Cather's Collected Short Fiction 1892–1912*. Lincoln: University of Nebraska Press.

Commissioner of Immigration. 1887. *Resources of Dakota*, Advance Sheet, p. 12.

Crothers, P. R. no date. 'Memories of a pioneer', unpublished ms, condensed from an undated account in the *Arlington Sun*, 1920s or 1930s.

Cunfer, G. 2005. *On the Great Plains: Agriculture and Environment*. College Station: Texas A&M University Press.

Endfield, G. 2008. *Climate and Society in Colonial Mexico*. New York: Wiley.

Gates, P. W. 1968. *History of Public Land Law Development*. Washington, DC: Public Land Law Review Commission.

Hanson, J., Marx, S. M. and Weber, E. U. 2004. *The Role of Climate Perceptions, Expectations, and Forecasts in Farmer Decision Making: The Argentine Pampas and South Florida*, IRI Technical Report 04–01. Palisades: International Research Institute for Climate Prediction.

Kepfield, S. S. 1998. 'They were in far too great want: Federal Drought Relief to the Great Plains, 1887–1895', *South Dakota History* **28**: 244–270.

Kingsbury County Independent, 23 February 1893.

Kingsbury County News, 14 and 28 September 1888.

Laskin, D. 2004. *The Children's Blizzard*. New York: Harpers.

Linville, P. W. and Fischer, G. W. 1991. 'Preferences for separating and combining events: a social application of prospect theory and the mental accounting model', *Journal of Personality and Social Psychology Bulletin* **60**: 5–23.

Marx, S. M., Weber, E. U., Orlove, B., Leiserowitz, A., Kranz, D., Roncoli, C. and Phillips, J. 2007. 'Communication and mental processes: experiential and analytic processing of uncertain climate information', *Global Environmental Change* **17**: 47–58.

Morris, R. C. 1926. 'The notion of a great American desert east of the Rockies', *Mississippi Valley Historical Review* **13**: 190–200.

Parton, W.J., Gutmann, M.P. and Ojima, D. 2007. 'Long-term trends in population, farm income, and crop production in the Great Plains', *BioScience* **57**: 737–747.

Popper, K. 1934. 'Scientific method', in Miller, D. (ed.) 1985. *Popper Selections*. Princeton: Princeton University Press.

Powell, J.W. 1878. *Report on the Lands of the Arid Region of the United States*. Washington, DC: US Geological Survey.

Riebsame, W.E. 1991. 'Sustainability of the Great Plains in an uncertain climate', *Great Plains Research* **1**: 133–151.

Robinson, D. 1904. *History of South Dakota,* vol. 1. DeSmet: B.F. Bowen.

Schell, H.S. 2004. *History of South Dakota*. Pierre: South Dakota State Historical Society Press.

Turner, F.J. 1893. 'The significance of the frontier in American History', *American Historical Association, Annual Report for 1893*: 199–227.

Weber, E.U. 1997. 'The utility of measuring and modeling perceived risk', in Marley, A.A.J. (ed.) *Choice, Decision, and Measurement: Essays in Honor of R. Duncan Luce*. Mahwah: Lawrence Erlbaum, pp. 45–57.

Weber, E.U. 2006. 'Experience-based and description-based perceptions of long-term risk: why global warming does not scare us (yet)', *Climatic Change* **77**: 103–120.

Wilder, L.I. 1940. *The Long Winter*. New York: Harper Collins.

Wilder, L.I. 1941. *Little Town on the Prairie*. New York: Harper Collins.

Wilder, L.I. 1962. *On the Way Home*. New York: Harper Trophy.

19

Climate change and adaptive human migration: lessons from rural North America

Robert McLeman

Introduction

When considering human capacity to cope with or adapt to stressful climatic conditions and events, an important consideration is that human populations are not stationary. During periods of climate-related stress, one possible response by members of a vulnerable household is to migrate away from the area at risk (McLeman and Smit, 2006). Currently, the popular discussion of climate change and migration tends to occur around themes of environmental refugees, forced migration and worst-case scenarios of future climate change impacts. While wholesale abandonment of many currently inhabited areas would indeed be a likely outcome in most worst-case scenarios, unless and until worst-case scenarios come to pass, the more likely and immediate outcome is that climate change-related migration will occur as one of a range of possible adaptations taken by exposed populations. It is therefore necessary to consider the ways in which vulnerability to climate change, adaptive capacity, migration and demographic change are connected, and consider where migration is situated within the broader range of adaptation strategies.

This chapter sets out by describing some general conceptual issues regarding the role of migration in the context of climate change adaptation. This is followed by the introduction of specific questions of how migration decisions are reached under conditions of climatic stress, the relative importance of migration vis-à-vis other possible forms of adaptation, and how the movement of households and individuals in and out of exposed regions affects the adaptive capacity of the community as a whole. Exploration of such questions is then carried out via empirical findings from two specific case studies of rural populations in North America. The chapter concludes with a discussion of how such cases provide new and pertinent insights that may advance our understanding of how migration is utilized as an adaptation

Adapting to Climate Change: Thresholds, Values, Governance, eds. W. Neil Adger, Irene Lorenzoni and Karen L. O'Brien. Published by Cambridge University Press. © Cambridge University Press 2009.

to climatic stresses, and how demographic change through changing migration patterns has feedback effects on the adaptive capacity of communities.

Migration as an adaptation to climate change

It has long been recognized that changes in environmental conditions can influence human migration patterns and behaviour, but the specific relationship between natural and human processes in this context remains poorly understood (McLeman and Smit, 2006). While migration is one of any number of potential adaptive strategies a household experiencing stressful climatic conditions may employ, it is rarely the adaptation of first resort. This is because even over relatively short distances or for brief periods of time, migration entails direct costs of movement or transportation to a new destination, and of finding accommodation and an income or livelihood opportunity at the destination. There may also be transaction costs associated with crossing jurisdictional boundaries (legally or otherwise), as well as opportunity costs to the household the migrant leaves behind resulting from lost labour and social support. As a result, large-scale migrations, especially over long distances, are not typically stimulated by modest or short-term fluctuations in climatic conditions. Historically, large climate-related migrations have tended to be stimulated by catastrophic extreme events, such as cyclonic storms or flooding, and adverse conditions that are experienced on a frequent or ongoing basis, particularly drought (see McLeman and Smit, 2006 for examples).

The forms that climate-related migrations take vary from one event to another, and a number of different migration responses may emerge from a single event. In dryland areas of West Africa, for example, complex migration patterns have emerged as rural populations adapt to both seasonal variability in precipitation and extended periods of drought (Swinton, 1988; Findley, 1994; Nyong and Kanaroglou, 1999; Rain, 1999; Mortimore and Adams, 2001; Roncoli et al., 2001; De Haan et al., 2002; Hampshire, 2002). In many sedentary agricultural communities, dry-season migration to regional urban centres by young men and, in some cultural groups, also by young women has become commonplace. During extremely dry periods, children may be moved out of drought-stricken areas to the homes of extended family members. Pastoralist populations modify their movements during extreme dry periods by moving larger numbers of animals into more southerly (wetter) parts of their range for longer periods of time, a practice that may bring them into conflict with sedentary farmers (Nyong et al., 2006; Brown et al., 2007). It is often in the period following a *favourable* growing season that young men attempt long-distance migration abroad, taking advantage of the temporary surplus in household resources – behaviour that is somewhat at odds with the popular supposition that inter-regional climate-change-related migration will be greatest under conditions of distress.

This simplified description of West African migration is offered merely to highlight the complexity of potential migration responses to climatic stresses, responses that are nested within the much larger range of possible adaptive responses households might employ. Such complexity is not unique to West Africa, but has been identified elsewhere (for example Meze-Hausken, 2000; Adger et al., 2002; McLeman, 2006). Given that climate change will likely increase the climate-related stresses faced by many populations, greater research is warranted into the socio-economic and political conditions and processes that give rise to differential adaptive migration. There are numerous specific research questions for which there currently exists limited information in scholarly literature. Why do some households (or individuals within households) adapt through migration while others adapt in other ways? What socio-economic conditions predispose some groups to migration as adaptation? How are migration destinations selected, and what influences the duration of migration? How does climate change feed into existing climate-related migration behaviour and how may it influence new migration trends? In what ways does migration affect the adaptive capacity of source and receiving communities?

While migration is usually a household-level adaptive response, in tackling such questions it is also important to recognize how adaptive migration is influenced by governance structures and processes, the nature of social networks within and between communities, and differential household access to capital in its various forms. Answering these and related questions inevitably requires giving a watchful eye to forces that work across societal, spatial and temporal scales.

The remainder of this chapter describes and compares two case studies of rural populations in North America that have experienced or are experiencing significant social and economic stress due to rapidly changing climatic conditions. In each case, discussion is based on findings from empirical research specifically designed to develop insights into questions such as those listed above. The first case considers household adaptation decision-making and migration outcomes during a period of extreme drought in Eastern Oklahoma in the 1930s. Specific findings from this multi-year research project have been reported elsewhere (McLeman 2006, 2007; McLeman et al., 2007); what follows below represents a synthesis of the project as whole and what it says about the role of migration vis-à-vis other forms of adaptation. The second case is based on very recent findings from a study of current adaptation needs in Addington Highlands, Ontario, an area of small communities where the economy is dominated by seasonal, resource-based activities. While this population has historically demonstrated a strong capacity to adapt to extreme climatic conditions, demographic changes due to migration flows in and out of the area are rapidly eroding the overall adaptive capacity of the population.

Drought adaptation and migration in rural Eastern Oklahoma in the 1930s

Between 2004 and 2007, empirical research was carried out in rural Eastern Oklahoma (Figure 19.1) to understand the adaptation strategies and migration behaviour during a period of extreme drought in the mid-1930s. This research included the systematic collection of secondary reports, census data, government agricultural and economic reports, meteorological data, and the identification and interviewing of survivors of that period, including inter-state migrants who presently reside in California and non-migrants who reside in Oklahoma (McLeman, 2006, 2007; McLeman et al., 2007).

Eastern Oklahoma has a generally rolling landscape that becomes increasingly rugged as one moves toward the Arkansas border. In the 1930s, the region had a predominantly rural population. Farms tended to be small, operated by tenants with modest amounts of capital. Heavy work was done with draft animals. Most farms produced a mixture of cotton, a cash crop, and corn, a feed crop for livestock. By the mid-1930s, Eastern Oklahoman farmers were coping with a harsh combination of climate- and non-climate-related stresses. Commodity prices had been falling steadily for many years, a decline accelerated by the Great Depression that began in 1929. After several years of generally below-average precipitation, in 1934 and 1936 most of Eastern Oklahoma experienced severe drought conditions during the growing season, destroying crops and making water for livestock and draft animals scarce (Figure 19.2). In 1935 and 1937, severe summer rainstorms led to flooding in many counties.

Three general migration patterns emerged in Oklahoma during the 1930s, each of which was closely interconnected with the drought adaptation of rural households in Eastern Oklahoma. The first of these, which began with the onset of the

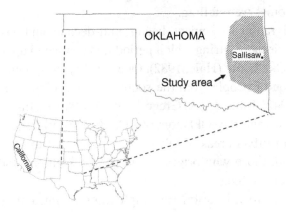

Figure 19.1 Location of Eastern Oklahoma study area.

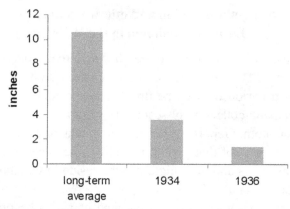

Figure 19.2 Annual precipitation, Sallisaw, Eastern Oklahoma.

Great Depression at the start of the decade, consisted of two groups moving into rural Eastern Oklahoma (McMillan, 1936; US House of Representatives, 1940; Hale, 1982; Whishenhunt, 1983). One group consisted of people displaced from the wage-labour economy of the region, particularly from urban centres and areas where oil production and mining took place. These sectors of the economy had contracted with the onset of the Depression, and many unemployed turned to subsistence farming to support their families. This back-to-the-land movement temporarily reversed a multi-decade decline in the proportion of Oklahoma's population living in rural areas. The other group consisted of migrants from semi-arid, grain-producing regions in western Oklahoma who had lost their farms through inability to meet mortgage or lease payments. Both groups were attracted to rural Eastern Oklahoma by its large number of small tenant farms, where a mix of cash-crop and subsistence farming could be taken up with modest amounts of economic capital. This in-migration had the effect of driving up demand for farmland in the area even as farm incomes were falling.

A second migration movement began mid-decade and lasted through the remainder of the decade, during which period it is estimated up to 300 000 people migrated out of Oklahoma (Hale, 1982). One-third of these joined the large inter-regional migration of people from the southern Great Plains states to California (McWilliams, 1942; Hale, 1982; Gregory, 1989). Two distinct sub-groups have been identified within the Oklahoma-to-California migrant stream: migrants who originated in urban areas of Oklahoma and tended to settle in urban areas in California; and, those who originated in rural Oklahoma and settled in rural California (Gregory, 1989).

The decade's third migration saw high rates of intra-state movements of people within rural areas and from rural areas to urban centres. Such movements reflected downward trajectories in the socio-economic status of rural households, as families tried to cope with the double stresses of economic recession and increasingly severe drought conditions. Few such moves resulted in improvements in the socio-economic lot of the migrant household (Tauber and Hoffman, 1937; McMillan, 1943). Eastern Oklahomans figured significantly in these latter two migrations.

The range of potential adaptations varied considerably among Eastern Oklahoman households struggling to adapt to this cocktail of climate- and non-climate-related stresses. Many possible forms of rural adaptation suggested in current climate change literature (Smit and Skinner, 2002) were either not available or had already been exhausted by the time the droughts struck in the mid-1930s. Most farmers had already maximized the potential benefits of modifying on-farm practices during their attempts to cope with tumbling commodity prices. Farmers had already long arranged corn and cotton plantings on their lands such that corn, the less drought-tolerant of the two crops, was planted on land with the highest potential soil moisture content. Although it is not very drought-tolerant, farmers could not switch away from corn because it was needed to feed livestock. Grains or other similarly drought-tolerant crops could not be grown profitably given the farm sizes and the quality of land in Eastern Oklahoma, and in any event, cotton was the cash-crop farmers knew and grew best.

Most farmers continued using the same methods and equipment that had been in use for decades; few attempted to innovate or adopt new technologies proven elsewhere. There are two straightforward reasons. First, since most farm operators did not own their land, few were willing to invest money or labour to improve it. Second, when crops failed due to drought, farmers received no revenue to invest in new technologies. This was an era before crop insurance programmes were implemented. Most lending institutions in the region had already gone out of business due to the Depression, and so farmers had few sources of credit beyond merchants, who charged prohibitive rates of interest.

Rural families attempted a host of other adaptations while trying to maintain themselves on the land. Friends and families pooled labour for heavy tasks such as bringing in crops or slaughtering hogs, which in non-drought years might have led to a casual labourer being hired. Some sold eggs and cream to raise cash; others engaged in riskier activities such as trapping skunks for pelts and operating illegal whisky stills. Households expanded their diets to consume plants and wild animals that were not regularly consumed in non-drought times.

As the range of available on-farm adaptations shrank, families increasingly sought income opportunities off the farm. Because non-agricultural sectors of the economy had contracted severely during the Depression, the supply of jobs was small and demand was considerable. Many looked for work on federal government programmes, such as infrastructure projects funded under the Works Progress Administration (WPA). The WPA enabled local governments across the USA, rural and urban, to apply for funding for projects to build or upgrade infrastructure. WPA projects provided hourly wages and paid residents to supply draft teams and trucks to work sites. Demand for WPA work was high, and participants rarely obtained more than 20 hours of work per week. In and of itself, income from WPA projects was insufficient to support a household, but such wages served as a critical income supplement for many rural families, allowing many to remain on the land.

Others looked beyond their immediate area. For decades, many Eastern Oklahomans had made it a practice to travel elsewhere in the region to work as casual agricultural labourers after their own crops had been harvested. Once harvests had been completed in those destinations, farmers would return home to ready their own farms for the next planting season. In the mid-1930s, however, traditional seasonal migration destinations were also suffering through extended droughts, with no employment to be had. This left the option of migration beyond the drought regions to find employment. California was the destination of choice for several reasons. California agriculture was characterized by large commercial farms that used extensive irrigation systems and were less sensitive to drought. While Californian farms were heavily mechanized in comparison with those of Eastern Oklahoma, there was then as now considerable demand for skilled farm labourers at critical stages of the growing season. This demand increased when the US government restricted seasonal immigration from Mexico on the pretext of protecting jobs for unemployed Americans.

This inter-regional differential in employment supply and demand does not in itself, however, explain why large numbers of Oklahomans moved west, for migration over such long distances was difficult, expensive and speculative. Migration to California was not a flood of destitute drought refugees. Although poor by Californian standards, those who migrated tended to be young, landless families who possessed some economic means and access to a motor vehicle, and transferable skills. They migrated as intact families, drawing on pre-existing networks of extended family and social connections in California for information and settlement assistance.

Although hundreds of thousands left drought-stricken Oklahoma during the 1930s, many more did not. Residents with economic capital invested in land ownership or business ownership tended to resist migration. Many of those who lost their farms or jobs due to drought-related crop failures and lacked the capital necessary to migrate out of the drought region fell into a downward socio-economic trajectory of poverty and transience. Thousands ended up living in conditions we today associate with distress migration in the developing world: squalid shack camps built along railways, riverbanks and on garbage dumps, with their attendant horrors of disease, child abuse and malnutrition.

In many cases, the migrants who left Oklahoma found themselves poor and landless in California, but there was little incentive to return to the economic and environmental disaster they left behind. Large, tight-knit communities of expatriate Oklahomans developed in central and southern California, communities that are still visibly identifiable to this day. Although initially received with hostility by many established Californian residents and their local governments, the migrants were overwhelmingly young, able, resilient families: the essential ingredients of successful communities. Many children of 1930s Oklahoman migrants today hold positions of economic and social influence in their Californian communities. Oklahoma's loss was California's gain.

Impacts of recent demographic change on community adaptive capacity: Addington Highlands, Ontario

A recently concluded study of current barriers, constraints and potential opportunities for climate change adaptation in Addington Highlands township of eastern Ontario, Canada (Figure 19.3), found that rapid demographic change caused by coincident flows of in- and out-migration is having significant impacts on the adaptive capacity of the population (McLeman and Gilbert, 2008). Addington Highlands is home to 2500 permanent residents living in a series of small communities linked by a single main highway. It is rough, forested country with many lakes and rivers used by vacationers, cottagers and seasonal residents. The area was cleared and settled in the late nineteenth century by family farmers who operated

Figure 19.3 Location of Addington Highlands, Ontario.

small, family farms in summer and worked in the logging industry in winter. Poor soils, a short growing season and distance from markets subsequently led to steady abandonment of intensive farming in all but the most suitable pockets of the township by the 1950s.

Today, those who live there year-round (referred to here as 'permanent residents') base their livelihoods on small-scale forestry, providing accommodation and services to visitors, or migrating away from the area for temporary periods to obtain wage employment. Household incomes are well below the national and provincial averages, even though the number of homes occupied by married couples is well above average. Employment within the township remains highly seasonal, and in many households the adults work a number of different jobs throughout the year. A significant number of households supplement wage income through subsistence activities such as hunting, trapping and fishing. Although just two hours' drive from the national capital, Addington Highlands lacks many services urban Canadians take for granted. There is no emergency medical clinic, no pharmacy and most residents have no access to a family doctor. There is no cell-phone service and, with the exception of two villages, no broadband service.

Through a combination of historical weather station data, indirect measures of climatic conditions (for example, historical snowpack records, maple syrup production records) and systematic collection of oral accounts from lifelong residents, significant changes in prevailing climatic conditions have been identified in this region. Winters are becoming shorter, less snowy and generally milder in temperature, and fewer periods of extreme cold occur. Spring arrives several weeks earlier than in the past. Average summer temperatures are becoming considerably less variable than in the past, and now hover consistently at the top end of their historical range. It is generally windier in summer, and extreme wind events are becoming more common (see McLeman and Gilbert, 2008 for more details on observed changes in this region). These observed changes have become manifest since the 1970s, and deviate significantly from observed trends since the late nineteenth century. It has not yet been established whether such trends are consistent with long-term variability in regional climate, are attributable to land use change in the region, or are the results of human-induced climate change. Sediment coring of lakes in the township is planned for 2009 in order to identify the likely causal origins of these observed climatic changes; but, whatever their origins, residents of Addington Highlands are presently obliged to adapt to them just the same.

These changes are having adverse impacts on the local economy. Lack of snow and shorter winters reduce the average number of snowmobiling and ice-fishing visitors to the township, two key generators of winter tourism revenues. Milder winter conditions also shorten the important winter season for local forestry, which is carried out when the ground and small waterways are frozen solid. Once the ground begins to thaw, foresters have difficulty moving equipment and logs through the bush, and local government restricts the maximum allowable axle-weight of vehicles to reduce road-surface damage (essentially precluding road use by heavily loaded logging trucks). The cost of maintaining and repairing roads, bridges and similar infrastructure is rising due to highly variable winter conditions that cause alternating freezing and thawing of materials. Warm and dry summers, continuous windiness and extreme windstorms increase the risk of damage to homes and property by falling branches, knock down electricity and telephone lines and lead to extreme dryness and consequent increased fire risks in the forest.

Even in the best of times, Addington Highlands has always been a hard place to earn a living, and so residents historically developed a robust capacity to cope with climate extremes that paralyse other Ontario communities. Households have historically been able to get along without electricity or telephones for extended periods, cope with closure of roads by extreme storms, hunt and fish for food, and extricate themselves and their neighbours from many emergencies. If a tree fell and blocked the road, one of the next drivers to come along invariably carried a chainsaw and began clearing the road; if a vehicle slid off a slippery road into the ditch, the next passing pick-up truck driver would break out a set of chains and pull the stranded vehicle out. Every home contained a log stove and months' worth of handsplit logs to supplement electrical and oil heaters and serve as an emergency replacement. Despite being geographically and economically marginalized from an increasingly urbanized Canadian population, a defining characteristic of the communities of Addington Highlands was a combination of ingrained self-reliance and unquestioned readiness and ability to assist fellow residents in time of need.

Recent demographic changes are rapidly eroding that capacity. Although the overall population level is the same as it was a century ago (countering the trend of steadily decreasing rural population numbers in Canada), the average age of the population has climbed to near 60. Where their parents' and grandparents' generations would often move away on a seasonal basis in search of temporary employment or make do by cobbling together several seasonal jobs, young residents now migrate away in large numbers to pursue higher education and advanced career opportunities, often never to return. The increasingly few who do remain often lack formal skills or training, and work at or near the minimum wage.

The overall population is stable because the out-migration of young, skilled individuals is being offset by in-migration of large numbers of retirees from urban areas, many of whom have been past summer visitors to the area. Those who 'retire to the cottage', as it is generally described, are typically urbanites who are wealthier and have higher levels of education than established residents. Given their age and past experiences, they tend to have different expectations and needs for services, and generally have a lower degree of self-reliance than established residents. They also have difficulty mixing socially with the permanent population because their economic, social and cultural backgrounds are quite different. This lack of social mixing is detrimental to both groups and to the community as a whole. Many new-comers have useful skills and expertise that are rare within the permanent popula-tion but are being underutilized, and any number of opportunities to enhance the economic and institutional infrastructure of the area is being missed. At the same time, the traditional knowledge of the natural environment and how to cope with its inherent variability, which has historically kept the vulnerability of the permanent population to climatic stresses extremes relatively low, is not being shared with the newcomers. As they age, the long-term residents are themselves increasingly less able to undertake many demanding physical activities upon which their historic self-reliance was based, such as hauling and splitting logs, digging out from heavy snows and operating machinery.

In aggregate, the population of Addington Highlands is quickly developing a higher degree of vulnerability to climate- and non-climate-related stresses than would exist if there were greater bridging of the two distinct social groups that increasingly make up their communities. A small number of service clubs and outdoorsmen's associations have been able to draw membership from both perma-nent residents and newcomers. One of the critical findings of the research study was that such organizations represent potential vehicles for rebuilding relationships of mutual support within the communities, for transferring skills and knowledge between groups, and for taking advantage of the expertise of newcomers to explore new potential directions for economic development and capacity-building that were not available previously. Should this bridging of social groups not improve, the combination of ongoing climatic stress and demographic change will continue to erode community adaptive capacity to the point where new out-migration and subsequent population decline will probably ensue.

Discussion and conclusions

There are a number of directions discussion of these case studies might take: I will concentrate here on three. The first is the question of the importance of migra-tion as an adaptation relative to other potential adaptations that may be under-taken by households exposed to climatic risks or stresses. In the first case study migration, particularly over great distances, was neither an adaptation of first nor last resort. For several years before the droughts hit, rural households in Eastern Oklahoma had been experiencing steady declines in income due to a broad-based economic contraction and falling commodity prices; but their means of adaptation tended to be those other than migration out of their home region. In fact, because the socio-economic configuration of rural areas was centered around small-scale, tenant-run agriculture, Eastern Oklahoma *attracted* migration from other areas, including from those to the West which had already begun experiencing drought. When the droughts came to Eastern Oklahoma, many possible forms of adaptation rural households might have otherwise used had already been implemented or had been rendered unavailable by the economic collapse. Increasing numbers of families were involuntarily displaced from the small tenant farms that had provided a large proportion of the rural population with livelihoods. It was at this point that large-scale migration of Eastern Oklahomans commenced.

Even so, the majority of Eastern Oklahomans did not migrate out of the region. Wherever possible, rural families attempted to access government assistance programmes as means of adapting. Migration to California was disproportionately undertaken by families with particular demographic characteristics. Landowning families typically did not migrate, the security of their tenure and the lack of portability of land as capital giving them few incentives to leave Oklahoma, regardless of the economic hardship. Those who quit the region for California and elsewhere were healthy young adults and their children, with transferable work skills, friends and family to help them settle in their chosen destination, and access to sufficient financial means to get there. Although poor, they were not absolutely destitute. Households or individuals that were destitute, or where members were elderly, infirm or unwell typically had fewer migration options than households that did not share these or similar traits.

I have elsewhere characterized this relationship in terms that household access to economic, social and/or cultural capital acts as a limiting factor on its range of potential adaptation options generally, and on the possibility of migration as a potential adaptation more specifically (McLeman and Smit, 2006). At any given point in time, a household's migration possibilities are a function of its available capital. Should adverse climatic conditions begin eroding the household's access to capital, its range of migration options, and especially the window for long-distance migration, may begin to close. This is seen in the Oklahoma case with the departure of landless young families before they fell into the final stages of absolute destitution and the social and physical debilitation that accompanies it.

The second point I wish to highlight is how the migration stimulated by Oklahoma droughts was so detrimental to the long-term success of the migrant source region and was an unexpected long-term benefit to the region that received the migrants. The tone in many current discussions about climate change and forced migration implies that adaptive migration is something that needs to be prevented. Such was the case in the 1930s. The government of California posted billboards along west-bound highways on the Great Plains to discourage migrants. The Los Angeles County Police actually instituted for several months a 'bum blockade' along the California–Arizona border, where officers attempted to intimidate and prevent would-be migrants from enter-ing the state, in spite of this being a blatantly illegal and unconstitutional act. John Steinbeck's 1939 fictional account of the migration, *The Grapes of Wrath*, which casts drought migrants in a sympathetic light and Californian and Oklahoman socie-ties in a less-than-favourable one, stimulated loud denunciations, library bans and occasional book-burnings in California and in Oklahoma.

It is true that newly arrived migrants often face many social, economic and cultural challenges as they attempt to establish themselves in a new place. It is also true that such arrivals will in the short run generate demands for services and resources that may not have existed before their arrival, although Black (1994) has documented that even the poorest of environmental migrants may integrate quickly and successfully if given freedom of mobility when they arrive at their destination. Oklahoma migrants did generate new demands for housing, schooling and health-care that had not previously existed when California's agricultural labour market was dominated by seasonal migrants from Mexico. But, within a few years these new arrivals from Oklahoma had successfully established themselves and begun moving up the socio-economic ladder. A lesson to be taken, therefore, is that when contemplating the potential for future migration due to climate change, scholars should continue to avoid value-laden judgements that might vilify those who adapt through migration.

My third and final observation relates to the transient nature of adaptive capac-ity, particularly when it is heavily dependent on social capital or individual physical well-being. In both cases discussed above, social capital emerges as an important factor in adaptation. The first case shows that strong localized concentrations of social capital facilitated adaptation by means other than migration; extended social networks and access to social capital in other places facilitated migration over long distances. The Addington Highlands case emphasizes how social capital is not static over time, and how demographic changes through migration can serve to erode or potentially enhance a community's social capital and consequently its adaptive capacity. Consistent with scholarship generated elsewhere, the Addington Highlands case demonstrates there is an important need to incorporate the creation and maintenance of community social groups, networks and activities within the context of climate change adaptation strategies and capacity-building. A simple message to be taken from this case is that robust and mutually supportive commu-nities can and will adapt to a tremendous spectrum of future climatic changes.

The Addington Highlands case further highlights how the adaptive capacity of communities changes over the life course of its members. The capacity to make a livelihood in the challenging natural environment of Addington Highlands has historically been based on physical robustness and skills that are acquired over time. As the young members of that community leave in increasing numbers and remaining members become older, the adaptive capacity of individual households and the community in aggregate has begun to decline. To stem this erosion requires innovation and to draw upon new reservoirs of social capital that arrive in the form of new arrivals from elsewhere. Migration, at least in this case, brings both losses and potential gains for the future adaptive capacity of the community.

It is an open question how universal the insights from these two cases may be or the extent to which they may be transferable to other regions around the world. I would suggest that many of the observations made here, particularly the emphases on the importance of social capital in adaptation, the recognition that climate-related migration is often undertaken by particular socio-economic groups, and the potential for erosion of adaptive capacity by demographic change, may resonate with researchers elsewhere. I would conclude by suggesting that researchers take care to avoid straight-line assertions that the impacts of climate change will lead to indiscriminate, large-scale population displacements as has occasionally been done in the past. The relationship between human migration behaviour and climatic conditions is a complex one, and future climate change will only serve to make it more so.

Acknowledgements

The research described in this chapter was supported financially by the Social Sciences and Humanities Research Council of Canada and Natural Resources Canada. Barry Smit is gratefully acknowledged for his support for much of the research described here.

References

Adger, W.N., Kelly, P.M., Winkels, A., Huy, L.Q. and Locke, C. 2002. 'Migration, remittances, livelihood trajectories and social resilience', *Ambio* **31**: 358–366.

Black, R. 1994. 'Forced migration and environmental change: the impact of refugees on host environments', *Journal of Environmental Management* **42**: 261–277.

Brown, O., Hammill, A. and McLeman, R. 2007. 'Climate change: the new security threat', *International Affairs* **83**: 1141–1154.

De Haan, A., Brock, K. and Coulibaly, N. 2002. 'Migration, livelihoods and institutions: contrasting patterns of migration in Mali', *Journal of Development Studies* **38**: 37–58.

Findley, S.E. 1994. 'Does drought increase migration? A study of migration from rural Mali during the 1983–1985 drought', *International Migration Review* **28**: 539–553.

Gregory, J.N. 1989. *American Exodus: The Dust Bowl Migration and Okie Culture in California*. New York: Oxford University Press.

Hale, D. 1982. 'The people of Oklahoma: economics and social change', in A.H. Morgan and Morgan, H.W. (eds.), *Oklahoma: New Views of the Forty-Sixth State*. Norman: University of Oklahoma Press, pp. 31–92.

Hampshire, K. 2002. 'Fulani on the move: seasonal economic migration in the Sahel as a social process', *Journal of Development Studies* **38**: 15–36.

McLeman, R. 2006. 'Migration out of 1930s rural Eastern Oklahoma: Insights for climate change research', *Great Plains Quarterly* **26**: 27–40.

McLeman, R., 2007. 'Household access to capital and its influence on climate-related rural population change: lessons from the dust bowl years', in Wall, E., Smit, B. and Wandel, J. (eds.) *Farming in a Changing Climate: Agricultural Adaptation in Canada*. Vancouver: University of British Columbia Press, Vancouver, pp. 200–216.

McLeman, R. and Gilbert, G., 2008. *Adapting to Climate Change in Addington Highlands: A Report to the Community*. Ottawa: University of Ottawa.

McLeman, R. A. and Smit, B. 2006. 'Migration as an adaptation to climate change', *Climatic Change* **76**: 31–53.

McLeman, R., Mayo, D., Strebeck, E. and Smit, B. 2007. 'Drought adaptation in rural Eastern Oklahoma in the 1930s: lessons for climate change adaptation research', *Mitigation and Adaptation Strategies for Global Change* **13**: 379–400.

McMillan, R. T. 1936. 'Some observations on Oklahoma population movements since 1930', *Rural Sociology* **1**: 332–343.

McMillan, R. T., 1943. *Migration and Status of Open-Country Families in Oklahoma*, Report No. T-19. Stillwater: Oklahoma Agricultural Experiment Station.

McWilliams, C. 1942. *Ill Fares the Land: Migrants and Migratory Labor in the United States*. Boston: Little, Brown.

Meze-Hausken, E. 2000. 'Migration caused by climate change: how vulnerable are people in dryland areas?', *Mitigation and Adaptation Strategies for Global Change* **5**: 379–406.

Mortimore, M. J. and Adams, W. M. 2001. 'Farmer adaptation, change and "crisis" in the Sahel', *Global Environmental Change* **11**: 49–57.

Nyong, A. O. and Kanaroglou, P. S. 1999. 'Modelling seasonal variations in domestic water demand in rural northern Nigeria: implications for policy', *Environment and Planning A* **34**: 145–158.

Nyong, A., Fiki, C. and McLeman, R. 2006. 'Drought-related conflicts, management and resolution in the West African Sahel: considerations for climate change research', *Die Erde* **137**: 223–248.

Rain, D. 1999. *Eaters of the Dry Season: Circular Labor Migration in the West African Sahel*. Boulder: Westview Press.

Roncoli, C., Ingram, K. and Kirshen, P. 2001. 'The costs and risks of coping with drought: livelihood impacts and farmers' responses in Burkina Faso', *Climate Research* **19**: 119–132.

Smit, B. and Skinner, M. 2002. 'Adaptation options in agriculture to climate change: a typology', *Mitigation and Adaptation Strategies for Global Change* **7**: 85–114.

Swinton, S. M. 1988. 'Drought survival tactics of subsistence farmers in Niger', *Human Ecology* **16**: 123–144.

Taeuber, C. and Hoffman, C. S., 1937. *Recent Migration from the Drought Areas*. Washington, DC: US Resettlement Administration, US Farm Security Administration, US Bureau of Agricultural Economics.

US House of Representatives 1940. *Testimony of Wheeler Mayo, Editor of Sequoyah County Times, Sallisaw, Okla*. Washington, DC: Select Committee to Investigate the Interstate Migration of Destitute Citizens, Sixty-Seventh Congress, Oklahoma City Hearings, pp. 2122–2128.

Whisenhunt, D. W. 1983. ' "We've got the Hoover blues": Oklahoma transiency in the days of the Great Depression', in K. E. Hendrickson (ed.) *Hard Times in Oklahoma: The Depression Years*. Oklahoma City: Oklahoma Historical Society, pp. 101–114.

Part III

Governance, knowledge and technologies for adaptation

20

Whether our levers are long enough and the fulcrum strong? Exploring the soft underbelly of adaptation decisions and actions

Susanne C. Moser

Give me a lever long enough,
and a fulcrum on which to place it,
and I shall move the world.

Archimedes (287 BC – c. 212 BC)

Introduction

Much of the focus in social science adaptation research in the climate and global change arena to date has focused on the international policy framework, national-level to place-based strategies in general, technological options and specific adaptation actions that can be employed by actors at any level of decision-making. By far the strongest emphasis has been on what potentially could be done, whether the necessary capacity for adaptation is in place, and – if wanting – how to enhance it (Smit et al., 2001; Adger et al., 2007). A much smaller number of investigators has looked at historical examples of successful or failed adaptation (from the cultural to local levels). Only in the last few years has there been a growing interest in the degree to which adaptation to anthropogenic climate change is already occurring, and if not, why not (see Adger et al., 2007). This line of work in particular, together with critical analyses of the socio-economic and political power dynamics and inequity issues underlying vulnerability and adaptive capacity (for example Adger et al., 2001, 2006) has brought us to the serious investigation of barriers and limits to adaptation that is the focus of this book and the conference preceding it.

A number of investigators have argued that an honest discourse and more critical examination of our true ability to adapt to climate change is warranted and

Adapting to Climate Change: Thresholds, Values, Governance, eds. W. Neil Adger, Irene Lorenzoni and Karen L. O'Brien. Published by Cambridge University Press. © Cambridge University Press 2009.

overdue, including for highly developed countries (for example Easterling et al., 2004; O'Brien et al., 2006; Pielke Jr et al., 2007; Moser and Luers, 2008; Adger et al., 2009). This urgent call stems from a number of relatively recent developments (several of which are also exemplified in chapters in this book), including:

- Concern with the fast pace of climate change (and the spectre of abrupt climatic shifts);
- A persistent and growing gap between the rich and poor, in any country;
- Recognition of high societal vulnerability to climate extremes even in developed countries (for example, the heat wave of 2003 in Europe, Hurricane Katrina in the United States in 2005, the extreme drought in Australia in 2007);
- A growing understanding of lags in social systems;
- Impatience with the rather slow response of national and local governments to climate change impacts to date (i.e., planned adaptation, preparedness);
- A critique (or at least questioning) within the scientific community of the almost exclusive emphasis on adaptive *capacity* while neglecting the question whether this capacity is actually being used and fully realized in actual adaptation actions.[1]

Factors commonly cited as determining the development or enhancement of adaptive capacity include the availability and equitable distribution of economic resources, technology, information and skills, infrastructure and institutions. Commonly, they vary widely across space, time, sectors and social groupings (Smit et al., 2001; Yohe and Tol, 2002; Adger et al., 2007). These factors can be viewed as constituting the *levers* one might apply to advance adaptation to climate change in practice, and this theme explores just two of them: knowledge and technology.

But what about the *fulcrum* – the support or pivot point on which the levers turn – which is deemed equally important in Archimedes' famous quote? This third theme also explores the governance structures and processes that underpin the use of adaptation levers, in short: the fulcrum. Some core questions to explore here include:

- In what ways do governance structures and processes matter in principle for the enhancement and realization of adaptive capacity?
- To what extent are governance structures and processes in any specific instance in place, functional and effective in supporting the use of economic, technological, policy and informational levers one could apply to affect change?
- When and how do governance structures enable and channel adaptive actions?
- When and how do governance structures and processes delay or render ineffective adaptive actions?
- When do these structures and processes function as (mutable) barriers or even as de facto (non-negotiable) limits to adaptation?

---

[1] This emphasis may well be an artefact of how the climate change community conventionally defines *vulnerability*, namely as a function of exposure, sensitivity and coping/adaptive capacity. Enhanced adaptive capacity (along with lowered exposure and sensitivity), thus, would make for less severe negative impacts or a greater ability to take advantage of potential positive impacts of climate change.

This chapter addresses some of these and related questions. Its primary goal is not to answer them, but to offer a systematic approach to examine governance issues in the context of adaptation decision-making, and to open up a wide enough thematic umbrella over the range of chapters that follow on this challenging topic.

A decision-centred, diagnostic approach

What do we mean then by 'governance'? Governance can be broadly conceived as the set of decisions, actors, processes, institutional structures and mechanisms, including the division of authority and underlying norms, involved in determining a course of action. One important implication of this definition is that governance is not just about institutions and clearly not restricted to public-sector actors, but involves all players involved in decision-making. A second important implication of this definition is that governance is constantly held in the dialectic tension between structure and agency. To do so, this perspective places decision-makers – as active agents embedded in particular institutional, normative, and political contexts – at the centre of governance. As a consequence the dynamics between those involved in decision-making, the qualities of the decision processes themselves, as well as the host of factors that influence decisions and decision-makers come into sharper focus. Figure 20.1 sketches the approach suggested here: the outermost area of influence on our decisions constitutes the values that are embedded and reflected in our culture, science, politics and the economy. In turn they determine the sectors and communities in which adaptation is considered, and influence the selection of organizations and institutions within which adaptation decisions get made by specific decision-makers and interested or affected stakeholders. These influences are

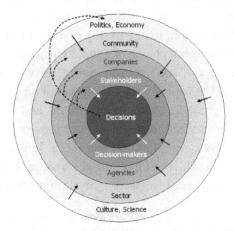

Figure 20.1 Layers of influence on adaptation decision-making.

enabling and constraining on governance decisions, which in turn influence future contextual circumstances.

One may begin then an exploration of governance as potential barrier to adaptation by identifying the arena (sector, industry, community, location, etc.) for which adaptation is being considered and, to the extent already known, more specifically, the (potential) adaptation decision(s) that may be made.[2]

In which arena are adaptation decisions to be made?

At the risk of pointing to the all-too-obvious, it is worth asking why we consider some sectors, industries or communities worthy of consideration for adaptation – and not others. Such choices can have very deep and difficult-to-disentangle roots. In the deepest sense, they are value-driven, i.e. reflective of what we value. (I refer the reader to the contributions under the second theme in this book for a fuller discussion.) These deeply held values may be formally and informally codified in social norms and expectations, enabling legislation, research programmes or management responsibilities. Equally importantly, they will colour the political and social context of adaptation governance. As a result, we can expect values to affect the relative importance different areas of interest gain via scientific investigation and assessment, and – underlying these – possible structural constraints such as available technical expertise, research funding, or disciplinary 'stove-piping' and blind spots (for a discussion of the importance of these science-related choices see for example Moser, 2005; Kiparsky et al., 2006). Choices as to what is and what is not being considered for adaptation planning and action may also result from (and result in) civic mobilization and public discourse (or the lack thereof), the presence or absence of a leader or champion, the occurrence of a focusing or distracting event (such as a disaster, an election or the passing of a law), prevailing economic goals and priorities and the selective valuation of certain aspects of our lives, or from political (lack of) attention, manoeuvring or even silencing.

These influences on the initial choice of adaptation arenas bound and sometimes constrain, in important ways, the decision context in which adaptation is being considered. They reflect the timing in the evolution of an issue domain, affect the information (about the problem, risks, opportunities and possible solutions) that is available for decision-making, determine the range of parties interested, and partaking, in adaptation decision-making, and reveal important aspects about the larger political context in which adaptation governance occurs. As we will see below, while easily ignored or dismissed as irrelevant context for the adaptation

---

[2] Note that the conscious or unconscious decision *not* to act toward greater preparedness for climate change impacts, *not* to plan ahead, or *not* to include any considerations of climate change in ongoing management and policy decisions is of equal relevance here. This is recognized here by the word 'potential'.

issues in question, these initial and often invisible choices deeply affect governance and thus the degree to which (normative, institutional, etc.) structures can enable or hinder effective adaptation actions (see also Edwards and Steins, 1999; Risbey et al., 1999; Adger et al., 2003; French and Geldermann, 2005). Once the adaptation arena has been identified and at least initially characterized, the door is now open to the identification of the institutions and people involved in making the decision.

Who initiates or has responsibility for developing adaptation options? Who has (potentially) decision-making and implementation authority?

The decision-making entity

The centrality of the decision-maker cannot be overstated, albeit not because it gives a priori primacy to agency over structure or because it allows an outside observer to assign responsibility (or potentially blame) to particular individuals for the (in)effectiveness of adaptation decisions. Rather, beginning with the decision-maker(s) provides a revealing diagnostic entry point into the structural governance context (i.e., the agency, company or organization: see Figure 20.1) that channels adaptation planning and implementation.

Key questions one may ask here to get at the structural context of adaptation governance include but are not limited to the following:

- In which government agency/ies or private-sector entity/ies are the decision-makers located who will be involved with the issue in question?
- At which level(s) of governance are they placed (for example, at a local or national level of government, at the top executive level in a multinational company)?
- Through what mechanisms is this level of decision-making linked to other levels, sectors, communities, and constituents that will be affected by the decision (for example, through top–down mandates, regulations, voluntary information-sharing mechanisms, unpredictable and inadvertent impacts on others)?
- What types of mandate and authority do the decision-making entities have (for example, strategic policy and planning, operational management and/or monitoring)? How far in space and time do their decisions reach (affected area, time horizons)?
- What laws, rules and regulations guide, demand and/or constrain these entities' actions, and in what ways may they allow, enable, support or hinder adaptation actions?
- In what ways may the decision-making entities be held accountable for their decisions (for example, legally, financially, politically), and by whom (voters, shareholders, current and/or future stakeholders, etc.)?
- What are the financial resources available for adaptation planning and implementation (for example, sources, levels, reliability and permanence of funding; are funds additional to, transferred from elsewhere to, or already included in, existing budgets)?

- For the specific decision-making entity in question, is there a specific mandate to engage in adaptation planning and action or is it a non-mandated initiative? What is the level of support – regardless of mandate – from the leadership, staff and constituents for such an effort?
- What is the general historic experience with the entity's/ies' functionality and accountability?

Answers to these and related questions shape the action space within which individual decision-makers can manoeuvre (see for example Klein and Möhner; Finan and Nelson; or Drieschova et al., Chapters 29, 21 and 24, this volume). How tight or flexible this action space ultimately is, however, is largely a question of the individuals acting within it. There is empirical evidence from coastal management in California, for example, where individual coastal planners and public works engineers are highly aware of and concerned with the risks of climate change, and even in principle ready and willing to tackle the adaptation question, but lack the financial and technical resources, staff, and leadership and guidance to begin the hard work of educating themselves about, planning for and dealing with the impacts of climate change (Moser and Tribbia, 2006/2007; Tribbia and Moser, 2008). Other evidence suggests that these structural constraints certainly serve as barriers but not necessarily as immutable limits to adaptation. Young (2007), for example, describes a variety of strategies, including networking, leadership support and creative financing which local-level government officials employed in cities around the United States to break through these perceived and real barriers to climate action.

Individual decision-makers

After sketching out the decision context (Q1) and circumscribing the action space within which actors work (Q2a), the diagnostic approach proposed here leads next to the decision-makers themselves (Figure 20.1).

At least until recently there have been almost no individuals anywhere in the private or public sectors of society with the 'luxury' of an exclusive directive or sole responsibility to deal with adaptation. And even if there are any, none would get to make adaptation decisions alone or in isolation of other non-climatic considerations, implement them single-handedly, or face the consequences of these choices by herself. Far more commonly, decision-makers hold responsibilities far greater than just for climate-sensitive decisions in a particular arena, and the question is where, when and how they can build climate change considerations into their day-to-day work (an individualized perspective on the process commonly referred to as 'mainstreaming') (for example Huq et al., 2003; Klein et al., 2005; Bouwer and Aerts, 2006; Klein et al., 2007), and how to balance attention to climate change with other core responsibilities. Thus, for example, a city manager cannot only concern himself with shoreline protection as sea level rises, but will also have to determine how to pay for the renovation of the local elementary school and the replacement of the community's fire truck fleet. In short, for any one individual involved in adaptation decisions, the mission, mandate, authority, job description, and even 'softer' factors such as professional identity, personal ambition and work ethic may be broader or narrower than would allow for easy integration of climate change concerns. For many, consideration of long-term climate conditions are not required or simply do not yet enter decision-making today (countless examples could be cited here; see Pulwarty and Redmond, 1997; Jones et al., 1999; Mote et al., 2003; Pulwarty, 2003; Jacobs et al., 2005; Morss et al., 2005; McNie, 2007). Frequently, the mandate to explicitly assess current decision protocols for their robustness under different climate regimes (trends, extremes, variability and uncertainty) is not yet in place.

In the absence of such requirements, it may not be obvious, easy or even desirable for individual actors to discern where and how to make room within their pre-existing professional constraints for climate change (for example Rayner et al., 2005). And again, these constraints are not absolute limits to adaptation. A change in the job descriptions or incentive structures for individuals may go a long way toward overcoming these barriers. It would be foolish, however, to underestimate how real and big these ingrained procedures, habits, norms, lines and pressures of accountability, and day-to-day realities are in imposing structural impediments to adaptation-related decision-making.

A different but related possible constraint on individual decision-makers arises if the questions of responsibility, mandate, workload or job description are removed. Assume for a moment a 'willing' and authorized decision-maker, one who is ready to address climate change impacts. For this individual, questions of (economic, information and even political) resource availability, access and rights or entitlements may well become the constraining factors – and if not *formal* rights or entitlement (i.e. actual structural realities), a *sense* of entitlement, capacity and empowerment at the individual or household levels (for example Leach et al., 1999; Kelly and Adger, 2000) or within the context of governance at the community or regional levels (for example Pelling, 1998; Jentoft, 2005; Naess et al., 2005; Tribbia and Moser, 2008).

Interaction among decision-makers

As already mentioned, decision-makers rarely if ever act in isolation from others on adaptation-related matters and thus may well seek out, or be forced, to collaborate with others. At the same time it is important to acknowledge that 'stove-piping' is equally pervasive in policy, administrative and management contexts. Thus several questions emerge which help elucidate how the identification and inclusion of the full range of relevant actors, as well as their interactions around adaptation-related decisions may be structurally constrained. Among those questions are the following:

- Who initiates the adaptation planning process?
- Who leads the adaptation planning process? How experienced is the leader in facilitating group processes? And how is the leader perceived by others involved (respect, trust, benevolence, power sharing, etc.)?
- Was a decision made to have an elite/exclusive or broad/inclusive decision process, involving only selected individuals or a broader set, including a possibly wide range of stakeholders? What are the benefits and drawbacks of this choice?
- How systematic, strategic or arbitrary was the process of identifying relevant participants in the adaptation planning and decision-making process? How much freedom or choice did initiators/leaders have?
- To the extent that the boundaries of the regions, ecosystems, sectors, industries, communities or portions within them for which adaptation is being considered do not coincide with the boundaries of influence of responsible decision-makers, do neighbouring decision-makers with similar responsibilities communicate or collaborate on the adaptation question? If so, is the interaction formally required or voluntary?[3]
- For decision-makers working within the same context (for example, forest management, public health services, food security, a single community, a company) for which adaptation is being considered but who have complementary responsibilities (for example, strategic planning, operational management, financial management, different services), do they communicate and collaborate on the adaptation question? If so, is the interaction formally required or voluntary?
- For decision-makers working within a given sector or industry but at different scales or levels of organization, do they communicate, consult or collaborate with each other? If so, is the interaction formally required or voluntary?
- In any given arena (sector, industry, region), are all relevant stakeholders aware of, invited to, and engaged in, the planning and decision-making process? If so, how are they engaged, and what procedural rules guide their involvement?
- What kinds of authorities and responsibilities do the involved actors have? Are there potential overlaps or gaps in authority?
- To what extent do the involved players have complementary skills, missions and goals? Do any have contradictory missions and goals?
- What are the formal and informal 'rules of engagement' among decision-makers?

---

[3] This and the following two bullets are asked to account for the challenges with cross-scale interactions and common mismatches between physical or ecological phenomena on the one hand and the extent and reach of the social processes and institutions that affect and govern them (for example, Lee, 1995; Cox, 1998; Cash and Moser, 2000; Berkes, 2002; Young, 2002; Adger et al., 2005; Lebel et al., 2005; McCarthy, 2005; Cash et al., 2006; Cumming et al., 2006).

- Is the collaboration around adaptation a novel interaction, or are there past experiences and historical legacies that may either help or hinder in the adaptation planning and decision-making process?
- What is the quality, form and frequency of communication and interaction among all involved?

In many instances, this list of questions will reveal a host of structural impediments to effective communication and collaboration as well as help characterize the 'institutional culture' of the entities involved (see for example Brockhaus and Kambiré; Eakin et al., Chapters 25 and 13, this volume). They can pose sometimes formidable barriers if not de facto limits to adaptation planning and implementation. The questions may, for example, uncover different levels, or lack of clarity around the division, of decision-making authority among those involved. Or they may reveal lack of clarity, skill and experience with stakeholder engagement processes. They may also uncover that different decision-makers are accountable to agencies with contradictory missions (and underlying laws) leading to opposite objectives. Alternatively, accountability may be to agencies with overlapping missions, which can lead to turf issues or power struggles. Participants may historically have had infrequent or complete lack of interaction across management divisions and lack clarity or authority to negotiate and assign roles, share duties, costs, responsibilities and benefits. In a case comparison of coastal management for sea level rise in the two US states Maine and Hawaii, such structural impediments between involved agency representatives were found to negatively impact managers' interaction, trust and shared learning; it also hindered information sharing around sea level rise and related coastal hazards and undermined the possibility for joint decision-making until regular interagency forums for communication and planning were set in place (Moser, 2005, 2006).

Individuals from historically separate and independently working decision-making entities may also be linked (and accountable) to different authorities and stakeholders, and obtain their preferred, needed and trusted information from different information sources, giving them access to different, or even contradictory, information. In such cases, the availability of face-to-face forums or other mechanisms for sharing knowledge, eliminating knowledge gaps and misunderstandings, and accelerating collective learning could be essential, and the lack thereof can constitute yet another structural impediment to effective adaptation governance. Such a situation is particularly problematic when there is little scientific consensus (yet) on the issue at hand; where long-held interests (sometimes codified in laws, customary rights or constitutional guarantees) are being contested; when values differ fundamentally; or power struggles over the issue at hand or an unrelated one are carried out in the process of the adaptation decision process (for example Barker and Peters, 1993; Slovic, 1993; Cvetkovich et al., 2002; Poortinga and Pidgeon, 2004; Sarewitz, 2004; Patt, 2007).

Finally, the individuals interacting in the process of adaptation governance are quite likely to have different levels of access to financial resources, technical resources, and political leadership and power. If not addressed deliberately, these structural inequalities may well be reproduced and perpetuated within or without broadly participatory processes (for example Adger et al., 2001, 2005; Cooke and Kothari, 2001; Hickey and Mohan, 2005; Few et al., 2006). The absence of forums for dialogue and influence, and appropriate, functional mechanisms for redress and conflict resolution in such situations can prove highly problematic for adaptation governance (Ostrom 1990; Gunderson et al., 1995; Folke et al., 2005; Humphreys 2005; Richardson 2005; Sidaway, 2005; Adger et al., 2006). At the same time, it must be recognized that stakeholder processes can often be lengthy, heated and involved, and may not necessarily lead to 'better' or more 'effective' adaptation choices. In fact, our democratic ideals may very well – in the face of growing urgency and future humanitarian disaster – be compromised, thereby creating political legacies that further undermine the fulcrum of governance (for example NRC, 1996; Beierle, 1998; Rowe and Frewer, 2000; Cooke and Kothari, 2001; Hickey and Mohan, 2004).

As suggested, structural impediments can produce direct and indirect, yet equally challenging constraints on effective adaptation governance, particularly through negative impacts on the trust, mutual understanding, and level of sharing of information and decision-making power among those involved. These 'soft' constraints hamper the efficiency and expediency of interactions and thereby delay but maybe not ultimately prevent adaptive action. The critical issue here, however, is that important 'windows of opportunity' for adaptation decisions (such as periodic planning or budgetary processes, the regular replacement of infrastructure or equipment, or more urgent occasions such as in the aftermath of disasters, and before or after elections) may be missed. Clearly, such impediments will colour the quality of human interactions and deeply affect perceptions of effectiveness of the adaptation decision processes and outcomes among all involved.

What influences adaptation decisions?

The discussion so far has drawn concentric, consecutively narrower circles around the adaptation decision process – beginning with the larger societal context (for example, political, social, cultural, economic), then – within that – the institutional context, including laws, norms and organizational structures, and within that the individuals who perform their duties and interact with each other (Figure 20.1). Each of these contextual factors demarcates the action space and intimately affects the interactions among decision-makers within it. Here now, the focus is more specifically on the interplay between characteristics of the decision-maker and those of the adaptation decisions, exploring the many motivating and constraining influences that ultimately lead to a decision (or non-decision). Clearly, this step in the diagnostic approach (a bottom–up focus on individuals' characteristics affecting decision-making) is complementary to Step 1 (a top–down focus on the arena of decision-making): both are faces of the same coin.

For this stage in the diagnostic approach, questions emerge that hinge primarily on matters of motivations or incentives (writ large) to take adaptation actions as well as on perceived and real barriers or constraints to doing so. The framework suggested for this discussion is derived from a multidisciplinary exploration of motivations and barriers to climate change action (Moser and Dilling, 2007a, 2007b).

Personal values, attitudes, beliefs and capabilities

The first set of questions reveals influences rooted in the deeply held values, beliefs and self-perceptions of the decision-maker and may include the following:

- For any individual decision-maker, what is the level of awareness and concern with climate change science and impacts?
- What underlying, more general environmental values and perceptions do decision-makers hold?
- What is the personal level of knowledge and understanding of the current state of the climate and the environment, of projections of future states of relevant aspects of the environment, and related social trends?
- What is the decision-maker's ability and inclination to acquire necessary information and knowledge, to learn, and to use this information in decision-making (even if current decision protocols do not require it)? What is the decision-maker's adaptive learning capacity?
- What are the decision-maker's personal interests, ambitions, and goals, his/her duties and obligations vis-à-vis his/her work?
- What is the decision-maker's perception in terms of perceived power and capacity, his role in his work environment and in society at large? What positional power and personal abilities/capacities does (s)he bring to make or influence decisions or to take specific actions, and is (s)he inclined to bring them to bear on adaptation questions?
- What social norms and expectations, peer pressures and peer support does the decision-maker experience?

In the absence of a formal obligation to begin adaptation planning and implement various actions, answers to these questions may well determine whether or not adaptation planning will commence at all. Yet whether the adaptation decision-making process is mandated or not by some higher authority, a decision-maker will always come to decision-making with his/her deeply held personal values, and long-established self-perceptions, beliefs, inclinations and capabilities (see for example Wolf et al., Chapter 11, this volume). They will influence negotiation positions, prolong or help overcome institutional inertia, prevent or prompt creative solutions thinking when seemingly impenetrable obstacles may slow progress, and shape every aspect of the decision process (from participation to transparency to power sharing to efficiency). While maybe outside the specific norms that directly underlie governance, in a decision-maker-centred approach, it would be foolish to disregard the wider set of social norms and values that influence the contributions (s)he will make in the adaptation process. These personal norms and values constitute the grease or grit in the mills of what is likely to be a difficult governance process.

Decision support resources: information and knowledge

Aside from the personal motivations or hesitancies to begin adaptation planning, a decision-maker may have or lack access to critical information and knowledge. Having such information may provide additional motivation to plan ahead, be necessary in the formulation of appropriate responses and adaptation options, and inform an adaptive management process over time as the outcomes of adaptation decisions are monitored, evaluated and decisions potentially revised.

A rapidly growing body of literature too large to synthesize here is concerned with the question of what type of information, tools and other knowledge resources would best aid decision-makers in adaptation planning (for example National Council for Science and the Environment, 2000; Cash et al., 2003, 2006; McNie, 2007).[4] Importantly, however, information alone is never enough. Often training in interpreting and using such information is equally if not more important (for example Tribbia and Moser, 2008). Moreover, the integration of even appropriately downscaled, usefully formatted and timely delivered climate change information into existing decision processes is not trivial at all. Frequently, information is uncertain, knowledge contested, and by itself not predictive of a particular course of action (for example Funtowicz and Ravetz, 1993; Bradshaw and Borchers, 2000; Ravetz, 2004; Reichert and Borsuk, 2005; Richardson, 2005; Moss, 2007; see also the chapter by Dessai et al., Chapter 5, this volume).[5]

---

[4] Decision support for climate change adaptation is already a well established activity of the UK Climate Impacts Programme and has been relatively neglected in the Climate Change Science Program of the United States. The US National Research Council has developed recommendations for the effective provision of decision support for mitigation and adaptation decisions (NRC, 2009).

[5] It is beyond the scope of this chapter to do justice to the challenges associated with decision-making under uncertainty or to discuss adequately the issues related to adjudicating between different knowledge claims, however important they often are, especially when lay publics or indigenous groups play critical roles as stakeholders in the adaptation governance process (see also Finan and Nelson, Chapter 21, this volume).

Nonetheless, questions that elucidate what forms of decision support are available or may be needed will reveal important structural constraints on adaptation governance. They may include the following:

- What climate change, environmental, economic and social information is available at present that informs problem formulation, understandings of the current state of affairs, causation, and future trends, as well as possible response options?
- What additional information is required to deepen understanding, bound or characterize uncertainties, assess the possible impacts of the set of available adaptation choices?
- What efforts are made to increase or enhance the collective knowledge of the problem and possible solutions (for example, periodic scientific assessments, literature reviews, surveys of traditional ecological knowledge)? Is there competition among different bodies of knowledge (range of disciplines and practical experiences) and different knowledge claims?
- What tools are available or needed to analyse, interpret, display and communicate relevant information, and to transform it into 'useable information' in the decision process?
- At what time(s) in the decision process is such information and knowledge needed? What role does it play in the decision-making process?
- Is knowledge about climate change (and any other environmental, social and technological conditions) viewed as fluid and evolving? And therefore, is an adaptive management process in place (for example Gunderson et al., 1995; Lee, 1993; Pulwarty and Melis, 2001; Tompkins and Adger, 2004; Arvai et al., 2006)?

Decision support is essential in enabling adaptation decisions, and resource managers frequently deplore the lack of needed information at present. At the same time, no claim is made here that the provision of such information would automatically lead to adaptation decisions. Aside from the many institutional and more personal constraints, what are viewed to be available and feasible solution choices in any given sectoral, political, economic and social context is at least as important.

The 'solution' options

Motivation is only part of the equation that determines whether specific actors plan for adaptation and enact different strategies or not. The other critical part of the equation is made up of actors' resistance and perceived or real barriers to making such decisions. Research shows time and time again that when actors do not know of solution options, do not view them as feasible, cannot enact them, or if they do not feel that the potential actions make any real difference in the problem, they are extremely unlikely to even want to try (for a review and synthesis, see Moser and Dilling, 2004, 2007a).

Here the focus is only on the availability and feasibility of technological solutions. All too often, the search for 'solutions' begins and ends with a question about technology. In such a context, the question of technological solutions is deeply intertwined with financial affordability (two common levers considered under the rubric of 'adaptive capacity'). The reasons for this self-imposed constraint is rooted in the larger cultural context of Western societies (be they highly developed countries and richer communities with more ample resources, or developing nations or poorer communities, often in a less fortunate position to afford technological solutions), where technological optimism not only favours such options, but where normative ideals about individual freedom, the role of government vis-à-vis the individual, and opposition to unpopular policies or social engineering may resist deeper social changes to address global change problems.

Technology thus holds a distinctly privileged position in the search for adaptation options, but, of course, is not necessarily available, accessible, affordable, advisable on environmental grounds, or socially acceptable to all stakeholders. On a case-by-case basis, it must be decided how mutable or absolute (at least at the time of decision-making) these constraints around technological solutions are (see for example Reeder et al., Chapter 4, this volume). For example, it is theoretically conceivable to place thousands of miles of dikes and seawalls around ocean-bordering countries in an effort to protect their shorelines from sea level rise and associated coastal erosion and flooding. The impacts on coastal ecosystems, fisheries, tourism and recreation, not to speak of that nation's treasury, would be unspeakable. The economics, in fact, are likely to dictate practical limits on how much, how high, and how frequently such structural protection could even be considered. Complementary technologies would be needed to deal with related impacts, such as salt water intrusion. None will guarantee 'safety' or cost-effective returns on the investment. Moreover, real feasibility limits would have to be confronted. It is more common to consider partial protection of high-value areas and retreat elsewhere, which entails intense political debates about fairness and justice, as witnessed for example at present in the UK (for example Few et al., 2004). These cursory examples from the coastal sector illustrate the many constraints that limit the feasibility of technological solutions options for adaptation. The perceived or real constraints around technological or any other type of 'solution' counteract any possible motivation individual actors may have to plan for adaptation, and what they see as their realistic choice set. Moreover, these perspectives also influence – even before a decision is ever made – how effective actors believe their adaptation strategies may ever be. To complicate things even further, proposed technological solutions will mobilize some stakeholders, while disengaging others, thereby playing an important role in shaping the political context and power dynamics around adaptation decisions.

What are the outcomes of a decision once made, and how do the decision-makers and affected stakeholders live with them?

Let us assume a choice between different adaptation options has been made, decisions are now being implemented. It is in the nature of these choices and in the nature of climate change that some impacts of these decisions can be seen almost immediately, others cannot be observed or evaluated (and hence adjusted, if necessary) until many years hence. Adaptation governance therefore cannot be viewed as complete with the first or one-time decision, but extends deeply into the future. To understand the long-term nature of adaptation governance, a number of questions must be asked that reveal how well governance institutions are prepared for adaptive management, including:

- Are mechanisms available to trace the impacts of decisions and those resulting from specific implementation actions (or behavior change)? Such mechanisms ideally should monitor for near- and longer-term environmental impacts, and differential social impacts.
- Are institutional mechanisms available to revisit previous adaptation choices?
- Are social/political mechanisms available to address social justice concerns, deal with power imbalances?
- How is outcome effectiveness defined and measured? According to whom? These are among the big questions behind adaptation and resilience (for example, is an adaptation strategy effective when current stakeholders are satisfied, an initially set goal is achieved, or when (un)anticipated benefits are reaped and (un)anticipated risks and disasters avoided, or is the measure of effectiveness ultimate survival or an emancipation to a higher level of well-being?
- How are differential outcomes weighed and compared, if they cannot be measured by the same metric? How do we adjudicate among different goals and embedded values?
- How reversible or irreversible are the decisions and (non)actions once made?
- What are the ethical obligations and actual legal (financial) liabilities to future generations, people and companies elsewhere?
- What forums are available for the expression of public discontent and redress? What mechanisms are available for 'just' compensation?
- How much flexibility is there (socially, institutionally, politically) to respond and correct (if necessary and/or possible) any prior actions?

The answers to these questions and the outcomes of the decisions made or not made ultimately affect the concentric circles of the larger decision context (Figure 20.1), the institutional action space, the motivations and barriers perceived and experienced by decision-makers, and the practical feasibility of any one adaptation option. As a result, we find ourselves in an ever-changing decision space for adaptation governance, where chances for 'success' or 'failure', however defined, constantly shift. Importantly, however, whatever may be perceived as 'failure' at a given time, may well be important information for learning over time. Thus, despite the changing context accountability mechanisms should allow for risk-taking and potential mistakes so as not to undermine a system's long-term ability to learn and change. This is precisely why political vigilance and skilled leadership will be required if adaptation planning and actions are to be advanced. While certain conditions conducive to adaptation decision-making may be identified, the multi-factorial movement of adaptation governance will remain difficult to predict, tricky to steer, and foolish to dismiss as the usual muddling through.

Conclusion

In this chapter, I have tried to ask questions about the governance of adaptation, and to what extent and in what ways the day-to-day realities of decision-making in given decision contexts affect our ability to translate adaptive capacity into real actions. The chapter tries to make the case for a shift in focus in adaptation research. In short, future studies of adaptation must go beyond laundry lists of potential options and (even) constraints. They also need to go beyond simplistic assertions that technology, more and better information, sufficient money, and the availability of democratic institutions will save the day, i.e. that these are the principal drivers and determinants of not just adaptive *capacity* but adaptive *success*. Clearly, it is time to acknowledge that there is little room for complacency (O'Brien et al., 2006) and little grounds to assume that – failing to significantly mitigate the extent and pace of climate change – we can easily or surely 'cope our way out of climate change' (Adger et al., 2007).

Instead, there is a growing need to understand and empirically test our understanding of the social dynamics that underpin (motivate, facilitate and constrain) on-the-ground adaptation strategies and actions through existing governance structures and mechanisms. As this chapter sketches out, the fulcrum cannot a priori be assumed to be a strong basis on which to turn our wishful levers for adaptation. Instead, we need to investigate whether these structures and mechanisms allow us to effectively use in practice what we understand in principle. Do they allow us to be in dialogue with one another about the present and future, i.e., about the kind of world we want to live in? Do they allow us to make explicit, and address, the value judgements and power dynamics embedded in adaptation decisions? Do they allow us to confront social challenges and potential conflicts, and recognize early on the signs of large-scale discontent (chronic or acute)? Finally, do they provide us with solution-oriented, socially just and non-violent forums and mechanisms for addressing these discontents and conflicts? Over the long term, as we may have to confront not just a little, but drastic climate change, we must also ask, whether these governance structures and mechanisms are robust enough to handle urgent decision-making, perpetually stressful times of change, and potentially socially and politically highly unpopular choices (for example Dodds et al., 2003; Lempert et al., 2003; Anderies et al., 2004; Dessai and Hulme 2007). Only to the extent that we can answer any of these questions in the positive can we have any confidence that the fulcrum is indeed strong.

If on the other hand we find the fulcrum to be deficient, we may ironically also find ourselves with levers that are too short. Why? Because the decision-making structures and processes that ultimately lead to adaptation actions, as the chapters in this book illustrate, may lead to inadequate resource allocation or inappropriate choices among policies, technologies, and institutional arrangements. This is not to say that there is some elusive 'perfect governance' approach that promises 'perfect adaptation'. In a world of rapid change, multiple stresses, ambiguous signals, significant uncertainty, and people who differ in values, attitudes, beliefs, capacities and goals, there can never be such a guarantee. However, as several chapters in this book illustrate with examples, shortcomings in the governance structures, processes and mechanisms can lead to technological 'solutions' that are known to involve significant negative environmental side effects, or to insufficient dedicated funds to finance adaptation, to reinforcement of social power imbalances and inequities, or deficient infrastructure investment. In short, inadequacies in governance can lead not just to poor use of the existing adaptive capacity, but also to inadequate furthering or even diminishment of adaptive capacity, leaving us less well prepared for future adaptation to change.

This all leads to the sobering conclusion that we must stop hand-waving about adaptive capacity, especially in richer, highly developed countries and communities, and increase our understanding of, and our ability to use or create, more effective governance structures to realize it. We must ask how to foster and employ the human and social capital necessary to actually and effectively move technical, economic, institutional, and policy levers for adaptation to climate change (Moser et al., 2008). Rapid environmental change may well confront us with a varied and ample menu of social, environmental and economic disruptions, and if history is any guide, we must admit that a society under multiple sources of stress will be challenged to handle them graciously or even peacefully.

Coda: An ethical imperative for adaptation researchers

It is for all the reasons laid out above that I see a moral imperative embedded in the challenge of strengthening and realizing our adaptive capacity, namely for (social) scientists to move more frequently and courageously out of the ivory tower, acknowledge these challenges, and engage more effectively with decision-makers and stakeholders in addressing them (Vogel et al., 2007; Moser, 2008). We have much knowledge that can help decision-makers be more effective, learn from others' experience, and help avoid governance traps and inadequacies. A socially relevant, ethically informed, and practically engaged science can offer its insights so that they may be applied toward an empowering and emancipatory interpretation of adaptation and resilience in an increasingly challenging world.

References

Adger, W.N., Benjaminson, T.A., Brown, K. and Svarstad, H. 2001. 'Advancing a political ecology of global environmental discourses', *Development and Change* **32**: 681–715.

Adger, W.N., Brown, K., Fairbrass, J., Jordan, A., Paavola, J., Rosendo, S. and Seyfang, G. 2003. 'Governance for sustainability: towards a 'thick' analysis of environmental decision-making', *Environment and Planning A* **35**: 1095–1110.

Adger, W.N., Arnell, N.W. and Tompkins, E.L. 2005a. 'Successful adaptation to climate change across scales', *Global Environmental Change* **15**: 77–86.

Adger, W.N., Brown, K. and Tompkins, E.L. 2005b. 'The political economy of cross-scale networks in resource co-management', *Ecology and Society* **10**: 9.

Adger, W.N., Paavola, J., Huq, S. and Mace, M.J. (eds.) 2006. *Fairness in Adaptation to Climate Change*. Cambridge: MIT Press.

Adger, W.N., Agrawala, S., Mirza, M.M.Q., Conde, C., O'Brien, K., Pulhin, J., Pulwarty, R., Smit, B. and Takahashi, K. 2007. 'Assessment of adaptation practices, options, constraints and capacity', in Parry, M.L., Canziani, O.F., Palutikof, J.P., Van der Linden, V.J. and Hanson, C.E. (eds.) *Climate Change 2007: Impacts, Adaptation and Vulnerability. Contribution of Working Group II to the Fourth Assessment Report of the Intergovernmental Panel on Climate Change*. Cambridge: Cambridge University Press, pp. 717–743.

Adger, W.N., Dessai, S., Goulden, M., Hulme, M., Lorenzoni, I., Nelson, D., Naess, L.-O., Wolf, J. and Wreford, A. 2009. 'Are there social limits to adaptation to climate change?', *Climatic Change* **93**: 335–354.

Anderies, J.M., Janssen, M.A. and Ostrom, E. 2004. 'A framework to analyze the robustness of social-ecological systems from an institutional perspective', *Ecology and Society* **9**: 18.

Arvai, J., Bridge, G., Dolsak, N., Franzese, R., Koontz, T., Korfmacher, K.S., Sohngen, B., Tansey, J. and Thompson, A. 2006. 'Adaptive management of the global climate problem: bridging the gap between climate research and climate policy', *Climatic Change* **78**: 217–225.

Barker, A. and Peters, B.G. (eds.) 1993. *The Politics of Expert Advice: Creating, Using and Manipulating Scientific Knowledge for Public Policy*. Pittsburgh: University of Pittsburgh Press.

Beierle, T.C. 1998. *Public Participation in Environmental Decisions: An Evaluation Framework Using Social Goals*, Discussion Paper No. 99–06. Washington, DC: Resources for the Future.

Berkes, F. 2002. 'Cross-scale institutional linkages for commons management: perspectives from the bottom up', in Ostrom, E., Dietz, T., Dolsak, N., Stern, P., Stonich, S. and Weber, E. (eds.) *The Drama of the Commons*. Washington, DC: National Academies Press, pp. 293–321.

Bouwer, L.M. and Aerts. J.C.J.H. 2006. 'Financing climate change adaptation', *Disasters* **30**: 49–63.

Bradshaw, G. A. and Borchers, J. G. 2000. 'Uncertainty as information: narrowing the science–policy gap', *Conservation Ecology* **4**: 1.

Cash, D. W. and Moser, S. C. 2000. 'Linking local and global scales: designing dynamic assessment and management processes', *Global Environmental Change* **10**: 109–120.

Cash, D. W., Clark, W. C., Alcock, F., Dickson, N. M., Eckley, N., Guston, D. H., Jäger, J. and Mitchell, R. B. 2003. 'Knowledge systems for sustainable development', *Proceedings of the National Academy of Sciences of the USA* **100**: 8086–8091.

Cash, D. W., Adger, W. N., Berkes, F., Garden, P., Lebel, L., Olsson, P., Pritchard, L. and Young, O. 2006. 'Scale and cross-scale dynamics: governance and information in a multilevel world', *Ecology and Society* **11**: 8.

Cooke, B. and Kothari, U. (eds.) 2001. *Participation: The New Tyranny?* London: Zed Books.

Cox, K. 1998. 'Representation and power in the politics of scale', *Political Geography* **17**: 41–44.

Cumming, G. S., Cumming, D. H. M. and Redman, C. L. 2006. 'Scale mismatches in social-ecological systems: causes, consequences, and solutions', *Ecology and Society* **11**: 14.

Cvetkovich, G., Siegrist, M., Murray, R. and Tragesser, S. 2002. 'New information and social trust: asymmetry and perseverance of attributions about hazard managers', *Risk Analysis* **22**: 359–367.

Dessai, S. and Hulme, M. 2007. 'Assessing the robustness of adaptation decisions to climate change uncertainties: a case study on water resources management in the East of England', *Global Environmental Change* **17**: 59–72.

Dodds, P. S., Watts, D. J. and Sabel, C. F. 2003. 'Information exchange and the robustness of organizational networks', *Proceedings of the National Academy of Sciences of the USA* **100**: 12 516–12 521.

Easterling, W. E., Hurd, B. H. and Smith, J. B. 2004. *Coping with Climate Change: The Role of Adaptation in the United States*. Arlington: Pew Center on Global Climate Change.

Edwards, V. M. and Steins, N. A. 1999. 'A framework for analysing contextual factors in common pool resource research', *Journal of Environmental Policy and Planning* **1**: 205–221.

Few, R., Brown, K. and Tompkins, E. L. 2004. *Scaling Adaptation: Climate Change Response and Coastal Management in the UK*, Tyndall Centre Working Paper No. 60. Norwich: Tyndall Centre, University of East Anglia.

Few, R., Brown, K. and Tompkins, E. L. 2006. *Public Participation and Climate Change Adaptation*, Tyndall Centre Working Paper No. 95. Norwich: Tyndall Centre, University of East Anglia.

Folke, C., Hahn, T., Olsson, P. and Norberg, J. 2005. 'Adaptive governance of social-ecological systems', *Annual Review of Environment and Resources* **30**: 441–473.

French, S. and Geldermann, J. 2005. 'The varied contexts of environmental decision problems and their implications for decision support', *Environmental Science and Policy* **8**: 378–391.

Funtowicz, S. O. and Ravetz, J. R. 1993. 'Science for the post-normal age', *Futures* **25**: 739–755.

Gunderson, L. H., Holling, C. S. and Light, S. S. (eds.) 1995. *Barriers and Bridges to the Renewal of Ecosystems and Institutions*. New York: Columbia University Press.

Hickey, S., and Mohan, G. (eds.) 2005. *Participation: From Tyranny to Transformation? Exploring New Approaches to Participation in Development*. London: Zed Books.

Humphreys, M. 2005. 'Natural resources, conflict, and conflict resolution: uncovering the mechanisms', *Journal of Conflict Resolution* **49**: 508–537.

Huq, S., Rahman, A. M., Konate, M., Sokona, Y. and Reid, H. 2003. *Mainstreaming Adaptation to Climate Change in Least Developed Countries (LDCs)*. New York: United Nations.

Jacobs, K., Garfin, G. and Lenart, M. 2005. 'More than just talk: connecting science and decisionmaking', *Environment* **47**: 6–21.

Janssen, M. A. and Anderies, J. M. 2007. 'Robustness-tradeoffs for social-ecological systems', *International Journal of the Commons* **1**: 77–99.

Jentoft, S. 2005. 'Fisheries co-management as empowerment', *Marine Policy* **29**: 1–7.

Jones, S. A., Fischhoff, B. and Lach, D. 1999. 'Evaluating the science–policy interface for climate change research', *Climatic Change* **43**: 581–599.

Kelly, P. M. and Adger, W. N. 2000. 'Theory and practice in assessing vulnerability to climate change and facilitating adaptation', *Climatic Change* **47**: 325–352.

Kiparsky, M., Brooks, C. and Gleick, P. H. 2006. 'Do regional disparities in research on climate and water influence adaptive capacity?', *Climatic Change* **77**: 363–375.

Klein, R. J. T., Eriksen, S. E. H., Næss, L. O., Hammill, A., Robledo, C., O'Brien, K. L. and Tanner, T. M. 2007. 'Portfolio screening to support the mainstreaming of adaptation to climate change into development assistance', *Climatic Change* **84**: 23–44.

Klein, R. J. T., Schipper, E. L. F. and Dessai, S. 2005. 'Integrating mitigation and adaptation into climate and development policy: three research questions', *Environmental Science and Policy* **8**: 579–588.

Leach, M., Mearns, R. and Scoones, I. 1999. 'Environmental entitlements: dynamics and institutions in community-based natural resource management', *World Development* **27**: 227–247.

Lebel, L., Garden, P. and Imamura, M. 2005. 'The politics of scale, position, and place in the governance of water resources in the Mekong Region', *Ecology and Society* **10**: 18.

Lee, K. N. 1993. *Compass and Gyroscope: Integrating Science and Politics for the Environment*. Washington, DC: Island Press.

Lee, K. 1995. 'Greed, scale mismatch, and learning', *Ecological Applications* **3**: 560–564.

Lempert, R. J., Popper, S. W. and Bankes, S. C. 2003. *Shaping the Next One Hundred Years: New Methods for Quantitative, Long-Term Policy Analysis*. Santa Monica: RAND Corporation.

McCarthy, J. 2005. 'Scale, sovereignty, and strategy in environmental governance', *Antipode* **37**: 731–753.

McNie, E. C. 2007. 'Reconciling the supply of scientific information with user demands: an analysis of the problem and review of the literature', *Environmental Science and Policy* **10**: 17–38.

Morss, R. E., Wilhelmi, O. V., Downton, M. W. and Gruntfest, E. 2005. 'Flood risk, uncertainty, and scientific information for decision making: lessons from an interdisciplinary project', *Bulletin of the American Meteorological Society* **86**: 1593–1601.

Moser, S. C. 2005. 'Impact assessments and policy responses to sea-level rise in three US states: an exploration of human-dimension uncertainties,' *Global Environmental Change* **15**: 353–369.

Moser, S. C. 2006. 'Climate change and sea-level rise in Maine and Hawai'i: the changing tides of an issue domain', in Mitchell, R. B., Clark, W. C., Cash, D. W. and Dickson, N. (eds.) *Global Environmental Assessments: Information, Institutions, and Influence*. Cambridge: MIT Press, pp. 201–239.

Moser, S.C. 2008. 'A new charge: engaging at the science–practice interface', *IHDP Update* **1**: 18–21.

Moser, S.C. and Dilling, L. 2004. 'Making climate hot: communicating the urgency and challenge of global climate change', *Environment* **46**: 32–46.

Moser, S.C. and Dilling, L. 2007a. 'Toward the social tipping point: Conclusions', in Moser, S.C. and Dilling, L. (eds.) *Creating a Climate for Change: Communicating Climate Change and Facilitating Social Change*. Cambridge: Cambridge University Press, pp. 491–516.

Moser, S.C. and Dilling, L. (eds.) 2007b. *Creating a Climate for Change: Communicating Climate Change and Facilitating Social Change*. Cambridge: Cambridge University Press.

Moser, S.C. and Luers, A.L. 2008. 'Managing climate risks in California: the need to engage resource managers for successful adaptation to change', *Climatic Change* **87**: 309–322.

Moser, S.C. and Tribbia, J. 2006/2007. 'Vulnerability to inundation and climate change impacts in California: coastal managers' attitudes and perceptions', *Marine Technology Society Journal* **40**: 35–44.

Moser, S.C., Kasperson, R.E., Yohe, G. and Agyeman, J. 2008. 'Adaptation to climate change in the Northeast United States: opportunities, processes, constraints', *Mitigation and Adaptation Strategies for Global Change* **13**(5–6): 643–659.

Moss, R.H. 2007. 'Improving information for managing an uncertain future climate', *Global Environmental Change* **17**: 4–7.

Mote, P.W., Parson, E.A., Hamlet, A.F., Keeton, W.S., Lettenmaier, D., Mantua, N., Miles, E.L., Peterson, D.W., Peterson, D.L., Slaughter, R. and Snover, A.K. 2003. 'Preparing for climatic change: the water, salmon, and forests of the Pacific Northwest', *Climatic Change* **61**: 45–88.

Naess, L.O., Bang, G., Eriksen, S. and Vevatne, J. 2005. 'Institutional adaptation to climate change: flood responses at the municipal level in Norway', *Global Environmental Change* **15**: 125–138.

National Council for Science, and the Environment. 2000. *Recommendations for Improving the Scientific Basis for Environmental Decisionmaking: A Report from the First National Conference on Science, Policy, and the Environment*. Washington, DC: National Academy of Sciences.

National Research Council (NRC) 1996. *Understanding Risk: Informing Decisions in a Democratic Society*. Washington, DC: National Academies Press.

National Research Council (NRC) 2009. *Informing Decisions in a Changing Climate*. Washington, DC: National Academies Press.

O'Brien, K.L., Eriksen, S., Sygna, L. and Naess, L.O. 2006. 'Questioning complacency: climate change impacts, vulnerability, and adaptation in Norway', *Ambio* **35**: 50–56.

Ostrom, E. 1990. *Governing the Commons: The Evolution of Institutions for Collective Action*. Cambridge: Cambridge University Press.

Patt, A. 2007. 'Assessing model-based and conflict-based uncertainty', *Global Environmental Change* **17**: 37–46.

Pelling, M. 1998. 'Participation, social capital and vulnerability to urban flooding in Guyana', *Journal of International Development* **10**: 469–486.

Pielke Jr. R.A., Prins, G., Rayner, S. and Sarewitz, D. 2007. 'Lifting the taboo on adaptation', *Nature* **445**: 597–598.

Poortinga, W. and Pidgeon, N.F. 2004. 'Trust, the asymmetry principle, and the role of prior beliefs', *Risk Analysis* **24**: 1475–1486.

Pulwarty, R. S. and Redmond, K. 1997. 'Climate and salmon restoration in the Columbia River basin: the role and usability of seasonal forecasts', *Bulletin of the American Meteorological Society* **78**: 381–397.

Pulwarty, R. S. 2003. 'Climate and water in the West: science, information and decision-making', *Water Resources Update* **124**: 4–12.

Pulwarty, R. S. and Melis, T. S. 2001. 'Climate extremes and adaptive management on the Colorado River: lessons from the 1997–1998 ENSO event', *Journal of Environmental Management* **63**: 307–324.

Ravetz, J. 2004. 'The post-normal science of precaution', *Futures* **36**: 347–357.

Rayner, S., Lach, D. and Ingram, H. 2005. 'Weather forecasts are for wimps: why water resource managers do not use climate forecasts', *Climatic Change* **69**: 197–227.

Reichert, P. and Borsuk, M. E. 2005. 'Does high forecast uncertainty preclude effective decision support?', *Environmental Modeling and Software* **20**: 991–1001.

Richardson, T. 2005. 'Environmental assessment and planning theory: four short stories about power, multiple rationality, and ethics', *Environmental Impact Assessment Review* **25**: 341–365.

Risbey, J., Kandlikar, M., Dowlatabadi, H. and Graetz, D. 1999. 'Scale, context, and decision making in agricultural adaptation to climate variability and change', *Mitigation and Adaptation Strategies for Global Change* **4**: 137–165.

Rowe, G. and Frewer, L. J. 2000. 'Public participation methods: a framework for evaluation', *Science, Technology and Human Values* **25**: 3–29.

Sarewitz, D. 2004. 'How science makes environmental controversies worse', *Environmental Science and Policy* **7**: 385–403.

Sidaway, R. 2005. *Resolving Environmental Disputes: From Conflict to Consensus*. London: Earthscan.

Slovic, P. 1993. 'Perceived risk, trust, and democracy', *Risk Analysis* **13**: 675–682.

Smit, B., Pilifosova, O., Burton, I., Challenger, B., Huq, S., Klein, R. J. T. and Yohe, G. 2001. 'Adaptation to climate change in the context of sustainable development and equity', in MacCarthy, J. J., Canziani, O. F., Leary, N. A., Dokken, D. J. and White, K. S. (eds.) *Climate change 2001: Impacts, Adaptation and Vulnerability. Contribution of Working Group II to the Third Assessment Report of the Intergovernmental Panel on Climate Change*. Cambridge, UK: Cambridge University Press, pp. 877–912.

Tompkins, E. L. and Adger, W. N. 2004. 'Does adaptive management of natural resources enhance resilience to climate change?', *Ecology and Society* **9**: 10.

Tribbia, J. and Moser, S. C. 2008. 'More than information: what coastal managers need to plan for climate change', *Environmental Science and Policy* **11**: 315–328.

Vogel, C., Moser, S. C., Kasperson, R. E. and Dabelko, G. 2007. 'Linking vulnerability, adaptation and resilience science to practice: players, pathways and partnerships', *Global Environmental Change* **17**: 349–364.

Yohe, G. and Tol, R. 2002. 'Indicators for social and economic coping capacity: moving toward a working definition of adaptive capacity', *Global Environmental Change* **12**: 25–40.

Young, A. 2007. 'Forming networks, enabling leaders, financing action: the Cities for Climate Protection campaign', in Moser, S. C. and Dilling, L. (eds.) *Creating a Climate for Change: Communicating Climate Change and Facilitating Social change*. Cambridge: Cambridge University Press, pp. 383–398.

Young, O. R. 2002. 'Institutional interplay: the environmental consequences of cross-scale interactions', in Ostrom, E., Dietz, T., Dolsak, N., Stern, P., Stonich, S. and Weber, E. (eds.) *Drama of the Commons*. Washington, DC: National Academies Press, pp. 263–292.

21

Decentralized planning and climate adaptation: toward transparent governance

Timothy J. Finan and Donald R. Nelson

Introduction: governance and adaptation to climate change

The emergent and growing literature on climate change and adaptation seems to have laid aside the effort to define a standardized, universally accepted set of concepts and theoretical frameworks and now acknowledges the coexistence of context-specific vulnerability and adaptation models (Eakin and Luers, 2006; Füssel and Klein, 2006). In effect, the hazard-based, economic-based and climate-based approaches resemble variations in linguistic dialect – similar enough to be mutually intelligible but different enough to create highly nuanced 'theory communities'. The reality is, however, that hazards (and risk), poverty and global climate change are intimately intermingled in both theory and practice (Halsnaes and Verhagen, 2007; Vogel et al., 2007) – and so must be the models that explain the related dimensions of vulnerability and adaptation. This chapter seeks to contribute to both the scholarly understanding and the praxis of vulnerability and adaptation in specific socio-economic, physical and political contexts by demonstrating how systems of governance act to integrate these variant dimensions of adaptation. The specific context of a drought-affected state in north-east Brazil provides the empirical basis for our argument.

There are four underlying themes present in the variant models of vulnerability and adaptation that seem critical to how the praxis of adaptation unfolds in the face of climate variability and change. The first of these is that the process of adaptation necessarily entails the articulation of different levels of scale. The stresses on the natural and social systems are global in origin and scope, while the adaptive responses are local or national in scale. Second, adaptation has a strong distributive dimension related to both socio-economic inequality (including poverty rates) and

Adapting to Climate Change: Thresholds, Values, Governance, eds. W. Neil Adger, Irene Lorenzoni and Karen L. O'Brien. Published by Cambridge University Press. © Cambridge University Press 2009.

the nature of the local and national adaptive responses (Kelly and Adger, 2000). The third theme focuses on the interactive nature of natural stressors (for example, drought, flood) and other sources of system stress, such as globalized market trends and conflict (Leichenko and O'Brien, 2001; O'Brien et al., 2004), which suggests that adaptation is a response not only to changes in the physical environment but also to other non-climate dynamics that can exacerbate the impact of climate change per se. The final theme is that successful adaptation to climate change will require the active participation of local communities in the process and that institutional adjustments will inevitably focus on community reorganization and initiative (Adger, 2003; Folke et al., 2005; IPCC, 2007).

Taken together, these cross-cutting themes reinforce the complexity of a climate change adaptation process that incorporates non-climate influences at multiple scales as well as historically determined socio-economic patterns manifest in the distribution of assets and power. The dynamics of adaptation are negotiated within local contexts as households and communities adjust to rhythms and changes in the natural system over time (Finan and Nelson, 2001). But the range of response is established in part by the form of articulation with higher-level systems, primarily those which constitute what we often call the 'public sector'.

Public entities and their policies play a fundamental role in most societies as the orchestrators of the adaptation process. Policy is best understood as a set of public decisions supported by sanctioned institutions of power (law, coercion, etc.), and governance is the structures and processes by which such decisions are made and implemented – i.e. how power is shared throughout the social system (Lebel et al., 2006). Policy sets and systems of governance define how local communities articulate with external forces in the dynamics of adaptation to climate variability and change. The influence of policy and governance is felt not only directly but also indirectly. Non-climate global factors which affect local communities are often filtered through public decisions (for example, through domestic pricing policies), and even the historic patterns of the socio-economic distribution of resources and power are also the outcomes of past and current policy decisions.

Consistent with the literature, we seek to demonstrate that the nature of governance is a major determinant of the success of an adaptation process to climate extremes and climate change. In their insightful review, Folke et al. (2005) conclude that ecological knowledge systems, including local knowledge, must be enhanced and built into local adaptive management practices and that 'polycentric' nodes of governance are necessary to maintain system flexibility and diversity. Lebel et al. (2006) further elaborate this approach and identify participation, deliberation, decision-making diversity, justice and accountability as the key attributes of an 'adaptive' governance system.

Where forms of governance preclude effective community participation and discourage co-management practices, local resilience tends to be low and adaptive capacity limited. On the other hand, a more resilient social–ecological system operates in multi-nodal, well-articulated decision-making context where knowledge production and learning are dynamic and stocks of social capital generate bonds of trust (Gaventa, 2002). Given this key relationship between governance and successful adaptive management, it appears logical that the support of appropriate governance institutions would constitute a priority element in an overall adaptation strategy.

There is, however, a general lack of focus on the institutional resistance of governance structures rooted in centuries of structural inequity. In the many parts of the world where democratic traditions are less prevalent and where rigid forms of human interaction have evolved to maintain power differences, changes in governance systems can be neither willed nor mandated. The process by which governance systems can be prodded in a direction of greater adaptive capacity – in the absence of radical change – is little documented but is certainly non-trivial. This chapter discusses how governance factors have constrained adaptive capacity among rural populations in north-east Brazil and reports on an experiment in governance change based on the principles of participation, polycentrism, transparency, accountability and justice. It began as an attempt to evaluate the impacts of a sophisticated climate forecasting system on rainfed farmers, evolved into community-based vulnerability assessments and, finally, into a methodology for decentralized planning that challenges the traditional model of governance. This approach integrates community participatory research tools (inspired by the PRA of Chambers, 1994) with a GIS-based mapping platform to engage the population in the planning process through the creation of locally generated vulnerability assessments, to serve as the basis for the development of mitigation and adaptation plans, and provide the ability to continually monitor progress. The ultimate outcome is a local level development plan with a dual purpose – to anticipate the next, inevitable climate crisis (drought) and to redress the underlying structural factors that make this population so vulnerable. To achieve the polycentric governance goal, the elaboration of local development plans is articulated with the state government and becomes part of the state's overall development strategy. The implications of this methodology with regard to adaptation to climate change are amply discussed throughout this chapter.

Ceará and its *sertão*

Ceará is one of the nine states that constitute the north-east of Brazil, a region famous for its contribution to the history of the country and infamous for its poverty and record of catastrophic droughts. The vast majority of this semi-arid landscape is called the *sertão* (translated in popular literature as the 'backlands') and is characterized by irregular rainfall distributed over a short four-month period, high rates of evapotranspiration, thin soils, and a unique, thorny savanna ecosystem known as *caatinga*. The area of the state is 144 000 km^2, 90% of which is sertão, and the population exceeds 8 million, a quarter of which lives in the rural regions.[1] There is no naturally flowing surface water in the state, although a network of dams and reservoirs has perennialized the flow in two major water basins.

After the discovery of Brazil in the north-eastern state of Bahia in 1500, the Portuguese established a sugarcane economy along the north-eastern coast and, for around 100 years, enjoyed a monopoly in the sugar trade (Furtado, 1963). The principal source of labour for this production system was slavery – first provided by the local indigenous groups (which quickly perished) and later brought in with the African slave trade. In the sertão, the Crown had distributed land to favoured settlers in large tracts that were ruled as fiefdoms by their owners. With the development of the sugar economy, a secondary economy arose throughout the sertão based on provisioning foodstuffs, animal power and leather (for ropes in the sugar mills). Whenever the coastal sugar economy entered a down cycle, the sertão hinterland retreated back into a subsistence mode, producing corn, beans and manioc on a sharecropper regime. The patterns established during this colonial dynamic are still operative today, with the sertão traditionally controlled by a small *latifundista* elite engaged in cattle-raising while their resident sharecroppers produce subsistence food crops. The seeds of this historical process have generated one of the highest levels of wealth and income concentration in the world, with 75% of the rural population of Ceará now living beneath the World Bank's poverty line.

The first official drought was recorded in the north-east in 1552 (Villa, 2000), and up until the middle of the last century droughts occurred on an average of one every decade. Since the 1970s, this frequency has increased to one every three years. Specific droughts, however, have seared the cultural memory of the *sertanejo* (resident of the sertão) giving the region its reputation for a supernatural capacity for suffering. During the drought of 1877–78, an El Niño episode, over 500 000 *nordestinos* died of starvation, thirst and illness, prompting the federal authorities to launch a national drought mitigation programme which, under one form or another, continues to this day. Since the 1970s, no fatalities have been officially attributed to drought, but during the 1998–99 El Niño drought in Ceará, nearly 500 million dollars were spent to sustain more than 300 000 families. Ominously, the north-east of Brazil is projected to experience increased aridity and an increase

---

[1] In 1970, 65% of the population in Ceará was rural. The state, as throughout the north-east region, has experienced dramatic levels of rural exodus into the urban centres of the country.

in the frequency and severity of drought under climate change (Marengo, 2007). These changes will place additional stress on a system already functioning at near maximum level.

The impacts of droughts are intimately linked to rural livelihoods and the lack of reliable water supplies. Most of the rural population in Ceará practice rainfed agriculture, and only around 15% of cultivated land is irrigated. The dryland production of corn, beans and manioc is highly sensitive to the distribution and amount of rainfall, which can be erratic and uncertain in any given year. The typical farm family is either a smallholder or sharecropper with several hectares of the staple food crops – corn and beans. Most farms also support a small number of animals, and a certain reliance on off-farm employment is common but irregular. Older residents (about 30% of households) have access to the federal rural retirement programme and receive a small monthly pension (Finan and Nelson, 2001). In a drought year, food production is decimated, forcing households to sell or slaughter their animals. *Fazendas*, the large landholdings, face the crisis of providing their livestock with fodder and water, and are also often forced to sell or move herds. Perhaps the most urgent impact of drought in the sertão of Ceará is the lack of drinking water. Households normally obtain water from alluvial wells or from small reservoirs located near settlement clusters (communities); however, during a drought, wells dry up and all but the largest reservoirs become saline, forcing households to depend on water distribution networks maintained by the local government.

Traditional governance in rural Ceará

The inequalities embedded in the economic structure of rural Ceará are further elaborated in the power structure. In traditional society, the flow of social interaction is mediated and channelled by the currents of patronage and clientilism. The patron–client relationship is the most pervasive form of social organization in rural society and the basis upon which the elites and the poor define one another. Economically marginalized households develop dyadic relationships with wealthier members of society as a strategy that serves to provide a form of safety net protection against the unforeseen risks that threaten daily survival. The poor households reciprocate this economic support, particularly in times of crisis, with political loyalty and often a preferential labour source. These relationships are often reinforced and cultivated in such cultural institutions as *compadresco* (co-godparenthood), which establishes a kin-like linkage between families. From a broader optic, then, economically dependent families seek to create wide networks of reliable 'patrons', while the wealthier households maintain their 'clients' as a source of political power (and a pool of manual services).

The formal political structure in Ceará also incorporates the patronage principle as the major strategy of alliance-building and, consequently, determines the distribution of public goods and resources. The smallest politico-administrative unit in Brazil is the *município*, roughly similar in size to the US county, and is constituted of an urban centre and numerous settlement concentrations, called *comunidades*, scattered across the rural landscape. There are 184 municípios in Ceará. The political institution that administers the município is the *prefeitura*, the head of which is the *prefeito(a)*, the major elected official. The prefeito occupies an extremely powerful position, since the prefeitura administers virtually all state and federal development programmes and services, including local schools and hospitals. The other elected post at the município level is the *Câmara dos Vereadores*, an elected board of supervisors which is meant to represent the interests of different regions and interest groups in the município. There are several political parties, but political philosophies are generally unimportant, since the political process more often than not is articulated as a struggle for power between elite families who have chosen competing party affiliations. Throughout the sertão, município politics are fiercely and sometimes violently contested because political power is a major avenue to economic capture.

Governance in rural society is thus a formal political structure – with its outward set of state-sanctioned rules, roles and responsibilities – built on an informal but unshakeable foundation of patronage and clientilism. Local município leaders 'deliver' a certain number of votes in exchange for economic or political favours, and public goods and resources are distributed along channels built on political loyalty and family. Perhaps an example from the field is illustrative. In 1997, the authors were visiting a large remote município in Ceará and for lack of alternative options, stayed at the house of the prefeita. At daybreak, there was considerable commotion outside the house, and we observed a line of impoverished individuals – some old and disabled, others in the tattered clothing of the poor – waiting for an audience with the prefeita. She received each one in turn, talking quietly with each, then from her own purse handing out small amounts of money, a note for the pharmacy or the doctor, a bag of food, or other minor benefices. The authors were struck by the manner in which the formal responsibilities of the office of prefeito had been personalized and recast in a patron–client relationship. These poor did not approach the prefeita in a spirit of exercising their rights as citizens, but rather as disadvantaged individuals asking a 'woman-in-power' for a favour. The reciprocal expectations went unspoken, for they were well understood by both sides.

The patronage principles of governance at the município level tend to be replicated at the state and federal levels. Most municípios have a negligible tax base and depend upon a flow of resources from the state and federal governments. State legislators curry favour and votes from municípios in exchange for the promise of resource flows, and prefeitos are expected to 'deliver' a certain number of votes to a specific legislator in return for greater access to state and federal programmes. In addition, populations in municípios in which the prefeito does not actively support the state governor's party are resigned to the knowledge that state funds will be reduced and more difficult to access for the duration of the political term.

It is important for our argument to note, however, that if the traditional patronage practices are still dominant, the narrative of governance has changed significantly in Ceará over the last two decades. With the end of the military dictatorship in the mid-1980s, Brazil embraced democracy and a commitment to a new sense of citizenship based upon individual effort and merit. In Ceará, the power base of the state shifted from the traditional landed elites to the more cosmopolitan commercial and industrial leaders (Tendler, 1997). As a result, the public narrative came to emphasize professional management in public administration (as in industry), needs-based and problem-solving investment, local participation, and, foremost, the decentralization of public decision-making (generating the neologism *municipalização*). The professionalization of local governance was further supported by the World Bank, which invested heavily in rural development in Ceará through locally based associations of producers.

Governance and drought: the 'drought industry'

We have argued elsewhere that the public response to climate extremes and climate change in Ceará has been palliative in nature without addressing the fundamental nature of livelihood vulnerability, which is rooted in historical causes and past public decision-making priorities (Nelson and Finan, 2009). Since the horrendous 1877–78 drought, the federal government sought to 'solve' the drought problem by building a large network of dams and reservoirs that would secure a water supply for the rural population. In the early 1900s, the government created a federal agency, today called DNOCS (National Department for Works against the Drought), headquartered in Fortaleza, the capital of Ceará. This agency has supervised the construction of basin earthworks and irrigation perimeters throughout the state. A state agency for meteorology and water resources management (FUNCEME) was instituted in 1972 to address the problem of drought through technology applications (including cloud-seeding), and in the 1990s it developed a sophisticated climate forecasting capacity. Despite these efforts to neutralize climate extremes through engineering and technology, the underlying vulnerabilities remain as major constraints to a process of adaptation.

The advent of a drought continues to be received as an unexpected event, and there is virtually no planning or preparation in place prior to its arrival. A typical drought year begins with an initial rainfall that moistens the fields sufficiently to plant (conventional wisdom is about 22 cm). Crops and native pasture species germinate but in the absence of further rainfall eventually wither without production, and the lower thorny shrubs and trees fail to produce leaves. The sources of water for animal and human consumption are not replenished and soon drinking water becomes scarce. For the rural household, the experience of drought is a lack of basic staple foods, lack of water and feed for animals, and a lack of water for household use. At the same time, market merchants respond to drought conditions by increasing prices, and many off-farm employment opportunities (such as working for a large cattle rancher) are drought-sensitive and disappear during a crisis.

Under traditional governance systems, the level of human suffering provides the trigger for public action. When drought-afflicted families begin to blockade the federal highways or invade public and private food stocks, a state of emergency is declared, and a chain of events is unleashed that mobilizes state action, then federal action which results in the appropriation of relief resources. Relief is then channelled back through the state to the município, where the prefeitura is generally responsible for distribution (Figure 21.1). Relief resources come in several forms, including food baskets, public work front employment, and drinking water distribution with tanker trucks. The actual registration of drought families and the types of public work projects are traditionally the responsibility of the prefeitura.

This governance system provides tremendous opportunity for capture by local elites, including the prefeito. The local term, the 'drought industry' encapsulates a collective cynicism regarding the overall impacts of past relief programmes, and accounts of corruption are both legion and legendary. For example, in the drought of 1983, it is widely acknowledged that the public work registration included names from cemetery headstones, names of newborns and even the names of domestic

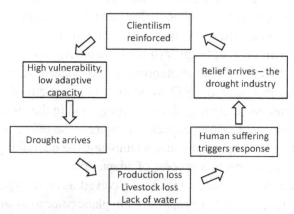

Figure 21.1 Drought relief: governance by clientilism.

pets. Public work projects often consisted of the construction of fence or private reservoirs on the lands of large landowners. Regardless of the veracity of these narratives, the critical characteristic of this form of relief governance is that it appears as personalized relief – removing it from the realm of one's right as a citizen to that of a personal favour from the political leader. From our field research, it is clear that scarce relief resources have often been distributed according to political allegiance and loyalty rather than according to need. We are confident in our conclusion that the extent to which the drought industry enhanced the coffers of local leaders is far surpassed by the extent to which clientilistic political power based on patronage has been reinforced during times of crisis.

It is precisely in this way that the traditional governance system fails to provide a solution to the continued crisis of regular drought and to the ability to adapt to projected increases in desiccation and drought frequency as a result of global warming. It becomes a self-reinforcing feedback loop. The patronage-based governance system in rural Ceará in effect benefits from environment crisis by increasing the resources that local authorities can distribute to enhance their power base and the adaptive capacity of the population is diminished in direct relation to this increase.

MAPLAN: a subversive methodology for local planning

Our research on climate vulnerability in Ceará began in 1997 with a survey of 484 rural households designed to assess the impacts of government policies on farm livelihoods, particularly the use of climate forecast information (Lemos et al., 2002).[2] This study revealed the deep and pervasive level of vulnerability among rainfed households and paucity of technological options available to them. The quest to understand the usefulness of climate information led to a focus on municípios, the minimal decision-making unit for public resource management, based on the logic that if individual farmers lacked the adaptive capacity to take advantage of climate information, perhaps the prefeitura was positioned to do so. After several pilot studies, the research team developed a methodology that could incorporate both the output and the perspective of science into a localized planning process to mitigate the impacts of drought and permanently reduce the levels of household vulnerability.

In the professionalizing political narrative of the state (described above), two key elements of the political discourse were decentralization and participation. The research team thus selected pilot municípios where the local prefeitos had publicly

---

[2] This research was funded by NOAA (National Oceanic and Administrative Administration) from 1996 to 2005 and was carried out in partnership with FUNCEME, the Government of Ceará and the Federal University of Ceará.

embraced and defended this discourse. The prefeitos were invited to discuss the ideas and process of the methodology with the state government (Secretariat of Cities – formerly the Secretariat of Regional and Local Development – SDRL), including the cost-sharing obligations of the município. Through this self-selection process several municípios withdrew. The methodology, known as MAPLAN (*Mapeamento para Planejamento Participativo* – 'Mapping for Participatory Planning'), integrates community participatory research tools with the development of a GIS. Community and wealth mapping are widely accepted tools for participatory research (Chambers, 2006), and the mapping of climate vulnerabilities is institutionalized in UN agencies such as the World Food Program. Participatory GIS is also well established as an innovative qualitative methodology (Rambaldi et al., 2006). What makes MAPLAN unique is that its outcome – a set of asset, vulnerability and priority problem maps – is less interesting in terms of its knowledge value but critically important in terms of the dynamic by which the maps are generated.

The implementation of MAPLAN is conducted in several stages, and the precise methodology is described elsewhere (Nelson, 2005). Here we briefly outline the methodology and then focus on several key aspects that are cogent to our present argument. The unit of analysis is the *comunidade* (community), a nucleated settlement that varies from 3–200 households in size. Any given município might have from a dozen to several hundred communities depending on its area and resources. Working with self-identified local assistants, and with the full support of the município governments, a series of community meetings were held to identify factors that constrained the ability of participants to adapt and respond to drought events and to identify local assets that could be mobilized in the future. Meetings were not held in each of the communities – conducting meetings in dozens of communities with 15 residents is logistically infeasible – but all communities were represented at the focus group meetings which were geographically spaced throughout the municípios. These data were incorporated into a series of attribute-specific maps as well as summary maps that visually represented the situation of each community. The maps were validated by participants and then served as the basis to discuss priority areas and possible actions. This information was then discussed at the município level and incorporated into drought planning and long-term vulnerability reduction planning. The adaptation plans are then negotiated with the state government to create an investment strategy that directly corresponds to each adaptation plan.

Nearly all of the data included in the GIS were provided by the local participants. During the community meetings individuals had the opportunity not only to characterize the various attributes of their regions, but more importantly, to identify what attributes should even be considered for discussion. Common to many of the discussions were issues of access to water, healthcare and electricity, transportation infrastructure, educational opportunities and others. Because all data in a GIS are geo-referenced and can therefore be represented on a map, it is possible to readily aggregate the data to the level of regions and the state, without losing any of the richness of the local detail.[3] The maps serve as vehicles of communication between local individuals and the larger public sector. The use of IT provides the outputs with an instant legitimization that policy-makers do not usually accord to non-experts and locally generated knowledge.

The original objective of MAPLAN predecessors was to contribute specifically to drought mitigation planning. This narrow drought focus shifted as the participants repeatedly suggested that the ability to adapt to frequent droughts is contingent on a multiple of socio-economic factors operative at local, national and global scales. Thus, community discussions rarely made use of the word 'drought', and participants explicitly stated that with sufficient access to resources and opportunities droughts would cease to be an issue of concern (as it has, for example, in the south-western USA). They identified the types of resources and opportunities specific to each community. As the integral partner in the development of MAPLAN, the SDRL committed itself to funding and implementing a number of the priority projects that were identified in community meetings and later elaborated within the município. This was a first step in the process of changing the way in which resource allocation has taken place within the state. Breaking with traditional practices, SDRL bypassed local power-brokers to distribute funds in a transparent manner for projects that were locally appropriate. Absent from this process is the political bargaining and patronage that have defined the management of public goods and resources throughout the state.

For 20 years now, 'participation' has been in the lexicon of policy-makers in Ceará. Yet what we consider to be true participation has often been lacking. Rather, participation included community consultations prior to the implementation of a (already designed and funded) project, or the evaluation of a project that was designed with no local input. In this model, participation never challenged the dominant governance structure. True participation, which MAPLAN was designed to facilitate, occurs in a context in which individuals have the opportunity to define and characterize the problem(s), identify opportunities and solutions, and contribute to the planning process. This requires not only the physical presence of people, but a process designed to engage with the specific people that are involved. The continued public discussion around the maps as the planning process unfolds assures that local community participation remains active and vital. The visual impact of

---

[3] Geo-referencing refers to defining an attribute in physical space by assigning map coordinates to that attribute. There are many different types of attributes that may be geo-referenced including, for example, a well, a community, or an area of similar livelihoods. Geo-referencing defines the spatial relationships between different attributes and allows for the scaling up and down of data based on map coordinates.

maps in bright colours (especially for the illiterate residents) serves as a basis of the monitoring and accountability for a given political administration. Community forums meet on a regular basis, which provides the opportunity to track change through time. If, for example, a community remains in the red condition (indicating the least favourable situation) for the four years of a prefeito's mandate, it is clear to all that needs have not been met, and the administration must justify the basis for their resource allocation.

We have called this methodology 'subversive' fully acknowledging that it seeks to undermine a traditional form of governance that has maintained extremes of inequality throughout the history of the region. MAPLAN has been designed to co-opt the political discourse and to demonstrate a process by which such terms as 'decentralization' and 'participation' are operationalized in practice rather than revered as distant ideals. This approach obliges a process of effective community participation and creates a dialogue between prefeito and people based more on the rights of citizenry than on personal favour. It is clear that MAPLAN would not function if it were implemented blanket fashion across the state. The entrenched forms of governance would quickly undermine any possible gains. Rather, the role of MAPLAN is to first fill a gap in experience and practice in situations where the public and public officials are already searching for a way to move beyond traditional political processes. Second, once the process takes root, the initial successes demonstrate to residents of neighbouring municípios that alternatives exist.

Finally, the impact of MAPLAN has been to erode some of the long-standing feedbacks in the patronage system and to reverse the planning process at the state level (Figure 21.2). Removing the incentives (individualized access to resources) for individuals to make themselves beholden to patrons the project methodology helps set the stage for initiating desired, appropriate and equitable adaptation. Characteristically, solutions for the rural areas are developed in the state capital by experts then exported to the municípios that are able or wish to participate. Thus, resources have traditionally been attached to project activities. Under MAPLAN, part of state and federal resources now meet the priorities of locally devised plans based on local needs assessments. There are many challenges to the institutionalization of MAPLAN, due to the resilience of traditional governance systems and their elites; however, the potential for change has been recognized by individuals at the state and local levels. The state government has now internalized the MAPLAN methodology and has applied it in over 20 municípios as part of the state rural development strategic plan. Locally, residents of municípios neighbouring those involved in MAPLAN recognize the benefits of the methodology and have petitioned to be involved in the next roll-out. Additionally, the GIS and maps are communal property of the participants. In several municípios, residents have

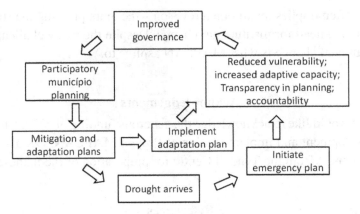

Figure 21.2 Governance for adaptation: new process through removal of incentives.

begun to use the data and information independently of MAPLAN. Holding their own community focus groups they are negotiating independently with município representatives and seeking development funds from outside sources.

Conclusions: governance and adaptation

The process of adaptation in Ceará is one that seeks sustainable forms of accommodation with a semi-arid environment given to devastating droughts. The mitigation of drought under the traditional governance system has served to protect the interests of the local power elites and has done little to reduce the chronic vulnerability or enhance the adaptive capacity of the rural population. We have attempted to demonstrate the inextricable link between forms of governance and adaptation. This link highlights some of the challenges for successful adaptation. Adaptation is a complex process that is intricately related to many non-climate factors. As the IPCC community and others have pointed to the critical need for local management of resources and local participation in the adaptation process, we have sought to present an experiment that is currently being implemented in Ceará. Promoting active, participatory adaptation may require significant changes in tradition and shifts in power relationships. For this reason, MAPLAN is not ideally suited to all contexts, even in Ceará. Rather it is rolled out on an incremental basis, expanding as the local demand for inclusion and voice increases.

MAPLAN is ambitious in scope and target, since it seeks profound change in rural society, change designed to assure that the marginalized voice finds space in the public arena. It addresses the complexity inherent in adaptation by bringing together theory and practice to tackle the issues of scale, redistribution, multiple stressors, and the augmentation of voice through comprehensive participation.

Adaptation often implies institutional revision – the more pressing the stressors the more urgent the need for institutional change. Despite the many challenges ahead, such institutional change is what MAPLAN aspires to.

Acknowledgements

The authors would like to acknowledge the generous funding of NOAA for supporting the development and implementation of Projeto MAPLAN. DN also acknowledges the support from the Tyndall Centre for preparation of the manuscript.

References

Adger, W. N. 2003. 'Social capital, collection action, and adaptation to climate change', *Economic Geography* **79**: 387–404.

Chambers, R. 1994. 'Participatory rural appraisal (PRA): challenges, potentials and paradigm', *World Development* **22**: 1437–1454.

Chambers, R. 2006. 'Participatory mapping and geographic information systems: whose map? Who is empowered and who disempowered? Who gains and who loses?', *Electronic Journal of Information Systems in Developing Countries* **25**: 1–11.

Eakin, H. and Luers, A. M. 2006. 'Assessing the vulnerability of social–environmental systems', *Annual Review of Environment and Resources* **31**: 365–394.

Finan, T. J. and Nelson, D. R. 2001. 'Making rain, making roads, making do: public and private responses to drought in Ceará, Brazil', *Climate Research* **19**: 97–108.

Folke, C., Hahn, T., Olsson P. and Norberg J. 2005. 'Adaptive governance of social–ecological systems', *Annual Review of Environment and Resources* **30**: 441–473.

Furtado, C. 1963. *The Economic Growth of Brazil*. Berkeley: University of California Press.

Füssel, H.-M. and Klein, R. J. T. 2006. 'Climate change vulnerability assessments: an evolution of conceptual thinking', *Climatic Change* **75**: 301–329.

Gaventa, J. 2002. 'Six propositions on participatory local governance', *Currents* **29**: 29–35.

Halsnaes, K. and Verhagen, J. 2007. 'Development based climate change adaptation and mitigation: conceptual issues and lessons learned in studies in developing countries', *Mitigation and Adaptation Strategies for Global Change* **12**: 665–684.

IPCC 2007. Parry, M. L., Canziani, O. F., Palutikof, J. P., Van der Linden, V. J. and Hanson, C. E. (eds.) *Climate Change 2007: Impacts, Adaptation and Vulnerability. Contribution of Working Group II to the Fourth Assessment Report of the Intergovernmental Panel on Climate Change*. Cambridge: Cambridge University Press.

Kelly, P. M. and Adger W. N. 2000. 'Theory and practice in assessing vulnerability to climate change and facilitating adaptation', *Climatic Change* **47**: 325–352.

Lebel, L., Anderies, J. M., Campbell, B., Folke, C., Hatfield-Dodds, S., Hughes, T. P. and Wilson, J. 2006. 'Governance and the capacity to manage resilience in regional social–ecological systems', *Ecology and Society* **11**: 19.

Leichenko, R. M. and O'Brien, K. L. 2001. 'The dynamics of rural vulnerability to global change: the case of southern Africa', *Mitigation and Adaptation Strategies for Global Change* **7**: 1–18.

Lemos, M.C., Finan, T.J., Fox, R., Nelson, D.R. and Tucker, J. 2002. 'The use of seasonal climate forecasting in policymaking: lessons from Northeast Brazil', *Climatic Change* **55**: 479–507.

Marengo, J.A. 2007. *Caracterização do clima no Século XX e cenários no Brasil e na América do Sul para o Século XXI derivados dos modelos de clima do IPCC*. São Paulo: Ministério do Meio Ambiente, Secretaria de Biodiversidade e Florestas, Diretoria de Conservação da Biodiversidade.

Nelson D.R. and Finan T.J. 2009. 'Weak rains: dynamic decision making in the face of extended drought in Ceará, Brazil', in Jones, E.C. and Murphy, A.D. (eds.) *The Political Economy of Hazards and Disasters*. Walnut Grove: Altamira Press.

Nelson, D.R. 2005. The public and private sides of persistent vulnerability to drought: an applied model for public planning in Ceará, Brazil. PhD dissertation, University of Arizona.

O'Brien, K.L., Leichenko, R., Kelkar, U., Venema, H., Aandahl, G., Tompkins, H., Javed, A., Bhadwal, S., Barg, S., Nygaard, L. and West, J. 2004. 'Mapping vulnerability to multiple stressors: climate change and globalization in India', *Global Environmental Change* **14**: 303–313.

Rambaldi, G., Kwaku Kyem, P.A., McCall, M. and Weiner, D. 2006. 'Participatory spatial information management in developing countries', *Electronic Journal for Information Sharing in Developing Countries* **25**: 1–9.

Tendler, J. 1997. *Good Government in the Tropics*. Baltimore: Johns Hopkins University Press.

Villa, M.A. 2000. *Vida e Morte no Sertão*. São Paulo: Editora Atica.

Vogel, C., Moser, S.C., Kasperson, R.E. and Dabelko, G.D. 2007. 'Linking vulnerability, adaptation, and resilience science to practice: pathways, players, and partnerships', *Global Environmental Change* **17**: 349–364.

22

Climate adaptation, local institutions and rural livelihoods

Arun Agrawal and Nicolas Perrin

Introduction

To understand the role of institutions in future adaptation of rural livelihoods to climate change, especially by poorer and more marginal groups, it is essential to attend to the historical repertoire of strategies used by rural populations. Natural resource-dependent rural households are likely to bear a disproportionate burden of the adverse impacts of climate change – droughts, famines, floods, variability in rainfall, storms, coastal inundation, ecosystem degradation, heat waves, fires, epidemics, and even conflicts. In some parts of the world, these effects may already be in play with potentially disastrous consequences for the poor (Adger et al., 2007). Many households in vulnerable regions could periodically be driven into destitution and hunger and find it difficult afterwards to recover.

Even as it is clear that poorer and disadvantaged groups around the world will suffer greatly from climate change, it bears remembering that the rural poor have successfully faced threats linked to climate variability in the past. History, as the cliché goes, may be a poor guide, but it is the only available guide. Even if future climate-related threats might appear prospectively to be historically unprecedented, analysing past impacts and responses is undoubtedly important in understanding the feasibility of future initiatives. After all, the only alternative to adaptation is extinction unless the world strictly and immediately limits its future emissions, an outcome surely in doubt given the record of the past decade.

The success of historically developed adaptation practices among the rural poor depends crucially on the nature of prevailing formal and informal rural institutions. Our chapter focuses on how rural institutions can help shape and enhance the adaptation practices of the rural poor in relation to climate change-induced risks and

Adapting to Climate Change: Thresholds, Values, Governance, eds. W. Neil Adger, Irene Lorenzoni and Karen L. O'Brien. Published by Cambridge University Press. © Cambridge University Press 2009.

how external interventions can help strengthen the functioning of rural institutions relevant to adaptation. It presents a brief typology of rural institutions using the familiar distinction between public, civic and private sectors, surveys some important recent work on adaptation, and then outlines an analytical framework through which to view the relationship between rural institutions, adaptation due to climate change, and livelihoods of the rural poor. It applies this analytical framework to 118 cases of adaptation practices drawn from the United Nations Framework Convention on Climate Change (UNFCCC) coping strategies database.[1] Using the basic finding from this analysis – that rural institutions are ubiquitous in framing and facilitating adaptation to climate change – it examines the adaptation projects and initiatives discussed in the 18 National Adaptation Programmes of Action (NAPAs) prepared by national governments with the support of the UNFCCC.[2] The analysis of the NAPAs shows the relatively limited attention national governments have paid to institutions even as the scholarly literature views institutions and governance as the centrepiece of future adaptation efforts. The chapter ends with an urgent call for action in three broad institutional domains if adaptation efforts are to lead to successful outcomes.

Rural institutions and livelihoods

Institutions are humanly created formal and informal mechanisms that shape social and individual expectations, interactions and behaviour. They can be classified as falling into public (bureaucratic administrative units, and elected local governments), civic (membership and cooperative organizations) and private sectors (service and business organizations) (Uphoff and Buck, 2006, p. 47). Table 22.1 below provides some indicative examples, but by no means a comprehensive listing, of the type of formal and informal institutions that may be present in rural areas and play a role in the ways rural households respond to climate change.

Livelihoods comprise the capabilities and material and social assets necessary for a means of living (Chambers and Conway, 1992). A sustainable livelihood includes the idea of coping with and recovery from external stresses so as to maintain or enhance existing capabilities and assets – a notion central to the definitions of resilience being discussed in relation to climate change.

Institutions influence the livelihoods and adaptation of rural households in three important ways.

---

[1] The UNFCCC database can be accessed at http://maindb.unfccc.int/public/adaptation/ (see UNFCCC, 2008a). For the analysis presented in this paper, the database was accessed and the relevant data downloaded in January 2008.

[2] The available NAPA documents for the different countries can be accessed at the UNFCCC website (see UNFCCC, 2008b). The analysis in this paper is based on documents available in January 2008.

Table 22.1 *Indicative examples of formal and informal rural institutions relevant to adaptation (I, informal institutions; F, formal institutions)*

	Public (state)	Private (market)	Civic (civil society)
Types of institutions	Local agencies (F)	Seed banks (I)	Labour exchanges (I)
	Local governments (F)	Service organizations (F)	Collective gatherings (I)
		Private businesses (F)	Membership organizations (F) Cooperatives (F)

(1) They structure the distribution of climate risk impacts. How particular social groups and populations will be affected by climate hazards is in part a function of the physical and structural characteristics of the hazard. It is also in part a function of the way macro- and micro-level institutions in a variety of domains affect distribution of risks related to climate hazards.

(2) They constitute and organize the incentive structures for household- and community-level adaptation responses that shape the nature of these responses. Institutional incentives are key in determining whether adaptation responses will be organized individually or collectively because institutions affect the emergence of leadership in different contexts, costs of collective action and the extent of transactions costs.

(3) They mediate external interventions into local contexts, and articulate between local and extra-local social and political processes through which adaptation efforts unfold. External interventions in the shape of finances, knowledge and information, skills training, new institutional inputs, and technological support can assume many different forms. Local institutions shape the acquisition and distribution of these interventions in fundamental ways, thereby affecting the degree of success of such interventions.

These basic points about the role of institutions in adaptation are summarized in the 'Adaptation, Institutions and Livelihoods' framework represented in Figure 22.1 (Agrawal, 2008).

Although much work on climate change and social responses to climate risks recognizes the relevance of institutions to adaptation, existing work on the subject has tended either to focus on highly specific case studies of local adaptation, or to examine national-level policies around adaptation. Comparative analyses of adaptation strategies that provide broadly generalizable insights into the role of different kinds of institutions, show how institutions link local responses to external interventions, and examine the institutional articulation at the local level among different kinds of institutions are sorely needed. Such comparative studies could play a significant role in developing middle-range generalizations about the role of institutions in adaptation – necessary both for deepening the theoretical understanding

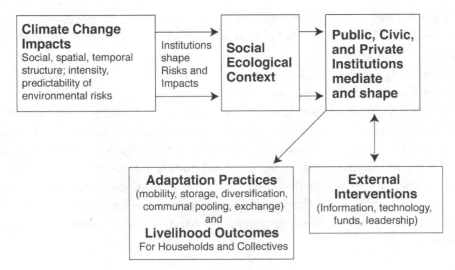

Figure 22.1 Adaptation, institutionsand livelihoods framework.

of the role of institutions in the context of climate change, and using such theoretical understanding to guide policy debates and discussions. Accordingly, the ensuing discussion examines especially the right-hand half of the above framework in relation to the role of institutions in adaptation by poor and marginal rural households in a variety of locations.

Climate change and adaptation

Given the nature of climate change hazards – droughts, heat waves, flooding and storms, among others – the stresses they create for rural livelihoods will manifest in two major ways: reduction of existing livelihood options, and perhaps more importantly in the short to medium run, greater volatility and unpredictability in streams of livelihoods benefits, especially in semi-arid, mountainous, polar and coastal ecological environments. Given the major uncertainties in how specific micro-locations and groups will experience and be affected by climate change, it is likely more fruitful for policy interventions to focus on improvements in adaptive capacity of disadvantaged rural populations rather than on identifying specifically how a given group of rural poor in a particular village or district will be affected by climate change.

Development strategies and institutional interventions that focus simply on improving total benefits to households without taking into account how households can address fluctuations in their livelihoods seem ill-suited to address the impacts of climate change. On the one hand, they ignore the most important characteristics of climate-related stresses – increased riskiness of livelihoods. On the other hand, they ignore the very real concerns of the rural poor about preventing hunger and destitution. Given that many rural households have only limited access to markets – for reasons both of ill-developed infrastructure and of limited purchasing power, high levels of riskiness in the environment cannot in a vast number of cases be ameliorated by engaging in market exchange.

To strengthen the adaptive capacity of the rural poor, therefore, governments and other external actors need to strengthen and take advantage of already existing strategies that many households and social groups use collectively or singly. Examining the environmental risks that rural populations have historically faced, their cultural responses to these risks, and the institutional configurations that facilitate individual and collective adaptation strategies is therefore a fruitful area of inquiry and policy analysis for generating effective coordination with external interventions.

Forms of adaptation in response to climate risks

A policy-relevant framework for examining adaptation practices in the context of rural institutions and livelihoods needs to be sufficiently general to cover the many empirical examples of adaptation practices used by different social groups, but also needs to be based on an analytical approach that takes into account the most important characteristics of the impacts of climate change on rural livelihoods – likely increases in environmental risks, reduction in livelihoods opportunities and stresses on existing social institutions. Few existing writings on adaptation, focused as they typically are on specific case studies or national-level policy concerns regarding appropriate interventions, present the kind of middle-range theories necessary to understand historical adaptation responses comparatively.

Climate change is likely to manifest around increased risks to rural livelihoods. These risks can be classified into four different types: across space, over time, across asset classes and across households. The basic coping and adaptation strategies in the context of livelihoods risks can correspondingly be classified into a set of four analytical types: mobility, storage, diversification and communal pooling. In addition, where households and communities have access to markets, market-based exchange can substitute for any of the four classes of adaptation strategies above (Halstead and O'Shea, 1989; Agrawal, 2008). Where successful, these responses either reduce spatial, temporal, asset-related and/or community-level risks directly, or reduce them by pooling uncorrelated risks associated with flows of livelihoods benefits from different sources.

Mobility is perhaps the most common and seemingly natural responses to environmental risks. It pools or avoids risks across space, and is especially successful in combination with clear information about potential precipitation failures.

Storage pools/reduces risks experienced over time. When combined with well-constructed infrastructure, low levels of perishability and high level of coordination across households and social groups, it is an effective measure against even complete livelihood failures at a given point in time.

Diversification reduces risks across assets owned by households or collectives. Highly varied in form, it can occur in relation to productive and non-productive assets, consumption strategies and employment opportunities. It is reliable to the extent benefit flows from assets are subject to risks and risks have different impacts on the benefit streams from different assets.

Communal pooling refers to adaptation responses involving joint ownership of assets and resources; sharing of wealth, labour or incomes from particular activities across households, or mobilization and use of resources that are held collectively during times of scarcity. It reduces risks experienced by different households.

Exchange is perhaps the most versatile of adaptation responses. Usually it is viewed as a means to promote specialization and increase revenue flows. But it can equally substitute for the first four classes of adaptation strategies to reduce risks when the poor have access to markets. As a means to reduce risks it can go together with high levels of specialization and institutionalization of exchange relations: consider as an example, buying insurance to cover risks of crop failure. Resorting to exchange or promoting exchange-based adaptation to address climate risks needs however to be treated with some caution given the highly unequal access to markets across different social groups, especially those who are in marginalized situations.

The success, and more generally the prospects of adaptation practices, depends on specific institutional arrangements – adaptation never occurs in an institutional vacuum. Thus, all adaptation practices require property rights, norms of trust are necessary for exchange, storage requires monitoring and sanctions, mobility cannot occur without institutions that provide information about the spatial structure of variability, and agricultural extension institutions can facilitate diversification.

The adoption of adaptation practices by specific households and communities is more or less likely depending on their social and economic endowments, networks of relationships and access to resources and power. For example, the poor are more likely to migrate in response to crop failure; the rich more likely to rely on storage and exchange.

There are natural affinities and incompatibilities among the broad classes of practices above. Storage and mobility tend not to go together. Other combinations complement each other: storage and exchange can play off temporal variability against spatial variability (Halstead and O'Shea, 1989, p. 4).

Finally, the effectiveness of adaptation can be institutionally enhanced by external interventions and local collective action: provision of information to reduce unpredictability associated with climate-related events and trends; technical advances leading to higher crop or resource productivity; financial and investment supports that make the adoption of technological changes more widespread; and leadership interventions that reduce costs of collective action (see Figure 22.1 above).

Case evidence on adaptation practices and rural institutions

Although there is a large case literature on adaptation and adaptive responses, there are few comparative studies of cases of adaptation. In this context of limited comparative work, the cases collected in the UNFCCC database on coping strategies form a useful empirical basis for assessing the usefulness of the framework represented in Figure 22.1, and for examining the relationship between different classes of adaptation practices and institutional types. The database includes cases from a large number of different countries as part of the effort undertaken by UNFCCC to explore the nature and distribution of adaptation responses and adaptive capacities in the poorer countries of the world.

Although the UNFCCC refers to the information collected on coping strategies as a database, the actual information conforms less to what is typically imagined as a structured database, more to a compendium of cases and different kinds of documents pertaining to a specific case. For each case, the UNFCCC website provides some basic information – ecological context, nature of hazard, types of impacts, location and the name and location of the case – and a brief description of the adaptation practice. For a majority of the cases included in the database, the UNFCCC also provides additional links to other documents (not all links are active) from which additional information about the case can be gleaned. Much of this information has probably been supplied by personnel in the relevant environment ministry or agency in a country, or non-government organizations involved in consultations around climate. As a result, it is likely that most of the cases on which information is available in the database are ones that have reached some level of official notice.

The cases included in the database are widely distributed around the world, covering 42 different developing nations in Africa, Latin America and Asia. It is evident that the UNFCCC database on coping strategies constitutes perhaps the most comprehensive effort worldwide to collect information on how different social groups around the world have attempted to cope with environmental variability.

It is worth mentioning that although the UNFCCC cases focus on coping strategies, analytically it is difficult to distinguish between coping and livelihoods strategies. Adaptation strategies are viewed by some scholars as being prospective in nature in contrast to coping efforts which are seen as being retrospective and in response to specific experiences of variability. However, given that many climate hazards are recurrent, strategies adopted as responses to experiences of climate risks are also prospective in terms of future experiences of scarcity. Further, historical efforts to cope with production failures associated with some kinds of risks can have significant utility in relation to other kinds of risks as well.

The ensuing discussion of adaptation cases builds upon an analysis of 118 distinct cases in the UNFCCC database.[3] The cases in the UNFCCC database all cover a single time period rather than describing changes over multiple time periods. The different cases include both customarily evolved examples of adaptation as well as more recently created or externally introduced examples. The information in the database is presented as case description rather than as an Excel spreadsheet. For the included cases, the database also typically provides links to additional documentation. The analysis of the UNFCCC case data involved reading each case description to identify the different variables of interest – particularly to gather information from the case descriptions on the nature of adaptation being examined, the types of institutions involved, and the nature of external inputs where these were available to facilitate adaptation. The case information was thus used to generate an Excel spreadsheet which formed the basis of the tables presented below.

Table 22.2 provides information about the distribution of different kinds and combinations of adaptation practices, and finer distinctions within the five classes of adaptation practices mentioned earlier. The evidence in the cases also indicates some interesting patterns. Perhaps the most interesting points concern the absence of mobility in the examined cases (see Table 22.2), and the occurrence of exchange typically only in combination with at least one other type of adaptation practice.

The absence of mobility among the case examples included in the database is interesting to say the least. It most likely pertains to biases in the way UNFCCC data on adaptation practices was collected. Mobility is a widespread adaptation strategy, particularly in dry and semi-arid areas by pastoralist and agro-pastoralist populations as a routine response to erratic rainfall or by agricultural populations moving in search of wage labour opportunities. It is also an emergency response to drastic and unpredictable hazards where existing social and economic systems fail – as would be the case for severe storms, hurricanes or floods. However, in

---

[3] In using 118 cases from the UNFCCC database, we dropped cases where the two cases were repeated as examples of different themes, where the information in a given case was a duplication of that from another case, or where the information was sketchy and incomplete. Duplication for example can refer to a situation where two cases each talked about an early warning system in two different locations in the same terms, described it as serving the same function, and focused on the same institutions in examining the implementation of the system.

Table 22.2 *Frequency distribution of major classes of adaptation practices*
(N=118)

Class of adaptation practice	Corresponding adaptation strategies	Frequency
Mobility	1. Agro-pastoral migration 2. Wage labour migration 3. Involuntary migration	0
Storage	1. Water storage 2. Food storage (crops, seeds, forest products) 3. Animal/live storage 4. Pest control	11
Diversification	1. Asset portfolio diversification 2. Skills and occupational training 3. Occupational diversification 4. Crop choices 5. Production technologies 6. Consumption choices 7. Animal breeding	33
Communal pooling	1. Forestry 2. Infrastructure development 3. Information gathering 4. Disaster preparation	29
Market exchange	1. Improved market access 2. Insurance provision 3. New product sales 4. Seeds, animal and other input purchases	1
Storage and diversification	Examples of combinations of adaptation classes are drawn from the strategies listed above	4
Storage and communal pooling		4
Storage and market exchange		6
Diversification and communal pooling		4
Diversification and market exchange		26

many discussions of mobility in the context of adaptation to climate change and
variability, it is viewed in a negative light, and often as a maladaptation. This is not
to say that the UNFCCC database is underpinned by a similar view of mobility –
but to point to one possible reason why contributors to the database may not have
focused on mobility as an adaptation practice.

For exchange to occur, the UNFCCC case data suggests, households and communities need to resort to at least one other type of adaptation practice as well. Table 22.2 suggests that the most common classes of adaptation responses are diversification and communal pooling on their own, and diversification and exchange as a pair.

The above patterns at a minimum can be taken as being informative about the more than 100 cases included in the UNFCCC database, in itself an advance over the state of the field which has tended to focus typically on single cases. But the conjunction of exchange with at least one other class of adaptation practice may also be representative of adaptation practices more broadly – it makes analytical sense that households will pursue exchange typically when they have some surplus to exchange – and such surplus is likely generated when households are also involved in some other classes of adaptation. The limited representation of mobility in the data seems an artefact of reporting bias – agro-pastoral and wage labour groups have used mobility as an adaptation to environmental variability for generations – indeed, mobility often also occurs in conjunction with other adaptation strategies such as diversification. Some form of the common official bias against mobility, often visible in climate change discussions in the invocation of climate refugees, may be at play in the underrepresentation of mobility as an adaptation strategy in the UNFCCC database as well.

The UNFCCC data also show other interesting patterns. In nearly all cases, local institutions are necessary enablers of the capacity of households and social groups to deploy specific adaptation practices (see Table 22.3). In 70 cases, the primary structuring influence of adaptation stems from local institutions without external interventions. In 41 other cases, local institutions work in conjunction with external interventions. The inference is evident – without local institutions, rural poor groups will find it far costlier to pursue any adaptation practice relevant to their needs. Table 22.3 also indicates that when rural institutions work in conjunction with external interventions, it is more likely that benefits from adaptation practices will be shared more widely in the collective.

Table 22.4 uses the data collected by the UNFCCC to examine how different kinds of institutions are associated with different types of adaptation practices, using the broad classification of private, public or state-sponsored, and civil society institutions as the relevant categories. The table indicates that civil society institutions play a striking role in adaptation. In contrast, market-based, private institutions seem to play a far more limited role in existing cases of adaptation in the UNFCCC database.

Although the UNFCCC database does not provide enough information to make a detailed assessment of the subdivisions within the broad categories of public, private and civic institutions, it does suggest that public institutions are only

Table 22.3 *Types of institutions and distribution of benefits adaptation (N=118)*

	Individually oriented benefits from adaptation practices	Collectively oriented benefits from adaptation practices	Total
Local institutions functioning in conjunction with an external intervention	15	26	41
Local institutions without external interventions	55	22	77
Total	70	48	118

Source: UNFCCC coping strategies database.

Table 22.4 *Association of adaptation practices with institutional arrangements (N=118)*

	Public	Civic	Private	Public and civic	Private and civic	Total
Storage	0	8	0	3	0	11
Diversification	0	**19**	1	12	1	33
Communal pooling	4	11	0	**14**	0	29
Storage and diversification	0	2	0	2	0	4
Storage and exchange	0	4	0	1	1	6
Diversification and exchange	0	**13**	4	5	4	26
Other	2	4	0	3	0	9
Total	6	61	5	40	6	118

Source: UNFCCC coping strategies database.

infrequently associated with market exchange processes promoting adaptation; and that when market actors are involved in adaptation practices, it is likely that they would assist exchange-based efforts.

Given the overall distribution of institutional arrangements through which adaptation is facilitated at the local level, it is not surprising that much of the institutional action is focused around civic and a combination of public and civic institutions. A few points are still worth highlighting from the information in this

Table 22.5 *Local institutions and their mediation of external interventions to promote adaptation (N=41)*

	Public	Civic	Public and civic	Civic and private	Total
Information	0	2	8	0	10
Technical inputs	2	4	1	0	7
Financial support	2	0	6	1	9
Information/ technical inputs	0	4	2	0	6
Technical inputs and financial support	0	4	1	0	5
Other	0	2	2	0	4
Total	4	16	20	1	41

Source: UNFCCC coping strategies database.

table (the relevant cells have the numbers in bold in Table 22.4). The first is that civic institutions and partnerships between civic and public institutions seem to occur more frequently to promote diversification and communal pooling. There are relatively few instances of civic institutions promoting storage or mobility, or for that matter a combination of different adaptation strategies. In contrast, much of the involvement of private institutions and the partnership between civic and private institutions seems to focus on the promotion of diversification and market exchange. This is an expected finding in many ways – one expects market actors and processes to be most suited for exchange-based activities, and indeed this is also the finding in the data.

Table 22.5 provides a summary overview of how public, civic and private institutions mediate external interventions to promote adaptation. It focuses on the 41 out of the 118 cases in Table 22.3 that clearly show the involvement of external actors in promoting adaptation. The total number of cases is too small, therefore, to make broad generalizations, but in looking at the distribution of the specific cases based on the main patterns in the data, some useful lessons can be derived.

The information in the table above suggests that the major external interventions to support local adaptation efforts have focused on providing information and financial support. There are fewer cases in which a variety of external interventions have been combined to facilitate adaptation, and in no case have external actors provided strong leadership or attempted local institutional reconfiguration to support adaptation. A closer look at the data explains these patterns. The vast majority of cases of information provision and financial support concern adaptation practices related to disaster preparedness, early warning systems about failure of rains, and private or public infrastructure that can withstand climate hazards such as floods and storms. Certainly, the role of external interventions in promoting adaptation is not exhausted by these three types of adaptation to the threat of climate change. As indicated by the list of specific adaptation strategies in Table 22.4, itself only a subset of the different types of adaptation practices that rural populations have already been attempting, there are many more ways in which external support can reinforce adaptation efforts and support institutions that are shaping, facilitating and reinforcing local adaptation efforts. The conclusion is inescapable that external forms of support focus on an incredibly small slice of the huge diversity of adaptation mechanisms that local actors and institutions are inventing and attempting.

In summary, the discussion above of the information on adaptation and institutions in the UNFCCC data brings out four important points. One, institutions are ubiquitous in local rural efforts to adapt to climate variability. It is important to highlight this point because of the nature of much policy debate on institutions. Many such discussions tend to focus on the institutions at the national and international levels that would be necessary to facilitate adaptation, missing the point that adaptation is inherently local and therefore it is critical to attend to local institutions in thinking about effective adaptation. The cases in the UNFCCC database were collected without reference to whether institutions were involved in the adaptation practices being described. The fact that institutions are relevant in all the cases of adaptation included in the UNFCCC data indicates with a very strong probability that local institutions are crucial, perhaps even necessary to adaptation.

The second point to be highlighted from the data is the absence of mobility as an important class of adaptation. It is quite likely that the absence of this adaptation strategy is the result of reporting bias, but it also indicates that official discourses around climate adaptation need to refer to and analyse mobility more carefully because different forms of mobility are undoubtedly one of the major ways in which social groups are likely to adapt to climate change.

The third important issue that the UNFCCC data bring up is the relative importance of civil society institutions in adaptation, either on their own, or in combination with public institutions. Civil society institutions are not only active in facilitating different kinds of adaptation practices; they are also very commonly associated with the mediation of external interventions for adaptation.

Finally, it is worth highlighting that unlike the situation for climate change mitigation, private and market institutions have been relatively absent in facilitating adaptation in rural areas. This absence constitutes an important arena of interventions for public policy to begin to craft incentives that can draw private institutions more centrally in facilitating adaptation.

Examining the National Adaptation Programmes of Action

It is interesting to compare the information from the analysis of the data brought together by the UNFCCC with the types of adaptation priorities identified in the National Adaptation Programmes of Action (NAPAs) which different national ministries of environment have prepared with the support of the United National Development Programme (UNDP). Information on only 18 NAPAs was publicly available at the time of writing this chapter (February 2008). Each of these NAPAs identifies the projects in which national ministries would like to invest adaptation funds if and when such funds become available. Unlike the data in the UNFCCC coping strategies database which indicate the type of historical adaptations that have been pursued in specific cases, the adaptation projects described in the NAPAs constitute expressions of what different countries would like to do around climate adaptation in the future. The most interesting parts of these NAPA documents for this chapter are their lists of priority areas and activities.

A review of the NAPA documents from the 18 countries listed on the UNFCCC website suggests that in all cases there was widespread participation by a cross-section of national government agencies and non-government organizations in their preparation. In many ways the NAPAs provide a comparable set of national level statements by official agencies about what they view as adaptation priorities, and how they expect to go about pursuing adaptation. It is therefore an appropriate set of documents to examine to assess the extent to which national level planning around climate adaptation has taken local rural institutions into account.

Figure 22.2 presents basic information about NAPA adaptation projects concerning their thematic focus and numbers. The figure shows that the largest number and proportion of adaptation projects are focused on sectoral issues related to improvements in natural resources-related activities such as in agriculture, forestry, water conservation and irrigation, and in the development of infrastructure and disaster relief. Very few of the countries have identified urban impacts of climate change or new research on the best means of adaptation as high priority areas needing support.

Although a detailed analysis of these projects along the lines of the UNFCCC database is not possible given the nature of information presented in the NAPA documents, it is still possible to make basic comparisons that point to the ways the policy process has worked in different countries to engage with local institutions

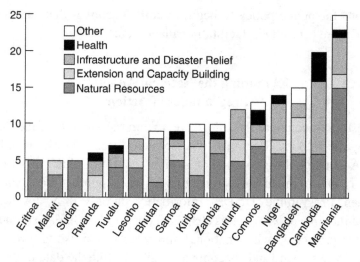

Figure 22.2 Sector-wise NAPA adaptation projects.

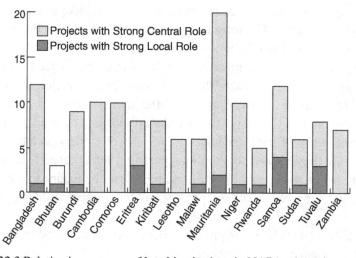

Figure 22.3 Relative importance of local institutions in NAPA adaptation projects.

in the urgent issues surrounding adaptation in the context of climate change. First, in contrast to the actual instances of adaptation described and enumerated in the UNFCCC database, most of the projects in the NAPA documents seem aimed far more at building the capacity of national governments and agencies to coordinate adaptation, to provide services to the general population, or to create infrastructure rather than to strengthen the capacity of local actors and institutions to undertake adaptation. Figure 22.3 provides information on two areas of concern for this chapter – the extent to which selected high-priority projects focus on communities, and the role they identify in the project design for community or local level public, private or civic institutions.

Thus, local institutions are incorporated as the focus of adaptation projects in just about 20% of the projects described in the NAPA documents. The limited focus on local actors is especially striking when it comes to the anticipated role of local-level institutions in adaptation. Only 20 of the 173 projects described in the NAPA reports identify local-level institutions as partners or agents in facilitating adaptation projects. Indeed, given this minimal level of attention to local institutions – even for projects that are focused on agriculture, water, forest management, fisheries, small-scale infrastructure and capacity-building for which local institutions can be viewed as basic components of an adaptation strategy – it is perhaps unnecessary to develop a refined argument about local institutions and adaptation. Despite widespread consultations that went into the production of the NAPA documents, it appears that the process was attentive in only a limited manner to the historical experiences of adaptation, indigenous or local adaptation strategies, and forms of local and cross-scale vulnerabilities of marginal peoples. In any further efforts to develop national adaptation plans the potential role of local civic institutions and institutional partnerships both at the local level and across multiple scales must receive much greater attention than it has hitherto received.

The analysis of the information on high-priority projects selected by relevant ministries in the least developed countries as identified by the UNFCCC brings home the enormous ground that still has to be covered by the national planning process in relation to adaptation and local institutions. Despite an explicit commitment to grass-roots processes and institutions articulated in the NAPA process, the actual documents and projects have paid relatively limited attention to rural institutions. Not only do most projects not incorporate local communities and institutions in adaptation plans, little evidence of consultation and coordination between the local and national level can be seen in the descriptions of the selected high-priority projects. Given that only a small proportion of all NAPA documents have been finalized at present, there is an opportunity both to redress this gap in the process by identifying how rural institutions can play a more defining role in projects targeted toward rural areas, and to provide guidelines for other attempts to develop territorially based adaptation plans in which interactions among institutions would be important to analyse and understand. It is especially useful to point to possible ways in which the analysis of NAPAs above may be useful – by calling for greater institutional coordination across levels, involvement of local institutional actors in project design and selection, and integration of different projects so as to promote a more holistic vision of adaptation in the context of climate-related threats to rural livelihoods.

Conclusion

This chapter identifies a framework through which to view the relationship between rural institutions, adaptation owing to climate variability and change, and livelihoods of the rural poor. Using the existing literature on risks and livelihoods, the chapter proposes five major classes of adaptation practices available to the rural poor in varying measures depending on their social networks, access to resources and asset portfolios: mobility, storage, diversification, communal pooling and exchange. Using data from the UNFCCC's local coping strategies database, the chapter identifies empirical patterns in the incidence and compatibility of these strategies. A comparison of these patterns with priority activities in NAPAs suggests that the NAPA documents – presenting the most widespread current national policy statements around adaptation – have paid relatively little attention to civil society or micro-level institutions in crafting national responses to climate change. It is quite likely that this inattention to local institutions is partly the result of the inadequate thought that has gone into the NAPA process, despite hundreds of thousands of dollars being spent on the NAPA process in each country. If adaptation is local, attention to local institutions is critically important in the design of adaptation projects and policies. Further, a close integration of different institutional arrangements is also likely critical for enhancing the effectiveness of adaptation practices. Without greater attention to local institutions and their role in adaptation efforts of different kinds, and the ways in which local and external institutions can be articulated in the context of adaptation, it is unlikely that adaptation interventions and investments will achieve much success.

Acknowledgements

We thank Andrew Norton, Robin Mearns, Mafalda Duarte, Gretel Gambarelli, Catherine McSweeney, Paul Siegel, Maria Lemos, Jesse Ribot, Neil Adger, Erin Carey, Rachel Kornak, Simon Batterbury, Julien Labonne, Daniel Miller, Nilufar Ahmad and participants at the 'Are there limits to adaptation' conference in February 2008 in London, and at the 'Social dimensions of climate change' conference in Washington DC in March 2008.

References

Adger, W. N., Agrawala, S., Mirza, M. M. Q., Conde, C., O'Brien, K., Pulhin, J., Pulwarty, R., Smit, B. and Takahashi, K. 2007. 'Assessment of adaptation practices, options, constraints and capacity', in Parry, M. L., Canziani, O. F., Palutikof, J. P., Van der Linden, V. J. and Hanson, C. E. (eds.) *Climate Change 2007: Impacts, Adaptation and Vulnerability. Contribution of Working Group II to the Fourth Assessment Report of the Intergovernmental Panel on Climate Change*. Cambridge: Cambridge University Press, pp. 717–743.

Agrawal, A. 2008. 'The role of local institutions in adaptation to climate change', paper prepared for the Social Development Department, World Bank, Washington, DC, 5–6 March 2008.

Chambers, R. and Conway, G.R. 1992. *Sustainable Rural Livelihoods: Practical Concepts for the 21st Century.* London: Defra.

Halstead, P. and O'Shea, J. (eds.) 1989. *Bad Year Economics: Cultural Responses to Risk and Uncertainty.* Cambridge: Cambridge University Press

United Nations Framework Convention on Climate Change (UNFCCC). 2008a. Database on Local Coping Strategies. Available at http://maindb.unfccc.int/public/adaptation/ (accessed 10 January 2008).

United Nations Framework Convention on Climate Change (UNFCCC) 2008b. National Adaptation Programmes of Action. http://unfccc.int/adaptation/napas/items/2679. php (accessed 10 January 2008).

Uphoff, N. and Buck, L. 2006. 'Strengthening rural local institutional capacities for sustainable livelihoods and equitable development', paper prepared for the Social Development Department, World Bank, Washington, DC.

23

Adaptive governance for a changing coastline: science, policy and publics in search of a sustainable future

Sophie Nicholson-Cole and Tim O'Riordan

Introduction

The risks of coastal flooding and erosion in the UK are changing in response to the likely effects of climate change, natural isostatic readjustment and the consequences of hard coastal defence initiatives which have lead to coastal instability elsewhere. In addition to the considerable uncertainty brought about by these factors, there has been a significant, strategic shift in national coastal management policy in England away from investing in expensive 'hard' engineered defence, toward designing a more naturally functioning coastline.

This policy change (Defra, 2005a) has come about in light of questions being asked about the physical sustainability (and increasing cost) of a reliance on engineered defences. This means that many coastal communities of varying size in England are now facing a situation of great unease and anxiety about their future; new policy preferences for retreat and realignment mean no future guarantees of protection. Coastal governance arrangements at national to local scales have not yet adequately responded to the new strategic outlook; the national shift of priority has not been matched by any compensation package or appropriate initiatives to promote the development and delivery of associated adaptation requirements in such locations.

Drawing on Tyndall Centre research, this chapter presents an analysis of the governance setting for the management of England's changing coastline (O'Riordan et al., 2006, 2008; Milligan and O'Riordan, 2007; Milligan et al., 2009). Taking the East of England region and in particular the county of Norfolk as a case study, the chapter examines why governance for a future sustainable coastline is so dysfunctional at present and why it is failing to create the conditions for adaptation to coastal change. It suggests that the present scientific evidence base is not

Adapting to Climate Change: Thresholds, Values, Governance, eds. W. Neil Adger, Irene Lorenzoni and Karen L. O'Brien. Published by Cambridge University Press. © Cambridge University Press 2009.

adequate for ensuring political and electoral consensus over decisions being made for the future planning of the coast amongst the many stakeholders involved. A mismatch and tension between national coastal management policy and strategy, and the cautious efforts to adapt at regional and local scales is highlighted.

This chapter contributes to the evolving policy arena in coastal management in Britain and further afield by clarifying the key underlying challenges facing adaptation to coastal change, which contribute to current coastal governance arrangements in England not yet being adaptive and therefore unable to deliver adaptation. These challenges operate at all scales of governance and account for the considerable conflict that characterises current coastal management in the East of England. They include different expectations and understandings surrounding the notion of a sustainable coast, strong sentiments over maintaining familiar and much-cherished 'safe' coastlines, huge dilemmas over appropriate planning for future settlement and infrastructure, an uncoordinated aggregate of governance arrangements and other institutional barriers to establishing a more adaptive coast for the middle of this century and beyond. These key challenges reflect possible limits to the prospects for adaptation to coastal change in England.

This chapter concludes that coastal governance itself must become more adaptive, and outlines a number of conditions that are necessary for coastal governance in England to deliver adaptation to coastal change at local to national scales. The term adaptive governance is used to reflect a coordinated and integrated approach to coastal governance (for example, between different parts of government, institutions and local communities, and across timescales). Adaptive coastal governance must incorporate learning and the flexibility to respond to examples of good and bad practice as well as the uncertainty and dynamism inherent in coastal systems. Additionally, at the heart of adaptive governance must lie a number of 'building blocks' or foundations of good governance – careful preparation, robust scientific information, common visions, coordination of responsibility, financing and responsive public engagement.

Key challenges facing adaptation
to a changing coast

In responding to an increasingly uncertain coastline which faces potentially significant and rapid change in the future, an adaptive governance framework is critical for ensuring the resilience and adaptive capacity of the region's environment, society and economy (for example Folke et al., 2005). Eight key factors currently inhibiting effective adaptation are presented in Figure 23.1. They summarise the difficulties being experienced in the management of the coastline of Eastern England and constitute key challenges to adapting to this dynamic coastline.

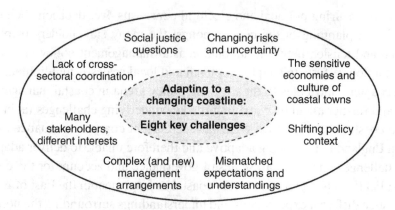

Figure 23.1 Eight key challenges facing adaptation to a changing coast.

Changing risks and uncertainty

The geologically soft coastline of Eastern England is highly dynamic and diverse in both its geomorphology and ecology, and it has a complex legacy of coastal defence. It encompasses eroding cliffs, shingle banks and sand dunes, which protect a low-lying interior. Much of this is at or below sea level, including the internationally famous wetland habitat, the Norfolk Broads (Bridges, 1998). Devastation caused by the North Sea storm surge of 1953 resulted in the construction of hard-engineered sea defences along much of this coast offering piecemeal protection from erosion and inundation. However, these alterations to natural patterns of sediment supply, transport, deposition and defence have led to an array of new coastal instabilities (see Leafe et al., 1998; French, 2004).

At a time when many of the defence structures are nearing the end of their life (or require substantial maintenance), climate change is becoming more of a reality. Sea level rise and the prospect of more extreme storm events threaten to exacerbate the already rapidly escalating costs of defending to higher standards and extremes (Hulme et al., 2002; Evans et al., 2004a, 2004b; IPCC, 2007). The south-eastern part of Great Britain is also subject to isostatic readjustment following the last ice age, exaggerating the impacts of a mean sea level rise (Shennan, 1989). Given the likelihood of future significant sea level rise and the physically vulnerable characteristics of parts of the UK coastline (for example Taylor et al., 2004), the UK Foresight Programme commissioned a study of flooding and coastal risk in the UK to inform long-term policy for the period 2030–2100 (Evans et al., 2004a, 2004b). It concluded that if flood management policies and expenditure continued unchanged, annual costs of protection would increase by the 2080s under every scenario considered, with the risk to coastal areas growing to unacceptable levels. The

Foresight study highlighted the need for key decisions to be taken, to invest more in sustainable approaches to flood and coastal management, and to learn to live with increased flooding (see also Pitt, 2007). This collection of changing physical drivers and the legacy of past management decisions present significant problems for coastal managers and communities today (for example Brennan, 2007).

McFadden (2007) argues that there is a need to improve the basic scientific knowledge that underpins policy-making at the coast; that the present governing framework does not give sufficient emphasis to this fundamental need. In this regard, the Tyndall Centre is developing an advanced integrated model to simulate the possible future coast under different climate and management scenarios (Nicholls et al., 2005, 2007). The aim of this Coastal Simulator is to support decision-making and dialogue about the future of our coastline (Brown et al., 2006; Jude et al., 2006, 2007; Milligan et al., 2009). What future coastlines actually become, however, lies in the realm of social and political engagement.

The sensitive economies and cultures of coastal towns

Changing physical risks and uncertainty are only one part of a more complex set of contributors to the vulnerability of coastal communities and their uncertain futures. A recent parliamentary inquiry (House of Commons Community and Local Government Committee, 2007a) presented a thorough overview of some of the sensitive social and economic aspects that contribute to the vulnerability of many coastal communities. The inquiry concluded that many English coastal towns share common, unsettling characteristics. These include:

- physical isolation (poor road networks and public transport provision);
- high deprivation levels and lack of affordable housing;
- inward migration of older people;
- high levels of transience (short-term residency);
- outward migration of younger people;
- poor-quality housing;
- fragile, isolated coastal economies.

One recommendation by the inquiry was for research into the specific social and economic challenges facing coastal communities. Another was to establish a coastal areas network to be supported by national and regional government in order to coordinate action to reduce the vulnerability of many people who live on the coast (House of Commons Community and Local Government Committee, 2007b). However, at present there is little understanding of how vulnerable coastal communities may be enabled to prosper given a 'naturally functioning', yet highly uncertain future coast.

A shifting policy context

The policy setting and governance arrangements for coastal management in England are in a state of transition. The recent official policy statement, entitled *Making Space for Water* (MSW) (Defra, 2005a) seeks to implement a more holistic and risk-driven approach to managing flood and coastal erosion risks. It represents a strategic shift in policy away from hard defence in favour of naturally functioning protection arrangements, for example through managed realignment. As part of the strategy, the government has widened the role of the Environment Agency (the leading public body for protecting and improving the environment in England and Wales) to include primary responsibility for coastal risk assessment and coastal management delivery, with a remit to strategically oversee the whole coastline. This means a power shift away from local government (Defra, 2005a), which does not equate with a comfortable arrangement for effective adaptation, particularly where realignment is emerging as a preferred option in the short to medium term (present to 50 years in the future). This is because there are, as yet, no mechanisms for enabling or coordinating the development of adaptation initiatives across organisations, governance scales or across sectors (O'Riordan et al., 2008). Part of the MSW work programme includes the development of a portfolio of possible options for facilitating the adoption and delivery of sustainable flood and coastal erosion risk management approaches. This is currently under way. However, there is no guarantee that any of the possible solutions considered through this project will be implemented (Defra, 2005b).

Shoreline Management Plans (SMPs) are the official mechanism for guiding decision-making about coastal protection for subsections of the coastline. They provide assessments of the risks associated with coastal processes and present a strategic and long-term policy framework to reduce these risks in a sustainable manner. They are nominally reviewed at five-yearly interviews to incorporate new scientific research results and revised regional and national policy guidance in line with national strategy (Leafe et al., 1998; Cooper et al., 2002; Defra, 2006). In Norfolk there has been difficulty in coming to a consensual acceptance of the proposals contained within the county's second-generation SMP (Anglian Coastal Authorities Group, 2006). Reflecting the national strategic redirection, a number of coastal sub-units now face recommended policy options such as 'no active intervention' in the long, medium or even short term (100, 50 and 25 years) to replace previous 'hold the line' policies. In these locations, this means that current coastal defences will not be maintained, and the coast will be allowed to retreat.

Despite the change in national coastal defence strategy, the government makes clear that compensation for people whose homes, land and businesses may soon be lost to the sea will not be examined as an option. Additionally, there exist no adaptation strategy, policy, funding or mechanisms to enable a transition to a different coastline by supporting people whose lives and properties will be affected by coastal change. Consequently, several communities in Norfolk and elsewhere are contending with a sudden expectation that they may sooner or later lose their homes and other valued community assets to coastal erosion or flooding. This is a frightening prospect in the absence of any adaptation or financial support options or assistance. Unsurprisingly, the change in recommended policy and its potential implications for individuals, households and communities has led to much local uncertainty, mistrust, anger and anxiety (Milligan et al., 2006; Nicholson-Cole et al., 2007).

Complex management arrangements, many stakeholders and lack of cross-sectoral coordination

Despite the Environment Agency having a strategic role, there is no coordinated or integrated approach to managing the changing coast between the elected bodies and statutory agencies (within and across scales) (O'Riordan et al., 2006), or to streamlining policy or funding for adaptation. Coastal governance arrangements are fragmented and there is little sectoral integration. In an audit of coastal activity in the East of England, CoastNet (2007) highlighted the complexity of the current coastal management and governance arrangements, finding three central government departments, four regional bodies, five statutory agencies, four ad-hoc groupings, seventeen local authorities and five forums all with an interest in coastline planning, but not necessarily working together (also, non-governmental organisations, other lobby groups, service providers, insurers, businesses, members of the public, various partnerships, coastal fora, and so on). There emerged five sets of overlapping plans, fourteen designations of coastal sites and landscapes, a mix of management bodies, many organisational cultures, un-coordinated organisational activity at different scales, and overlapping jurisdictions, responsibilities and functions. These were operating at different institutional scales, with different remits and interests, and spanning public, private and civil sectors.

The sectoral fragmentation of England's present coastal governance arrangements is particularly problematic. Public, private and voluntary organisations are differentially concerned with conservation and biodiversity, transport and other infrastructure, development planning, marine interests, agriculture and other rural land use, utilities provision, insurance, etc. On the whole, they do not have the mechanisms or capacity to work in an integrated manner to plan for coastal adaptation. Taking the example of development planning: the policy setting of coastal flood and erosion risk sits uneasily with opportunities for comprehensive, resilient land use planning and appropriate mechanisms for controlling development in future high-risk areas. Local planners work to a short-term 20-year time horizon, which is mismatched with the longer-term thinking that climate change impacts demand (Few et al., 2007a). They also face much political pressure to encourage the regeneration of coastal economies through development in coastal areas, even where the coastline cannot be permanently defended in its current configuration. In the past, this has meant allowing new housing and commercial development in areas of flood risk or partially defended coastal locations. This is also because of the difficulty associated with adjusting to a longer-term planning outlook given the unpredictability of coastal risk, and the lack of interaction and collaboration between planners and other coastal managers (Lee, 1993; Leafe et al., 1998; Taussik, 2007). This is not a problem specific to England; Moser and Tribbia (2006) point out a similar dilemma in California.

An effort to establish a Regional Coastal Initiative in the East of England (Government Office for the East of England, 2007–8) is promising; it aims to agree a long-term strategic vision for the coast, provide a forum to consider strategic coastal issues and policy areas, and enable joint working and action. It is hoped that the initiative will provide a mechanism for connecting stakeholders, policy and delivery, address common concerns through joined up action, and provide a focus for a more integrated and coordinated approach to managing the region's coast by public, private and civic stakeholders (for example Baker, 2002). However, demonstrable progress is slow, partly because the initiative is set against a broader institutional backdrop which is resistant and ill-equipped to enable adaptation efforts to go beyond research and into practice. More integration and coordination is necessary at national to local scales for coastal management to lay down the foundations for a more sustainable future coast. This is not a new concept, as is clear through European and national efforts and arguments made in the literature to pursue the realisation of Integrated Coastal Zone Management (Defra, 2007).

Mismatched expectations and understandings

Coastal residents and business owners who love and depend upon their coastline are almost unanimously in favour of demanding the status quo; losing any of it is a cultural icon of failure. A recent uprising of direct action in Eastern England is illustrative of this, as well as of the growing undercurrent of local concern, anxiety and dissatisfaction about decision-making processes and proposals for the future management of the coast. There is widespread public feeling that meetings and consultations run by decision-makers about the future of the coast have been inconsequential, tokenistic and dismissive of local knowledge and interests (Milligan et al., 2006, 2009; Milligan and O'Riordan, 2007; Nicholson-Cole et al., 2007). This highlights the importance of Few et al.'s (2007b) assertion of the need to avoid 'the illusion of inclusion' in order to ensure meaningful public participation processes in coastal management (as an illustration of climate change adaptation).

Figure 23.2 Campaigners to save the Suffolk's Blythe Estuary gather on Walberswick beach. (Source: www.mike-page.co.uk.)

A number of coastal protection and concern groups have consequently organised and mobilised,[1] and momentum continues to grow, with local Members of Parliament becoming actively involved. A 2008 protest against proposals to retreat in the County of Suffolk's Blythe Estuary area attracted nearly 2000 people (see Figure 23.2).

Tyndall Centre research findings point to the need for civic recognition of coastal change as part of a wider process in the quest for a more economically, socially and geo-ecologically sustainable coastline (O'Riordan et al., 2006). The present impasse and public expectation that if enough fuss is made they will be defended cannot continue (Milligan and O'Riordan, 2007). However, imminent public acceptance of potentially significant coastal change is highly unlikely because deeply held values, cultures and identities are at stake (for example Tunstall et al., 1998). These are strongly linked to people's attachments with familiar coastlines and the fear brought about by considering potentially very different and unfamiliar coastal futures. The contention in Eastern England is not unique. Other SMP processes are under way elsewhere in England, demonstrating similar problems, where the Environment Agency and other coastal management stakeholders are starting to consider realignment and retreat options but where local communities are unsurprisingly expressing strong wishes to see the status quo maintained, i.e. continued coastal protection. In some places, individuals are taking matters into their own hands, investing privately to defend small stretches of coastline from

[1] For example, www.happisburgh.org.uk; http://scratbycoastalerosion.org.uk.

erosion. There are no legal arrangements to stop this from happening, even when they are strategically inappropriate. In some cases, people have embarked upon legal battles with statutory agencies over their rights to defend land and property. The current governance arrangements are anything but adaptive, and currently ill-equipped to respond to this type of action.

Some effort is being made to push the boundaries of present coastal governance arrangements in search of more adaptive approaches to managing coastal change. One local authority in particular (North Norfolk District Council) has responded to the discontent surrounding the SMP process by developing a Coastal Management Plan, which guides investments in short-term coastal defence measures to 'buy time' for adaptation options to be developed and implemented. It has also initiated more community involvement in the development of options for the future management of the coast. These measures are designed to overcome the 'injustice' of adaptive inaction where retreat is now, but was not previously anticipated, and to give local communities some degree of confidence about their futures.

Social justice questions

In the absence of any clear adaptation policy, funding, mechanisms or participatory processes to ensure a managed transition to a new future coastline, the concept of social justice has become central and contested in coastal management debates at national to local scales. Many communities are adopting social justice arguments, and are questioning the fairness of SMP policy recommendations that have been developed on the basis of cost–benefit calculations (amongst other methods). For example, they perceive an underestimation of the full social and economic value of community activity which goes well beyond property values.

Where long-established investment in coastal defence measures may not be continued, and in the absence of any compensatory or adaptation arrangements, it is inevitable that questions about fairness of treatment and the implications for the future well-being of coastal communities are being asked. Cooper and McKenna (2008) highlight a tension, however, between social justice arguments at different spatial and temporal scales in the context of managing coastal change. Coastal defence and compensation claims tend to be the basis of current social justice arguments at local levels, but these are made in the context of a narrow view of 'fairness', rather than being set within any wider philosophical context of either social justice or sustainability. For example, arguments to withdraw public spending on inappropriate coastal protection, allowing the coast to resume its natural functioning and funding that transition instead.

Paavola and Adger (2006) identify four key social justice dilemmas associated with the necessity to adapt to climate change (in an international context). One of them is fair participation in planning and making decisions on adaptation which demonstrate adherence to social justice. Accordingly, Stringer et al. (2006) argue that adaptive practices have the potential to make environmental management more democratic through the involvement of different stakeholders, via a range of participatory mechanisms to create conditions for social learning and favourable outcomes for diverse stakeholders. Lubell (2005) asserts that public learning and cooperation from grass-roots stakeholders is one important element of adaptive governance, and that adaptive governance processes should therefore strive to create a 'vanguard of cooperation' (consisting of a core policy network of trusted government officials and leaders representing relevant groups of grass-roots stakeholders). However, while the quest for common goals and visions is paramount if there is to be effective adaptation, there are presently no established mechanisms to enable coastal stakeholders jointly to visualise, negotiate and define coastal futures, and no clear framework for supporting coastal partnerships or cooperation (CoastNet, 2007; Milligan et al., 2009).

It could also be deemed unjust to embark upon a course of action or policy decision when there are potential injustices that can be predicted to emerge in consequence. In this coastal case, the threat to communities of coastal erosion and flooding risk has been newly created by policy change. However, the implications of this were not considered fully in the policy process. Consequently, in many locations a fundamental discontent and lack of trust in those responsible for coastal decision-making has developed. This is hindering progress on adaptation to coastal change in Eastern England even where it aims to be participatory and inclusive.

Limits and barriers to adapting to coastal change in the UK

The challenges presented above are acting to delay and even limit the development and implementation of adaptation to coastal change in Eastern England. Gunderson and Light (2006) suggest that in settings where the environment has been highly managed, the legacy of technological fixes, engineering solutions and control over resources, which have come about to meet shifting social objectives, create a social trap; arguably a barrier to adaptation. As corroborated in this chapter, they assert that this trap is characterised by governmental mandates, planning paradigms and vested interests all interacting to inhibit the resolution of chronic environmental issues – where in fact experimentation and learning embraced by adaptive management is needed. Figure 23.3 demonstrates some constraining factors that are acting as barriers and even limits to the success of attempts at coastal adaptation, reflecting the propositions about the possible limits to adaptation proposed by Adger et al. (2009).

There exist some discernible attempts nationally and in the East of England to move toward more adaptive governance at different scales and these must be

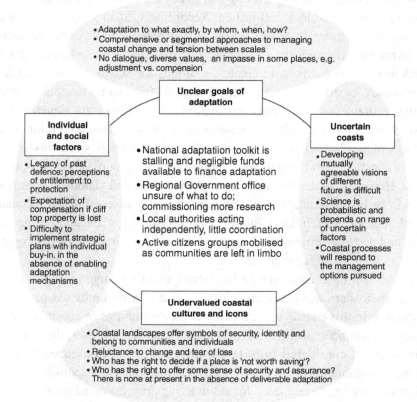

- Adaptation to what exactly, by whom, when, how?
- Comprehensive or segmented approaches to managing coastal change and tension between scales
- No dialogue, diverse values, an impasse in some places, e.g. adjustment vs. compension

Unclear goals of adaptation

Individual and social factors

- Legacy of past defence: perceptions of entitlement to protection
- Expectation of compensation if cliff top property is lost
- Difficulty to implement strategic plans with individual buy-in. in the absence of enabling adaptation mechanisms

- National adaptatiion toolkit is stalling and negligible funds available to finance adaptation
- Regional Government office unsure of what to do; commissioning more research
- Local authorities acting independently, little coordination
- Active citizens groups mobilised as communities are left in limbo

Uncertain coasts

- Developing mutually agreeable visions of different future is difficult
- Science is probabilistic and depends on range of uncertain factors
- Coastal processes will respond to the management options pursued

Undervalued coastal cultures and icons

- Coastal landscapes offer symbols of security, identity and belong to communities and individuals
- Reluctance to change and fear of loss
- Who has the right to decide if a place is 'not worth saving'?
- Who has the right to offer some sense of security and assurance? There is none at present in the absence of deliverable adaptation

Figure 23.3 Adaptation to coastal change: barriers and limits across scales.

credited (O'Riordan et al., 2008). One must be sympathetic to the difficulties facing policy-makers and executive agencies in the course of their efforts; there is no blueprint for adaptation to coastal change and no route map for delivering improved cooperation of policy, funding, justice and participation, in what are highly unfavourable and constraining institutional conditions. Additionally, there exists no clear policy or political momentum, no supportive financing for coastal adaptation, and no clear and agreed goal or vision about what adaptation is and what it will require. What we observe in the East of England, and more broadly in the UK, is an aggregate of governance arrangements and a highly uncertain future for an environmentally, socially and economically vulnerable coastline. Presently, there is no clear capacity to assess what adaptation to a changing coast really means and requires in practice.

As stakeholders at different scales attempt to embark upon understanding what adaptation does mean and require, it is becoming clear that the scope for adaptation is larger than what is politically and institutionally deliverable in the short term. Additionally, the goals of adaptation are neither clear nor agreed. No one knows what a sustainable coastal society and economy could look like, let alone how it might be achieved. Scientifically we cannot be sure how any coast facing climate change will react to different management policy options and possible new configurations of viable coastal adaptation can never be scientifically certain. Ways of looking at future coasts are also limited by the lack of vision that existing governing arrangements impose. This means that stakeholders are hard pushed to work towards joint visualisations of an acceptable future coast (Milligan et al., 2009). On the other hand, as Brunner and Steelman (2005) point out, the task may be less to predict the future than to shape it to advance common interests.

We are witnessing the early stages of progress toward adaptive coastal governance; there is a growing realisation that existing coasts cannot stay the same, and recognition that adjustment is possible via debate, dialogue and creative approaches to adaptation. But this will be a long and difficult process, and it comes at a price. We may need to think of completely different forms of future coasts. However, the reality of doing this means serious consideration of coastal retreat and realignment. These are notions which challenge people's powerful identities, senses of place and their deep attachments to their surroundings and the landscapes in which they presently live. The alternatives are unfamiliar; even considering them has been shown during the course of Tyndall Centre research to bring about extreme anxiety, as well as an aggressive fighting spirit.

Communities cannot be expected to adapt to very different coastlines until there is fair treatment through a supported process of adjustment. There needs to be a credible adaptation package in place, which is supported by a governance structure that is itself adaptive. Ultimately what people cannot conceive of is to abandon their existing homes and communities until there is fair treatment and a supported transition by way of a comprehensive and trustworthy adaptation process. But it is clear that there is no capacity to do this at present. Adaptation to coastal change, when it is urgently needed, can not proceed.

Adaptive governance for a sustainable future coast

We propose that eight key conditions need to be in place before coastal governance can progress toward being more adaptive (see Figure 23.4).

Planned adaptation to inevitable coastal change is absolutely critical. Adaptation needs to be managed in the short term as well as 50 or 100 years into the future. This requires enabling policy, funding, tools and mechanisms for delivering long-term adaptation, incorporating options for those people at most risk in the next 20 years. Foremost, policy-making and delivery needs to be much more integrated and

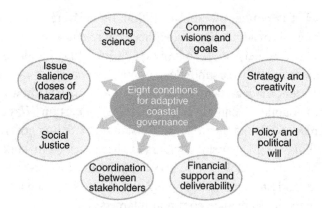

Figure 23.4 Eight conditions for adaptive coastal governance.

inclusive with clear and common objectives. This will require a partnership model of governance combining public, private and civil society into new coordinating arrangements, which will help to address the tension between national strategic frameworks, and local flexibility for delivery. It will also help in the development of new and positive approaches to looking at options for the future of our coast within a framework which proactively seizes the opportunities arising through change.

The transition has begun but it will be a long and difficult process to overcome government reluctance to fund adaptation, and for present governance arrangements to transform and become integrated and coherent. The quest to establish common adaptation goals and visions for a sustainable future coast will undoubtedly take even longer, and be more difficult. Unless there is adaptive governance, there can not be meaningful participation, and presently there is not the government, science or economic capacity to do this. This, for the moment at least, leaves coastal communities in a state of limbo and still searching for a sustainable future.

Acknowledgements

The views expressed in this article are those of the authors and are not necessarily those of the organisations to which the authors are affiliated.

References

Adger, W. N., Dessai, S., Goulden, M., Hulme, M. and Lorenzoni, I. 2009. 'Are there social limits to adaptation to climate change?', *Climatic Change* **93**: 335–354.
Anglian Coastal Authorities Group 2006. *Kelling to Lowestoft Ness Shoreline Management Plan: First Review*, Final Report, November 2006. Available at www.northnorfolk.org/acag/

Baker, M. 2002. 'Developing institutional capacity at the regional level: the development of a coastal forum in the north west of England', *Journal of Environmental Planning and Management* **45**: 691–713.

Brennan, R. 2007. 'The North Norfolk coastline: a complex legacy', *Coastal Management* **35**: 587–599.

Bridges, E. M. 1998. *Classic Landforms of the North Norfolk Coast*. Sheffield: The Geographical Association.

Brown, I., Jude, S., Koukoulas, S., Nicholls, R., Dickson, M. and Walkden, M. 2006. 'Dynamic simulation and visualisation of coastal erosion', *Computers, Environment and Urban Systems* **30**: 840–860.

Brunner, R. D. and Steelman, T. A. 2005. 'Toward adaptive governance', in Brunner, R. D., Steelman, T. A., Doe-Juell, L., Cromley, C., Edwards, C. and Tucker, D. (eds.) *Adaptive Governance: Integrating Science, Policy and Decision Making*. New York: Columbia University Press, pp. 269–304.

CoastNet 2007. *Scoping Study Regarding Current Coastal Activity and the National ICZM Programme, and Implications for East of England*, prepared by CoastNet for the Sustainable Development Round Table – East. Available at http://library.coastweb.info/

Cooper, J. A. G. and McKenna, J. 2008. 'Social justice in coastal erosion management: the temporal and spatial dimensions', *Geoforum* **39**: 294–306.

Cooper, N. J., Barber, P. C., Bray, M. J. and Carter, D. J. 2002. 'Shoreline management plans: a national review and engineering perspective', *Proceedings of the Institution of Civil Engineers: Water Maritime Engineering* **154**: 221–228.

Defra 2005a. *Making Space for Water: Taking Forward a New Government Strategy for Flood and Coastal Erosion Risk Management in England*, First Government response to the autumn 2004 *Making Space for Water* consultation exercise. London: Department for Environment, Food and Rural Affairs.

Defra 2005b. *Making Space for Water: Developing a Broader Portfolio of Options to Deliver Flooding and Coastal Solutions*. London: Department for Environment, Food and Rural Affairs.

Defra 2006. *Defra guidance on Preparing Shoreline Management Plans*. Available at www.defra.gov.uk/environ/fcd/policy/smp.htm (accessed 9 July 2007).

Defra 2007. *Summary of Responses to the Consultation: Promoting an Integrated Approach to Management of the Coastal Zone (ICZM) in England*. London: Department for Environment, Food and Rural Affairs.

Evans, E., Ashley, R., Hall, J., Penning-Rowsell, E., Saul, A., Sayers, P., Thorne, C. R. and Watkinson, A. R. 2004a. *Foresight: Future Flooding – Scientific Summary*, vol. 1, *Future Risks and Their Drivers*. London: Office of Science and Technology.

Evans, E., Ashley, R., Hall, J., Penning-Rowsell, E., Sayers, P., Thorne, C. R. and Watkinson, A. R. 2004b. *Foresight: Future Flooding – Scientific Summary*, vol. 2, *Managing Future Risks*. London: Office of Science and Technology.

Few, R., Brown, K. and Tompkins, E. L. 2007a. 'Climate change and coastal management decisions: insights from Christchurch Bay, UK', *Coastal Management* **35**: 255–270.

Few, R., Brown, K. and Tompkins, E. L. 2007b. 'Public participation and climate change adaptation: avoiding the illusion of inclusion', *Climate Policy* **7**: 46–59.

Folke, C., Hahn, T., Olsson, P. and Norberg, J. 2005. 'Adaptive governance of social–ecological systems', *Annual Review of Environment and Resources* **30**: 441–473.

French, P. W. 2004. 'The changing nature of, and approaches to, UK coastal management at the start of the twenty-first century', *Geographical Journal* **170**: 116–125.

Government Office for the East of England 2007–8. *Transforming Lives Transforming Places: GO-East Business Plan 2007–08*. Cambridge: Government Office for the East of England.

Gunderson, L. and Light, S. S. 2006. 'Adaptive management and adaptive governance in the everglades ecosystem', *Policy Sciences* **39**: 323–334.

House of Commons Community, and Local Government Committee 2007a. *Coastal Towns: Second Report of Session 2006–07*. London: The Stationery Office.

House of Commons Community, and Local Government Committee 2007b. *Coastal Towns: The Government's Second Response*. London: The Stationery Office.

Hulme, M., Jenkins, G. J., Lu, X., Turnpenny, J. R., Mitchell, T. D., Jones, R. G., Lowe, J., Murphy, J. M., Hassell, D., Boorman, P., McDonald, R. and Hill, S. 2002. *Climate Change Scenarios for the United Kingdom: The UKCIP02 Scientific Report*. Norwich: Tyndall Centre for Climate Change Research, University of East Anglia.

IPCC 2007. 'Summary for Policymakers', in Solomon, S., Qin, D., Manning, M., Chen, Z., Marquis, M., Averyt, K. B., Tignor, M. and Miller, H. L. (eds.) *Climate Change 2007: The Physical Science Basis. Contribution of Working Group I to the Fourth Assessment Report of the Intergovernmental Panel on Climate Change*. Cambridge: Cambridge University Press, pp. 00–00.

Jude, S., Jones, A. P., Andrews, J. E. and Bateman, I. J. 2006. 'Visualisation for participatory coastal zone management: a case study of the Norfolk coast, England', *Journal of Coastal Research* **22**: 1527–1538.

Jude, S. R., Jones, A. P., Watkinson, A. R., Brown, I. and Gill, J. A. 2007. 'The development of a visualisation methodology for integrated coastal management', *Coastal Management* **35**: 525–544.

Leafe, R., Pethick, J. and Townend, I. 1998. 'Realizing the benefits of shoreline management', *Geographical Journal* **164**: 282–290.

Lee, E. M. 1993. 'The political ecology of coastal planning and management in England and Wales: policy responses to the implications of sea-level rise', *Geographical Journal* **159**: 169–178.

Lubell, M. 2005. 'Public learning and grassroots cooperation', in Scholz, J. T. and Stiftel, B. (eds.) *Adaptive Governance and Water Conflict: New Institutions for Collaborative Planning*. Washington, DC: Resources for the Future, pp. 174–184.

McFadden, L. 2007. 'Governing coastal spaces: the case of disappearing science in integrated coastal zone management', *Coastal Management* **35**: 429–443.

Milligan, J. and O'Riordan, T. 2007. 'Governance for sustainable coastal futures', *Coastal Management* **35**: 499–509.

Milligan, J., O'Riordan, T., Watkinson, A., Amundsen, H. and Parkinson, S. 2006. *Implications of the Draft Shoreline Management Plan 3b on North Norfolk Coastal Communities*, Final Report to North Norfolk District Council. Norwich: Tyndall Centre for Climate Change Research, University of East Anglia.

Milligan, J., O'Riordan, T., Nicholson-Cole, S. A. and Watkinson, A. R. 2009. 'Nature conservation for sustainable shorelines: lessons from seeking to involve the public', *Land Use Policy* **26**: 203–213.

Moser, S. C. and Tribbia, J. 2006. 'Vulnerability to inundation and climate change impacts in California: coastal managers' attitudes and perceptions', *Marine Technology Society Journal* **40**: 35–44.

Nicholls, R. J., Richards, J., Bates, P., Dawson, R., Hall, J., Walkden, M., Dickson, M., Jordan, A. and Milligan, J. 2005. *Regional Assessment of Coastal Flood Risk*. Norwich: Tyndall Centre for Climate Change Research, University of East Anglia.

Nicholls, R. J., Watkinson, A., Mokrech, M., Hanson, S., Richards, J., Wright, J., Jude, S., Nicholson-Cole, S., Walkden, M., Hall, J., Dawson, R., Stansby, P., Jacoub, G. K., Rounsvell, M., Fontaine, C., Acosta, L., Lowe, J., Wolf, J., Leake, J. and Dickson,

M. 2007. 'Integrated coastal simulation to support shoreline management planning', Paper presented at *Flood and Coastal Erosion Risk Management Conference*, 3–5 July 2007, University of York, UK.

Nicholson-Cole, S. A., Milligan, J. and O'Riordan, T. 2007. *Investigating the Scope for Community Adaptation to a Changing Shoreline in the Area between Caister and Hemsby*, Report for Great Yarmouth Borough Council. Norwich: Tyndall Centre for Climate Change Research, University of East Anglia.

O'Riordan, T., Watkinson, A. R. and Milligan, J. 2006. *Living with a Changing Coastline: Exploring New Forms of Governance for Sustainable Coastal Futures*. Norwich: Tyndall Centre for Climate Change Research, University of East Anglia.

O'Riordan, T., Nicholson-Cole, S. A. and Milligan, J. 2008. 'Designing sustainable coastal futures', *Twenty First Century Society* 3: 145–157.

Paavola, J. and Adger, W. N. 2006. 'Fair adaptation to climate change', *Ecological Economics* 56: 594–609.

Pitt, M. 2007. *Learning Lessons from the 2007 Floods: An Independent Review by Sir Michael Pitt*. London: Cabinet Office.

Shennan, I. 1989. 'Holocene crustal movements and sea-level changes in Great Britain', *Journal of Quaternary Science* 4: 77–89.

Stringer, L. C., Dougill, A. J., Fraser, E., Hubacek, K., Prell, C. and Reed, M. S. 2006. 'Unpacking "participation" in the adaptive management of social–ecological systems: a critical review', *Ecology and Society* 11: 39.

Taussik, J. 2007. 'The opportunities of spatial planning for integrated coastal management', *Marine Policy* 31: 611–618.

Taylor, J. A., Murdock, A. P. and Pontee, N. I. 2004. 'A macroscale analysis of coastal steepening around the coast of England and Wales', *Geographical Journal* 170: 179–189.

Tunstall, S. M. and Penning-Rowsell, E. C. 1998. 'The English beach: experiences and values', *Geographical Journal* 164: 319–332.

24

Climate change, international cooperation and adaptation in transboundary water management

Alena Drieschova, Mark Giordano
and Itay Fischhendler

Introduction

The recently published report of the Intergovernmental Panel on Climate Change (IPCC, 2007) suggests clear evidence of climate change impacts. Global average air and ocean temperatures are rising. As a result, snow and ice are melting, leading to a rise in sea levels. Extreme weather events and hazards, such as flooding, heat waves and cyclones, are becoming more frequent, and the geographical and temporal clustering of precipitation patterns is shifting. In short, our hydrologic and ecological systems are changing. While an extensive body of literature has already developed around the question of international cooperation as a necessary prerequisite for mitigating climate change (Dunn and Falvin, 2002; Haas, 2004; Caney, 2005), other research indicates that some of its effects are already irreversible (IPCC, 2007). Thus scholars are also increasingly examining the possible role of international cooperation in the adaptation to climate change impacts.

There are at least three distinct roles for international cooperation in climate change mitigation and adaptation, which differ from each other in terms of their global communitarian basis. First, international cooperation can lead to *positive-sum outcomes* for the particular states involved. For example, the pooling of resources for the provision of public goods such as data, information and know-how, if shared in a non-exclusive way, can produce positive externalities. Second, *confidence-building measures* between two or more states can help to avoid conflicts over the use of increasingly uncertain stocks and flows of some internationally shared resources. This additional environmental uncertainty may require deviations from 'business as usual' interaction between states and can thus lead to a potential alienation and conflict. Confidence building measures in the form of increased communication or the development of international institutional structures can help to countervail or forestall possible negative consequences. Finally,

Adapting to Climate Change: Thresholds, Values, Governance, eds. W. Neil Adger, Irene Lorenzoni and Karen L. O'Brien. Published by Cambridge University Press. © Cambridge University Press 2009.

a feeling of *international solidarity* manifested for instance in the form of aid or assistance in case of emergency is at the heart of the concept that there is a shared responsibility for the current state of the Earth, especially in face of the uneven distribution of and responsibility for climate change effects.

Each of these possible roles for international cooperation in climate change adaptation has in fact already been formalized within numerous of the many treaties designed to manage variability in internationally shared water resources. While climate change is adding a considerably higher degree of urgency to the problem of flow variability, the difference is one of magnitude and not a qualitative change. As a result, the specific analysis of existing flow variability management in international water treaties provides an interesting, and possibly unique, opportunity to consider the potential of international cooperation in climate change adaptation in general.

As we have elaborated in Drieschova et al. (2008), many countries are dependent upon water that originates outside their borders. In fact basins shared by more than one country cover almost half of the Earth's land surface (UNEP, 2002) and often require regulative frameworks to coordinate riparians' actions. The importance of flow variability management in transboundary waters has long been acknowledged. As early as 1863, the Netherlands and Belgium made allocation of the Meuse's water conditional on annual availability. However, little work has been done to identify those variability management mechanisms that have been employed in environmental agreements, water treaties in particular, and the frequency of their use. A few have conceptualized the role of climate uncertainty in the development of treaty regimes and tested their hypothesis on case studies (Fischhendler, 2004), but to our knowledge no large N-studies have been undertaken to conceptualize the issues raised by flow variability.

The present study aims to partially fill this gap and uses the results to provide insights into international cooperation and climate change adaptation in general. It does this by first exploring the three distinct ways in which international cooperation can facilitate climate change adaptation and examines how these address the issue of flow variability in particular. It then uses the resulting typology to frame an analysis of the commonality of mechanisms, and their variation, to address flow variability in 50 treaties signed between 1980 and 2002. Particularly noteworthy is an apparent trade-off between flexibility and enforceability of treaty mechanisms, a finding which can be of relevance for international cooperation beyond the water sector.

The theoretical potential for international cooperation: water and beyond

Allegedly cooperation is least controversial when it leads to *positive-sum outcomes,* potentially because of the existence of comparative advantages and complementarities (Keohane, 1984). In the case of transboundary water resources, mechanisms to jointly develop infrastructure or transfer technology to either increase supply (in the case of shortage) or decrease it (in the case of flooding) can mitigate changing supply or demand conditions. Alternatively, mutually developed infrastructure, such as dams and barrages, can be used to more equally distribute water through time, without changing the overall amount of water available for human use. According to functional theories of international relations, such technical cooperation, as a form of 'low politics' is especially useful as a starting point for more comprehensive cooperation (Haas, 1961; Weinthal and Marei, 2002).

However, it would be naive to think that all situations lend themselves to positive-sum outcomes. Under certain conditions it is necessary to divide an existing resource pie and the resulting distributional issues can lead to conflict. This may especially be the case if historical use patterns and the institutional arrangements associated with them are disturbed due to changing environmental conditions, as will invariably be the case with climate change. Two types of *confidence-building measures* can help forestall undesirable outcomes. The first is clear but adaptable rules on how resources will be shared between states. Clear rules create a predictable environment conducive for cooperation, an order in the anarchical international system (Bull, 1977). As Kratochwil (1991, p. 87) claims: 'only explicitly formulated rules make it possible to discuss diverging expectations, and to justify deviations and exceptions without calling into question the nature of the social relationship as such'. The second is sufficiently well-developed communication channels to increase contact between affected states, reducing threat perceptions and creating epistemic communities across borders able to address contentious issues in a cooperative and technical manner (Adler, 1991; Haas, 1992).

In transboundary water management rules regulating the allocation of shared waters are a widely discussed topic and represent one of the most controversial aspects of treaty negotiations (Wolf, 1999). In fact, codified inflexibility in allocation has been cited as one factor behind the failure of agreements to withstand variability shocks. Other allocation mechanisms, such as flow percentages, have been advocated as an alternative (Wolf, 1998). Developing formalized communication between parties through the establishment of, for example, joint management institutions can overcome the rigidity of water treaties and serve as a venue for solving water conflicts (Feitelson and Haddad, 1999). Establishing conflict resolution mechanisms and encouraging data exchange are two other means through which communication channels can be established. Conflict resolution mechanisms provide an agreed forum for the discussion of changes in resource conditions not envisioned within initial agreements. Data exchange can reduce the potential impacts of flow variability by facilitating early identification of future trends, and offsetting the problem of asymmetrical information between riparians.

Lastly, when states and their inhabitants are driven by a concern for human security, rather than national security, they can provide each other with assistance in cases of emergencies, for instance when natural hazards occur. Obviously, international assistance does not have to be codified in treaties nor exclusively related to climate change inflicted problems. The international Asian tsunami (December 2004) relief efforts as well as the international support provided to Kashmir in the aftermath of its earthquake (October 2005) are but two cases in point. However, legal codification of international assistance is qualitatively different from ad hoc approaches, and there is a legitimate rationale for its use. Humanity in general, but the developed world in particular, bears responsibility for climate-change-inflicted natural hazards. The codification of assistance in international treaties makes it then obvious that such assistance is not a philanthropic gesture dependent upon the good will of particular governments, but that it is an obligation of some and a right for others. Codified assistance provides a form of a social security network, which many less developed countries badly need. Such assistance can for instance take the form of a fund towards which countries contribute and from which they withdraw money in emergency situations. The operation of the UN Food Program in the case of famine provides a general example whose structure could also be adopted in international schemes for climate change adaptation. Within water treaties this assistance is usually codified to refer to obligations for assistance in situations of floods, or at the opposite extreme, during periods of prolonged drought.

Institutional design: the need for a trade-off between flexibility and enforcement

From this discussion a suggestion emerges that two conditions should be fulfilled at once for codified international cooperation to be most effective in facilitating adaptation to climate change. First, ecosystem management requires an adaptive approach that, instead of aiming to maintain a fixed management regime, includes management rules that are sufficiently flexible to meet unexpected conditions (Walker et al., 2002). This is particularly relevant in the case of climate change. Dietz et al. (2003) introduced the concept of adaptive governance to expand adaptive ecosystem management to broader social contexts including human and ecological uncertainties. One of the variables often stressed as a factor in the capacity to adapt to variability is the degree of flexibility incorporated in governance systems (see Chapter 23). Flexibility can mean either the ability to change the rules of the game, for example in order to allow for the incorporation of new scientific knowledge (Boockmann and Thurner, 2006), or the option to apply a variety of policies in the face of changing conditions (Arvai et al., 2006).

Since transboundary environmental agreements are typically rigid instruments that are modified only under exceptional circumstances, the need for some degree of flexibility in their design is clear (McCaffrey, 2003). In addition to playing a direct role in resource management, flexibility in treaty content can also reduce the sovereignty costs of negotiations and allow regime creation to move forward, even if issues of uncertainty about the future state of the world are not fully resolved (Mitchell and Keilbach, 2001). It was in this context that flexible criteria were used to overcome political obstacles in the formation of the Kyoto Protocol (Thompson, 2005). Once agreements are in place, flexibility can also allow countries to deviate from a treaty when unexpected change occurs while still maintaining the overall benefits of the agreement (Fischhendler, 2004).

Second, there must be some way to ensure that agreement content will actually be followed. Enforcement mechanisms within treaties provide a degree of certainty to the parties involved, which is of particular importance under conditions of distrust – a situation which can potentially become more prevalent with climate change. To be sure, international law in its current state does not and cannot provide the degree of enforcement typical at the national level. Some scholars have therefore proclaimed the irrelevance of international law (Gilpin, 1983), while others have argued in favour of the incorporation of mechanisms making non-compliance costly (Oye, 1985). However, an increasing number of scientists point to the fact that states tend to respect international law to a considerable degree, because they are concerned about their reputation and even derive their own legitimacy from that very same law (Kratochwil, 1991). This logic of appropriateness makes it then obvious that enforcement in international law is to some extent a function of clarity in the overarching purpose and language of a treaty.

The following section provides an analysis of how the three roles of international cooperation for climate change adaptation have already been embodied in transboundary water treaties over the last quarter century. It then provides an examination of the content of these treaties and their attempts to broach the apparently conflicting demands for flexible and enforceable treaty provisions.

The state of variability management

In order to ascertain if and how transboundary water treaties address resource variability, a content analysis of available agreements signed since 1980 was undertaken and placed within the framework presented above. For the analysis, a treaty is considered to be 'an international agreement concluded between States in written form and governed by international law, whether embodied in a single instrument or in two or more related instruments and whatever its particular designation' (Vienna Convention of the Law of Treaties, 1969, Article 2).

To limit the analysis to those of primary relevance to flow variability, only treaties concerning 'water as a scarce or consumable resource, a quantity to be managed, or an ecosystem to be improved or maintained' are included in the analysis, while those dealing 'only with boundaries, navigation or fishing rights' were excluded (Hamner and Wolf, 1998, p. 158). The Transboundary Freshwater Dispute Database (TFDD) is the most comprehensive source of agreements related to these definitions and criteria. For the period 1980 to present, the TFDD contained 50 basin-specific agreements that met the above criteria and were included in our analysis.

Flow variability is explicitly mentioned in the text of 34 of the 50 treaties. For instance Article 40(3(d)) of Annex III of the Israeli–Palestinian interim agreement on the West Bank and the Gaza Strip provides for 'adjusting the utilization of the resources according to variable climatological and hydrological conditions'. The fact that 68% of the agreements explicitly mention flow variability strongly indicates the degree to which the issue is an important element in transboundary water management. The following sections examine how agreements are actually designed to handle variability issues.

The search for positive-sum outcomes

The creation of positive-sum outcomes is allegedly the least controversial method of international cooperation; in the case of transboundary water management this can sometimes be achieved by changing the level and timing of resource availability. Some of the surveyed agreements (14%) provide for the common construction of infrastructure to increase available water supplies or to disperse water supply throughout time. Another set of agreements address the opposite force of variability, i.e. flooding. The establishment of joint flood control mechanisms (30%) and warning systems (18%) to manage unexpected high flows in fact forms a major theme in recent international water treaties. Such agreements generally have a high degree of enforceability because of high opportunity costs, which make non-compliance particularly expensive. On the other hand the existing sunk costs make these agreements less adjustable and their degree of flexibility is therefore relatively low (Pahl-Wostl, 2005).

An additional 16% of agreements vaguely mention that the riparians would like to take joint measures to increase water supply, but without specifying the form that cooperation would take. For example, Bangladesh and India signalled in the 1996 treaty their intent to increase water supply at the Farakka barrage during dry periods but did not specify how or when such work would take place. Further, 42% of the agreements in the sample included provisions for the transfer of technology, which might be used to address variability. However, it is not generally clear from the treaty text to what extent the technology will be used for variability management. Such ambiguous language employed in treaty texts provides a relatively high degree of flexibility, whereas the degree of enforceability is relatively low.

The creation of confidence-building measures

As discussed above, the provision of clear rules as well as the establishment of formalized channels of communication are two distinct methods of creating certainty and developing confidence. Since allocation is a key topic in transboundary water management, the manner in which it is codified can have significant implications for the resilience of agreements as resource conditions vary. The mechanisms through which allocation may be addressed can be divided into three general categories. First, Direct Allocation mechanisms can be used to explicitly divide waters between co-riparians. Second, Indirect Allocation mechanisms can be used to establish the processes through which allocation will be determined, but without codifying the specific quantities or proportions to be shared. Consultations as a step to determine later allocations, a requirement for co-riparians to consent to any increased water use and a prioritization of water uses can all be considered as Indirect Allocation mechanisms. Finally, Principles for Allocations can establish the broader ideas or concepts for determining how water should be allocated now or in the future. These principles include concepts discussed in the 1997 UN Convention such as equitable and reasonable use, rational use, sustainable use, the requirement not to cause significant harm and the protection of existing uses.

At least one of the three allocation mechanisms was included in 60% of the agreements and 26% of the agreements included one or more Direct Allocation methods. More than half of those agreements that contained Direct Allocation mechanisms tied water rights to a certain degree to water availability (16% of the whole treaty sample). Some treaties allocate percentages of flow (6%), others allocate fixed quantities which themselves vary depending on water availability (10%) and still others allocate fixed quantities with the provision that in the case of insufficient water the deficit will be recouped in the following period. Generally Direct Allocation mechanisms have a high degree of enforceability, while their degree of flexibility varies.

For example the 1944 Treaty between the United States of America and Mexico relating to the waters of the Colorado and Tijuana Rivers, and of the Rio Grande, states in Article 4 that the USA has a right to a minimal contribution of 350 000 acre–feet of water annually from the Rio Grande, unless extraordinary drought or serious accident to the hydraulic systems make it difficult for Mexico to provide the required minimal quantity. In this case the deficiency shall be made up in the following five-year cycle. Despite the relative flexibility of the agreement, the treaty could not cope with ten consecutive years of low flows in the 1990s. The result was a growing water debt for Mexico and calls on both sides to renegotiate the treaty (Fischhendler, 2004). Thus the treaty revealed itself to be insufficiently flexible. Allocation based exclusively on percentages probably would have helped to avoid the tensions which arose.

Indirect Allocation mechanisms were used in almost half of the treaties (48%). However, in some cases, the indirect mechanisms operated only as complements to a direct mechanism. For example, in Article 2 of the Additional Protocol to the Convention on Cooperation for the Protection and Sustainable Use of the Waters of Portuguese–Spanish Hydrological Basins, the quantity of water the lower riparian should receive is given but subject to availability. When the flow is less than 65–70% (dependent on the individual river) of normal conditions, the parties to the agreement should inform the Commission, which then divides the available water on the basis of a set of Allocation Principles stated in Article 1 of the Additional Protocol. Thus the agreement provides a relatively high degree of flexibility, but also a certain degree of enforceability, because it makes clear which uses are prioritized.

Interestingly, all of the agreements which allocate water also incorporate at least one Principle of Allocation, and Principles of Allocation are rarely employed independently. Thus the often acclaimed lack of enforceability of these principles is actually offset by other mechanisms incorporated in the treaty and the advantage of their flexibility can be utilized to at least some degree. As a result, in situations of unexpected flow variability, the parties can still maintain the spirit of agreements by focusing on the relatively vague overarching principles. In disputes, tribunals can make their decisions so as to maintain the primary objectives of an agreement. For example, in the dispute related to the 1977 treaty signed between Czechoslovakia and Hungary on the construction of the Gabcikovo–Nagymaros multipurpose project, the International Court of Justice stated inter alia that Slovakia undertook a disproportional retaliatory measure and violated Hungary's right to an equitable and reasonable share of the Danube waters, when it started filling its part of the reservoir as a reaction to Hungary's non-compliance with the agreement (ICJ, 1997). Of the various possible general principles, equity in allocation is employed most frequently; it appears in 22% of the studied agreements. It is followed by rational use (18%), no significant harm (16%), protection of existing uses (8%) and sustainability (8%).

The provision of clear rules is one method of creating certainty and developing confidence, the establishment of formalized channels of communication is another. The specific communication mechanisms identified in the sample treaties include the formation of joint management institutions (88%), regular political consultations (46%), consultations as conflict resolution (90%), data exchange (86%) and arbitration (42%). While the incorporation of these mechanisms within treaties does not provide any guarantee that the parties will effectively address flow variability, it provides at least an institutional environment conducive for the search of cooperative solutions. Ultimately, in any concrete situation of variability, it depends upon the political will and agency of states or political entrepreneurs to use the existing institutional environment for the mutual benefit of each riparian.

International solidarity

International cooperation based on a feeling of overarching solidarity may be the highest degree of cooperation that can be obtained. Not surprisingly, this form of cooperation does not occur particularly frequently. Nonetheless, the fact that in 14% of agreements, riparians consented to assist each other in the event of unforeseen flow variability, indicates that intentions of such forms of cooperation are not a utopia. For instance the Convention on Cooperation for the Protection and Sustainable Use of the Danube states in Article 17 that 'in the interest of enhanced cooperation and to facilitate compliance with obligations of this Convention, in particular where a critical situation of riverine conditions should arise, Contracting Parties shall provide mutual assistance upon the request of other Contracting Parties'. Such clauses suggest that the riparians respect a shared responsibility to keep treaty provisions in the face of flow variability. The clarity of the language employed provides a certain degree of enforceability and such clauses are per se flexible, since their application depends upon the actual environmental conditions.

Evaluation: governance strategies to address variability

Three general conclusions emerge from this analysis. The first is that resource variability has clearly been perceived as an important issue in transboundary water treaties, both before and after climate change became a major issue of discussion. The second is that transboundary water treaties have already developed and employed a range of mechanisms which can assist riparian states in adapting to flow variability. Some of these mechanisms are explicitly related to variability management, while others are more general but with clear applicability. Finally, it is clear that the vast majority of flow variability management mechanisms are either flexible or enforceable, but not both. The only mechanism with a very high degree of both flexibility and enforceability is the allocation of waters based on percentage of flow. However, the use of this mechanism in the treaty sample was in fact very low. The mechanisms most often adopted in practice are those which provide a high degree of flexibility and have a low degree of enforceability including consultations as conflict resolution and principles of allocation.

The finding that only a minority of mechanisms adopted are both flexible and binding suggests certain barriers to the use of this 'ideal' management form. Some of these barriers may be political; for example, competition and power struggles between institutions have often been found to block innovation at the individual level (Adger and Kelly, 1999). The finding also suggests a trade-off between flexibility and enforceability; i.e. the risk of too much flexibility, which increases the likelihood of a treaty breach, against the risk of entering an agreement so constraining that it impedes state actions in regime implementation. While it is true that there is a certain preponderance of flexible mechanisms, we also found a high frequency of mechanisms reflecting a compromise between flexibility and enforceability such as Indirect Allocation mechanisms or data exchange. The trade-off between flexibility and enforceability can also explain the commonality of informal, broad commitments and institutions prevailing in international climate change negotiations (Karkkainen, 2004).

Variability of water flows can create risks for the longevity of agreements, because it is a change of circumstances which may cause states to change preferences, thereby reducing incentives to follow agreements signed in the past. In addition, unpredicted variability is also arguably a *force majeure* which could allow states to use a legal vacuum in general international law to abrogate treaty obligations and thereby escape the logic of appropriateness. For instance, Article 23 of the Draft Articles on Responsibility of States for Internationally Wrongful Acts states that 'the wrongfulness of an act of a State not in conformity with an international obligation of that State is precluded if the act is due to force majeure, that is the occurrence of an irresistible force or of an unforeseen event, beyond the control of the State, making it materially impossible in the circumstances to perform the obligation'. Lower riparians in particular may be interested in filling this vacuum on a case-by-case basis in order to guarantee a degree of certainty about the behaviour of their upstream neighbours as those neighbours themselves react to uncertainty in water availability. In contrast, upper riparians may not want to make concrete commitments related to an unpredictable natural world which might impose on them unknown future costs. The emergence of a flexible and not fully binding agreement can be interpreted as a compromise between these two extreme positions.

Another interesting finding relates to the prevalence of each of the three forms of international cooperation for climate change adaptation. Interestingly, most mechanisms in international water treaties have as their objective the creation of confidence-building measures (see Table 24.1), rather than the development of positive-sum outcomes as most neo-liberal theories would predict. This could suggest that the more important role of international cooperation lies in the function of communication, socialization and general interaction between states and

Table 24.1 *Flexibility and enforceability in three types of international cooperation*

	Positive-sum cooperation	Creation of confidence-building measures	International solidarity
Flexible	Ambiguous water supply increase (8), Technology transfer (21)	Principles of Allocation (30), Consultations as conflict resolution (45)	
Enforceable	Infrastructure changing supply (7), Joint flood control mechanisms (15), Early warning systems (9)	Fixed allocation (10), Arbitration (21)	
Both		Indirect Allocation (24), Data exchange (43), Percentage allocation (3)	Mutual assistance (7)

societies on one shared globe, rather than in the search for economic benefits from cooperation. On the other hand it is not particularly surprising that the least amount of mechanisms is based on a feeling of international solidarity, which requires a very high degree of a communitarian feeling at the international level.

Moreover Table 24.1 shows that within positive-sum cooperation a larger proportion of mechanisms has a high degree of enforceability than is the case for confidence-building measures, where a larger proportion of mechanisms is rather flexible. This is not too surprising, as almost by definition, positive-sum cooperation involves the economic calculation of costs and benefits. Compliance as well as non-compliance are more clearly based on rational choice assumptions and therefore highly enforceable mechanisms can make deterrence more costly. By contrast, confidence-building measures are based more on a constructivist understanding, on the development of norms and rules as well as peaceful interaction. Thus such measures follow rather a logic of appropriateness, than a logic of consequences. The enforceability of these mechanisms might be less important and the role of flexibility comes to the fore.

Conclusion

In the face of climate change, it is of crucial importance to identify both the factors that make society vulnerable and how society can physically and socially adapt

(Adger, 2005; Füssel, 2007). While earlier research focused on technical solutions, more recently researchers have increasingly focused on the human dimensions of global environmental change, and concepts of resiliency, vulnerability and adaptability have started to gain prominence (Walker et al., 2002; Cutter et al., 2003).

Some studies have analysed these issues on a case-by-case basis (Fischhendler, 2004; Conway, 2005). However, there have been few opportunities for large N-studies to analyse institutional designs for climate change adaptability. This study took the opportunity to examine how societies have adapted to one aspect of expected climate change, namely increased variability in river flows and how transboundary water treaties address the issue in their design. The study showed that flow variability can be and has been governed using a variety of mechanisms. Some mechanisms, such as allocation of waters based on a percentage of flows, explicitly address variability while the majority of mechanisms use less direct approaches that create open-ended rules for regulating water. Establishing communication channels and adopting indirect allocation mechanisms are two examples for this approach.

Flexibility and enforceability in rules regulating transboundary waters are often stressed as key positive attributes for governing shared water resources. However, flexibility can reduce the certainty around the actual flows of water parties will receive from an agreement, and enforceability can increase negotiation costs and may impinge on sovereignty. In international water treaties countries have generally tried to find a compromise between these two extreme and possibly contradicting positions. This insight can be of general relevance for the development of international institutions for climate change adaptability. The apparent demand for the trade-off between flexibility and enforcement should encourage researchers and policy-makers to seek hybrid governance structures that can combine the two qualities into single institutional designs which still meet environmental and political feasibility requirements.

At the same time, the general applicability of the results of content analysis has clear limits. First there is a difference in scale: water is not a global pool resource and transboundary water cooperation typically takes place at a regional level, which arguably makes cooperation and coordination easier to obtain. The number of players is smaller and free-riding becomes more difficult.

Second, as always, content analysis has its advantages as well as disadvantages and should therefore ideally be complemented with case studies. For example, we have to remember that a treaty becomes operative as a whole and while the study clearly found that flow variability mechanisms are rarely employed in isolation but rather as part of larger treaty packages, it did not examine in detail how various mechanisms interact, and possibly counteract each other. Third, since countries not only have to sign treaties but also comply with them, further study is required on the effectiveness of particular treaty regimes. Finally, attention is also needed on informal agreements not concluded in written form. Even at the transboundary level, institutions and governance structures do not have to be explicitly codified to have impact in general or for flow variability management in particular.

This study is an attempt to identify the broad range of mechanisms available for climate change adaptability in the concrete case of flow variation. The next step is to develop a better understanding of the relevance of our findings for other aspects of climate change adaptability, of the function of particular mechanisms, alone and in concert, and to ensure that the mechanisms most appropriate for particular conditions are in fact incorporated in future international agreements.

References

Adger, W.N. and Kelly, P.M. 1999. 'Social vulnerability to climate change and the architecture of entitlements', *Mitigation and Adaptation Strategies for Global Change* **4**: 253–266.

Adger, W.N., Arnell, N. and Tompkins, E.L. 2005. 'Adapting to climate change: perspectives across scales', *Global Environmental Change* **15**: 75–76.

Adler, E. 1991. 'Cognitive evolution: a dynamic approach for the study of international relations and their progress', in Adler, E. and Crawford, B. (eds.) *Progress in Postwar International Relations*. New York: Columbia University Press, pp. 43–88.

Arvai, J., Bridge, G., Dolsak, N., Franzese, R., Koontz, T., Luginbuhl, A., Robbins, P., Richards, K., Smith Korfmacher, K., Sohngen, B., Tansey, J. and Thompson, A. 2006. 'Adaptive management of the global climate problem: bridging the gap between climate research and climate policy', *Climatic Change* **78**: 217–225.

Boockmann, B. and Thurner, P.W. 2006. 'Flexibility provisions in multilateral environmental treaties', *International Environmental Agreements* **6**: 113–135.

Bull, H. 1977. *The Anarchical Society*. London: Macmillan.

Caney, S. 2005. 'Cosmopolitan justice, responsibility and global climate change', *Leiden Journal of International Law* **18**: 747–775.

Conway, D. 2005. 'From headwater tributaries to international river: observing and adapting to climate variability and change in the Nile basin', *Global Environmental Change* **15**: 99–114.

Cutter, S.L., Boruff, B.J. and Shirley, W.L. 2003. 'Social vulnerability to environmental hazards', *Social Science Quarterly* **84**: 242–261.

Drieschova, A., Giordano, M. and Fischhendler, I. 2008. 'Governance mechanisms to address flow variability in water treaties', *Global Environmental Change* **18**: 285–295.

Dietz, T., Ostrom, E. and Stern, P.C. 2003. 'The struggle to govern the commons', *Science* **302**: 1907–1912.

Dunn, S. and Flavin C. 2002. 'Moving the climate change agenda forward', in Worldwatch Institute (ed.) *State of the World 2002*. Washington, DC: Worldwatch Institute, pp. 24–50.

Feitelson, E. and Haddad, M. 1999. *Identification of Joint Management Structures for Shared Aquifers: A Comparative Palestinian–Israeli Effort*, World Bank Technical Paper No. 415. Washington, DC: World Bank.

Fischhendler, I. 2004. 'Legal and institutional adaptation to climate uncertainty: a study of international rivers', *Water Policy* **6**: 281–302.

Füssel, H.-M. 2007. 'Vulnerability: a generally applicable conceptual framework for climate change research', *Global Environmental Change* **17**: 155–167.

Gilpin, R. 1983. *War and Change in World Politics*. Cambridge: Cambridge University Press.

Haas, E. B. 1961. 'International integration: the European and the universal process', *International Organization* **15**: 366–392.

Haas, P. M. 1992. 'Introduction: epistemic communities and international policy coordination', *International Organization* **46**: 1–35.

Haas, P. M. 2004. 'Addressing the global governance deficit', *Global Environmental Politics* **4**: 1–15.

Hamner, J. and Wolf, A. 1998. 'Patterns in international water resource treaties: the transboundary freshwater dispute database', *Yearbook of the Colorado Journal of International Environmental Law and Policy* 1997: pp. 157–177.

IPCC 2007. *Climate Change 2007: Synthesis Report. Contribution of Working Groups I, II and III to the Fourth Assessment Report of the Intergovernmental Panel on Climate Change*. Geneva: IPCC.

International Court of Justice (ICJ) 1997. *Judgement in the Case Concerning the Gabcikovo–Nagymaros Project*. Available at www.icj-cij.org/docket/files/92/7375.pdf?PHPSESSID=1d5b3126d354b2c0c026ad4540a232f6 (accessed 26 February 2008).

Karkkainen, B. C. 2004. 'Post-sovereign environmental governance', *Global Environmental Politics* **4**: 72–96.

Keohane, R. O. 1984. *After Hegemony: Cooperation and Discord in the World Political Economy*. Princeton: Princeton University Press.

Kratochwil, F. V. 1991. *Rules, Norms and Decisions: On the Conditions of Practical and Legal Reasoning in International Relations and Domestic Affairs*. Cambridge: Cambridge University Press.

McCaffrey, S. C. 2003. 'The need for flexibility in freshwater treaty regimes', *Natural Resources Forum* **27**: 156–162.

Mitchell, R. B. and Keilbach, P. M. 2001. 'Situation structure and institutional design: reciprocity, coercion, and exchange', *International Organization* **55**: 891–917.

Oye, K. A. 1985. 'Explaining cooperation under anarchy: hypothesis and strategies', *World Politics* **38**: 1–24.

Pahl-Wostl, C. 2005. 'Transitions towards adaptive management of water facing climate and global change', Paper presented at the *International Conference on Integrated Assessment of Water Resources and Global Change: A North–South Analysis*, 23–25 February, Bonn, Germany.

Thompson, A. 2005. 'The rational choice of international institutions: Uncertainty and flexibility in the climate regime', *Proceedings of the Annual Meeting of the American Political Science Association*, 1–4 September, Washington, DC.

TFDD. Transboundary Freshwater Disputes Database. Available at www.transboundary-waters.orst.edu

UNEP 2002. *Vital Water Graphics*. Available at www.unep.org/dewa/assessments/ecosystems/water/vitalwater/03.htm

United Nations 1969. Vienna Convention of the Law of Treaties, Treaty Series, vol. 1155, 331. Washington, DC : United Nations.

Walker, B., Carpenter, S., Andereis, J., Abel, N., Cumming, G. S., Janssen, M., Lebel, L., Norberg, J., Peterson, G. D. and Pritchard, R. 2002. 'Resilience management in social–ecological systems: a working hypothesis for a participatory approach', *Conservation Ecology* **6**: 14.

Weinthal, E. and Marei, A. 2002. 'One resource, two visions: the prospects for Israeli–Palestinian water cooperation', *Water International* **27**: 460–467.

Wolf, A. 1998. 'Indigenous approaches to water conflict resolution and implications for international waters', Paper presented at *Conference on Water and Food Security in the Middle East*, April 20–23, Nicosia, Cyprus.

Wolf, A. 1999. 'Criteria for equitable allocations: the heart of international water conflict', *Natural Resources Forum* **23**: 3–30.

25

Decentralization: a window of opportunity for successful adaptation to climate change?

Maria Brockhaus and Hermann Kambiré

Introduction

In times of fundamental changes and shifts in institutional, political and economic structures at various scales, climate change and variability also exert a strong pressure on the resilience of social–ecological systems (IPCC, 2007). The IPCC states that African countries will be affected most by future climate change, since among other factors, widespread poverty, demographic changes, constrained institutional realities and inadequate political strategies are significantly limiting local adaptation capabilities (DFID, 2006; World Bank, 2006). In West Africa, livelihoods are highly dependent on forest ecosystem goods and services (FEGS), often in interplay with agricultural and livestock production systems. To reduce the growing risk of vulnerability under climate change, technical and societal adaptation is needed. Revised governance structures may enable adaptation at multiple levels and layers.

This chapter examines the opportunities and barriers for successful adaptation to climate change and variability in the context of an ongoing decentralization process, by examining the relatively understudied relationship between adaptive capacity and features of governance and culture. Here we present a case study on forests, climate change and aspects of adaptive capacity under a changing institutional landscape in two municipalities in the south-west of Burkina Faso. Adaptive capacity, in this chapter, is also understood as a function of governance features – such as institutional governance structures – and the individual understandings of the actors involved in decision-making processes related to FEGS. The chapter concludes by highlighting the importance of knowledge to overcome resource dependency and of two key features of governance

Adapting to Climate Change: Thresholds, Values, Governance, eds. W. Neil Adger, Irene Lorenzoni and Karen L. O'Brien. Published by Cambridge University Press. © Cambridge University Press 2009.

essential for technical and societal adaptation to climate change: (1) individual understandings and (2) institutional flexibility dependent upon close links with local realities.

Governance and adaptive capacity

Theoretical frame

Among other scholars, Huq and Burton (2003) recommend that the research community advance the theory and practice of 'adaptation science' and clearly identify the elements for enhancing adaptive capacity as well as improving the assessments of potential climate change impacts. Following the IPCC (2001, p. 6) definition, adaptive capacity can be understood as 'the ability of a system to adjust to climate change (including climate variability and extremes), to moderate potential damages, to take advantage of opportunities, or to cope with the consequences'. In literature there are different approaches to determine, and various concepts to measure, adaptive capacity, but these tend to be scale dependent (Brooks et al., 2005; Mukheibir and Ziervogel, 2007; Vincent, 2007). Furthermore, scholars increasingly recognize the importance of governance issues in the success or failure of adaptation efforts. In this chapter we refer to the UNDP (1997, Ch. 1) definition of governance as the complex 'mechanisms, processes and institutions through which citizens and groups articulate their interests, exercise their legal rights and obligations, and mediate their differences'.

Adaptive capacity is affected by the ongoing change of institutional management of natural resources. Institutions do not exist, develop and act independently, neither do they operate only at the local level. For instance, institutions governing the use of natural resources at a particular place in a particular time are also affected by institutions that operate on a regional, national or global level. Herrmann and Hutchinson (2005) emphasize that system-inherent adaptive capacity is weakened by inappropriate policies and programmes or by a total lack of them. Research needs to be done on the governance–adaptation linkages, including 'soft' issues such as culture, history and psychology into policy development (ICSU, 2005) related to adaptation. Folke et al. (2002) stress the need to develop adaptive, flexible and learning institutions at all levels to respond to the non-linear dynamics of natural resource and human systems. Tompkins et al. (2004) argue that for sustainable ecological management under climate change the building of social resilience is a key feature. Attributes of governance and individual, organizational or community capacities for adaptation determine the success of adaptation to climate change, and learning and flexibility are seen as key features for adaptation (Pelling et al., 2005).

Decentralization and adaptation in Burkina Faso

Following the ongoing decentralization processes in Burkina Faso, the institutional landscape is changing, administration and local governance are reorganizing, and a transfer of resources to the local municipalities is ongoing. These processes may offer new windows of opportunity to integrate adaptation into development policies but will simultaneously create new challenges and obstacles. According to various scholars (Agrawal and Ribot, 1999; Colfer and Capistrano, 2005; Ribot et al., 2006; Tacconi, 2007), decentralization can be considered as a promising approach for increased adaptive capacity at the local level. Although theoretical examinations tell of positive outcomes of decentralization practice still seems to be reluctant in providing evidence for these.

Decentralization

In Burkina Faso the political reform process started with the new constitution on 2 June 1991, which aimed at the reorganization of territorial administration. The *Code Général des Collectivités Territoriales* was adopted by the National Assembly in December 2004. In 2006, country-wide democratic elections were held at municipal level. The transfer of authority to the new local authorities continued creating the *conseil municipal* which is part of the new local governance structure that elects the mayor. The law stipulates the establishment of three permanent commissions in urban municipalities, one of these is in charge of local development and environment. At the village level, members of the elected CVDs (village commissions for development) form part of the *conseil municipal*. The administrative structures like the units for agriculture, livestock and environment at regional (Regional Department: DR) and provincial (Provincial Department: DP) level are intended to play an advisory role in this new configuration. The transfer of resources, competencies and planning authority to the local level should enable a highly adaptive management of the specific assets of natural resources in the established municipalities. Figure 25.1 provides an overview of these actors/ institutions at different levels.

Climate change and adaptation

At national level, the National Commission for Environment and Development with its permanent secretariat (SP/CONEDD) is the executive agency for issues related to climate change and adaptation. Burkina Faso submitted a National Communication to the United National Framework Convention on Climate Change (UNFCCC) in 2002 and a National Adaptation Programme of Action (NAPA) in December 2007. These documents outline that the main focus needs to be on technical adaptation

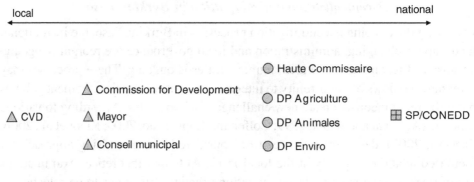

Figure 25.1 Adaptation and the administrative and decentralized structures. CVD (Commission Villageois de Développement): village commission for development; SP/CONEDD (Secrétariat Permanent du Conseil National pour l'Environnement et le Développement Durable): Permanent Secretariat of the National Commission for Environment and Development; DP Agriculture (Direction Provinciale de l'Agriculture, de l'Hydraulique et des Ressources Halieutiques): Provincial Department of Agriculture and Water; DP Animales (Direction Provinciale de Ressources Animales): Provincial Department of Animal Production; DP Enviro (or DP Environnement) (Direction Provincial d'Environnement et de Cadre de Vie): Provincial Department of Environment; Conseil municipal: City Council (elected); Commission for Development: permanent municipality commission for development and environment (different from SP/CONEDD); Haute Commissaire: High Commissioner; Forestry Brigade: a newly created body under the Ministry of Environment with executive and sanctioning powers.

of agricultural production systems including water management in the country. Governance issues are not addressed explicitly. This exemplifies the tendency to focus on technical problems at national level.

Research context and methods

The research area

The two urban municipalities researched are located in the south-west of Burkina Faso where annual precipitation rates are relatively high (900–1200 mm). The region is part of the dry forest area; vegetation is mainly made up of savanna wood-lands and gallery forests along the rivers. Over the last three decades the region has faced strong immigration, particularly of herders, mainly due to droughts in the north and the unstable political situation in the bordering Ivory Coast. Conflicts between farmers and herders are a widespread phenomenon in the region and of growing concern (Brockhaus, 2005).

Table 25.1 *Key characteristics of the two municipalities studied in Burkina Faso in 2005*

	Municipality 1	Municipality 2
Size	972 km²	874 km²
Population	26 260	41 505
Density	27 inhabitants/km²	47 inhabitants/km²
Components of the municipality (sectors and villages)	61 (5 + 56)	64 (8 + 56)
Tax revenues	1 434 865 FCFA	23 286 535 FCFA

Source: Adapted from Burkina Faso Info Route (2008).

The two neighbouring municipalities have similar geographical and physical conditions, as well as a partially shared history of settlement. The strongest ethnical groups in the region are the Lobi, the Dagara and the Birifor; these are present in both municipalities. Agricultural production practices are also very similar: extensive agriculture based on the use of the traditional hoe and only in few cases draught oxen.

However, apart from those shared features, the two municipalities manifest significant differences related to the economic and social characteristics of the communities, the state of the natural resources, as well as the experience with administrative processes and the decentralization in Burkina Faso. Table 25.1 shows the huge differences between the two municipalities especially regarding economic activities related to internal tax revenues, as well as population density. Another major difference is the municipalities' administrative and political development. Municipality 2 has a long history as a town hosting administration. Since 1995 an elected mayor has led the town together with the conseillers municipals. Municipality 1 had its first elections as a municipality in 2001 (now also led by a mayor and conseillers municipals). Additional differences are documented in the Burkina Faso Info Route (2008) regarding local development in terms of access to electricity and or communication, but also in terms of economic activities other than agriculture and livestock production, and the activities of civil society organizations between Municipality 1 (limited in all categories) and Municipality 2 (more diverse organizational landscape, presence of various development projects and other stakeholders).

The two research sites were chosen due to their specific characteristics and previous experience by the authors in those areas. The authors have worked in the two municipalities since 2001 on conflict and conflict management in the context of decentralization and land reform, thus building a solid base of trust during the various research periods. This enabled the research reported here to be embedded with elements of action research, which necessitates a deeper understanding of the local processes of change as well as deep 'trust' between interviewees and research-ers. Furthermore, climate change and variability was experienced and anecdotally reported in the two areas over recent decades as growing risk of flooding caused by heavy rainfalls, higher variability, uncertainty regarding length of the growing season, and changing tree species compositions, although quantified observations (statistics) were not available. The consequences of these changes, exacerbated by the droughts in the 1970s and 1980s, have also been experienced by immigrating herders from northern parts of West Africa.

Research methods and analyses

This section illustrates findings from comparative research focusing on governance, forests and adaptation in two municipalities in the south-west of Burkina Faso. In-depth interviews (from June 2007) were conducted with 16 actors from govern-ment, municipalities and the environmental, agricultural, and animal production extension services as well as with representatives of development projects active in both municipalities (see Table 25.2), representing decentralized and adminis-trative structures and development projects involved in the region. The interviews covered:

- role and contribution of FEGS for livelihoods (focus was on the sectors/topics: energy; water; and non-timber forest products and the use and management of trees and forests);
- perception/experience of climate change and extreme events and needed adaptive responses, and the envisaged challenges/threats under ongoing climate change/extreme events pressure;
- the roles and responsibilities for adaptation of the different actors in the arena;
- qualities of a 'good adapter' at individual and organizational level;
- actors' networks of information and influence.

Interviews were recorded and partially transcribed, and analysed following a scheme related to the above questions (see Table 25.3), specifically focusing on information given related to institutional flexibility in the governance system and the individual understandings of the interviewees.

Beside the formal interviews, informal discussions with the interviewees con-tinued during the entire research period and the process took the form of ongoing learning and knowledge sharing between and among the interviewees and the researchers. Demand for a workshop with all interviewees and other interested actors from different levels and with different organizational background was expressed and will take place in late 2008.

Table 25.2 *Summary of elites interviewed in both Burkina Faso municipalities, according to their role and position in different departments and activities*

	Municipality 1	Municipality 2
Decentralized structure		
Mayor	Mayor	Mayor
Commission for Environment and Local Development	President (interim)	President
Administrative structure		
Haute Commissaire	NA (not applicable)	High Commissioner
DP Agriculture (provincial governmental department for agriculture, technical assistance/extension service)	Director	Director
DR Animales (regional governmental department for livestock production, technical assistance/extension service)	NA	Director
DP Animales (provincial governmental department for livestock production, technical assistance/extension service)	Director	Not available at time of research
DP Environnement (governmental department for environment, technical assistance/extension service)	Director	Director
DP Environnement (governmental department for environment, technical assistance/extension service)	Not available at time of research	Chef de Service Departmental
Development projects		
PDR/SO (State project, focus on rural development, technical assistance, capacity building)	NA	Director (interim) (supports projects in M1, but organization operates from M2)
PDA/GTZ (Development cooperation project (German financial and technical assistance), focus on rural development, capacity building, technical assistance)	NA	Director (supports projects in M1, but organization operates from M2)
PROGEREF (State project, focus on sustainable management and valorization of environmental goods and services)	Director	Director

Table 25.3 *Analytical scheme applied to interviews in both Burkina Faso municipalities*

Actors' individual understandings	Institutional flexibility
Risk perception: complexity understandings, controllability, responsibility, severity, probability	Management, use of FEGS
Policy preferences and responsibilities	Historical failures and successes in management
Information	
Incentives	
Regulative	
Role understandings	Networks of influence and information ally/opponent/neutral information flows formal/informal structure
Attributes, factors and variables influencing abilities for adaptation	Knowledge about existing adaptation policy strategies

The research in Burkina Faso presented in this chapter also serves as a local starting point to identify structural gaps in the national policy arena for FEGS, and brokers and bridges across the different levels and layers from local to global. This work is part of the TroFCCA research project (Tropical Forests and Climate Change Adaptation) undertaken by CIFOR (Center for International Forestry Research). Research activities are ongoing in different regions and countries (West Africa, Asia and Central America) since 2006. Overall aim of TroFCCA is to support mainstreaming of forests and adaptation to climate change into development policy. TroFCCA emphasizes a policy–science dialogue integrated in decision-making processes rather than offering technocratic solutions (Forner et al., 2006). As part of the TroFCCA work undertaken in Africa, existing policies are evaluated using network and multi-stakeholder analyses.

Actors, institutions, and the role of forest ecosystem goods and services

Livelihoods and forests in the two municipalities

Forest and trees play a central role in the life of the local population even if these contributions are often non-monetary. Our interviewees identified the functions of trees and forests as economic, ecological, social and cultural. In this context the interviewees mentioned that high dependency on FEGS and the lack of knowledge about, and access to, alternative goods or services (for example the use of gas instead of wood as energy source) clearly limits the number of feasible adaptation responses. And, as was explained by some informants, the lack of valorization of

FEGS and the emphasis on the relevance of the agricultural sector (for example the expansion of cotton production) for development in Burkina Faso leads to a prioritization of the latter. Trade-offs despite the interdependencies and the cooperative potential between forestry, agricultural and livestock production systems lead to a focus on technical adaptation (for example the development of new seeds) in the agricultural sector thereby neglecting the importance of forests and trees for sustainable natural resource management in the region. Deforestation in the region and reduced availability of the above mentioned FEGS was reported by the informants. The interviewees also mentioned a variety of reasons when asked for the main factors affecting this availability, including complex relationships between poverty, population growth, economic interests in conflict with conservation efforts, inefficiency of agricultural and livestock production systems and the lack of participation of the local population in the management of forest resources. Climate change, mostly defined or perceived as changing rainfall patterns and shorter growing seasons, affecting the appearance of specific tree species, was also mentioned, but identified by only two interviewees as a main driver for changing availability of FEGS. For most informants this changing availability was a result of human activity in the region and less an outcome of biophysical processes related to climate change. But according to nearly all informants climate change and extreme events can become a major driver for reduced availability of FEGS in the long term.

Institutions in the governance structure

Use and management of FEGS

Trees and forests are managed through a variety of parallel rules and regulations – a well-known phenomenon not only in this region. Elaborated rules for the cultural use of FEGS, and the right of planting trees as an indicator of tenure property, are examples of traditional natural resource management. However, formal rules regulating the access and use of forests were introduced during the colonial period and are currently enforced with limited success by the forestry administration. Nearly all informants from all administrative and decentralized institutions in the two municipalities emphasized the importance of a shift to more participation in the decision-making processes for sustainable natural resource management. In particular, this need was highlighted by interviewees from the various development projects but also from the environmental administration units in both municipalities.

Failure and success of managing FEGS under decentralization

With the decentralization process the control over forest resources was transferred to the local level and the provincial environmental administration (DP

Environnement) is supposed to play an advisory role in this management. This has not always taken place in practice, and according to our interviewees in Municipality 1 this situation has lead to the 'sell-off' of a teak forest without considering any advice regarding environmental and economic aspects of this transaction. But informants from Municipality 2 used this as an example of 'lessons learned' and emphasized the importance and value of knowledge and advice from the technical service of the provincial and regional environmental administration (DP and DR Environnement) as a base for future decisions concerning forest resources.

Networks of information and influence

The three different technical institutions that participated in this study showed remarkable differences in approaching the new decentralized configuration in the established municipalities. The agricultural institution was well placed due to strong human resources even at village level, and members of the institution in both municipalities were president or acting president of the commission for local development and environment and well aware of the changing roles in the institutional landscape. On the other hand, the environmental institution is actively attempting to change its image from a control and enforcement agency to an advisory one. A Forestry Brigade was introduced with executive and sanctioning powers, while the provincial and regional departments (DP and DR) for the environment act as advisors and redesign their image, aiming to gain the trust of the population. According to interviewees from the livestock production service (DP Animales) even under changing institutional conditions, 'business as usual' continued, without any new strategies enacted as a result of the new decentralized structures. Regarding power relations among the different institutions in the governance structure, it became obvious during the interviews that the DP and DR for livestock production were isolated, whereas the relationship between agricultural and environmental units is still dominated by differences: as one informant put it, 'agriculture and we [from the environmental administration] do not always speak the same language', even if the need for more cooperation among them was highlighted by the informants. Beside these formal and informal networks between the actors, networks of information related to FEGS and adaptation needs or actions were not evident at time of the research, as we discuss in the next section.

Knowledge about existing global, regional and national
adaptation policy processes

Remarkably, for all informants access to information on climate change and adaptation was limited to public media. Only in one case (DP Environ in Municipality 2) the informant was aware of the ongoing NAPA preparation process. Nobody else mentioned participation or any information exchange about the national policy process for adaptation. Additionally, inside the administrative and decentralized structures, as well as for the development projects, it appears that no exchange of information has taken place across the hierarchical levels. Especially, the mayor of Municipality 2 complained about the uninformed technical services and the non-use of existing platforms for exchange and advanced training (cadre de concertation). One exception was efforts for technical adaptation to climate variability (introduction of new drought-resistant seeds) in the agricultural extension unit in interplay with local adaptation efforts, induced by local demand.

As Figure 25.2 shows, the situation resulting from the situation as just discussed is dominated by limited connectedness of the actors in the arena of FEGS, climate change and adaptation. Even if the networks of influence are changing and new allies or partners are identified (actively by the environmental unit), regarding information flows most actors remain unconnected. The SP/CONEDD at national level is only connected via a very weak tie to the DP Environ in Municipality 2; the local level was mentioned as part of the adaptation networks with the DP of agriculture in both municipalities but is not fully integrated in any other network. Furthermore, civil society organizations received no mention by any interviewee, except by one part of the development project. This may be either because civil society organizations do not yet operate in the domain of FEGS or, more probably because integration of civil society and its organizations in the political processes of decentralization has not occurred until now. This could also explain

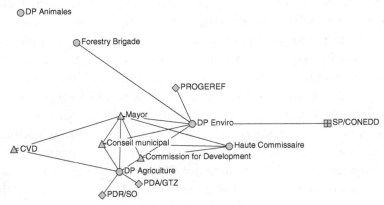

Figure 25.2 Actors and their connectedness in the local arena for adaptation to climate change and forest ecosystem goods and services. PROGEREF (Projet de gestion des ressources forestier): project for the management of forest resources; PDA/GTZ (Projet de développement d'agriculture): project for agricultural development of the German technical cooperation (GTZ); PDR/SO (Projet de développement rural au Sud Ouest): project for rural development in the southwest (of Burkina Faso).

why especially in the smaller and 'in decentralization terms less experienced' Municipality 1 there was no mentioning of civil society as a future key player in decision-making at local level but as a passive actor, perceived as 'the population [which] needs to have trust, confidence and patience with its elected'. However, similar views and perceptions of civil society were expressed by the mayor of Municipality 2.

Perceptions of roles and responsibilities for adaptation

Awareness, risk perceptions and adaptation responses

Due to local experiences of climate change and extreme events – for example droughts in the north with secondary consequences (migration, land use changes) in the south-west – built over long timescales, the existence of climate change was not questioned by interviewees. However, climate change was not perceived as a major driver of change. Particularly the agricultural units emphasized their belief in the controllability of the problem through technical responses. This perception was shared by the mayor of Municipality 2 and partially by the mayor and his deputies in Municipality 1. They described climate change as one risk among many, 'a normal situation for a municipality'. Informants of the development projects however indicated concern about the uncontrollability of climate change and the severity of forthcoming negative impacts due to climate change for achievement of development goals. However, the strongest barriers to the implementation of successful adaptation strategies and responses were identified (especially by the staff in the agricultural units in both municipalities) as unchanged systems of use and management of natural resources due to lack of techniques or access to techniques, or the lack of capacities to implement new production techniques (for example in agro-business). Strongest opportunities for successful adaptation were seen by informants from the development projects in capacity-building, individual behaviour changes and stronger involvement and participation of civil society in planning and implementation of adaptive responses. A positive (and provocative) view of climate change as a positive driving force for change came from an informant from the more environmentally oriented development project. For him, under climate change, changes in awareness and behaviour would become necessary and unavoidable. These would then lead to changes in use and management of natural resources, a better future. In other words, he felt that success of adaptation would result from the growing pressure of climate change.

Most informants gauged that climate change will bear significant impacts in the future, as a multiplier of the negative impacts already induced by human activity. In relation to future scenarios for the region in general, most mentioned 'the arrival of a second Dori in the region'.[1] In response to the possibility of further droughts in the north, informants mentioned a growing risk of conflict between different land users like farmers and herders due to strong migration induced by climate change in other parts of West Africa. However, the animal production extension service in both municipalities was indifferent to these concerns and seemed to be less worried about a future scenario with growing conflict potential.

Policy responses and understandings

The proposals for policy responses to realize adaptation at multiple layers varied across the informants. Because pressure and coercion has led to policy failure in the past, most informants preferred policy instruments that promoted information, awareness raising, and changes in economic incentives (for example, PES). The national government (together with a global climate regime) was seen as the responsible entity to set a policy framework to finance and regulate adaptation actions at the local level. However, local governance including the participation of all actors involved in use and management of ecosystem goods and services was considered as the most important structure in identifying needed adaptive responses and in planning and implementing these. Therefore, decentralization was seen by all informants as a strong opportunity to achieve participatory governance and adaptation, given that the process seemed to offer new institutional flexibilities and 'short distances' to local realities which should result in adapted and highly responsive planning.

Attributes, factors and variables influencing adaptation

Most actors emphasized the role of knowledge for successful adaptation. Motivation for learning at the individual level was mentioned as the most important quality of a 'good adapter', whereas at the organizational level the internal structures (formal and informal) for information exchange across vertical and horizontal levels was seen as a prerequisite to enable successful action for adaptation. Interviewees also emphasized the importance of curiosity, open-mindedness and willingness to learn (at both individual as well as organizational levels) as key variables for being a 'good adapter'. Interviewees also preferred information policy options and education tools over coercion instruments. However, while the representatives from the development projects mentioned these variables in a context of an active and empowered actor or a reflective, less hierarchical organization, the representatives of the decentralized as well as part of those from the administrative structures associated these characteristics with a 'receptive', 'listening' or 'executive' actor. Although the latter seem to contradict the basic nature of these general shared

---

[1] Dori is a Sahel town located in the north of Burkina Faso which serves as a metaphor for 'desert and extreme drought and heat without any shade'.

attributes, the first group of interviewees identified them in a 'good adapter' as a 'self-critical, reflective innovator', while the other group identified and considered a 'well-educated follower'.

In addition, an interviewee from the decentralized permanent commission in Municipality 2 mentioned need and scarcity as the strongest external drivers of change, independent of other personal or organizational characteristics. This understanding of innovation or change as driver-induced (i.e. scarcity of input factors) is clearly grounded in informants' experiences in Municipality 2, where pressure on natural resources is higher compared to Municipality 1. These findings and statements underline the strong and complex relationship between individual understandings and policy preferences for the future.

Limits of and opportunities for adaptation

The findings of this comparative research show that formal governance systems, and the institutional settings and structural conditions within, determine the overall space for adaptive capacity of the system, and set limits of successful adaptation. To what degree this potential space is used depends highly on the actors involved in the policy and decision-making processes, and their individual understandings as well as the resource dependency they are confronted with.

One could argue that in Burkina Faso the decentralized governance system offers maximum space for adaptation due to the potential of governance at the local level. But the comparison between the two municipalities shows that its implementation is highly dependent upon individual experiences – experiences with climate change as the context-related challenge and the experiences with the new roles and responsibilities in the changing institutional environment as a structural challenge. Adaptation is hindered at the local level by recently achievable short-term economic benefits, underpinned by trade-offs between the development of the agricultural and livestock sectors and the environment, under conditions of a changing institutional landscape characterized by yet unclear roles and responsibilities and limited experience in the long-term management of resources by the newly introduced formal governance structure. The comparison between the two municipalities indicates that lack of knowledge or experience is one barrier (Municipality 1), but individual understandings of civil society as a 'good adapter' passive under a steering local force (the elected) could also indicate ignorance or abuse of power by the newly elected as another, even stronger, barrier for change and successful adaptation. However, this could be a transient phenomenon of the reform period.

Sharing of experiences across actors, institutions and legal entities can reopen the space for increasing adaptive capacity in the municipalities' decision-making processes. Actors' risk perception, awareness, communication structures and formal roles in the new configuration of local governance determine their adaptive capacity but also preferences for policy instruments. Furthermore, their connectedness to other layers and levels of decision-making limits or enables adaptation. As we have seen, none of the actors in the two municipalities was involved in the NAPA process, connection between national and local level was lacking, most obviously in Municipality 1. This lack of connectivity between national and local, worsened by a degradation of participation to pure consultation, as well as the gaps between locally elected representatives and civil society, may lead to further isolated political processes; a reduction of the adaptive capacity at the local level; further loss of availability of FEGS and the growing risk of vulnerability of the local population.

The variables explored in this study (namely resource dependency, institutional flexibility and individual understandings within the governance system) all have knowledge as a key common feature. If decentralization is the process which determines the governance space in which adaptation is negotiated, the 'adaptive space' is the overlapping optimum of those three variables and determines the degree to which opportunities and potentialities for adaptation in the arena of FEGS, climate change and adaptation can be realized (Figure 25.3).

Conclusion

The study and findings discussed in this chapter suggest that successful adaptation is strongly driven by structural conditions, underpinning the management of FEGS and the behaviour of actors involved in decision-making. Besides limitations posed by resource dependency, adaptive capacity can further be related to two key features

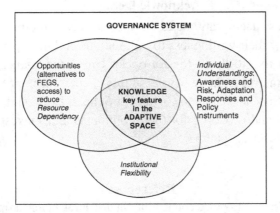

Figure 25.3 Knowledge as key feature in the adaptive space given by the three variables limiting or enhancing adaptive capacity in the governance system.

of governance which are essential for technical and societal adaptation to climate change: (1) individual understandings; and (2) institutional flexibility in governance structures, ensured by strong direct connections ('short distances') to local realities. Knowledge underpins all these variables and is the key to (1) the planning flexibility resulting from the introduction of decentralized local institutions with planning responsibility and (2) the opportunities provided by common learning and sharing of information among and across individuals and organizations.

Decentralization offers strong opportunities for the design and implementation of adaptation strategies due to growing institutional flexibility, higher responsiveness, and selective planning and implementation at local level. However, this chapter shows that decentralization is far from being a guarantee for successful adaptation. Whether it can fulfil its potential depends on a variety of factors. In the two case studies, success is hindered by the lack of structures for shared learning, lack of knowledge and a biased agenda setting for adaptation due to perceived trade-offs among various sectors. In order to use the window of opportunity decentralization offers, it has to be accompanied by capacity-building, knowledge transfer and the establishment of sound communication structures.

Local governance structures and the individual actors involved in those need to be more integrated into national political processes on forest ecosystem goods and services, climate change and adaptation. If 'participation' is nothing more than 'consultation', adaptation efforts designed at global, regional or national level are not informed by local experience and will not have an impact at the local level, where the adaptation is practised, needed and should be supported. On the other hand, participation in policy processes is needed, otherwise reform processes will remain isolated – far away from local realities and far away from local needs.

Acknowledgements

The authors express deep gratitude to all interviewees and other informants in the research region for their willingness to talk with us and their patience to do so for hours. We would like to thank the European Union for financing the TroFCCA project. We are grateful to the reviewers of this chapter for their helpful comments and suggestions. Special thanks to C. van der Schaaf, H. Djoudi, J. Sehring, H.-N. Bouda, D. Tiveau and C. Colfer for their comments and suggestions provided on earlier versions of this chapter.

References

Agrawal, A. and Ribot, J. 1999. 'Accountability in decentralization: a framework with South Asian and West African cases', *Journal of Developing Areas* **33**: 473–502.

Brockhaus, M. 2005. *Potentials and Obstacles in the Arena of Conflict and Natural Resource Management : A Case Study on Conflicts, Institutions and Policy Networks in Burkina Faso.* Göttingen: Cuvillier Verlag.

Brooks, N., Adger, W. N. and Kelly, P. M. 2005. 'The determinants of vulnerability and adaptive capacity at the national level and the implications for adaptation', *Global Environmental Change* **15**: 151–163.

Burkina Faso Info Route 2008. *Burkina Faso Info Route.* Available at www.inforoute-communale.gov.bf/monographies_cadre1.htm (accessed 20 July 2008)

Colfer, C. and Capistrano, D. (eds.) 2005. *The Politics of Decentralization: Forests, People and Power.* London: Earthscan.

DFID 2006. *Eliminating World Poverty: Making Governance Work for the Poor.* London: DFID

Folke, C., Carpenter, S., Elmqvist, T., Gunderson, L., Holling, C. S., Walker, B., Bengtsson, J., Berkes, F., Colding, J., Danell, K., Falkenmark, M., Gordon, L., Kaspersson, R., Kautsky, N., Kinzig, A., Levin, S. A., Mäler, K.-G., Moberg, F., Ohlsson, L., Olsson, P., Ostrom, E., Reid, W., Rockström, J., Savenije, S. and Svedin, U. 2002. *Resilience and Sustainable Development: Building Adaptive Capacity in a World of Transformations*, Report for the Swedish Environmental Advisory Council. Stockholm: Ministry of the Environment.

Forner, C., Nkem, J., Santoso, H., and Perez, C. 2006. *Setting Priorities for Forests in Adaptation to Climate Change*, TroFCCA's first year report. Bogor: CIFOR.

Herrmann, S. and Hutchinson, C. 2005. 'The changing contexts of the desertification debate', *Journal of Arid Environments* **63**: 538–555.

Huq, S. and Burton, I. 2003. *Funding Adaptation to Climate Change: What, Who and How to Fund?* Sustainable Development Opinion, RING Alliance No. 4. London: International Institute for Environment and Development.

ICSU 2005. *Harnessing Science, Technology, and Innovation for Sustainable Development*, A Report from the International Council for Science (ICSU)-ISTS-TWAS Consortium Ad Hoc Advisory Group. Paris: ICSU.

IPCC 2001. *Climate Change 2001:Impacts, Adaptation and Vulnerability. Contribution of Working Group II to the Third Assessment Report of the Intergovernmental Panel on Climate Change.* Cambridge: Cambridge University Press.

IPCC 2007. Parry, M. L., Canziani, O. F., Palutikof, J. P., Van der Linden, V. J. and Hanson, C. E. (eds.) *Climate Change 2007: Impacts, Adaptation and Vulnerability. Contribution of Working Group II to the Fourth Assessment Report of the Intergovernmental Panel on Climate Change.* Cambridge: Cambridge University Press.

Mukheibir, P. and Ziervogel, G. 2007. 'Developing a Municipal Adaptation Plan (MAP) for climate change: the city of Cape Town', *Environment and Urbanization* **19**: 143–158.

Ribot, J. C., Agrawal, A. and Larson, A. M. 2006. 'Recentralizing while decentralizing: how national governments reappropriate forest resources', *World Development* **34**: 1864–1886.

Tacconi, L. 2007. 'Decentralization, forests and livelihoods: theory and narrative', *Global Environmental Change* **17**: 338–348.

Tompkins, E. and Adger, W. N. 2004. 'Does adaptive management of natural resources enhance resilience to climate change?', *Ecology and Society* **9**: 10.

UNDP 1997. *Governance for Sustainable Human Development*, a UNDP policy document. Available at http://mirror.undp.org/magnet/policy/ (accessed 10 July 2008)

Vincent, K. 2007. 'Uncertainty in adaptive capacity and the importance of scale', *Global Environmental Change* **17**: 12–24.

World Bank Group Global Environment Facility Program 2006. *Managing Climate Risk: Integrating Adaptation into World Bank Group Operations*. Washington, DC: World Bank.

26

Adapting to climate change in Sámi reindeer herding: the nation-state as problem and solution

Erik S. Reinert, Iulie Aslaksen, Inger Marie G. Eira,
Svein D. Mathiesen, Hugo Reinert and Ellen Inga Turi

We have some knowledge about how to live in a changing environment. The term 'stability' is a foreign word in our language. Our search for adaptation strategies is therefore not connected to 'stability' in any form, but is instead focused on constant adaptation to changing conditions.

Johan Mathis Turi, Chairman of International Centre for Reindeer
Husbandry (ICR), Tromsø, UN Environmental Day, June 2007

Introduction

Climate change is likely to affect the Sámi regions in Norway, Sweden, Finland and Russia, with greater variability in temperature, precipitation and wind, and higher winter temperatures (ACIA, 2005; Tyler et al., 2007). These factors strongly affect snow quality and quantity, with snow quality as a crucial factor for reindeer herding. Considering their experience obtained through time and their traditional ecological knowledge, the pastoral practices of Sámi herders are inherently well suited to handle huge variations in climatic conditions. Reindeer herding and its natural environment have always been subject to large variability in weather patterns, and skilful adaptation to these past variations offers important insights on adaptation to climate change. In particular, it is crucial to recognize the importance of traditional ecological knowledge (Berkes, 2008).

The adaptation of Sámi reindeer herding to climate change is conditioned by its political and socio-economic environment (ACIA, 2005, p. 971; E. S. Reinert, 2006; Tyler et al., 2007). Important parts of the traditional adaptive strategies – the composition of herds and the flexibility to move reindeer herds between summer and winter pastures – are challenged by nation–state policies restricting herd

Adapting to Climate Change: Thresholds, Values, Governance, eds. W. Neil Adger, Irene Lorenzoni and Karen L. O'Brien. Published by Cambridge University Press. © Cambridge University Press 2009.

diversity and mobility and by rigid regulations. Management of Sámi reindeer herding in Norway is strongly conditioned by models of agricultural husbandry, not suitable for reindeer herding (ACIA, 2005, p. 978; E. S. Reinert, 2006; Tyler et al., 2007).

This chapter discusses the role of the nation–state and systems of governance and institutions as barriers and solutions to adaptation to climate change from the point of view of Sámi reindeer herders. Climate change is likely to affect the Sámi regions. The reindeer herders' adaptation to climatic variation, embodied in traditional herding practices with diversification of risk, based on the traditional ecological knowledge of biological diversity, is the key to improved resilience and successful adaptation to climate change. A crucial factor for enhancing the adaptive capacity to climate change is to modify the incentives in Norwegian official administration of reindeer herding management.

First we discuss traditional ecological knowledge and the anthropological concept of ecological niches, in terms of reindeer herding. We then outline the recent historical background for the governance structure of reindeer herding in Norway, and argue that adjustment of the governance structures is crucial for survival of the Sámi reindeer herding culture. Reindeer herding administration in Norway is characterized by detailed and inflexible government interference, from the structure of the herd to the movement of animals. In this chapter our focus is on reindeer herding in Finnmark, the northernmost area of Norway. Recapturing key aspects of the traditional organizational form – herd diversity and cross-boundary mobility – might be necessary in order for Sámi herding culture to adapt to climate change.

Traditional ecological knowledge and ecological niches

Traditional ecological knowledge is defined as the knowledge, practice and beliefs about dynamic relationships of living beings and the environment, a knowledge which has evolved in adaptive processes and been handed down from generation to generation (Ingold, 2000; Berkes, 2008). Combining traditional and scientific knowledge about ecological systems, and their interrelationships with cultural and economic systems, is crucial for understanding the resilience capacity of ecological and social systems and for identifying factors that can enhance the potential for sustainable development and self-sufficiency (Berkes et al., 2000; Berkes, 2008). The ACIA (2005) report recognizes that traditional ecological knowledge is important in order to supplement and enrich scientific data on climate change impacts. Sámi reindeer herders are unique observers of how changing weather patterns are altering grazing possibilities for reindeer and the sustainability of reindeer herding (Tyler et al., 2007). The EALAT project of the Sámi University College, which this chapter draws upon, represents an innovative approach to including traditional ecological knowledge into studies of adaptation to climate change.

Ecological niche is a well-known concept from ecology, describing the adaptation of an organism to a particular environment. The anthropological concept of ecological niche is also used in interdisciplinary studies of ecology, geography and anthropology, introduced by geographer Carl Troll (1899–1975) who studied parts of the world with extreme climatic variations (Troll, 1931, 1966). In the Andes mountains he described 'landscape belts' (*Landschaftsgürtel*) along the mountain ranges, not only with specific agricultural and pastoral activities, but with climate-related opportunities for creating a livelihood and human settlements.

Troll compared the high mountains in the tropics and the subarctic regions of the Sámi reindeer herds, where frequent freezing and thawing (*Frostwechselhäufigkeit*) was a common element. The most extreme climate zones of the planet are made inhabitable by the extreme variations – in terms of 'windows of opportunity' – that Nature presents to human beings, both in terms of ecological niches in which different plants and animals thrive, in terms of the possibility of migrating between such niches at relatively short distances, and in terms of temperature changes inside each niche, daily and annually. Humankind's main strategy to cope with and adapt to climatic extremes is in tune with Nature's own answer: maximizing variety in each Andean potato field and each Sámi reindeer herd reduces risk through principles similar to that of an insurance policy. These are principles alien to the practices underlying modern agricultural production. The basic foundation for a successful system of reindeer herding governance – particularly under climatic change – is the understanding of diversity and cyclicality, rather than stability, as the key feature of both the natural environment and the herders' response.

In the subarctic regions, the difference in snow quality is a key element of diversity, and topography is an important component determining how snow is distributed over an area. Winter grazing areas are determined by the time of the year and snow conditions (Sara, 2001, p. 46). In choosing grazing areas, the Sámi herders' first concern is what is called *guohtun* (the possibility for the reindeer to get to their food, i.e. grazing conditions through snow). Understanding and ability to handle changes in snow conditions are of vital importance for sustainable Sámi reindeer herding.

Troll's theories were further developed by anthropologist John Murra (1916–2006) in the context of the Andean cultures (Murra, 1975, 2002). In the Andes, Murra found a 'vertical archipelago' of micro-climates – ecological niches – and explained the pre-Columbian Andean cultures as based on sequential utilization of crops and animals found at these different ecological levels (*pisos ecológicos*). Reindeer herding can be understood in the context of Murra's 'archipelago' of ecological niches, but niches that are more horizontal than vertical. Herding is based on the sequential usufruct of a multitude of such ecological niches, moving the animals over large distances in annual cycles in order to find optimal grazing, ranging

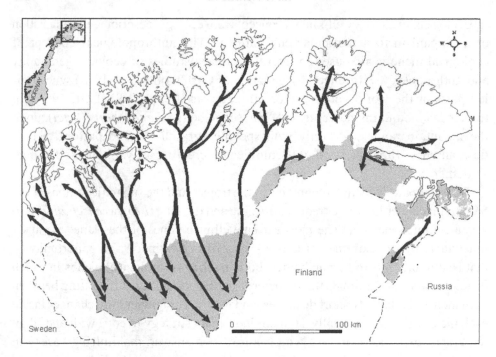

Figure 26.1 Reindeer herders' migratory pattern in Finnmark, Norway. (Source: Tyler et al., 2007.)

from the coast in summer to the inland areas with accessible snowpack in winter. The eight seasons of the reindeer herders – compared to the four in temperate climates – are but one reflection of the complexity of the system. Inside the niches controlled by the herders, there are also micro-niches that are used by the reindeer themselves. Dark-coloured animals – those that might suffer most from insects – find their way to the last patches of snow in summer, where insects are scarce. The annual treks normally cover several hundred kilometres and permanent human settlements and slaughtering facilities are often found at the passage points between summer and winter pastures.

Herds of mixed age and sex, varying in size from 100 to 10 000 animals, are maintained on natural pastures all year round and are typically moved between coastal summer pastures and inland winter pastures (Figure 26.1). The pattern of migration, from time immemorial, is clearly an adaptation to climatic conditions. Feeding conditions for reindeer in winter are chiefly a function of the availability of forage through snow, so the quality of the snowpack becomes a crucial variable. Weather variability with thawing and refreezing – during winter progressively increasing both the density and the hardness of the snowpack – makes it difficult for the animals to dig down to the plants beneath. Such conditions occur frequently at the coast, where winters are mild and precipitation high. Inland, however, conditions are colder, drier and temperatures have been more stable, creating favourable grazing conditions (i.e. snow conditions) – *buorre guohtun* – for the animals.

In herders' terminology the concept of *guohtun* is only used in connection with snow, i.e. the condition of the snow and the amount of snow (Magga, 2006). The term does not refer to grazing as such, to the availability of moss, lichen or plants etc. The herders' term *buorre guohtun* conveys how easy it is for the reindeer to dig through the snow in order to reach their food. If it is easy for the animals to reach the food through the snow, one says there is *buorre* (good) *guohtun*. This means that the snow is dry and grainy, and that the snow cover is not very thick. Under such conditions the reindeer do not have to use a lot of energy in order to get to their food. *Heajas* (bad) *guohtun* means that it is difficult for the animals to get to their food. The snow is hard and there are layers of ice both near the ground and at higher levels in the covering snow, and the animals will consequently use much energy to reach their food. Norwegian authorities have tended not to understand that feeding problems are often a result of lack of access to food through the snowpack, a problem which is overcome by the return of spring, rather than a permanent problem of 'overgrazing'. This is but one example of the lack of awareness of herders' traditional ecological knowledge.

One important possible effect of climate change is that the area with high *Frostwechselhäufigkeit* is extending further inland. If the ground surface of an area free from snow freezes in autumn, while wet, this produces *botneskárta* – an icy cover that will block access to vegetation until spring. Faced with this condition, reindeer herders say it is best to sell as many reindeer as possible: this means a winter catastrophe. This condition will not change until springtime comes and the hard ice-snow becomes grainy snow.

Institutions and risk-reduction mechanisms under extreme climatic risks

A comparative study of institutions in the Andes and among the reindeer herders exhibits many similarities (E. S. Reinert, 2007). The utilization of climatic niches created by climatic diversity forces both cultures into long annual treks. Both societies are organized in extended family groups, the *ayllu* in the Andes and the *siida* of the Sámi. The mechanisms developed both by the Andean potato farmers and the Sámi reindeer herders in order to reduce unpredictable climatic risks are based on the principle of insurance policies, on diversity. Even close to the present day in the Andes, farmers would grow up to 40 varieties of potatoes every year. One potato variety would survive severe frost in the growing season, another extreme drought, etc. Although not maximizing the yield in a 'normal' year, a diversity of varieties ensured survival under virtually any circumstance (Murra, 1975, 2002). Both the variety of the Andean potato fields and the traditional composition of a reindeer herd exemplify the diversity called for by a precautionary principle. When climate change adds to the already extreme weather conditions, it appears wise to understand and strengthen the very efficient coping mechanisms that have made survival possible for centuries.

In reindeer herding the challenges posed by unpredictable climate variation are met through herders' finely tuned skills in exploiting the options presented by the presence of a myriad of ecological niches, in other words, by the diversity of the landscape and the diversity of their herds. As one herder puts it:

The more landscape types one has – that is, alternatives with which to meet different situations – the more secure reindeer pastoralism will be over a longer period of time. Contrariwise, in a uniform landscape without alternatives, one is left helpless when faced with natural changes (within a season, between years) (Mikkel Nils Sara quoted in Paine, 1992).

Reindeer herders have traditionally maintained high levels of phenotypic diversity in their herds with respect, for example, to the age, sex, size, colour and temperament of their animals (N. Oskal, 2000; Magga, 2006). The Sámi concept of a 'beautiful' herd of reindeer (*čáppa eallu geallu eallu*) incorporates, therefore, a diversity which is the antithesis of the monoculture of homogeneity observed in a pure-bred herd of livestock developed by selection to suit the requirements of modern, high-yielding agricultural ruminant production systems.

The traditional diversity of the structure of the reindeer herds reflects a coping strategy aimed at reducing their vulnerability to the consequences of unfavourable – and unpredictable – conditions (for example Nilsen, 1998; A. I. Oskal, 1999). In this way apparently 'non-productive' animals have particular roles, which contribute to the productivity of the herd as a whole. For example, in the 1960s reindeer herds in Finnmark typically comprised between 25% and 50% adult males, many of which were castrated (Paine, 1994). Castrates were required for traction and to keep the herd gathered. They may also have lowered the general level of activity of the females, hence contributing to increased net energy gain in the herd. Modern agronomists have considered adult males unproductive and today few herds in Finnmark comprise more than 10% large bulls, but variation is large (Nilsen, 1998). With more temperature variability and difficult grazing conditions in winter, castrated male reindeer play a crucial role in breaking the ice cover, enabling the smaller animals to graze. The lack of herd diversity is a considerable threat to sustainability and adaptation to climate change.

Cyclicality and change

The climate in the Arctic is neither stable nor predictable. Great variability in weather patterns and the need for constant adaptation is the rule rather than the

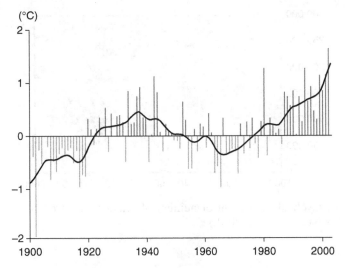

Figure 26.2 Cyclical behaviour of temperatures in the Arctic in the twentieth century: annual average change in near-surface air temperature from stations on land relative to the average for 1961–1990, for the region between 60° and 90° N. (Source: ACIA, 2005, p. 3.)

exception in Sámi reindeer herding. 'One year is not the next year's brother', goes a Sámi saying.

The study of climatic cycles in the Arctic goes back to Norwegian polar researcher Fridtjof Nansen (Nansen, 1926). Figure 26.2 shows the historic development from 1900 in average air temperatures in the Arctic. Today is not the first time the Arctic has seen rapid temperature change. Similar patterns were evident during the 1920s and 1930s. There are still reindeer herders alive who have knowledge about how reindeer husbandry coped and adapted in the past, knowledge that can be crucial to face a future with increasing climate change.

Historical records indicate large variations in the number of reindeer. Figure 26.3 shows the number of reindeer in Sweden from 1900 to 2000, displaying a distinctly cyclical variability. Between 1978 and 1998 the annual number of reindeer slaughtered in Norway, Sweden and Finland tripled, and then fell back again almost to the previous level. Despite very different governance regimes in the three countries, the production curves rise and fall in a remarkably parallel fashion (E. S. Reinert, 2002, p. 39).

By mitigating the negative effects of environmental variability, the skills of the herders create a protective cushion that minimizes the effect of this variability. A downward cycle in the number of reindeer may be triggered by a climatic shock to the herds so big that the herders are not able to compensate. Such events are characterized as *nealgedálv* ('the year when the reindeer starved') (Eira et al., 2008). The coping mechanisms involved in cushioning against variability are of the same

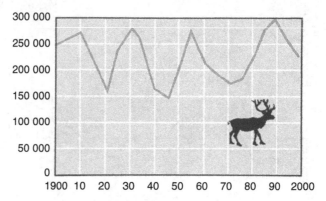

Figure 26.3 Cyclical movement of number of reindeer in Sweden, 1900–2000.
(Source: Statistics Sweden, 2001.)

kind that will be needed in order to understand, cope and adjust to permanent
change. The experiences from the cyclical patterns must be part of the analysis
of adaptation to climate change. The relationships that create cyclicality in the
number of reindeer are highly complex, and one of the goals of the EALAT project
is to investigate the factors involved and their internal relationships.

Reindeer herding and the nation–state

Early books on natural and political geography, as Giovanni Botero's *Relazioni
Universali* (originally published in 1596), list the Sámi regions of Lappia, today's
Sápmi, as an independent nation on a par with Norway and Sweden. When *Sápmi*
or *Lappia* later was absorbed into the modern nation–states, this was done in a
way that cut across traditional ethnic areas. Figure 26.4 shows how the dotted lines
marking the borders (from left to right) of Norway, Sweden, Finland and Russia cut
across the areas of the Sámi linguistic groups that also represented the traditional
migratory range of the herders. Comparing Figure 26.4 and Figure 26.1, which
shows the present-day migratory patterns of herders in Finnmark, renders the idea
of the importance of the traditional migration across present nation–state border,
within the domains of the linguistic groups.

Initially, even when Lappia later was absorbed inside the borders of Norway and
Sweden (at the time also including the Grand Duchy of Finland), the traditional free
movement of the Sámi reindeer herders in Northern Fenno-Scandia was maintained
and codified in *Lappecodicilen* of 1751. In this way the organizational units of the
Sámi herders – the *siida* – were preserved across national borders. Their sequential
usufruct of land continued across national boundaries – and indeed their whole economic system – continued unaffected by the consolidation of the nation–states for

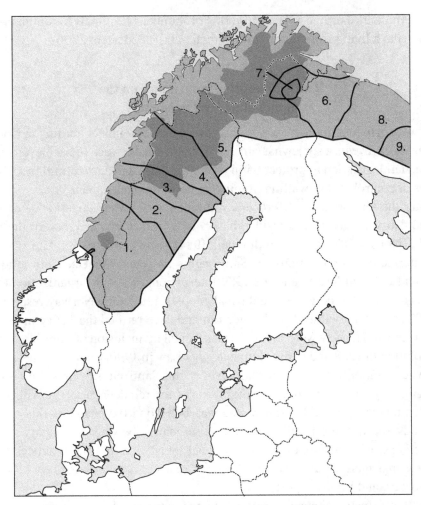

Figure 26.4 Sámi languages and nation–state borders. Geographical distribution of the Sámi languages: 1, Southern Sámi; 2, Ume Sámi; 3, Pite Sámi; 4, Lule Sámi; 5, Northern Sámi; 6, Skolt Sámi; 7, Inari Sámi; 8, Kildin Sámi; 9, Ter Sámi. Darkened area represents municipalities that recognize Sámi as an official language. (Source: Wikipedia, 2008.)

a long time (N. Oskal, 1999). Between 1809 and 1917 the Grand Duchy of Finland was part of Russia.

The most devastating effects on the reindeer herders' economy did not take place until 1852. The border between Russia/Finland and Norway had been drawn in 1826, but the free movement of reindeer and herders continued. In 1852 the border between Norway and Russia/Finland was closed to the herders. Herders' access to some of their key ecological niches was blocked, breaking off the main artery of the annual migration of the herders, seriously undermining the carrying capacity of the Finnmark herding system in terms of number of animals and humans that could make a living from herding (Bull et al., 2001; Pedersen, 2006).

Reindeer herding governance on the nation-state level in Norway

We now turn to look more closely at the governance of reindeer herding in Norway. Our analysis reveals a somewhat surprising mismatch between the achievements of Norwegian legislation to protect Sámi rights, and the rigid agricultural models and modes of control still prevailing in reindeer herding management.

After the long history of forced assimilation – 'Norwegianization' – came to an end, Norway has made considerable progress in achieving rights for the Sámi people (Norges Offentlige Utredninger, 1984, p. 18; 2007, p. 13). The two central international conventions on Sámi rights are the 1966 UN Convention on Civil and Political Rights and the 1989 International Labour Organization (ILO) Convention 169 on indigenous and tribal peoples, ratified by Norway respectively in 1972 and 1990 (neither Sweden nor Finland have ratified the ILO convention). Sámi rights in Norway were further strengthened by inclusion of para. 110a in the Constitution in 1988 and in the Human Rights Law in 1990. Central to these rights is the recognition that the material basis of the Sámi culture must be secured, strengthening the Sámi rights to the pastures for reindeer herding. Although a detailed analysis of the legislative changes is beyond the scope of this chapter, our study shows there is still a distinct discrepancy between the rights secured through extensive national and international legislation and the practices of the reindeer management authorities. These are recounted in E. S. Reinert (2006) in more detail, and briefly reported below.

In 1978, a new reindeer herding law was introduced in Norway. The principles of this legislation were, typically for the time, based on a strong faith in mass production and industrial agriculture, attitudes prevalent in the Norwegian Ministry of Agriculture (Lenvik, 1988, 1990). The Lenvik studies which informed this legislation were undertaken on particular areas only, in the southern reindeer districts in Norway, where annual meat production is stable, not showing the variability found further north. Moreover, the agricultural models for livestock feeding were not translatable to reindeer grazing on natural pastures. In particular, while the agricultural model does not attribute much value to an old wether, the situation is quite different in reindeer herding, where the castrated male reindeer (known by the herders as 'the gentlemen of the tundra') play an important role in breaking the ice so that the smaller animals may graze.

The 1978 reindeer herding law introduced an element of common pastures, without roots in previous Sámi legal tradition (where pastures traditionally were managed by the *siida*). Not surprisingly the reindeer herders themselves call the 1978 law the 'barnyard law', following which reindeer herds have increasingly been managed – almost exclusively – to maximize annual meat production, thereby seriously increasing the vulnerability of the herds. Thus the centrally-imposed management regime paid little attention to the traditional regulation of the use of pastures and to the role of natural cycles of production. The use of common pastures contributed to fuelling an official perception of 'overgrazing' recalling the classical 'tragedy of the commons' situation (Hardin, 1968). Herders were simultaneously accused of maximizing profits, by exploiting common resources for individual benefit, and of maximizing herd sizes. Either way, the policy conclusion was the same: too many reindeer.

As a consequence, the Norwegian Parliament (*Stortinget*) defined a non-revisable maximum number of reindeer in Finnmark, which warrants 'forced slaughtering' of reindeer by the Ministry of Agriculture. Officially, a year resulting in a large number of reindeer is a call for alarm, since according to governmental analysis heightened conditions of (re)production are interpreted as a threat to the 'sustainability' (statically defined) of the lichen the reindeer feed on, rather than as a natural cyclical change within a wider pattern of sustainability. This view has recently been challenged by a Sámi study of reindeer numbers in relation to ecological and traditional notions of sustainability (Joks et al., 2006).

The official consensus in Norway during the widely publicized crisis in reindeer herding in the late 1990s was that it was due to herder irresponsibility. During the 1990s Sámi reindeer herding was subjected to a fixed pricing regime. The volume of production was halved, with similar consequences to herders' income (E. S. Reinert, 2006). Other pieces of legislation, both national and international, contributed to further changes in the management of reindeer herds. Sanitary regulations enforced within the European Economic Area meant that the Norwegian Sámi lost control over the elements in the value chain where profits are made, in particular slaughtering and marketing (H. Reinert, 2007). In 2002 an estimated 80% of all reindeer in Sweden and Finland were slaughtered in establishments owned by the herders themselves. In Norway this number was around 20%, a striking difference from the neighbouring countries (E. S. Reinert, 2002). The herders lost control not only over a key stage in their productive cycle, but also an important aspect of herding culture to a monopsony (purchasing monopoly) controlled by non-Sámi economic interests. This economic regime also severely reduced the options to market reindeer meat in the luxury food market segment (Jensen, 1999), nationally and internationally, thus reducing herders' profitability even further (E. S. Reinert, 2006; H. Reinert, 2007).

In other words, the 1978 law denied Sámi reindeer herders access to key elements in the value chain. Individual purchasers were replaced by a system of 'target price' established through negotiations with the government. An important part of this regime of herding administration was a policy of equalization, whereby large reindeer owners through forced slaughter had to significantly reduce the size of their herds. At the same time, numerous new economically weak herding units with small herds were established (E. S. Reinert, 2006). This policy model from agriculture has attempted to reform Sámi reindeer herding to conform to monoculture practices of modern agriculture and to Fordist mass production (E. S. Reinert, 2006). A key goal in this policy has been to level out the natural cycles of production. Traditional ecological knowledge was not recognized as integral to sustainable reindeer management practice: 'It is as if everything we know has no value', a university-educated female herder pointed out (E. S. Reinert, 2006).

The detrimental effects of the 1978 law on Sámi reindeer herding in Norway have slowly been recognized, and it was supplanted by a new law in 2007 with the purpose of amending several of the unfortunate consequences of the 1978 law (Norges Offentlige Utredninger, 2001, p. 35). It remains to be seen how the new law will influence the practice of the reindeer herding administration and hence the sustainability of reindeer herding. As one reindeer herder put it: 'Before, we were used to work with an unpredictable nature, now we also have to work with an unpredictable government administration' (E. S. Reinert, 2006). Nevertheless, additional challenges remain to a more sustainable herding of reindeer, for example, the proposal that reindeer management should not be overseen by the Ministry of Agriculture (Norges Offentlige Utredninger, 2001, p. 35). It will be a formidable challenge to modify the structure of reindeer herding management and bring the governance practices in line with the improvements in legislation for Sámi rights obtained through the past 25 years; to secure the material basis for survival and continuation of Sámi culture; and to build adaptive capacity based on traditional ecological knowledge.

Some are calling for the traditional ecological knowledge of Sámi reindeer herders to be recognized and included in animal welfare legislation. There is fear that Sámi reindeer herders might lose their right to castrate male reindeer using their traditional knowledge and insights. As mentioned earlier, these animals provide vital support for the herds. Without them the vulnerability of the herds is increased, both as regards predatory animals, animals welfare, and reduced access to food through ice-covered pastures. Norway's National Committee for Ethics in Science and Technology (NENT) developed ethical guidelines, in effect giving support to the Sámi by recommending natural scientists to integrate and respect alternative sources of knowledge such as traditional knowledge. Recently, Vladimir Etylin from a reindeer herding family in Chukotka, Eastern Siberia, said in this respect:

'Being an indigenous representative and having been born on the tundra myself, I consider a ban on castration as a serious threat to all reindeer husbandry. [...] Castrated males do have their own place in the herd's structure too. Humans would not have been able to domesticate reindeer without castration. It is one of the corner stones of the domestication process. [...]Without castrations it is not possible to build up a controllable reindeer herd. Geldings have many functions in a reindeer herd. The first one is that they are the calmest animals of a herd. Which means that a reindeer herd with castrates quiets down easily. For example: In Chukotka it is impossible to survive without crushing ice during a so-called black ice period, when everything gets covered with a layer of ice. When this happens only castrates are strong enough to break such ice. [...] Reindeer cows follow after them and eat the fodder left over (Etylin, 2007).

Although the basis for Sámi legal rights is much stronger in Norway than in the other Nordic countries, a comparison of the practices of reindeer herding management in Norway, Sweden, Finland and Russia reveals striking differences, with important lessons for adaptive capacity for climate change. For example, in Russia, reindeer herders on the Yamal peninsula, organized in so-called brigades, have been allowed strong autonomy within their pasture areas. Within the borders of the pasture allocated to the municipal enterprise there are no formal limitations on flexible use of land, allowing brigades to 'trade snow' with neighbouring brigades, and even respond to severe climatic conditions, such as frozen pastures, by using seasonal pastures off season. An example is the winter of 2003/04 when rain in January caused locked pastures in the winter pastures south of the Yamal-Nenets Autonomous Okrog when some brigades responded to this by turning back towards summer pastures, other responded by not migrating as deeply into the winter pastures as they normally would (E. I. Turi, 2008).

The ability to self-organize according to their traditional knowledge is an important factor in strengthening reindeer herders' resilience to changes. The general secretary of World Reindeer Herders' Association stated that 'Nothing is liable to arouse more disturbances within reindeer husbandry than encroachments on its internal organization' (J. M. Turi, 2002, p. 71). Without a fluent and flexible organization reindeer pastoralists would lose the source of their greatest adaptive capacity. Institutional settings where reindeer pastoralists' traditional organization is restricted – as in Norway – represent a serious institutional constraint on adaptation.

Conclusion: building adaptive capacity for climate change

The crucial factor in adaptation to climate change is to improve ecological, cultural and economic resilience, through building adaptive capacity. To summarize, this chapter suggests four main elements for increasing the capacity of reindeer herding to adapt to climate change.

(1) Restructure herds to decrease vulnerability to climatic change. A *čáppa eallu* ('beautiful' herd of reindeer) is highly diversified, unlike a herd of livestock developed to suit the requirements of a modern, high-yielding agricultural production system. Modify government incentives that work against herd diversity. Include traditional ecological knowledge to enhance diversity, flexibility and adaptive capacity. Recognize that the complexity inherent in the reindeer herding and its traditional management is precisely the key to successful adaptation to the complexity presented by the uncertainties of climate change.

(2) Re-establish Fenno-Scandinavian reindeer herding across nation–state boundaries, allowing the *siida* to regain its transboundary character, through political processes involving the Sámi organizations in the countries in question.

(3) Limit the increasing permanent loss of ecological niches available to the herders, due to pastoral land being used for other purposes, including new infrastructure. This point differs from point 2 which refers to ecological niches that do exist, but are blocked from herders' use by national or international rules and regulations, whereas point 3 refers to ecological niches being permanently destroyed.

(4) Improve the economic basis of the reindeer herders by giving them back access and ownership to the most profitable activities in the value chain: slaughtering and marketing. A solid economic base will better enable herders to absorb the costs associated with climatic change.

Acknowledgements

This work is in part supported by Research Council of Norway, project IPY EALAT-RESEARCH: Reindeer Herders' Vulnerability Network Study: 'Reindeer pastoralism in a changing climate', grant number 176078/S30. The lead author thanks Reindriftens Utviklingsfond (The Reindeer Herding Development Fund) in Norway for a grant to study comparative reindeer herding governance in Finland, Norway and Sweden. We thank Professor Kirsti Strøm Bull for valuable comments.

References

ACIA 2005. *Arctic Climate Impact Assessment*. Cambridge: Cambridge University Press.

Berkes, F. 2008. *Sacred Ecology*. New York: Routledge.

Berkes, F., Colding, J. and Folke, C. 2000. 'Rediscovery of traditional ecological knowledge as adaptive management', *Ecological Applications* **10**: 1251–1262.

Botero, G. 1622. *Le relazioni universali*. Venice: Alessandro Vecchi.

Bull, K. S., Oskal, N. and Sara, M. N. 2001. *Reindriften i Finnmark: Rettshistorie 1852–1960*. Oslo: Cappelen.

Eira, I. M. G., Magga, O. H., Bongo, M. P., Sara, M. N., Mathiesen, S. D. and Oskal, A. 2008. *The Challenges of Arctic Reindeer Herding: The Interface between Reindeer Herders' Traditional Knowledge and Modern Understanding of the Ecology, Economy, Sociology and Management of Sámi Reindeer Herders*. Available at

http://arcticportal.org/uploads/Oe/1k/Oe1kPFlFGawnDaldpSRdCQ/Eira_127801.
    pdf, pp. 1–18 (accessed 20 June 2008)
Etylin, V. 2007. *On Reindeer Castration. Presentation to the Norwegian Ministries
    about the Work of the International Centre for Reindeer Husbandry.* Video
    podcast in Russian available at http://video.google.com/videoplay?docid=-
    3047476777473731395&hl=en (accessed 24 July 2008)
Hardin, G. 1968. 'The tragedy of the commons', *Science* **162**: 1243–1248.
Ingold, T. 2000. *The Perception of the Environment: Essays on Livelihood, Dwelling and
    Skills.* London: Routledge.
Jensen, R. 1999. *The Dream Society: How the Coming Shift from Information to
    Imagination will Transform your Business.* New York: McGraw-Hill.
Joks, S., Herriksen, I. M., Mathisen, S. D. and Magga, O. H. 2006. *Reintallet i
    Vest-Finnmark: Forskningsbasert vurdering av prosessen rundt fastsettelse av
    høyeste reintall i Vest-Finnmark.* Kautokeino: Samisk Høgskole/Nordisk Samisk
    Institutt.
Lenvik, D. 1988. *Utvalgsstrategier i Reinflokken.* Alta: Reindriftsadministrasjonen.
Lenvik, D. 1990. 'Flokkstrukturering – tiltak for lønnsom og ressurstilpasset reindrift',
    Rangifer **4**: 21–35.
Magga, O. H. 2006. *Diversity in Saami Terminology for Reindeer, Snow and Ice.*
    London: Blackwell Publishing.
Murra, J. 1975. 'El "Control Vertical" de un Máximo de Pisos Ecológicos en las
    Sociedades Andinas', in Murra, J. (ed.) *Formaciones Económicas y Políticas del
    Mundo Andino.* Lima: Instituto de Estudios Peruanos, pp. 59–115.
Murra, J. 2002. *El Mundo Andino: Población, Medio Ambiente y Economía.* Lima:
    Instituto de Estudios Peruanos/Universidad Católica.
Nansen, F. 1926. *Klima-vekslinger i Historisk og Postglacial Tid.* Oslo: Dybwad.
Nilsen, Ø. 1998. 'Flokkstrukturen i Varanger-reindrifta på Slutten av 1800-tallet og i
    dag', *Varanger Årbok* **1998**: 107–115.
Norges Offentlige Utredninger 1984. *Om Samenes Rettsstilling.* Oslo: Norges Offentlige
    Utredninger, (Nov 1984: 18).
Norges Offentlige Utredninger 2001. *Forslag til endringer i reindriftsloven.* Oslo: Norges
    Offentlige Utredninger, (Nov 2001: 35).
Norges Offentlige Utredninger 2007. *Den nye sameretten.* Oslo: Norges Offentlige
    Utredninger, (Nov 2007: 13).
Oskal, A. I. 1999. 'Tradisjonelle vurderinger av livdyr', paper presented at the 10th
    Nordiske Forskningskonferanse om Rein og Reindrift, 13–15 March 1998,
    Kautokeino. Rangifer Report No. 3: 121–124. Nordisk Organ for Reinforskning
    (NOR).
Oskal, N. 1999. 'Det Moralske Grunnlaget for Diskvalifiseringen av Urfolks
    Eiendomsrett til Land og Politisk Suverenitet', in Oskal, N. and Sara, M. N.
    (eds.) *Reindriftssamiske Sedvaner og Rettsoppfatninger om Land.* Manuscript.
    Kautokeino: Nordisk Samisk Institutt/ Samisk Høgskole.
Oskal, N. 2000. 'On nature and reindeer luck', *Rangifer* **2–3**: 175–180.
Paine, R. 1992. 'Social construction of the 'Tragedy of the Commons' and Sámi reindeer
    pastoralism', *Acta Borealia* **2**: 3–20.
Paine, R. 1994. *Herds of the Tundra: A Portrait of Sámi Reindeer Pastoralism.*
    Washington, DC: Smithsonian Institution Press.
Pedersen, S. 2006. Lappekodisillen i nord 1751–1859. Fra Grenseavtale og Sikring av
    Samenes Rettigheter til Grensesperring og Samisk Ulykke. PhD thesis, University
    of Tromsø.

Reinert, E. S. 2002. *Reinkjøtt: Natur, Politikk, Makt og Marked*. Tromsø, Verdiskapningsprogrammet for Rein, SND. Available at www.othercanon.org/uploads/native/Reinkjoett_natur_politikk_makt_og_marked.pdf (accessed 24 July 2008)

Reinert, E. S. 2006. 'The economics of Reindeer herding: Saami entrepreneurship between cyclical sustainability and the powers of state and oligopoly', *British Food Journal* **108**: 522–540.

Reinert, E. S. 2007. 'Institutionalism ancient, old, and new: a historical perspective on institutions and uneven development', in Chang, H.-J. (ed.) *Institutional Change and Economic Development*. New York: United Nations University Press, pp. 354–534.

Reinert, H. 2007. The corral and the slaughterhouse: knowledge, tradition and the modernization of indigenous reindeer slaughtering practice in the Norwegian Arctic. PhD thesis, Scott Polar Research Institute, Cambridge University.

Sara, M. N. 2001. *Reinen – et Gode fra Vinden*. Karasjok: Davvi Girji.

Statistics Sweden 2001. Number of reindeer at different times during the twentieth century. Diagram. Available at www.rennaringsstatistik.scb.se/diagram1.asp#A (accessed 24 July 2008)

Troll, C. 1931. 'Die geographische Grundlage der Andinen Kulturen und des Inkareiches', *Ibero-Amerikanisches Archiv* **5**: 1–37.

Troll, C. 1966. *Ökologische Landschaftsforschung und vergleichende Hochgebirgsforschung*. Wiesbaden: Steiner.

Turi, E. I. 2008. Living with climate variation and change: a comparative study of resilience embedded in the social organisation of reindeer pastoralism in Western Finnmark and the Yamal Peninsula. Master thesis in political science, University of Oslo.

Turi, J. M. 2002. 'The world reindeer livelihood: current situation, threats and possibilities', in Kankaanpää, S., Müller-Wille, L., Susiluoto, P. and Sutinen, M.-L. (eds.) *Northern Timberline Forests: Environmental and Socio-Economic Issues and Concerns*, Research Paper No. 862. Helsinki: Finnish Forest Research Institute.

Tyler, N. J. C., Turi, J. M., Sundset, M. A., Bull, K. S., Sara, M. N., Reinert, E., Oskal, N., Nellemann, C., McCarthy, J. J., Mathiesen, S. D., Martello, M. L., Magga, O. H., Hovelsrud, G. K., Hanssen-Bauer, I., Eira, N. I., Eira, I. M. G. and Corell, R. W. 2007. 'Saami reindeer pastoralism under climate change: applying a generalized framework for vulnerability studies to a sub-arctic social-ecological system', *Global Environmental Change* **17**: 191–206.

Wikipedia 2008. Figure on the geographic distribution of the Sámi languages. Available at http://en.wikipedia.org/wiki/Sami_people (accessed 24 July 2008)

27

Limits to adaptation: analysing institutional constraints

Tor Håkon Inderberg and
Per Ove Eikeland

Introduction

The academic literature has been biased towards depicting adaptation to climate change as a rational decision-making process, with constraints being mainly available resources and technologies. An ideal rational adaptation process would start by evaluating the problem of climate change (assessing exposure), mapping possible solutions (possible adaptive measures), and, through a cost–benefit approach, the best and most feasible adaptation measure(s) would be decided and simply implemented. Should organisational structure not favour the implementation of measures directed at the goals set, it would be accordingly altered.

Recent studies have focused on institutional barriers and limits to adaptation, however (for example Adger et al., 2007). We share their concern that neglecting institutional constraints may conceal true human limits to adaptation and also narrowly frame the scope of measures that could be taken in order to increase adaptive capacity. Society consists of formal and informal social structures (regulatory factors, values, norms and cognitive limits) influencing choice and behaviour. These factors have still not received the deserved attention in the adaptation literature. After all, resources and technology are of little use if such institutional factors hinder implementation of proper adaptive measures.

In this chapter, we develop an institutional approach and seek to apply it to analyse constraints to adaptive capacity in the national energy system, chosen due to its vital role as a hub linking together other societal systems and functions. Human life, welfare and security in modern societies are highly dependent on stable supply of energy. Energy-technical structures are exposed to the forces of nature and

Adapting to Climate Change: Thresholds, Values, Governance, eds. W. Neil Adger, Irene Lorenzoni and Karen L. O'Brien. Published by Cambridge University Press. © Cambridge University Press 2009.

are vulnerable to changes in these stemming from disturbances in the global climatic system. Keeping national energy systems intact under climate change impacts is therefore a paramount societal challenge.

Our framework views the national energy system as a complex *socio-technical system*, defined as a social network of agents interacting in a specific technology area under a particular *institutional infrastructure* to generate, diffuse and utilise technology (Carlsson and Jacobsson, 1997, p. 268). The system has a technical part consisting of physical installations directly exposed to climate change. A parallel social network of private and state organisations own, operate, maintain, plan, construct, regulate and use these technical energy structures. The social network is responsible for adapting to climate change; institutional barriers to adaptive capacity are present within this network. The two parts are interconnected and must be analysed together.

The framework also draws on institutional organisation theory, viewing the national energy system as a potential *organisational field* with interacting agents bound together by institutional factors (routines, norms, values) constraining adaptation, not only in single organisations but the entire system. The main questions addressed in this chapter are:

- How will an institutional theoretical framework add to our understanding of adaptive capacity and its limits?
- What are the most important institutional barriers or limits to adaptation in a national energy sector, illustrated by the Norwegian situation?

The section below addresses the first question, presenting the theoretical foundation for an institutional critique of the rational approach to adaptation. The section that follows addresses the second question, discussing how institutional factors would interfere with a rational scheme for adaptation in the energy sector. Some short concluding remarks are made in the final section.

The institutional approach to adaptation

The institutional critique of rational adaptation

Adaptation is a topical issue in the organisation theory literature with contrasting views on the degree to which organisations can change the way they work in response to changes in their environments. On the one hand is the idea that organisational actors can readily survey their environments, interpret what their problems are and change their organisation internally (i.e. what it does and how it does it) to promote organisational survival and performance. This rational approach to organisational life and behaviour dominated the early organisation theory literature (Hatch and Cunliffe, 2006). The paradigm gradually developed from a unitary rational organisational perspective to 'bounded rationality' (Simon, 1976) and to allow for conflicts between rational individual interests.

The 'institutionalist' reaction to the idealised rational perspective saw organisational actors as constrained, having difficulties in interpreting what changes meant to their organisation and opposing changes that would challenge entrenched views on how things already worked (March and Olsen, 1979). The approach highlights the role of institutional factors, stable rules, routines, norms and cognitive limits not only in constraining but also in empowering social action, by giving meaning to social life in spite of a changing external organisational environment. Such institutional factors exist not only at the individual and organisational levels but could also span an entire sector or the equivalent of what the organisational theory literature has coined an 'organisational field', defined as a recognised area of institutional life, including key suppliers, resource and product consumers, regulatory agencies and other organisations that produce similar services and products (DiMaggio and Powell, 1983). Agents within such functional fields share a common regulatory framework and tend over time to develop shared values, norms and cognitive scopes, beyond the technical requirements of the task at hand (Selznick, 1983, p. 17), limiting their available future options, which Krasner (1988, p. 71) has termed a 'path dependent' organisational room of decisional manoeuvrability.

Adaptive capacity as capacity to learn and act

The insights from institutional theory have implications for our understanding of societal adaptive capacity to climate change, notably within sectors showing the characteristics of an organisational field with strong regulatory constraints, norms, values and cognitive frameworks (Zucker, 1987).

The ideal rational approach to climate change adaptation would depict goal-oriented agents gathering and processing knowledge about expected climate-change-induced weather phenomena (and how these may produce second-order geo-physical effects), evaluating the potential vulnerabilities/opportunities of the technical system in light of these effects and choosing optimal strategy(ies) to reduce vulnerabilities/act on opportunities given the risks calculated. Adaptive capacity would be constrained by resources available (March and Olsen, 1979; Brunsson, 2003, p. 168). As noted in the introduction, this approach is common in the adaptive capacity literature, putting emphasis on formal factors, or 'determinants' of adaptive capacity in general models (IPCC, 2001; Yohe and Tol, 2002).

The institutional approach to climate change adaptation would look for constraints in the gaining of knowledge about climate change effects, vulnerabilities and opportunities, and in acting on this knowledge. Adaptive capacity would accordingly be constrained by limited *capacity to learn* and *capacity to act*, reflecting the parallel organisational systems of 'thought and idea' and 'action'.

Institutional constraints to learning

The institutional understanding of constraints to learning stems from a thorough understanding of the learning process. Learning within an organisation or organisational field is dependent on *access to, processing of* and *dissemination of* information or knowledge within the organisation and organisational field, and the capacity to store this knowledge and erase it when obsolete. An organisation learns through everyday routine-based activities (Young, 1991); and through active target-oriented creation, capture, transfer and mobilisation of knowledge (Levitt and March, 1988). In essence, the learning process thus involves *learning by doing* (Young, 1991), *learning by interacting* (Sabatier, 1988; Brown and Duguid, 1991) and *organisational memory* (Levitt and March, 1988). Learning by doing basically represents the daily routine improvement. This can be associated with *single-loop learning*, meaning incremental, technical modifications improving an organisation's ability to achieve its set goals – without changing norms, values, or questioning the goals (Nilsson, 2005, p. 11). Learning by interaction, on the other hand, reflects learning that involves cumulative processes where individual knowledge is combined to create knowledge that is more than its distinctive parts. Learning within organisational fields depends on the culture of cooperation between agents within the field and their interaction with agents outside the field with different experiences (Levitt and March, 1988, p. 329). Broad interaction would increase the chance of *double-loop learning*, which basically is learning that modifies or at least questions norms, priorities, goals and objectives – needed in climate change adaptation due to the sheer complexity of the issue and its potentially devastating effects. Double-loop learning tends, however, to be the bottleneck of organisational learning (Argyris and Schön, 1978, p. 21).

Institutional constraints to interactive learning are a topical issue in the 'systems of innovation' literature, which highlights how some systems have developed and others lack trust and a culture (routines) of cooperation (flow of knowledge). Some have extensive long-term cooperative ties, while others are narrower (Carlsson and Jacobsson, 1997; Carlsson and Stankiewicz, 1991; Lundvall, 1992). This literature also highlights the impact of formal institutional factors, notably governmental regulations, on the capacity to learn.

Institutional mechanisms inhibiting learning outside a particular *path* is another topical issue addressed in the literature. One of these has been coined the 'competency trap'. This relates to a situation where increased efficiency in one area, created by incremental refinement of routines, leads to specialisation and further refinement. A competency trap may occur if high performance with an inferior routine leads an organisation to gain further experience based on it, and by doing so stick to the inferior routine at the cost of not selecting better ones (Levitt and March, 1988, p. 322). The technology systems literature holds that networks in sectors experiencing increasing returns to capital investments in expansionary periods will run the risk of becoming 'blind' to alternative technological strategies needed when changes occur in their business environment. Past successes may limit adaptation to new circumstances, 'locking' the networks into 'old' technologies. In this context firms and networks search outside their traditional areas in a highly localized capacity (Carlsson and Jacobsson, 1997).

Other mechanisms can been grouped under the umbrella *ambiguity of interpretation* (March and Olsen, 1979). Factors like wishful thinking, preference for the status quo, and an organisational tendency to frame problems according to culturally accepted norms and cognitive schemas, will all contribute to hinder learning in the sense of deviating from the 'path'.

Organisational ability to act

Capabilities to learn are only half the story of what constitutes adaptation capacity. To effect measures of adaptation, one also needs organisational abilities to implement feasible actions by taking appropriate decisions, coordinating action, and generating and allocating the necessary resources. If organisational learning is exploration of new opportunities, action can be the exploitation of them (March, 1991).

The rational perspective sees organisational coordinated action as following automatically from decisions, and action effectiveness to be a matter of adapting structure and regulations to the goals set – a logic of consequence. Action can in general be understood to be decision-making and resource allocation (March and Olsen, 1999, p. 56). The institutional perspective holds that organisational decisions are more restricted and follow a logic of appropriateness – action as bound by norms, values and cognitive scopes, in which decisions made would be similar to those made in earlier comparable decision-situations (routines or standard operating procedures) (Cyert and March, 1992, pp. 120f). Action then is what is *expected* by an actor in a process where situation and roles are matched, thus fitting an informal 'rule' to a situation (March and Olsen, 1989).

Organisational ability to act is closely connected to the concept of governance; namely the process whereby societies or organisations make important decisions, determine whom they involve, and how they render account. Governance includes the administrative and process-oriented elements of state governing, both a loose steering role, management of incentives and issuing of commands through regulations (Christensen and Peters, 1999, p. 163). An institutional conception of *governance* emphasises the legitimisation of policy choices and the maintenance of norms (Christensen and Peters, 1999, p. 167) seen as necessary for ensuring successful adaptation (March and Olsen, 1989). Successful governance for climate change adaptation depends on a long-term normative and legitimacy basis to avoid resistance to changes in structures and practices (Næss et al., 2005, p. 129). The capacity to act on climate change adaptation measures is thus constrained by prevailing norms, values and cognitive scopes, and the abilities of actors to either align measures to these institutions or change the institutions *as such*.

Adaptation in the Norwegian energy sector: institutional barriers

How then may institutional factors act as barriers to adaptation in the national energy sector? This issue must be discussed on the premise that adaptation to climate change would need a very *long-term* and *broad* process of learning and acting, given the potential devastating effects that a faulty adaptation process could have, such as long-term energy blackouts with consequent harmful effects on human life and welfare.

A complete understanding of how climate change would affect national energy systems is on the other hand impossible, given that the decision-situation is one not only of high uncertainty but partly also ignorance. Regional and local climate change parameters are still highly uncertain despite progress made by IPCC projections about regional changes in temperature, precipitation and wind conditions. Another major information-processing task is the identification of how these parameters interact with other geo-physical conditions to produce second-order effects of relevance for the energy system, such as the probability of landslides, flooding and avalanches stemming from new patterns of precipitation; the probability of downfall of trees generated by higher wind-speeds; local climatic changes caused by changes in temperature, etc.

An ideal rational energy sector learning process would encompass participation of agents knowledgeable about the weaknesses of the system, climate change impacts and solutions to increase system robustness. Agents should give out information openly and act in a spirit of cooperation. This would be needed to ensure that new knowledge is generated based upon a comprehensive understanding of existing information. Next, appropriate actions based on knowledge would include flexibility in reaching agreement between energy system agents on reallocation of funding and on reorientation of investments. Finally, decisions made would be implemented.

However, national energy sectors are typical of technology systems with strong organisational field characteristics constraining such an ideal and broad approach to learning and flexibility in acting, as sketched in the previous section. They have typically evolved historically with institutionalised forms of interaction between agents. These put a strain on learning by interaction and favour technological trajectories that promote *learning by doing* in a single-loop way.

Development of the Norwegian energy sector

The Norwegian energy system evolved under a social contract with the government to provide abundant power supply for industrial development and economic growth in return for resources and legal basis enabling massive development of large-scale hydropower. Strong cooperative ties between agents in the system (producers, technology suppliers, large industrial consumers and the state energy agency) made Norwegian energy a success story. The abundant mountainous river systems were tamed and complex engineering projects carried out giving Norway an energy system highly dependent on electricity (nearly 100% hydropower), and with a web of transmission lines transporting power from hydropower-rich areas in the western and northern mountainous areas down to population-dense areas at the coast. Sector agents (predominantly engineers) developed a great sense of pride in belonging to one of the most prestigious national economic sectors.

The other side of the coin of this success story was an energy system highly specialised in hydropower and relatively 'vulnerable' due to its dependence on one source of supply and the need to transfer power over long distances, across mountainous areas susceptible to rough weather conditions. Until the 1990s, this was no major problem. A major surplus capacity had been developed, reflecting the many large-scale development projects carried out despite major overruns of initial cost estimates, and the sector had so far managed to attract high-quality engineers for maintenance.

This specialisation led the Norwegian energy system into a competency trap, with hydropower engineers sharing technical norms and routines about how things should be done to secure supply. Values concerning the societal benefits of continued large-scale hydropower development were also shared in the sector (Nilsen and Thue, 2006, p. 47). Alternative investments were seen as either too small or too different to fit in with the logic of rapid system expansion. Critiques of this logic – from scientists detecting negative ecological effects of large-scale hydropower; economists pointing to the societal economic losses from the engineering logic of large-scale development at any costs; energy system specialists pointing to the need for diversification and the economic rationale of giving focus to demand-side management rather than further expansion of supply – were fended off more or less in unison by hydropower sector agents. Even the state regulatory agency shared their norms and values, which led some scholars to denote the Norwegian energy system as a hydropower-corporatist regime (Midttun, 1987).

In 1991, the Norwegian Parliament introduced a new market-based electricity trading regulation; new formal institutions challenged existing informal institutions. Power suppliers were now allowed to compete for customers across former local monopoly areas. A functional split was made between monopoly activities (operation of the grid) and activities subject to competition (production and supply), with the former subjected to incentive regulations aimed at cutting costs of operating and maintaining the grid. Investments came to a halt due to market-based risks and falling prices resulting from massive existing surplus capacity. The engineering norm was replaced by a norm of economic rationalisation, cost minimisation and paying dividends to shareholders. Restructuring of the sector introduced private business logic with outsourcing of functions (engineering, construction, maintenance) and lay-off of personnel. In other words, it involved a substantial replacement of the goals of security of supply and maintenance with cost minimisation and maximisation of dividends to shareholders.

During the following decade, demand growth caught up with the surplus capacity and security of supply concerns grew after several incidences of particularly low seasonal rainfall. Higher prices increased the willingness to invest, but interest was primarily directed towards further hydropower development, despite its potential vulnerability to low precipitation incidences, thus indicating that old competencies and norms still prevailed in the system. Into the 2000s, those speaking out for greater diversification in energy carriers and energy sources found a greater stronghold among Norwegian politicians, with state support granted to investments in district-heating systems based on biomass and wind power. So far this has attracted little interest, partly because the support system is far less ambitious than that seen in other European countries.

More recent considerations: adaptation of the energy sector

Against this background, discussion about adaptation to climate change has started in the Norwegian energy sector. Our research is still in progress, but a few observations illustrate how institutional factors appear to influence the choice of strategies for learning about climate change, about its effects on the energy system and about future action.

The rational adaptation model would predict a broad evaluation by energy system agents of all major climate change effects, energy system weaknesses and possible solutions in response to the challenges of climate change. So far, the work carried out by the Norwegian Watercourses and Energy Directorate (NVE), the agency responsible for national long-term energy supply security, has been quite narrow in scope. The Directorate's Department of Hydrology together with the Norwegian Meteorological Institute carried out a major project in 1996–2000 on the effects of climate change on hydrological balances in Norway. This work was funded by the Norwegian Association of Power Suppliers and the Research Council of Norway. The main focus was how climate change would alter precipitation levels, patterns of water flow and affect the risk of flooding in rivers and security of dams in hydropower reservoirs. The work has resulted in flood maps being created for different regions in Norway, serving as input to further municipal planning work.

This cognitive focus on hydrology indicates institutional influences on the interpretation of climate change impacts on the Norwegian energy system – a *framing effect*. Hydrology and flood prevention is a main competency area of the NVE, in addition to its responsibility for long-term energy security more generally. Well-known problems like flooding have simply received more attention at the cost of more uncertain effects, like landslides or changes in wind patterns. Yet another potential cognitive bias comes from the main conclusion drawn from the work by the NVE: 'climate change will be beneficial for hydropower production in Norway and the rest of the Nordic area. More stable runoff to the reservoirs throughout the year and an increase in total runoff will give higher production' (NVE, 2006). This might be an expression of a 'focusing effect' – a prediction bias occurring when people place too much importance on one aspect of an event, causing error in accurately predicting the utility of a future outcome. Such cognitive biases could potentially be constraints for a far broader evaluation of climate change impacts on the Norwegian energy system.

An area where knowledge availability and gathering are still scant in the Norwegian system regards the potential effects of climate change on electricity infrastructure that stretches through landscapes already experiencing tough climatic conditions. To be sure, much of the infrastructure is already well built for rough weather conditions, and the scant attention could reflect complacency, given the still weak understanding of second-order climate effects. A recent report concluded that increased frequency of landslides and melting of permafrost in mountainous areas, affecting unstable landmasses, could be expected from the higher frequency and intensity of rainfall in parts of Norway (Førland et al., 2007). This knowledge has so far been poorly connected to the location of local and central energy infrastructure.

A recent March 2008 landslide in the town of Aalesund, resulting in several deaths, focused media attention on the general lack of mapping of areas in Norway exposed to landslides, a critique valid also for local energy infrastructure exposure. History has recorded catastrophic landslides affecting hydropower facilities. After heavy rains in 1963, landslides into the Vaiont reservoir in Italy caused the stored water to spill over the dam, sweeping away the village of Longarone and flooding nearby hamlets; some 2000 people drowned (Hendron and Patten, 1985).

A closer look at existing local energy planning documents in Norway[1] shows that few if any local energy companies by March 2008 had included any evaluation of their facilities' exposure to climate change. Nor had such an evaluation been ordered by the regulatory authorities. The NVE had started work aimed at increasing attention among local energy system operators to the possible need for new networks to endure greater strains from climate change-induced phenomena such as icing, potentially a problem in wintertime due to more days around freezing point, combined with greater air humidity. Ice on the installations frequently destroys wires, attachment points and pylons of Norwegian electricity infrastructure. Climate change has not yet figured as a prominent driver of decisions, as illustrated by the current heated discussion in Norway on whether electrical cables should be buried underground or not. When the NVE in late 2007 hosted its annual 'Energy Days Conference', the major annual gathering of energy industry representatives in Norway, burying of the grid was a topical issue, engaging also the Minister of Energy in her address to the conference. The conflicting stances are between industry seeking to avoid additional costs from burying cables against local groups protesting against the degradation of aesthetic nature values from overhead pylons and cables. The climate change exposure of the grid, with more icing and higher frequency of lightning affecting overhead facilities, has not received much attention – especially compared to other European countries (Commission of the European Communities, 2003).

Hence, at the start of 2008, adaptation work in the Norwegian energy sector was still characterised by scattered information gathering and no well-systematised, broad-scope learning processes. Those involved were first and foremost agents at national regulatory level and less often the local energy systems agents (i.e. those actually operating the exposed infrastructure). The lack of attention at the local level could be interpreted institutionally, as representing a perception of complacency that 'we have ridden off storms before' and are well prepared for new ones. Signs of imperfect preparedness are clear, however. Winter storms, icing and heavy snowfall in parts of Norway in recent years have crashed many grids and blacked out local communities, with some consumers experiencing cut-off of power for a prolonged period. For instance, the storm *Narve* affected parts of Norway badly in January 2006, caused major devastation of the central and local grids, leaving the entire north of Norway dependent on supply via the grid to Sweden and cutting power for thousands of people, many of them without power for more than a week. One argument heard about why the storm created such devastation was that it came in from a different direction than normal storms,

---

[1] Local energy companies are ordered by regulation to make local energy plans and power system plans.

showing that adaptation to past experience is no guarantee for preparedness for future climate change effects.[2]

Two reports published by Norwegian Defence Research Establishment (FFI) in 2000 and 2001 concluded that the Norwegian energy system had major flaws in robustness to external events, with formal institutional changes after the 1991 transition to a market-based regulatory regime partly to blame for increased vulnerability (Rutledal et al., 2000; Fridheim et al., 2001). The reform's division of grid and production was accompanied by highly reduced incentives for the still-regulated grid companies to keep up maintenance efforts. Continued increase in vulnerability is expected unless measures are taken, according to one of the reports (Fridheim et al., 2001). The other indicates that changes in the sector's professional demography (from an engineer-dominated to an economist-dominated industry) after the 1991 reform influenced culture and cognitive scopes in the sector. This noticeably altered the former prevailing culture and norm of 'overinvestment' in dimensioning of grids and level of maintenance towards 'social–economic feasibility' as the dominating norm, in reality leading to an 'underinvestment' problem in many companies (Rutledal et al., 2000: p. 25). This has implications for both learning and acting capacities. The 'learning by doing' routines of the sector have changed following significant outsourcing as short-term contracts of system robustness maintenance. The 'learning by interaction' routines have also changed, from engineers meeting engineers to economists interacting with economists; outsourcing has fragmented and delayed learning processes. While the grid seems to be a weak point of the system, far fewer resources are invested in maintenance than before deregulation, mirroring goal replacement discussed earlier (Rutledal et al., 2000). The challenges of climate change have so far not affected this picture, as illustrated by the plans presented for investments in the central, high-voltage grid between 2005 and 2020, which do not refer to climate change or changes in weather at all (Statnett, 2005).

Even though the regulatory bodies are aware of these institutional influences, they face problems in finding solutions – economic incentives given are too weak to prompt an adequate increase in maintenance. This, combined with the focus on higher financial surplus in the sector, contribute to an impression of an organisational field having taken on a culture for short time scope, reducing the capacity for learning and acting on climate change. The fragmentation caused by unbundling of the sector has the potential of further reducing possibilities for coordinated learning and action, use of holistic systems approaches, and alternative thinking 'outside the box'.

A final comment on institutional constraints follows from the still quite narrow focus on strategies to increase the robustness of the Norwegian energy system. From

---

[2] Inger Ha Hansen-Bauer, senior research fellow at the Norwegian Meteorological Institute, pers. comm.

a rational viewpoint, a range of technical investment alternatives exists that could assist in balancing the system at a higher level of robustness. *Retrofitting* would include proper maintenance and strengthening of existing technical structures on the supply side; scaling up dams, cables and pylons to withstand greater loads expected from precipitation, landslides, wind, ice, snow, lightning, etc. It could also include installing more energy-efficient gear to save energy and reduce the need for erecting new vulnerable structures for production/transmission. For instance, *decentralisation* would imply strengthening of production capacity close to or at consumption sites as an alternative to development of vulnerable long-distance transmission grids. *Diversification* in primary energy sources and energy carriers would entail taking a broader scope, choosing less vulnerable energy sources and carriers. *Relocation* of existing production and transmission facilities to safer locations could typically be burying cables underground to make them less vulnerable to wind, lightning, icing and downfall of trees. *Modification of nature conditions* would entail such things as cutting down vegetation and planting new species of vegetation to hinder downfall of trees on installation or modification of geological structures for the same purpose.

Presently, a clear cognitive bias appears in the adaptation measures discussed in the Norwegian energy sector, with most of the focus given to retrofitting on the supply side (production and transmission) and modification of the environment. Diversification in the Norwegian energy system proceeds slowly, without adaptation to climate change as a decision-premise either to speed up diversification or to influence the choice of alternatives based on evaluation of their robustness in case of higher frequency of more extreme climate conditions in Norway (for example the robustness of wind power towers erected). The slow speed of diversification is consistent with a lock-in effect on hydropower; many of the new development projects given a go-ahead in Norway in the past year have been small hydropower projects. Decentralisation lost focus after the 1991 market reform abolished the focus on local security of supply, when the new goal of cost minimisation at the national level gave priority to inter-regional trade in electricity. Decentralisation has so far not been relaunched as a climate change adaptation strategy. Demand-side measures have never been given any great priority among Norwegian energy system agents due to low energy prices and focus on the abundant opportunities for further development of hydropower. So far, no efforts have been made to include demand-side measures in a climate change adaptation strategy.

Concluding remarks

We have suggested that an institutional perspective adds crucial insights on constraints to adaptation in societal sectors. Formal and informal institutions (regulations, norms, values and cognitive scopes), and their impacts on the capacity to learn and act on climate change, are too often neglected in the adaptation literature. We argue that even with technological, financial and human resources in place, institutional factors may still hinder their wise deployment and use in climate change adaptation, due to strong path-dependencies. A conscious adaptation strategy needs to be aware of institutions, like norms and values, in order to either measure them for legitimacy purposes or enact processes of institutional change when they act as barriers to adaptation. Some institutions may pose absolute limits to adaptation but others will be malleable. Path-dependency does not mean that all other options are closed for the future in a deterministic fashion, but mean that decisions at one point in time can restrict future possibilities (Thelen, 1999, p. 394). Climate change may be a critical junction along the way, eventually prompting change.

We have used the institutional framework to discuss institutional constraints to adaptation in the Norwegian energy sector. Institutional factors help explain the interpretation or framing of climate change vulnerabilities and solutions discussed to date by sector agents as well as their overall scant attention to adaptation. Regulations, routines, norms, values and cognitive views appear with sometimes forceful effects in narrowing the scope of learning and acting on the problem and its solutions.

With empirical work to continue on adaptation in the Norwegian energy sector (national and local levels) and with similar comparative studies planned for Sweden, our research will continue refining the analytical framework, aiming at making it applicable as a tool for analysing institutional limits to adaptive capacity across not only energy systems but different socio-technical systems more generally.

References

Adger, W. N., Agrawala, S., Mirza, M. M. Q., Conde, C., O'Brien, K., Pulhin, J., Pulwarty, R., Smit, B. and Takahashi, K. 2007. 'Assessment of adaptation practices, options, constraints and capacity', in Parry, M. L., Canziani, O. F., Palutikof, J. P., Van der Linden, V. J. and Hanson, C. E. (eds.) *Climate Change 2007: Impacts, Adaptation and Vulnerability. Contribution of Working Group II to the Fourth Assessment Report of the Intergovernmental Panel on Climate Change*. Cambridge: Cambridge University Press, pp. 717–743.

Argyris, C. and Schön, D. A. 1978. *Organisational Learning: A Theory of Action Perspective*. Reading: Addison-Wesley.

Brown, J. S. and Duguid, P. 1991. 'Organizational learning and communities-of-practice: toward a unified view of working, learning, and innovation', *Organization Science* **2**: 40–57.

Brunsson, N. 2003. *The Organisation of Hypocrisy: Talk, Decisions and Action in Organisations*. Copenhagen: Copenhagen Business School Press.

Carlsson, B. and Jacobsson, S. 1997. 'Diversity creation and technological systems: a technology policy perspective', in Edquist, C. (ed.) *Systems of Innovation, Technologies, Institutions and Organizations*. London: Pinter, pp. 266–294.

Carlsson, B. and Stankiewicz, R. 1991. 'On the nature, function, and composition of technological systems', *Journal of Evolutionary Economics* **1**: 93–119.

Christensen, T. and Peters, B. G. 1999. *Structure, Culture, and Governance: a Comparison of Norway and the United States*. Lanham: Rowman and Littlefield.

Commission of the European Communities. 2003. *Undergrounding of Electricity Lines in Europe*. Brussels: Commission of the European Communities.

Cyert, R. M. and March, J. G. 1992. *A Behavioral Theory of the Firm*. Malden: Blackwell.

DiMaggio, P. J. and Powell, W. W. 1983. 'The iron cage revisited: institutional isomorphism and collective rationality in organizational fields', *American Sociological Review* **48**: 147–160.

Førland, E. J., Amundsen, H. and Hovelsrud, G. K. (eds.). 2007. *Utviklingen av naturulykker som følge av klimaendringer: Utredning på oppdrag fra Statens Landbruksforvaltning*, CICERO Report No. 2007:03. Oslo: CICERO.

Fridheim, H., Hagen, J. and Henriksen, S. 2001. *En sårbar kraftforsyning – Sluttrapport etter BAS3*. FFI/Rapport-2001/02381. Kjeller: Forsvarets Forskningsinstitutt.

Hatch, M. J. and Cunliffe, A. L. 2006. *Organization Theory: Modern, Symbolic, and Postmodern Perspectives*. New York: Oxford University Press.

Hendron, A. J. and Patten, F. D. 1985. *The Vaiont Slide*. US Corps of Engineers Technical Report No. GL-85–8. Washington, DC: US Government Printing Office.

IPCC 2001. *Climate Change 2001: Impacts, Adaptation, and Vulnerability. Contribution of the Working Group II to the Third Assessment Report of the Intergovernmental Panel on Climate Change*. Cambridge: Cambridge University Press.

Krasner, S. D. 1988. 'Sovereignty: an institutional perspective', *Comparative Political Studies* **21**: 66–94.

Levitt, B. and March, J. G. 1988. 'Organizational learning', *Annual Review of Sociology* **14**: 319–340.

Lundvall, B.-Å. 1992. *National Systems of Innovation: Towards a Theory of Innovation and Interactive Learning*. London: Pinter.

March, J. G. 1991. 'Exploration and exploitation in organizational learning', *Organization Science* **2**: 71–87.

March, J. G. and Olsen, J. P. 1979. *Ambiguity and Choice in Organizations*. Bergen: Universitetsforlaget.

March, J. G. and Olsen, J. P. 1989. *Rediscovering Institutions: The Organizational Basis of Politics*. New York: Free Press.

March, J. G. and Olsen, J. P. 1999. 'Institutional perspectives on political institutions', in March, J. G. (ed.) *The Pursuit of Organizational Intelligence*. Malden: Blackwell, pp. 52–72.

Midttun, A. 1987. *Segmentering, institusjonelt etterslep og industriell omstilling: Norsk kraftutbyggings politiske økonomi gjennom 1970- og 1980-årene* (Segmentation, institutional inertia and industrial transformation: the political economy of Norwegian electrical power development in the 1970s and 80s). Uppsala: Philosophical Department, University of Uppsala.

Nilsen, Y. and Thue, L. 2006. *Statens kraft 1965–2006: Miljø og marked*, vol. 3. Oslo: Universitetsforlaget.

Nilsson, M. 2005. *Connecting Reason to Power: Assessments, Learning, and Environmental Policy Integration in Swedish Policy*. Stockholm: Stockholm Environment Institute.

NVE. 2006. *Endringer i det globale klimaet og norsk vannkraft.* Norges Vassdrags- og Energidirektorat (NVE). Available at www.nve.no/modules/module_111/news_item_ view.asp?iNewsId=28671&iCategoryId=1523 (accessed 27 March 2008)

Næss, L. O., Bang, G., Eriksen, S. and Vevatne, J. 2005. 'Institutional adaptation to climate change: flood responses at the municipal level in Norway', *Global Environmental Change* **15**: 125–138.

Rutledal, F., Hagen, J., Nystuen, K. O. and Østby, E. 2000. *Kraftmarkedets føringer for sårbarheten i norsk kraftforsyning,* FFI/Rapport-2000/03451/Kjeller: Forsvarets Forskningsinstitutt.

Sabatier, P. A. 1988. 'An advocacy coalition framework of policy change and the role of policy oriented learning therein', *Policy Sciences* **21**: 129–168.

Selznick, P. 1983. *Leadership in Administration: A Sociological Interpretation.* Berkeley: University of California Press.

Simon, H. A. 1976. *Administrative Behavior: A Study of Decision-Making Processes in Administrative Organization.* New York: Free Press.

Statnett. 2005. *Kraftsystemutredning for sentralnettet 2005–2020,* revidert. Oslo: Statnett SF.

Thelen, K. 1999. 'Historical institutionalism in comparative politics', *Annual Review of Political Science* **2**: 369–404.

Yohe, G. and Tol, R. S. J. 2002. 'Indicators for social and economic coping capacity: moving towards a working definition of adaptive capacity', *Global Environmental Change* **12**: 25–40.

Young, A. 1991. 'Learning by doing and the dynamic effects of international trade', *Quarterly Journal of Economics* **106**: 369–405.

Zucker, L. G. 1987. 'Institutional theories of organization', *Annual Review of Sociology* **13**: 443–464.

28

Accessing diversification, networks and traditional resource management as adaptations to climate extremes

Marisa Goulden, Lars Otto Næss, Katharine Vincent and W. Neil Adger

Introduction

There is considerable evidence that rural economies in Africa use diverse strategies to adapt to environmental variability and change (Davies, 1996; Scoones, 1996; Mortimore, 1998; Mortimore and Adams, 2001; Eriksen et al., 2005). This finding is reiterated in to the context of climate change (for example Thomas et al., 2007). Much of the evidence so far, however, has been location-specific and has not drawn out commonalities between context-specific adaptations, particularly on how they are accessed, by whom, and their relevance for policies and planned adaptation responses to climate change. It is thus not clear to what extent location-specific findings provide useful lessons for adaptation policy and practice, and what opportunities and barriers may exist. We address this gap in the understanding of adaptation by comparing local adaptation responses to similar climatic events in rural societies across three case study locations in South Africa, Tanzania and Uganda. The chapter captures the diversity of adaptation actions in a wide range of livelihood contexts that include economies with arable farming, livestock, fisheries, forestry, and strong urban linkages through migration, arguing that there are common constraints to accessing adaptation strategies. In other words, it is wide resource access and appropriate governance structures that are the keys to successful adaptation.

Theorising adaptation

We draw on theories on entitlements and access (Sen, 1981; Leach et al., 1999; Ribot and Peluso, 2003) to conceptualise barriers to and opportunities for adaptation in

Adapting to Climate Change: Thresholds, Values, Governance, eds. W. Neil Adger, Irene Lorenzoni and Karen L. O'Brien. Published by Cambridge University Press. © Cambridge University Press 2009.

terms of access to resources mediated by institutions. Adaptation, in the context here, is an active set of strategies and actions taken in reaction to or in anticipation of change by people in order to enhance or maintain their well-being, broadly defined. Inevitably these strategies are location and context specific. Indeed, Morton (2007) suggests that factors such as the integration of agricultural and non-agricultural livelihood strategies and exposure to various stressors, ranging from natural stressors to those related to policy change, make adaptation difficult to model and predict for smallholder and subsistence agriculture. Faced with this complexity various suggestions and typologies are proposed for how adaptation progresses in subsistence economies. Thornton et al. (2007) suggest that intensification, diversification and increasing off-farm activities are the most common adaptations in pastoralist settings, while Mortimore and Adams (2001) and Eriksen et al. (2005) observe, in addition, the use of greater biodiversity within cropping systems and use of open access resources such as wild foods. Arun Agrawal and Nicolas Perrin, in Chapter 22 of this volume, suggests that all strategies in rural institutions in effect involve functions that pool and share risks through mobility, storage, diversification, communal pooling and exchange.

Here we observe adaptations as falling predominantly into three main categories. These are diversification of livelihoods, the use of local-based knowledge of resource systems, and the use of existing or extending social networks to share risk. Each of these types of adaptation relies on differing bundles of resources although there are interactions between them.

Livelihoods are composed of 'the assets (natural, physical, human, financial and social capital), the activities and the access to these (mediated by institutions and social relations) that together determine the living gained by the individual or household' (Ellis, 2000, p. 10). Households, or communities, that have access to natural capital assets are able to benefit from a range of ecosystems services, in particular the provisioning and regulating services of ecosystems upon which many rural livelihood activities depend (Millennium Ecosystem Assessment, 2005).

Livelihood diversification includes 'on-farm' diversification (for example, keeping livestock in addition to cultivating crops), other natural resource access diversification (for example, fishing and collecting forest products) and non-farm diversification (for example, engaging in migrant labour in cities) (Ellis, 2000). Access to natural resources and the benefits of ecosystems services, and hence the ability to diversify into natural-resource-dependent activities is mediated by institutions and also depends on households possessing other types of assets, for example human or social capital. Livelihood diversification has been recognised as having the potential to alleviate poverty as well as reduce vulnerability to shocks (Ellis, 2000; Ellis and Freeman, 2004).

Local knowledge is the unique knowledge developed over time and held by a given culture or society, and existing outside a scientific domain (Warren et al., 1995). It is often largely informal, largely tacit and closely linked to social institutions that maintain and evolve such knowledge over time. The role of local knowledge has received increasing interest in adaptation studies, particularly in the Arctic (for example Riedlinger and Berkes, 2001; Ford et al., 2006). At the same time, its status in relation to scientific knowledge (Agrawal, 1995) and its relevance in the face of future climate change remain contested. Here, we address the role of different forms of local knowledge providing assets for households that can be used for adaptation.

Social networks comprise links or relationships between individuals and exist within and between households, communities and institutions of governance. Social links can be described in terms of social capital assets since they can be used to facilitate productive activity (Coleman, 1988) or coordinated action (Putnam et al., 1993). Social links can provide households with access to social capital of different types: bonding social capital from strong links with relatives, close friends and neighbours, bridging social capital from weaker links with others in the village or local area and linking social capital with outsiders to the village who have superior political or economic power (Woolcock, 2001).

Having classified adaptation strategies into three types, the critical barriers to action include the access to these, and the institutional and social mediation of this access. We focus on this access issue in order to identify who adapts and, critically, who is left behind in such processes and the role of institutions and processes of governance in facilitating or hindering access to adaptations. According to Susi Moser in Chapter 20 there is a growing need to understand the social dynamics that act through governance structures and mechanisms to motivate, facilitate or constrain adaptation.

Changing governance: the context

There has been a continuous process of change in governance structures over the past decades, with trends towards decentralisation and the so-called hollowing-out of the state in terms of its core responsibilities in parallel with the rise of supra-national global governance structures. These trends are nowhere more apparent than in many African countries during the post-colonial period of the past 40 years. Decolonisation, political upheaval and civil war, liberalisation of trade and agricultural markets, economic reforms and the replacement of traditional forms of governance with centralised and latterly multi-level governance structures have all influenced changes in governance (Robinson and White, 1999).

In Uganda, a wide programme of reform has been undertaken by the government since 1987 with respect to the economy, the judiciary, administration and politics, an important component of which has been decentralisation (Francis and James, 2003). Under decentralisation, new institutions of local government have been formed at multiple levels from the community level to the district, with links to central government. As in Uganda, local government reforms have been undertaken in Tanzania over the last decades, especially since the implementation of the Local Government Reform Programme from the late 1990s, building on reform efforts started during the 1980s. The relevance of these reforms to adaptation strategies was pointed out in Tanzania's National Adaptation Programme of Action (NAPA), though without going into much detail on the nature of linkages (URT, 2007). In South Africa the transition to democracy in 1994 led to reintegration of the former homelands (including the areas covered in this chapter) with the rest of the country, and has been accompanied by the establishment of new governance structures and policy frameworks to which all South Africans have statutory access. There are now three levels of governance: national, provincial and local (including district municipality and municipality). However, the existence of traditional authority is also enshrined in the constitution, and there is, at times, an uneasy relationship between the two governance forms, leading to gaps and overlaps in responsibilities.

Methods and data

The chapter draws on three studies in rural communities that assess how social institutions influence access to adaptations to experienced climate impacts such as droughts, floods and climatic trends in South Africa, Tanzania and Uganda. The IPCC Fourth Assessment report highlights Africa as particularly vulnerable to the adverse impacts of climate change, reflecting both exposure and social vulnerability. Based on an aggregate index of social vulnerability to climate-change-induced changes in water availability that ranks countries in Africa (reported in Vincent, 2007b), Uganda is the sixth most vulnerable country, Tanzania the tenth most vulnerable, and South Africa the 45th most vulnerable country in the continent (see Figure 28.1). However, this vulnerability index explores national level determinants of vulnerability and thus masks sub-national variation. By comparing the mechanisms by which adaptation to the impacts of climate variability and extremes occurs in the policy, social and environmental contexts of these three African countries, we demonstrate similar patterns in location-specific determinants of the ability of households and communities to adapt to climate change.

Fieldwork in South Africa was conducted between October 2004 and February 2005, in Tanzania between July and December 2006, and in Uganda between

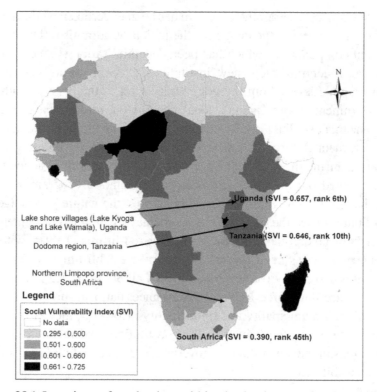

Figure 28.1 Locations of study sites within the landscape of vulnerability of African countries to climate change.

November 2003 and April 2004. Data on adaptation responses and the characteristics of the households and institutions in the studied communities were collected using household surveys, semi-structured interviews and participatory-style group exercises and meetings (Goulden, 2006; Vincent, 2007a). Eighty-five households were surveyed in South Africa, 140 in Tanzania and 80 in Uganda. Key characteristics of the samples and locations are shown in Table 28.1 and Figure 28.1. These exhibit differing natural resource and livelihood contexts. In all fieldwork locations, crop and livestock farming are common activities, and in the Tanzanian case forest resources are also important, whilst in Uganda the lakes and wetlands provide fishing-related and other livelihood activities. The South African village exhibits a diverse pattern of livelihood activities since non-natural-resource-based activities have become increasingly important. The cases experience some similarity in the climatic contexts, all three villages having suffered droughts in recent years and the villages in South Africa and Uganda having experienced the impacts of flooding events.

In order to investigate the hypothesis that barriers to adaptation arise from the inability to access resources and that access is mediated by social institutions, in the next section we examine the opportunities and barriers to adapting for the three cases by first focusing on the types of adaptation that are used, then by looking at access to these adaptations and finally by examining the role of governance institutions and policy in providing access to these adaptations. We then finish with a discussion of the implications of our findings.

Opportunities for adaptation and barriers to adapting

Table 28.1 shows examples of the adaptations that have been taken in response to climatic stresses for the different case studies, classified into three types: those based on livelihood diversification, local-knowledge-based adaptations and adaptations based on social links. Many of the observed adaptations relate to livelihood diversification. They involve a range of natural resource dependent livelihood activities and include aspects of flexibility in livelihoods over time and space and the mobility of people. Three different types of livelihood diversification are observed: concurrent, spatial and temporal. *Concurrent diversification* refers to doing different activities at the same time, *temporal diversification* involves changing activities over time and *spatial diversification* refers to doing activities in different locations (Goulden, 2006). Opportunities for non-natural-resource-dependent activities are limited in the Tanzanian case compared to the South African case. In the Ugandan case, many of the activities that are not directly dependent on natural resources, such as small businesses and trading, are indirectly related to natural resource abundance since the strength of the local economy depends on fish catches. In all three cases we see that knowledge of local natural resources such as wild plants is used for providing alternative foods in the case of drought. Traditional practices are also observed such as forecasting or spiritual rituals. Local knowledge also provides access to adaptations that require some skill or knowledge of traditional ways of doing things, for example salt-making and brewing.

Social links provide an important source of adaptations in all three of the cases. Links with relatives and friends are important for the adaptations used in the Tanzanian and Ugandan cases. Many of the social capital based actions happen at an individual or household level, for example borrowing food from friends or relatives, but some are collective actions involving many members of the community, for example the setting up of and participation in credit groups and fishing committees. Government assistance also plays a role, particularly in South Africa, where absolute vulnerability is being reduced by the increasing availability of long-term predictable cash transfers through social protection programmes, such as the old age pension and child support grant. In Tanzania, government assistance is restricted to emergency food aid and some limited

Table 28.1 *Location, livelihood, climate variability and examples of adaptation to climate stresses for study areas in South Africa, Tanzania and Uganda*

	South Africa	Tanzania	Uganda
Location	One village in a dryland region; former Venda homeland in northern Limpopo province; 85 households surveyed	Two villages in dryland areas, Bahi and Chamwino Districts in Dodoma Region; 140 households surveyed	Two villages on the shores of Lake Kyoga (Kamuli District) and Lake Wamala (Mityana District); 80 households surveyed
Natural resources and livelihoods	Northern foothills of the Soutpansberg mountains; legacy of arable and pastoral farming but now more diversified non-agricultural livelihoods	Savanna landscape with mainly hill forests. Mixed livelihoods based on arable and pastoral farming and forest resource use	Shallow lakes, wetlands and farmland that support mixed livelihood activities based on arable farming, livestock and fishing
Climatic trends and extreme events	High inter-annual variability punctuated by floods and droughts in recent memory	High inter-annual variability in rainfall. Long oral history of droughts and famines. Perception of decrease in rainfall	Flooding and drought events and cyclical variability in lake area, farmland area and fish catches
Livelihood diversification-based adaptations	*Concurrent diversification* Seeking employment/ setting up micro-enterprises Irrigation Brewing *Temporal diversification* Substituting foods Modifying planting dates *Spatial and temporal diversification* Purchasing fodder for livestock Gathering wild fruits/animals Migration	*Concurrent diversification* Change to cash crops and more drought-tolerant crops Charcoal making Selling firewood Selling water *Spatial diversification* Expand or move farm to new areas (incl. clearing forest lands) *Spatial and temporal diversification* Seasonal labour within and outside village Migration to towns	*Concurrent diversification* Adopting a range of natural resource (NR) dependent and non-NR dependent activities within the household to spread risk and increase income *Temporal diversification* Swapping to new livelihood activities as activities fail *Spatial diversification* Changing location (migration) or activities in several locations

Table 28.1 (cont.)

	South Africa	Tanzania	Uganda
Local-knowledge-based adaptation	Moving to better pasture Gathering wild fruits and animals	Consumption of forest fruits and vegetables Traditional wells Tree species protected Traditional crop varieties (change away from) Traditional weather forecasts and rainmaking Selling of forest products Salt-making, brewing	Harvesting wild swamp plants during famine Spiritual rituals when Lake Wamala receded Knowledge of previous recession and expansion of lake prevented some households from cultivating the lake bed
Social-links based adaptation	Borrowing food from friends and relatives Seeking government assistance Moving away to stay with relatives for work	Informal 'food loans' among villagers Barter trade (for example exchange tomatoes for grain) Exchange of livestock for food Informal credit Remittances Livestock trusteeship (for milk, ox ploughing) Government assistance and NGO support schemes	Seeking loans or gifts Remittances Seeking shelter or employment from other households Moving away to stay with relatives for work Participation in fishing or credit groups Spiritual rituals Village committees seek emergency food aid from district government

schemes, such as seed distribution, whilst in Uganda government assistance is generally only available in the form of food aid for exceptional emergencies and is often inadequate.

Table 28.1 shows examples where livelihood diversification, local knowledge and social links are able to enhance adaptation to climate impacts. However, there are some cases where these strategies are less effective in enhancing adaptation or even may hinder adaptation. For example, concurrent livelihood diversification does not reduce vulnerability if all activities are impacted by an extreme climate event, for example if all activities depend on variable natural resources. Some responses may reduce resilience to subsequent climate impacts, for example those people that used temporal diversification by swapping from fishing to cultivating the dried-up lake bed when Lake Wamala receded were then vulnerable to the rapid expansion of the lake and flooding of their crops a few years later. Another example is that of strong social links between fishers, which can mean that they resist attempts at co-management and fisheries regulations that are designed to promote sustainable fishing. In the Tanzanian case, several respondents attributed recent droughts to loss of traditional beliefs and the adoption of new religions. However, such beliefs do not appear to be a significant barrier to actions taken to adapt to droughts, as have been suggested in other cases (see Chapter 6 by Anthony Patt). However, this does emphasise the need to understand local perceptions of environmental changes and their causes.

In Table 28.2 we see that access to many of the adaptations is governed by a number of attributes of households and individuals including wealth, gender, age, health, available labour, skills and social links (or natural, physical, financial, human and social capital). We group the examples in Table 28.2 by those where gender and age influence access to adaptations and those where wealth, in its broad sense, influences access to adaptations.

Many adaptations based on livelihood diversification are difficult for the more vulnerable groups to access such as women, elderly and the poorest households. Nevertheless, even limited diversification of livelihoods, such as temporal diversification, can provide some resilience or coping ability to these more vulnerable households. Since women, the elderly and the poorest households have fewer alternatives, adaptations based on local knowledge or traditions can be important, but access to them can be reduced either because of gender constraints or because the activities of the wealthier households and social and environmental changes have reduced the availability of the necessary resources. We see that access to bridging and linking social capital can be important for access to several types of adaptation including, spatial diversification, government assistance and for the functioning of collective action groups but is less available to poorer or female-headed households. Bonding social capital is less able to provide access to adaptations, especially when climate impacts affect all activities and households in the village and the surrounding area. However, those with fewest alternatives often rely proportionately more on bonding social capital than other types of mechanism for adaptation, as a last resort.

Individuals' perceptions of changes that they have experienced can influence their choice of adaptations. For example, people living on the shore of Lake Wamala in Uganda had experienced a collapse in the fishery and fishing-based livelihoods when the lake receded between 1991 and 1994. Although people were not always in agreement as to the reason for the changes they saw in the lake, there was a perception that human activities had harmed the fishery and this motivated some of the fishers and government fisheries department officials to set up and participate in co-management of fisheries. Others responded by diversifying their livelihoods

Table 28.2 *Access to adaptations by type of adaptation depends on gender, age and wealth*

	Gender and age	Wealth
Access to livelihood-diversification-based adaptations	Women and elderly have less access to diversification options requiring good health, physical strength and mobility, for example, livestock keeping, wild fruits, hunting, charcoal making and migration for seasonal labour Stage in life of household head influences motivation for and access to diversification options	Access to diversification increases with wealth. Poorer households restricted to low-income natural resource based activities and *temporal* diversification Richer households more able to access *concurrent* and *spatial* diversification that requires capital and labour, such as irrigation, farm expansion and livestock keeping
Access to local-knowledge-based adaptation	Some traditional options are not accessible to all; for example, women depend on trusted men to be able to move to better pastures Local-knowledge-based options are undermined by social and environmental changes, affecting women and elderly who have fewer alternatives	Strategies accessible to richer households, such as forest clearing for sesame growing and charcoal making, restrict poorer households' access to traditional strategies such as use of wild plants and animals Some cases of traditions hindering access to options such as ox-ploughing
Access to adaptations based on social links	Younger households with migrant workers have more linking and bridging social capital than other households, enabling them to access government assistance Older members of the community more likely to be members of burial groups Older female headed households restricted to bonding social capital within the village, often insufficient for responding to covariate risks from climate	Richer households are more likely to have links with politically or economically influential people outside the village (linking social capital), giving improved access to adaptation options The poorest and displaced persons tend to have less access to strong social links, less education, less capital for collateral, less ability to repay favours or fewer kin, although they often rely on assistance through social links as a last resort

so as not to rely on the fishery because they had a perception that a collapse in the fishery might well happen again in the future.

In Tanzania, perceptions of drier weather have led to widespread change in crop varieties over the last few years. Not all farmers have access to alternative seeds, however. For example, government-run seed support ahead of the rainy season in 2006 was given on the condition that farmers ploughed and applied manure on their lands prior to receiving seeds. Due to labour and money shortages, some poorer farmers were unable to fulfil the conditions, and as a result did not collect or sow the drought-tolerant seeds.

In South Africa, access to formal policy frameworks and the assistance that they provide to households in adapting, such as the availability of drought recovery funds from the Department of Agriculture, are very much contingent on individuals being aware that such opportunities exist. Men tend to have very different human capital, through the legacy of better education, with also more linking social capital across a wider spatial area, due to their participation in labour migration. They are thus more likely to hear about such opportunities than elderly women who are less well educated and typically rely on social capital within the village where they have spent most of their lives.

In Table 28.3 we summarise how governance institutions and policies influence the types of adaptations that are available and which members of society have access to these. We find that aspects of governance can have a positive or negative influence on people's access to adaptations. The multi-level aspect of governance has played a role sometimes in facilitating and sometimes in hindering access to resources that constitute adaptive capacity.

Local- or village-level governance institutions such as local councils and committees, credit groups and fishing committees can facilitate access to livelihood diversification and adaptations based on social links. However, inadequate capacity and corruption or mismanagement in both central and local government institutions can limit the ability of government schemes or policies to strengthen livelihoods. This is particularly a danger in underfunded local government institutions.

In the Tanzanian and Ugandan cases certain government schemes or policies can limit adaptations. In the Tanzanian case, government initiatives were sometimes in conflict with traditional knowledge. In the Ugandan case, for fisheries co-management on Lake Kyoga, conflict between fishers and the national and regional state institutions appeared to occur because the goals of the state institutions did not match the goals of the local population. This difference in goals is not easily resolved due to the poverty and pressing livelihood needs of the population and therefore represents a considerable barrier or even a limit to successful co-management and arguably to long-term adaptation.

Table 28.3 *Governance institutions and policy influence on access to adaptations*

	Positive influence of governance institutions and policy on access	Negative influence of governance institutions and policy on access	Other factors influencing adaptation access
Livelihood-diversification-based adaptations	Village-based institutions, such as village local council committees, credit groups and fishing committees can facilitate access to livelihood diversification	Formal credit systems not available to poor groups Recent introduction of fisheries co-management and increase in enforcement of fishing regulations on Lake Kyoga reduced options for fishing based livelihoods for some Weak institutional capacity of government extension service impedes support of agricultural livelihoods	
Local-knowledge-based adaptation	Village-level governance committee asks village elders to perform spiritual rituals to pray for the expansion of the lake, drawing on traditional religion	Forest conservation rules in conflict with traditional forest use Government directives in conflict with traditional crop farming practices, ban on growing many local varieties Insufficient emergency food aid prompted harvesting of unpalatable swamp plants	As livelihoods diversify, they are less natural-resource-dependent and local knowledge of wild fruits and animals is diminishing

Table 28.3 (cont.)

	Positive influence of governance institutions and policy on access	Negative influence of governance institutions and policy on access	Other factors influencing adaptation access
Adaptation based on social links	Gaps and overlaps between the democratic and traditional governance structures leave space for exploitation of opportunities for those with sufficient human capital Existence of government assistance is reducing pressure on traditional systems of reciprocity	Government schemes of support to disadvantaged groups (seed, food aid) appear in practice to favour those with better social links During periods of social upheaval, conflict and breakdown or reorganisation of institutions (for example civil war, decentralisation) there is less access to collective action adaptations State-led policies can be at odds with the goals of the local population and limit adaptations, for example in fisheries management institutions on Lake Kyoga	The climate-induced collapse in Lake Wamala's fishery provided an incentive for collective-action-based adaptations

In South Africa, social, political and economic changes have opened up new opportunities and prompted a move away from natural-resource-dependent livelihoods, thus cushioning people from the unpredictability of the environment. However, gaps and overlaps between the responsibilities of traditional and multilevel democratic institutions of governance exist and failure to clearly articulate responsibility can make things confusing for individuals and households, and have implications for their access to adaptations. Also, whilst there are comprehensive national policies that promote adaptation to climate change in the realms of water, agriculture and disaster management, the failure to follow through on implementation through to grass-roots level means that not all groups in all places are able to benefit from them. Key knowledge-brokers tend to be responsible for transmitting knowledge on policy frameworks, and at the local level this includes the village chief, ward councillor and key government department personnel. In Limpopo the absence of an agricultural extension officer means that the community is effectively cut off from accessing agricultural policies and programmes. Only those with good linking social capital and contacts outside the village – typically young working-age men – get to hear about such policies and can pursue them directly through the local office.

Government assistance in South Africa means that people rely less on traditional reciprocity and are more likely to resort to both formal governance structures, such as the ward councillor, and to a lesser extent informal governance structures, through the village chief, in times of crisis. A lack of government assistance in Uganda means that social-capital-based adaptations are still very important but these can be limited where the climate impact is both widespread and serious and affects the majority of the population.

There are trade-offs and interactions between the different types of adaptation in this typology. An example of a positive interaction between social links and livelihood diversification is the ability of those with good social links (especially bridging and linking social capital) to be able to access distant markets, which facilitates livelihood diversification. In Tanzania, it seems clear that diversification by mainly male youths into non-traditional activities such as charcoal making and sesame cultivation, has recently led to significant reductions in the forest cover (especially large trees protected by traditional rules), at the expense of local-knowledge-based options such as gathering of wild fruits and vegetables. Farmers also perceive that loss of big trees due to increased charcoal making has led to reduced rainfall and loss of ability to predict the rainy season using traditional signs. In South Africa, whilst the access of the case study village inhabitants to formal governance structures and frameworks since 1994 has acted to reduce absolute vulnerability, a side effect of this has been the insertion of a 'cushioning layer' between households and variability in the natural environment. Whereas before they were largely agriculture-dependent and had to adapt their livelihood strategies to climate variability, now there is a shift away from the land, and with high unemployment there is dependence on pensions and child support grants. This acts to insulate households' economic livelihoods from climate variability, and can impede the adaptability of more traditional forms of resource-dependent livelihoods.

Conclusions

We have examined the opportunities and barriers to adaptation in rural Africa by first focusing on the types of adaptation that are used, by examining access to these adaptations, and by investigating the role of governance institutions and policy in providing access to these adaptations.

Our findings indicate that rural societies have diverse ways of adapting to the impacts of extreme climate events that rely on livelihood diversification, local knowledge and social links. However, the results also support the hypothesis that there are real barriers to adaptation as a result of uneven access to assets including natural resources, financial capital, human capital and social capital. It appears that this uneven access to adaptation results partly from the differing assets possessed by households. Gender, age and wealth have a strong influence on people's ability to access adaptation: women, elderly people and the poorest households having less options available to them. Adaptation is also influenced by institutions of governance and the context in which adaptation is occurring. Multi-level governance systems do provide some opportunities for adaptation. However, lack of capacity in governance systems, differing goals and a lack of recognition of different types of knowledge are all factors that contribute to barriers to adaptation. These findings have relevance to emerging national adaptation strategies.

We suggest that barriers to adaptation can be better understood by a broader recognition of the multiple drivers of vulnerability, such as poverty, gender and age and an awareness of how governance affects local adaptation. We argue, in line with Mortimore and Adams (2001), Thomas et al. (2007) and others, that the notion of African rural societies as powerless to act in the face of climate change inadvertently risks neglecting people's knowledge and experience of adaptations based on past climatic variability. Our findings indicate that there is therefore a need both to focus on promoting opportunities for livelihood diversification, building strong social capital and incorporating local knowledge into adaptation planning and also to remove barriers to the accessibility and success of these adaptations within structures of governance. Where policies and institutions can reduce vulnerability to climate change it is necessary to ensure that access is not differential based on gender, age and wealth, for example. However, we recognise that some households will continue to be vulnerable due to their limited ability to access many of these adaptations so that interventions need to be targeted at the most vulnerable. This should be accompanied by an improved understanding of how livelihoods, and the natural resources upon which many of them depend, vary with the climate and other aspects of local context.

Acknowledgements

The research presented in this chapter was supported by PhD studentships funded by NERC and ESRC in the UK and the Research Council of Norway. We thank Ed Allison, Katrina Brown, Declan Conway, Mike Hulme, Don Nelson and Jouni Paavola, colleagues in the Tyndall Centre, our research assistants in South Africa, Tanzania and Uganda, and the families who participated in the research.

References

Agrawal, A. 1995. 'Dismantling the divide between indigenous and scientific knowledge', *Development and Change* **26**: 413–439.

Coleman, J. S. 1988. 'Social capital in the creation of human capital', *American Journal of Sociology* **94**: S95–S120.

Davies, S. 1996. *Adaptable Livelihoods: Coping with Food Insecurity in the Malian Sahel*. London: Macmillan.

Ellis, F. 2000. *Rural Livelihoods and Diversity in Developing Countries*. Oxford: Oxford University Press.

Ellis, F. and Freeman, H. A. 2004. 'Rural livelihoods and poverty reduction strategies in four African countries', *Journal of Development Studies* **40**: 1–30.

Eriksen, S., Brown, K. and Kelly, P. M. 2005. 'The dynamics of vulnerability: locating coping strategies in Kenya and Tanzania', *Geographical Journal* **171**: 287–305.

Ford, J. D., Smit, B. and Wandel, J. 2006. 'Vulnerability to climate change in the Arctic: a case study from Arctic Bay, Canada', *Global Environmental Change* **16**: 145–160.

Francis, P. and James R. 2003. 'Balancing rural poverty reduction and citizen participation: the contradictions of Uganda's decentralization program', *World Development* **31**: 325–337.

Goulden, M. 2006. Livelihood diversification, social capital and resilience to climate variability amongst natural resource dependent societies in Uganda. PhD thesis, School of Environmental Sciences, University of East Anglia, Norwich.

Leach, M., Mearns, R. and Scoones, I. 1999. 'Environmental entitlements: dynamics and institutions in community-based natural resource management', *World Development* **27**: 225–247.

Millennium Ecosystem Assessment 2005. *Ecosystems and Human Well-Being: Synthesis*. Washington, DC: Island Press.

Mortimore, M. 1998. *Roots in the African Dust: Sustaining the Sub-Saharan Drylands*. Cambridge: Cambridge University Press.

Mortimore, M. J. and Adams W. M. 2001. 'Farmer adaptation, change and "crisis" in the Sahel', *Global Environmental Change* **11**: 49–57.

Morton, J. F. 2007. 'The impact of climate change on smallholder and subsistence agriculture', *Proceedings of the National Academy of Sciences of the USA* **104**: 19 680–19 685.

Putnam, R. D., Leonardi, R. and Nanetti, R. Y. 1993. *Making Democracy Work: Civic Traditions in Modern Italy*. Princeton: Princeton University Press.

Ribot, J. C. and Peluso N. L. 2003. 'A theory of access', *Rural Sociology* **68**: 153–181.

Riedlinger, D. and Berkes, F. 2001. 'Contributions of traditional knowledge to understanding climate change in the Canadian Arctic', *Polar Record* **37**: 315–328.

Robinson, M. and White, G. (eds.) 1998. *The Democratic Developmental State: Politics and Institutional Design*. Oxford: Oxford University Press.

Scoones, I. 1996. *Hazards and Opportunities: Farming Livelihoods in Dryland Africa: Lessons from Zimbabwe*. London: Zed Books.

Sen, A. 1981. *Poverty and Famines: An Essay on Entitlement and Deprivation*. Oxford: Oxford University Press.

Thomas, D., Twyman, C., Osbahr, H. and Hewitson, B. 2007. 'Adaptation to climate change and variability: farmer responses to intra-seasonal precipitation trends in South Africa', *Climatic Change* **83**: 301–322.

Thornton, P. K., Boone, R. B., Galvin, K. A., BurnSilver, S. B., Waithaka, M. M., Kuyiah, J., Karanja, S., Gonzalez-Estrada, E. and Herrero, M. (2007). 'Coping strategies in livestock-dependent households in East and Southern Africa: a synthesis of four case studies', *Human Ecology* **35**: 461–476.

URT (United Republic of Tanzania) 2007. *National Adaptation Programme of Action (NAPA)*. Dar es Salaam: Vice President's Office, Division of Environment.

Vincent, K. 2007a. Gendered vulnerability to climate change in Limpopo Province, South Africa. PhD thesis, School of Environmental Sciences, University of East Anglia, Norwich.

Vincent, K. 2007b. 'Uncertainty in adaptive capacity and the importance of scale', *Global Environmental Change* **17**: 12–24.

Warren, D. M., Slikkerveer, L. J. and Brokensha, D. (eds.) 1995. *The Cultural Dimension of Development: Indigenous Knowledge Systems*. London: Intermediate Technology Publications.

Woolcock, M. 2001. 'The place of social capital in understanding social and economic outcomes', *Canadian Journal of Policy Research* **2**: 11–17.

29

Governance limits to effective global financial support for adaptation

Richard J. T. Klein and Annett Möhner

Introduction

The United Nations Framework Convention on Climate Change (UNFCCC) is one of only a few global environmental treaties with near universal membership (192 Parties). It includes all main contributors to climate change as well as the majority of those that will need to adapt to the adverse effects of climate change. One of the commitments of developed countries under the UNFCCC is to assist developing countries in meeting costs of adaptation to these adverse effects (Article 4.4). The UNFCCC has therefore developed a dedicated system for providing global financial support for adaptation to climate change in developing countries.

As the need for such adaptation is becoming increasingly clear, the World Bank (2006) notes that the total amount of funding for adaptation projected to be available to developing countries by 2012 falls well short of the estimated amounts needed to cover its costs, and considers this shortfall to be the major impediment to adaptation funding:

> An assessment of the current financial instruments shows that, while they are technically adequate to respond to the challenge of achieving climate-resilient development, the sums of money flowing through these instruments need to be substantially increased.
>
> *(World Bank, 2006, p. 40.)*

However, developing countries have expressed the additional concern that the complexity of current funding arrangements constrains their access to money (Decision 3/CP.12; UNFCCC, 2006a, 2007a, 2007b). This would challenge the World Bank's view that the current financial instruments are 'technically adequate'. Without questioning the need for additional funding, this chapter assesses the financial instruments' responsiveness to the needs of developing countries.

Adapting to Climate Change: Thresholds, Values, Governance, eds. W. Neil Adger, Irene Lorenzoni and Karen L. O'Brien. Published by Cambridge University Press. © Cambridge University Press 2009.

It presents an assessment of technical adequacy from a governance perspective, based on an analysis of 42 adaptation projects. This chapter adopts the definition of governance as put forward by the Commission on Global Governance (1995). This Commission defines governance as:

> the sum of the many ways individuals and institutions, public and private, manage their common affairs. It is a continuing process through which conflicting or diverse interests may be accommodated and co-operative action may be taken. It includes formal institutions and regimes empowered to enforce compliance, as well as informal arrangements that people and institutions either have agreed to or perceive to be in their interest.

Governance system for adaptation funding under the UNFCCC

Figure 29.1 depicts the global governance system through which the financial instruments for adaptation have been developed, designed and implemented under the UNFCCC to date. This governance system is aimed at meeting developing countries' needs for adaptation by providing funding for adaptation projects in accordance with guidance developed for the respective financial instruments.

The main actors in the system are the developing countries, the Conference of the Parties (COP) of the UNFCCC, the Global Environment Facility (GEF) and its Implementing and Executing Agencies (IAs and EAs). The COP is the supreme decision-making body of the UNFCCC and meets annually to review and advance the implementation of the Convention. The GEF, as an entity entrusted with the operation of the financial mechanism of the UNFCCC, manages the instruments for the transfer of financial resources from developed to developing countries. It provides new and additional funding to meet the agreed incremental costs of projects to generate global environmental benefits in climate change and other areas (GEF, 1994). The GEF functions under the guidance of and is accountable to the COP. It operates the financial instruments by establishing operational programmes, providing programming documents and allocating resources.

During negotiations developing countries express their needs and concerns, and pursue their interests in adaptation funding. In response the COP establishes financial instruments with specific priorities (i.e. which activities are to be funded), eligibility criteria (i.e. who can receive funding) and policies, including disbursement criteria (i.e. what share of a project can be funded). Developing countries can further pursue their interest in adaptation funding and further negotiate operational modalities at meetings of the GEF Council, which take place twice a year to decide on the operation of the financial instruments. Once a financial instrument is operational, eligible countries can propose projects based on their adaptation

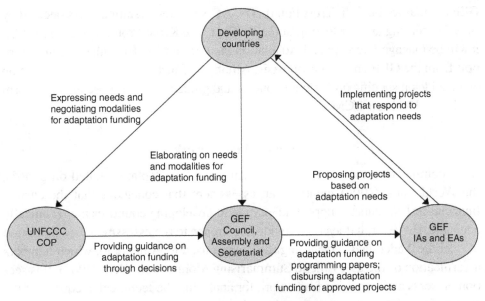

Figure 29.1 Governance system of financial instruments for adaptation under the UNFCCC and the GEF. (Source: Möhner and Klein, 2007.)

needs through one of the three IAs of the GEF: the United Nations Development Programme (UNDP), the United Nations Environment Programme (UNEP) and the World Bank. Seven additional EAs, including regional development banks, contribute to the implementation of GEF projects.

Current financial instruments for adaptation under the UNFCCC

The UNFCCC COP has adopted numerous decisions that contain guidance on the establishment and operation of various financial instruments for adaptation. At its first session in 1995 the COP laid out the initial guidance for the provision of financial support for adaptation from the GEF Trust Fund's climate change focal area. Relevant activities funded under the GEF Trust Fund include vulnerability and adaptation assessments in the context of National Communications, as well as capacity building. In 2001 the COP requested that support be extended to a range of adaptation-related activities, which prompted the GEF to establish under its Trust Fund the Strategic Priority 'Piloting an Operational Approach to Adaptation' (SPA).

In 2001 the COP established three additional funds for adaptation, each with its own guidance: the Least Developed Countries Fund (LDCF) and the Special Climate Change Fund (SCCF) under the UNFCCC, and the Adaptation Fund (AF) under the Kyoto Protocol. Both the LDCF and the SCCF are operational and managed by the

GEF (similar to the GEF Trust Fund). The AF is not yet operational. As decided by the COP serving as the Meeting of the Parties to the Kyoto Protocol (CMP) in 2007, it will be managed by a special Adaptation Fund Board (AFB), with secretariat support from the GEF on an interim basis. Among its functions, the AFB will develop and decide on specific operational policies and guidelines, and subsequently inform the CMP (Decision 1/CMP.3).

Analytical framework

In its comment on current financial instruments for adaptation (cited on p. 465), the World Bank (2006) mentions an assessment that concludes that the current financial instruments to support adaptation in developing countries are technically adequate. However, it does not provide a reference to the assessment or a definition of 'technical adequacy' (including the criteria used to evaluate it), which hampers a verification of the assessment. Summarising Möhner and Klein (2007), this section presents an alternative framework for analysing the technical adequacy of the current financial instruments for adaptation.

Expanding on Sagasti et al. (2005), technical adequacy reflects the match between financial instruments for adaptation and the needs of developing countries to be able to access funds and to be able to use them in line with their adaptation requirements. For the purpose of assessing technical adequacy this chapter considers that a match between financial instruments and the two needs exists if:

- the application and approval process is efficient;
- decision-making procedures are fair;
- the financial instruments respond to developing countries' needs.

The GEF Evaluation Office (EO) has assessed efficiency as part of its third overall performance study of the GEF (GEF EO, 2005) and as part of its evaluation of the GEF activity cycle and modalities (GEF EO, 2006). Qualitative analyses of the fairness of funding arrangements for adaptation have been conducted by Mace (2005), Paavola and Adger (2006) and Müller (2007). Therefore this chapter presents an assessment of the responsiveness of the existing financial instruments to developing countries' adaptation needs. It is based on an analysis of, first, the adherence by the GEF to guidance from the COP (as COP decisions) and, second, the adherence by IAs to guidance from the GEF (in the form of programming papers; see Figure 29.1). Full adherence to these guidances would mean that adaptation projects respond to developing countries' needs.

Adherence to guidance

Möhner and Klein (2007) present a detailed analysis of adherence using the three aforementioned elements of funding guidance as indicators: priorities, eligibility

Table 29.1 *Adherence by the Global Environment Facility and its Implementing Agencies to guidance on priorities*

	COP guidance	GEF adherence to COP guidance	IA adherence to GEF guidance
GEF Trust Fund	Funding for 14 adaptation-related activities, no distinction between ecosystem and development benefits	*Partial adherence* Programming for only one activity through establishment of the SPA, which focuses on ecosystem benefits	*Partial adherence* Support for pilot adaptation projects; however, some projects focus on development benefits
SCCF	Funding for adaptation, including for water and land management, agriculture, health, infrastructure development, fragile ecosystems, and integrated coastal zone management, no distinction between ecosystem and development benefits	*Partial adherence* Programming includes all activities included in the COP guidance; however, focus on development benefits	*Partial adherence* Support for adaptation projects; however, some projects focus on ecosystem benefits
LDCF	Funding for the preparation and implementation of National Adaptation Programmes of Action (NAPAs) for least developed countries (LDCs) and other elements of the LDC work programme	*Partial adherence* Programming for the preparation and implementation of NAPAs but not for other elements of the LDC work programme	*Full adherence* Support for the preparation and implementation of NAPAs

and disbursement. It is based on a review of publicly available documents, including COP decisions, documents from the GEF and its IAs, as well as documents of 34 GEF-funded adaptation projects. This chapter presents an update, based on all 42 GEF-funded adaptation projects as of November 2007 (UNFCCC, 2006a, 2007c; GEF, 2007a, 2007b, 2007c).

Non-adherence on *priorities* (Table 29.1) constrains developing countries in undertaking certain adaptation activities. For example, the GEF has yet to develop operational guidance (in the form of programming papers) for all but one of the 14 adaptation activities for which the COP decided support should be provided through the GEF in 2001 (Decision 5/CP.7). Without such guidance developing

Table 29.2 *Adherence by the Global Environment Facility and its Implementing Agencies to guidance on eligibility*

	COP guidance	GEF adherence to COP guidance	IA adherence to GEF guidance
GEF Trust Fund	Developing countries	*Not assessable* No details on eligibility in programming paper	*Partial adherence* Support for developing countries and one country with an economy in transition
SCCF	Developing countries	*Full adherence* Programming includes only developing countries	*Full adherence* Support for developing countries
LDCF	LDCs	*Full adherence* Programming includes only LDCs	*Full adherence* Support for LDCs

countries cannot apply for funding for activities such as systematic observation and monitoring networks, and early-warning systems. As another example, GEF guidance foresees projects under the SCCF to generate development benefits, whilst the focus of those funded by the GEF Trust Fund should be on creating ecosystem benefits (GEF, 2004, 2005). Yet, a number of proposed or approved projects under the SCCF have a primary focus on ecosystem benefits. For example, the project 'Design and Implementation of Pilot Climate Change Adaptation Measures in the Andean Region' is directed at mountainous ecosystems. Likewise there are projects under the GEF Trust Fund aiming to generate development benefits, such as the project 'Integrating Vulnerability and Adaptation to Climate Change into Sustainable Development Policy Planning and Implementation in Southern and Eastern Africa'.

Adherence to guidance on *eligibility* is almost complete (Table 29.2). Non-adherence on eligibility reduces the availability of funding to those countries that are eligible for adaptation support. For example, Hungary, an EU member state with an economy in transition, has received support under the SPA for its project 'Lake Balaton Integrated Vulnerability Assessment, Early Warning and Adaptation Strategies', even though the COP mandated support under the SPA for developing countries only (Decision 5/CP.7).

Adherence to guidance on *disbursement* (Table 29.3) is incomplete, especially with regard to the SPA. Non-adherence either increases the cost of adaptation projects for recipient countries or reduces the availability of funding to other

Table 29.3 *Adherence by the Global Environment Facility and its Implementing Agencies to guidance on disbursement*

	COP guidance	GEF adherence to COP guidance	IA adherence to GEF guidance
GEF Trust Fund	No explicit guidance	*Not assessable* Programming foresees disbursing a 'double increment': one for adaptation from the SPA and one for generating global environmental benefits from the GEF focal area in which the benefits are generated	*Partial adherence* Support for only five out of 15 projects is composed of funding from both the SPA and one or more other focal areas; the other ten projects receive funding from the SPA alone
SCCF	No explicit guidance	*Not assessable* Programming includes provision of additional (adaptation) costs, determined through a sliding scale, whereby smaller projects receive proportionally more GEF funding than bigger projects since they are assumed to have a larger adaptation component	*Full adherence* Supported projects either comply with the sliding scale or provide information on the adaptation costs as stipulated by GEF guidance
LDCF	Preparation: Full-cost funding Implementation: Full-cost funding to meet the additional costs imposed on LDCs to meet their immediate adaptation needs	*Full adherence* Programming includes full-cost funding for preparation For implementation: provision of additional (adaptation) costs, determined through a sliding scale, whereby smaller projects receive proportionally more GEF funding than bigger projects	*Full adherence* Full-cost funding support for NAPA preparation For implementation: supported projects either comply with sliding scale or provide information on the adaptation costs as stipulated by GEF guidance

eligible countries. For example, ten out of 15 SPA projects receive or are expected to receive their GEF funding solely from the SPA, whereas GEF guidance stipulates that part of their GEF funding should come from other focal areas (for example, bio-diversity, international waters, land degradation). Examples are the West African project 'Adaptation to Climate Change: Responding to Coastline Change and its Human Dimensions through Integrated Coastal Area Management' or Colombia's 'Integrated National Adaptation Plan: High Mountain Ecosystems, Colombia's Caribbean Insular Areas and Human Health'.

Discussion

It is important to understand the nature and reasons for non-adherence to guidance and how it limits developing countries' access to funds if the situation is to be improved. The analysis presented here (based on a more detailed presentation in Möhner and Klein, 2007) suggests that non-adherence is due to the complex governance system for adaptation funding, which leads to imperfect design and inconsistent implementation of guidance on the operation of the financial instruments.

The design of guidance can lead to non-adherence if guidance is unspecific or ambiguous. For example, with respect to disbursement, the COP has not defined adaptation costs in a way that allows the GEF to make a clear distinction between adaptation and development. As a result, the GEF has developed and applied the concepts of additional and incremental costs to determine its share of project funding. Calculations using these concepts invariably raise the question as to which part of a project concerns adaptation, to be funded by the GEF, and which part concerns development, which is the recipient country's responsibility. The introduction of the sliding scales for the LDCF and the SCCF have simplified the process of determining the GEF share of funding, but they have been designed in a way as to lead to different marginal values of GEF funding depending on the size of a project. This could lead to strategic behaviour by developing countries seeking to maximise GEF funding for their projects.

Even if guidance is relatively unambiguous, implementation of the guidance can still contradict the design intent. For example, with respect to eligibility, COP guidance stipulates that funding under the SPA be available only to developing countries (Decisions 5/CP.7 and 1/CP.10). However, a non-developing country (Hungary) has received funding in line with the GEF's own criteria that countries with economies in transition are eligible for support under the Trust Fund (GEF, 1994).

In other cases non-adherence to guidance results from a mix of design and implementation issues. For example, the COP requests complementarity between the different funds (Decision 7/CP.7), yet in Decision 5/CP.7 it requests support for the same adaptation activities (such as technologies for adaptation) from more than one fund (in this case from the GEF Trust Fund and SCCF). The GEF and its IAs attempt to establish complementarity by distinguishing between the development-focused SCCF and the ecosystem-focused SPA. However, the 'Coping with Drought and Climate Change' projects received funding for project preparation under the SPA (UNFCCC, 2006b), whilst project implementation is now funded under the SCCF (GEF, 2007a). Such shifting of projects between funds could lead to confusion amongst developing countries and create another incentive for strategic behaviour: to apply to those funds that have the most resources available. Instead of being provided with straightforward opportunities for adaptation funding, developing countries need to adapt their projects so as to secure support for their proposed adaptation activities.

Conclusions

As outlined above (p. 468), the technical adequacy of financial instruments is determined by efficiency and fairness as well as responsiveness to developing countries' needs (reflected by adherence to guidance). As far as efficiency is concerned, GEF EO (2006) concluded that the GEF activity cycle is not efficient, that the situation has grown worse over time and that GEF modalities have not made full use of trends towards new forms of collaboration that serve to promote efficiency. It is too early to judge if the implementation of the GEF reform agenda, which began in 2007, is improving efficiency (UNFCCC, 2007c). Mace (2005), Paavola and Adger (2006) and Müller (2007) conclude that there is room for improvement when it comes to the fairness of adaptation funding arrangements. This chapter concludes that the financial instruments are also technically inadequate in terms of responding to developing countries' needs.

Improvements are necessary for all instruments and by all actors in the governance system to enhance the funds' responsiveness to developing countries' needs. The COP could provide more explicit guidance: it could clarify the relationship between adaptation activities under the GEF Trust Fund and adaptation activities under the LDCF and SCCF, as well as the relationship between adaptation and development. The GEF, which has been requested by the COP to give due priority to adaptation activities (Decision 2/CP.12), could make operational all COP guidance on adaptation as part of its review of the SPA and its deliberations about the future use of the Trust Fund for adaptation. As requested by the COP, the Implementing Agencies should be better informed of the relevant Convention provisions and decisions to ensure that their support responds better to developing countries' needs (Decision 7/CP.13).

As part of the implementation of the 2007 Bali Action Plan (Decision 1/CP.13), the COP could improve access to adaptation funding by simplifying the governance system. This may be achieved by consolidating current funding arrangements for adaptation into a single financial instrument, with an unambiguous design in terms of priorities, eligibility and disbursement and straightforward implementation. Finally, the AFB, which will establish a new governance system for the AF with specific operational policies, guidelines and entities, has the opportunity to learn from experience with the SPA, the SCCF and the LDCF. Rather than adding to the complexity of global adaptation funding, it could benefit from its closer proximity to the CMP and its more direct link to developing countries and design a governance system that is technically adequate and no longer limits the effectiveness of global financial support for adaptation.

Acknowledgements

This chapter is based on SEI Working Paper *The Global Environment Facility: Funding for Adaptation or Adapting to Funds?* (Möhner and Klein, 2007). The reader is referred to the working paper for details omitted in this chapter due to the word limit. Richard Klein acknowledges funding from the European Commission under project number FP6–018476–2 (Adaptation and Mitigation Strategies: Supporting European Climate Policy; ADAM) and from the Swedish Foundation for Strategic Environmental Research (Mistra) under its Climate Policy Research programme (Clipore). Annett Möhner conducts PhD research whilst also being employed by the UNFCCC Secretariat as a Programme Officer. The views expressed in this chapter are those of the authors and do not necessarily reflect the views of the UNFCCC Secretariat and the United Nations.

References

Commission on Global Governance 1995. *Our Global Neighborhood: The Report of the Commission on Global Governance.* Oxford: Oxford University Press.

GEF 1994. *Instrument for the Establishment of the Restructured Global Environment Facility.* Report of the GEF Participants Meeting, 14–16 March 1994, Geneva, Switzerland.

GEF 2004. *Programming to Implement the Guidance for the Special Climate Change Fund Adopted by the Conference of the Parties to the United Nations Framework Convention on Climate Change at its Ninth Session.* Washington, DC: GEF.

GEF 2005. *Operational Guidelines for the Strategic Priority 'Piloting An Operational Approach to Adaptation' (SPA).* Washington, DC: GEF.

GEF 2007a. *LDCF and SCCF Programming Update.* Washington, DC: GEF.

GEF 2007b. *Joint Summary of the Chairs*, GEF Council Meeting, 14–16 November 2007. Washington, DC: GEF.

GEF 2007c. *Progress Report on the Least Developed Countries Fund (LDCF) and the Special Climate Change Fund (SCCF).* Washington, DC: GEF.

GEF Evaluation Office 2005. *OPS3: Progressing Toward Environmental Results*, third overall performance study of the GEF. Washington, DC: Office of Monitoring and Evaluation of the Global Environment Facility.

GEF Evaluation Office 2006. *Evaluation of the GEF Activity Cycle and Modalities*, Joint Evaluation of the GEF Evaluation Office and the Evaluation Offices of the Implementing and Executing Agencies of the GEF. Washington, DC: GEF.

Mace, M. J. 2005. 'Funding for adaptation to climate change: UNFCCC and GEF developments since COP-7', *Review of European Community and International Environmental Law* **14**: 225–246.

Möhner, A. and Klein, R. J. T. 2007. *The Global Environment Facility: Funding for Adaptation or Adapting to Funds?* SEI Working Paper. Stockholm: Stockholm Environment Institute.

Müller, B., 2007. *Nairobi 2006: Trust and the Future of Adaptation Funding.* Oxford: Oxford Institute for Energy Studies.

Paavola, J. and Adger, W. N. 2006. 'Fair adaptation to climate change', *Ecological Economics* **56**: 594–609.

Sagasti, F. R., Bezanson, K. and Prada, F. 2005. *The Future of Development Financing: Challenges and Strategic Choices.* Basingstoke: Palgrave Macmillan.

UNFCCC 2006a. *Report on the Latin American Regional Workshop on Adaptation.* Geneva: UN.

UNFCCC 2006b. *Report of the Global Environment Facility to the Conference of the Parties.* Geneva: UN.

UNFCCC 2007a. *Report on the African Regional Workshop on Adaptation.* Geneva: UN.

UNFCCC 2007b. *Report on the Expert Meeting on Adaptation for Small Island Developing States.* Geneva: UN.

UNFCCC 2007c. *Report of the Global Environment Facility to the Conference of the Parties.* Geneva: UN.

World Bank 2006. *An Investment Framework for Clean Energy and Development: A Progress Report.* Washington, DC: World Bank.

30

Organizational learning and governance in adaptation in urban development

Marte Winsvold, Knut Bjørn Stokke, Jan Erling
Klausen and Inger-Lise Saglie

Adaptation in urban development:
a complex coordination problem

In this chapter we focus upon the lack of coordination between different actors, which is a pronounced problem in urban planning and development, as a possible barrier to adaptation to climate change. This has to date received little attention in the literature. Our aim is to contribute to the development of a model for analysing how different coordination mechanisms can hinder or encourage adaptation. Different forms of coordination are normally attributed to different modes of governance, and in this chapter we distinguish between three modes (which are often also observed in urban planning and development): a *hierarchical* mode in which coordination of different actors is ensured through formal regulations, command and control; a *market* mode in which coordination of actors is ensured by the price mechanism of the market; and a *network* mode of governance in which coordination ideally happens through arguing and bargaining among the involved actors. We explore how the mode of governance and the corresponding form of coordination influences how actors interpret signals from their surroundings, how they search for solutions to perceived problems, how solutions are articulated and put into action and eventually how the implementation of solutions is evaluated. In short, the specific mix of governance modes at play will influence the actors' learning processes, their use of available knowledge, and hence their capacity to adapt to climate change. In order to identify limits and barriers to adaptation in an urban context, we therefore combine theories on governance modes with theories on organizational learning.

Adaptive capacity can be defined as the ability or potential of a system to respond successfully to climate variability and change (Brooks and Adger, 2005).

Adapting to Climate Change: Thresholds, Values, Governance, eds. W. Neil Adger, Irene Lorenzoni and Karen L. O'Brien. Published by Cambridge University Press. © Cambridge University Press 2009.

However, actual adaptation does not necessarily follow from successful learning, as adaptation requires both the capacity to learn and the capacity to act. We therefore also discuss the relation between governance and the capacity to act, and moreover the relation between learning and taking action.

The importance of urban areas in climate change politics seems to be increasingly recognized, and growing numbers of conurbations have put this issue high on their agendas. Whereas national governments have occupied a prominent role in climate change politics not least due to their participation in the international climate negotiations, urban areas are in a position to take the lead in terms of adaptation to climate change, given their size, their actual or potential vulnerability and not least their mandates and resources. About half the global population currently lives in urban areas, and in developed regions this figure approaches 75% (United Nations, 2007). Considerable amounts of any nation's social, political and economic life are concentrated in urban areas, and this is especially the case in modern knowledge economies. At the same time, many conurbations are vulnerable to climate change. As many cities are located in close proximity to water, they are particularly vulnerable to rising sea levels and increasing flood risk. Due to the 'urban heat island effect', global warming will be more pronounced in cities than in rural areas. The very limited surface permeability in urban areas poses a particular drainage problem and the capacity of drainage systems is often difficult to increase because of the high density of existing buildings. Governing bodies of urban areas have responsibility for a diverse portfolio, including spatial planning, housing, transportation, water and sewage. A key challenge is to include adaptation into established remits and activities.

Innovative approaches to adaptation are hampered by the complexity of this policy field. Consider firstly the complexity of assessing vulnerability. Although great advances have been made in recent years, existing knowledge on climate change and about future projections is still incomplete even on the global level. Downscaling these changes to the regional level adds layers of uncertainty. Latitude, topography and even geology are among the factors that produce regional variations. Knowledge deficiencies pertaining to such regional contingencies clearly represent grave impediments to adaptation. But the literature on vulnerability has identified a number of factors that affect actual impacts, such as soil composition, quality of buildings and infrastructure, adequacy of flood protection measures and numerous other region-specific physical features. Finally, business structure and the general composition of activities in society also constitute regional variations in vulnerability. Secondly, identifying vulnerability issues is one thing, taking effective action is another. The complexity of developing and implementing adaptation measures reflects the complexity of vulnerability assessment, in the sense that climate change affects such a wide variety of activities, actors and interests.

Effective adaptation will require collaboration between a large number of parties, across policy areas in the public domain and in the private sector as well as in civil society.

The perspective taken in this chapter is that these issues of knowledge and action can be analysed as rather complex problems of *coordination*. As indicated above, vulnerability assessment – the knowledge problem – requires the consideration of a composite body of knowledge, not just from several branches of sciences but also from the practical world of architecture, engineering, city planning and business life. No one actor is in the possession of all relevant knowledge, and from this fact arises the need for *knowledge transfer and collective learning*. A key element is how this knowledge flow may be managed and coordinated.

Furthermore, the 'action problem' is partly a matter of motivating the wide range of relevant actors to implement adaptation measures either unilaterally, in the context of their own properties and spheres of responsibility, or jointly in order to be effective, giving rise to further needs for coordination.

Generally speaking, there are a limited number of options available for societal coordination. Drawing on the extensive generalized literature on coordination (Ouchi, 1980; Williamson, 1985; Hood, 2000), this chapter considers the potentials and limitations of markets, hierarchies and networks in meeting the coordination needs that arise from the knowledge and action aspects of adaptation. Key questions this chapter theoretically seeks to answer are: To what extent will markets provide adaptation through autonomous self-adjustment? What are the limits to regulatory devices implemented by the hierarchical power of public governments? What are the potential strengths and weaknesses of networks and partnerships in managing knowledge transfer and taking effective and efficient action?

In order to address these questions we first introduce a model of organizational learning and elaborate on the characteristics of the different governance modes. Next, we combine these two into one model and discuss how the effect of governance modes on learning can present both potentials and barriers to adaptation. This approach provides the theoretical background of the empirical study on adaptation in urban development projects in four different Norwegian cities, currently taking place.

Urban governance and the capacity to learn and act

The knowledge problem: institutional learning for adaptation

Drawing on theories of organizational learning, Berkhout et al. (2004) have developed a simple model capturing the main aspects of an adaptation process. The

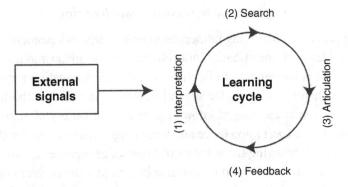

Figure 30.1 A model for institutional learning. (Source: Berkhout et al., 2004.)

basis is that adaptation requires learning, and learning involves the encoding in organizational routines of lessons learnt from experience. Learning hence leads to changes in organizational behaviour. This process is in fact a process of adaptation. Of particular relevance for discussions around climate change adaptation is the issue of uncertainty, given that organizations are being encouraged to change their organizational routines not on the basis of lessons learnt from past *experience*, but based on *scenarios* of future happening, and these scenarios have a high degree of uncertainty attached to them. Berkhout et al.'s model depicts learning resulting in adaptation as a process with four stages. The organization (institution/actor) receive signals from the outside world; these may be signals of actual or anticipated climate change, new knowledge about climate change, political signals (such as new regulations, claims from environmental organizations), economic signals (such as a change in the demand of a certain good or service, new available technology, changes in insurance policy). The first step in the learning process occurs when these signals are received and *interpreted* (1) and eventually acknowledged as a problem (see Figure 30.1). If a need for adaptation is identified at this interpretative step in the process, the organization will *search* (2) for possible solutions to the perceived problem. Next, a selected set of solutions will be *articulated* (3) and codified into new routines of the organization. Last, experience with the new routines will be evaluated and perhaps changed (*feedback*) (4).

These four phases will probably rarely occur in the tidy, linear sequence suggested by Berkhout et al.'s model, and we do not propose that it accurately reflects actual adaptation processes. In this chapter the four-stage conceptualization is used for elaborating on how learning and adaptation may take place under different governance arrangements. How do, for instance, the mechanisms for 'interpretation' differ in markets as compared to in hierarchies? Such issues are discussed below, following a brief overview of the three basic modes of coordination.

Governance for institutional learning

As learning processes involving numerous operationally independent actors, these introduce problems of coordination not reflected in the linear model presented in Figure 30.1. The achievement of effective, broad-scale adaptation requires coordination of the learning process, for several reasons. As noted in the introduction, adaptation involves processing of knowledge held by a variety of actors, including different strands of expert knowledge and knowledge on numerous local contingencies. We argue that bringing these forms of knowledge together in order to achieve adequate interpretations of – and potential solutions to – the problem needs to take place as part of a coordinated process because no one actor is in the possession of all relevant knowledge. And as effective adaptation involves the articulation of solutions in the routines of several operationally independent organizations, coordination is needed to motivate actors to implement new routines, to monitor the extent to which they actually do so and their effectiveness.

The concept of 'governance modes' captures the relations between the stakeholders as well as the rules guiding their interaction. It thereby offers a tool to analyse multi-actor constellations and processes of government, involving both public and private actors, which is exactly what characterizes most adaptation processes in urban areas: official authorities are responsible for the regulations, decrees etc., whereas most adaptive measures are implemented by private actors.

As briefly mentioned earlier, we focus upon three archetypes of governance modes: the hierarchical mode, the network mode, and the market mode. In a *hierarchical* governance mode, coordination takes place in the form of command and control. The key prerequisite for achieving hierarchical coordination is the existence of centrist authority with legitimate powers of decision-making and means to ensure compliance, including control over funding and means of coercion. This is the traditional modus operandi of representative government, in which decisions made in the representative body are implemented by public employees in subordinate departments and agencies through a consistent chain of command. Most business enterprises are however also organized as hierarchies, as are many organizations in civil society.

The market is a mode of coordination devoid of the centrist authority found in hierarchies. Coordination is achieved through the autonomous self-adjustment of numerous operationally independent actors. The sole means of communication between the actors is the price mechanism. In perfect markets, no one actor is able to affect prices. Instead, a product price will emerge that balances supply and demand, and the actors will adjust their behaviour accordingly.

Networks in several ways represent a point in between hierarchies and markets. Contrary to hierarchies, there is no centrist authority and no set patterns of

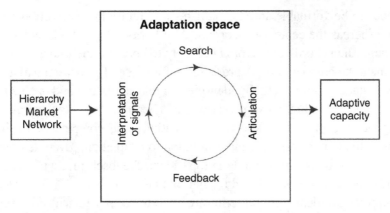

Figure 30.2 A conceptual model for adaptation, combining institutional learning and governance modes.

subordination. Participants in networks are operationally autonomous but contrary to market actors they are mutually interdependent, requiring direct communication. Neither hierarchies nor markets leave room for arguing and bargaining, which are the key modes of interaction in networks. Contrary to hierarchies, ideal-type networks are self-regulating and informal.

The different governance modes are ideal types, which will hardly be found empirically. Instead, for coordinating action, a mix of them, a 'governance arrangement', is usually in play (Heinelt et al., 2006). We hypothesize, as earlier mentioned, that the specific mix of governance modes will influence the learning processes and hence the adaptive capacity of involved actors. The relation between governance mode, learning process and adaptive capacity is pictured in Figure 30.2. We discuss below how the three modes of governance may create different potentials and limits for adaptation throughout the learning process, conceptualized as per Berkhout et al.'s model by the four components of interpretation, search, articulation and feedback.

Linking governance modes and learning

In a *hierarchical mode*, the responsibility for interpretation and search for solutions resides with the centre of authority. The goals defined as important at the top of the hierarchy will dictate what signals and what information is perceived as relevant and whether it necessary to take action (for example adaptive measures). The search for possible solutions is likely to take place within the organization, drawing from a limited repertoire of already tested solutions. Articulation takes place in the form of codified, binding prescriptions that are passed on to the hierarchically subordinate actors by means of legislation, regulations, instructions and so forth. Adherence to these needs to be verifiable and mechanisms for control and correction need to be in place. Whereas the centre of hierarchical authority takes all relevant knowledge into account, the subordinate actors need only to consider the knowledge required to implement the prescriptions. A prerequisite for successful hierarchical adaptation is however that the centre actually identifies all relevant issues and is able to codify these into binding and effective prescriptions. Hierarchy is in its generalized form not designed to exploit local knowledge, creativity and competence to their fullest extent. This may prove a serious limitation to adaptation, given its complexity. Moreover, as the hierarchical mode is top–down, feedback mechanisms are often weak and dependent on established formal routines.

In a *market mode*, learning and action are likely to occur primarily due to changing demands. Increasing awareness of potential economic losses due to climate change means that interpretation takes place among consumers. As a result, demand for adaptation may increase, raising prices for commodities such as climate-friendly energy solutions or buildings constructed to withstand more frequent and intense extreme weather events. Raising prices could induce suppliers of such goods to search for possible solutions, for example new construction methods. It is likely that the main focus of market governance would be on cost-effective and profitable solutions, based on assessments of risk and discount horizon. Articulation is partially a question of implementing such solutions in new projects as a routine matter, partially a question of consumers actually buying these solutions. Because learning is costly, new routines will only be developed and implemented to the extent that the costs of doing so are smaller or equal to the potential costs of climate-induced damage. In a market mode, feedback mechanisms are well developed and respond rather quickly to new signals among customers. Developers may also learn from one project to the next. However, such feedbacks may not result in appropriate adaptation measures because of a biased preference for profitable solutions.

In a *network* mode, learning and action is coordinated among operationally autonomous actors who recognize patterns of interdependence – the need to achieve coordinated action in order to handle common problems. Learning in ideal-type networks would take place quite differently than in hierarchies and markets. Interpretation and search would require direct communication between all relevant actors, on an equal basis. Professional networks, business organizations, conferences and workshops would be relevant arenas for such learning. Different signals will normally be absorbed due to the fact that the different stakeholders involved have different goals. In a horizontal network, a shared understanding of the situation is the key. The absence of a hierarchical authority implies that articulation is a matter of voluntary action, although actors may choose to forgo their independence by entering into binding agreements. The search for adaptation solutions will be discussed among the different actors in order to arrive at common solutions. It is likely that such a governance mode may result in a high capacity to find innovative and comprehensive solutions. On the other hand, networks are vulnerable because of their unstable nature and because of the absence of an authority that may enforce the implementation of solutions agreed on. For the same reasons, feedback mechanisms will not necessarily exist.

Linking governance modes and the capacity to act

We argue that the relevant dimensions for assessing the potentials and limits to the capacity to act, pertaining to the three governance modes outlined above, are: *implementation, range* and *democratic legitimation.* Systems of coordination need to ensure that action is actually taken, across the range of relevant dimensions and in accordance with accepted democratic standards.

Hierarchical coordination has proven a very powerful tool in a multitude of contexts, and remains ubiquitous in spite of the purported diffusion of networks, matrix organizations and informal, horizontal coordination. Most if not all formal organizations in all spheres of society retain elements of hierarchy – for instance, very few employed members of any workforce would find themselves not leading or being led by others, through formal structures, laws and regulations governing the actions of private companies and individuals. A hierarchical mode of governance may enhance or preclude adaptation through the following. First, by promoting implementation. In functioning hierarchies, the centre is enabled to use (coercive) power to enforce rules or regulations – for instance, removing leaders that fail to meet requirements or using legal means of enforcement over private actors who do not observe the rules. On the flip side of this coin there is however a limitation. Enforcement through coercion requires extensive resources devoted to control, straining the administrative capacity of the centre and diverting capacity from other tasks.

Second, by covering an extensive range. Although adaptation is a complex and multidimensional issue, where extensive bodies of regulation are already in place, they may govern activities relevant to adaptation. In other words, existing hierarchical coordination can probably be applied to most issues relevant for adaptation. However, as already mentioned a prerequisite for successful hierarchical adaptation is that the centre actually identifies all relevant issues and codifies these into binding regulations. If the centre fails to do so, important aspects may be overlooked and appropriate action will not take place.

Third, through democratic legitimacy. Democratic equality and representative accountability can only be ensured through formal procedures, some of which require a hierarchical mode – for instance, making governments accountable by means of elections, establishing political rule over the administrative domain and subjecting society in general to the rule of law. Democratic limitations related to hierarchy could be the loss of legitimacy due to government failure; if a representative government loses the trust of its electorate and is perceived as unable to deal with emerging challenges and opportunities – including adaptation – legitimacy and support cannot be secured through elections and legality alone.

Coordination through markets is based on specific assumptions concerning the actors involved. They are expected to maximize their own interests as effectively as possible. The only precondition – regarding the focus on adaptation in this chapter – is that market actors recognize climate change as potentially damaging to their own interests. Expectations of loss of profits or degradation of capital value will induce rational actors to adapt to climate change in order to serve their own interests. Herein lay the potentials as well as the limits of markets as coordinating mechanisms for adaptation. Firstly, regarding the implementation dimension: contrary to hierarchies, adaptation does not require central authorities to codify relevant aspects of the problem into specific and binding regulations. Furthermore, adherence to regulations need not be controlled and enforced. This is so because actors are expected to adapt following their own initiative, in order to serve their own interests. Clearly this indicates a potential for achieving adaptation without straining the administrative capacity of public government. The corresponding limitation is however that adaptation will only occur to the extent that costs of adaptation are offset by potential losses of profit and capital value. Adaptation through market-based coordination cannot by nature extend beyond this limitation.

Concerning range, many relevant aspects of adaptation will be taken care of by market mechanisms, but not all. Furthermore, markets are known to be vulnerable to failures. Investors may not be aware of the potential effects of climate change and so fail to take these into account in their future plans. They may act rationally but with a short time horizon: if they expect to sell a building shortly after or even before its completion, their profits or future capital value may not be jeopardized by long-term effects such as climate change. In some cases, collective-action issues may hamper resolution of complex issues: the situation may require a number of actors to implement cost-inducing measures, but a stalemate may be reached if actors free-ride, thus pursing a strategy of reaping the benefits without sharing the costs. Because market-based adaptation takes place outside the sphere of public policy, potentials and limitations related to issues of democratic legitimation are less relevant here than in the other two modes of coordination.

Finally, concerning governance networks, we argue that a prime motivation for using networks as a mode of coordination in adaptation is to offset the limitations and potential failures of hierarchies and markets. Contrary to hierarchies and markets, network coordination does not in itself contain specific assumptions concerning motivation for implementation of measures, such as securing profits or adhering to regulations. Under this mode of coordination, the range of adaptation issues will depend on local contingencies and situational aspects and is in the general sense unpredictable. As for the democratic legitimacy of governance networks, this is a contested and potentially difficult issue. The growing body of literature on democratic potentials and problems of network governance cannot be summarized here, but it should be noted that the absence of formal rules and procedures governing decision-making, guaranteeing of equal access, transparency and accountability, often are weak or absent in network governance. On the other hand, networks may provide added opportunities for direct participation.

However, we argue that networks can be regarded as a mode of coordination fit to circumvent the potential problems associated with enacting adaptation in markets and hierarchies. Networks can serve to disseminate knowledge about climate related vulnerabilities, and so induce market actors to implement these into their cost functions. They may also serve to enhance the knowledge of hierarchical authorities, channelling information on local contingencies into the rule-development process thus making the codes and regulations more accurate and effective. Finally, the elements of direct contact and dialogue between actors may make networks especially suitable for achieving coordinated action. Many adaptation measures need to involve a plurality of actors jointly, as they may involve synergies and externalities that must be considered. The following section explores the relation between learning and action in urban planning and development.

Capacity to learn and capacity to act: a non-linear relation in urban areas

In the previous section we argued that learning as well as coordination among actors is important and necessary for adaptation. Learning is necessary for changing practices and in the diversity of political, social and economic actors in any situation (including urban areas) requires some form of coordination. The complexity of relationships between the high numbers of actors involved in urban development calls for a flexible approach to modes of governance in urban political leadership. It has been argued that representative government will only deviate from a hierarchical government when driven by necessity (Schmitter, 2002), and it is further argued that successful political leadership is dependent on the ability to switch between the various governance modes in reacting to the challenges (Jessop, 2002). We can therefore expect that successful adaptation is dependent on the ability to move between modes depending on the issue faced.

Planning has been described as the link between knowledge and action (Friedman, 1987), well suited to a hierarchical government structure where experts provide the knowledge and suggest solutions from the top, which local authorities are expected to implement. An example of this approach in Nordic countries is where municipalities have the main responsibility for land use planning, but they place trust in the state with regards to adaptation measures (Berglund and Nergaard, 2008). However, there is an absence of a comprehensive national policy for adaptation to climate change in Scandinavian countries, contrary to mitigation (Granberg and Elander, 2007). In this case, the adoption of such a hierarchical model where national policy is wanting may be a severe barrier to adaptation. In addition, Næss et al. (2005) found in their study of responses to the flood in eastern Norway in 1995, that the coordination between the national, regional and local level of government could be characterized as mainly hierarchical, and that this was a barrier to the development of local solutions for adaptation based on local knowledge.

However, this theoretical notion of planning has been redefined, following what is known as the argumentative or communicative turn, according to which planning becomes relevant in a network governance mode (Fischer and Forester, 1993; Healey, 1997). Planning is understood as a combined process of knowledge and interaction with the aim of coordinating actors, interests and arguments. The inclusion of non-public actors is a prerequisite in this revised understanding of planning (Saglie and Vabo, 2004). Hierarchical command and control may not be sufficient when implementing specific measures, for instance related to site layout and design of buildings as recommended in guidance books (Land Use Consultants, 2006). Property developers respond to market signals from their customers and cooperation between non-public actors and public authorities may be required in a network governance mode.

However, a high capacity to learn combined with a high capacity to act through ability to switch between modes of governance will still present barriers to adaptation, such as conflicting goals, conflicting values and interests, as well as uneven power relations among actors. In urban development there are potentially situations of *conflicting goals* and *conflicting interests*. One particular conflict of goals, for instance, resides in measures for reducing emissions of greenhouse gases and adaptation in urban areas. Norwegian and European policy has been to promote a compact city concept over the last 10–15 years (Commission of the European Communities, 1990; St. meld. 31, 1992–3), which has been promoted as a solution for sustainable urban development, supported by transport research showing that urban structure matters with regard to transport volume (Newman and Kenworthy, 1999; Næss, 2006). This policy for a desirable urban form include increasing densities within existing urban area, develop high-density nodes in the transport system and avoid suburban and exurban sprawl. But it is also a contested policy (Breheny, 2002). The consequences of this policy may be that green areas are reduced and more land has sealed surfaces thus contributing to problems of flooding in urban areas. In addition, former harbour areas and industrial areas along rivers have been subject to urban regeneration in line with a compact city policy. This is problematic because of the exposure to increased risks with regard to urban flooding, fluvial flooding and sea level rise. In addition, densification means that more marginal sites, including slopes and hillsides, are used for construction of houses. These may be more vulnerable to risks for landslides. Thus a balance has to be found between these two potentially competing goals.

Value conflict is a well-known phenomenon in urban planning and another potential barrier to adaptation. Conflicts over values also mean conflict over the good policy. With regard to adaptation to climate change these value conflicts may include differences in risk acceptance. Waterfront housing may be so attractive that future risks of sea level rise are accepted by the individual, while for society at large, huge investments in infrastructure and built environments in risk areas may not be desirable. The way adaptation measures are implemented may also be the subject of diverging opinions.

Conflicting interests may also pose barriers for adaptation. A compact city policy with high-density development has been in line with market forces in Norway in the past decades. There has been a strong demand for centrally located housing, and high-density building is profitable particularly in attractive waterfront locations. However, adaptation measures may mean less profit for the developer. Local water management and more green (permeable surfaces) may not enable high-density housing. Thus the actors may have conflicting interests, resulting in power struggles affecting the outcome, as has taken place in rural communities in Norway (Næss et al., 2005); and there is no reason to expect this to be different with regard to urban communities. Conflicts between rationality and power have been discussed both theoretically and shown empirically in a number of studies of urban planning processes as well as in governance studies (Forester, 1989; Stoker, 1995; Flyvbjerg, 1998; Mäntysalo, 1999; Hillier, 2002). Finally, the uncertainty of future climate change scenarios and consequently the question 'what shall we adapt to?' is crucial when studying adaptation processes. Different opinions, assessments of future scenarios, hopes and beliefs will influence the attitude to the necessity as well as the way to adapt. Assessments and interpretations of uncertainty, alongside actors' preferences and planning time horizons should be a central element of any adaptation studies.

Conclusions

Adaptation to climate change can be facilitated by coordinated learning and action among a wide and diverse range of actors. Such coordination can be achieved in many ways. A key conclusion of this chapter is that the choice of coordination mechanism may impact substantially on the implementation of adaptation measures. There is however no universal solution, each generic mode of coordination seems to have its strengths and drawbacks. This chapter has outlined the potentials of and limits to adaptation encapsulated by each coordination mechanism with some specificity according to the following three questions.

(1) What are the potentials and limits to regulatory devices implemented by the hierarchical power of public governments?

The key potentials of hierarchical coordination include its ability to enforce the implementation of measures, across a wide range of relevant dimensions and with a high degree of democratic legitimation. The limits include the demands it puts on administrative capacity and its lack of ability to utilize local creativity and competence.

(2) To what extent will markets provide adaptation?

As for market-based coordination, its key potentials for adaptation seem to mirror its key limitations: effective adaptation will only occur to the extent that future climate change is perceived as a threat to future profits or capital value. While profit maximization certainly may be a powerful driver for adaptation, market failures such as free-riding and short-sightedness on behalf of investors pose serious limitations.

(3) What are the potential strengths and weaknesses of networks in managing knowledge transfer and taking effective and efficient action?

Coordination in networks can be considered as a mechanism for potentially alleviating the limitations of hierarchies and markets. For instance, the arenas for dialogue and information dissemination provided by networks may increase investors' awareness of potential risks, thereby making market-based adaptation more effective. They may also enhance communication between hierarchical authorities and local actors, allowing greater utilization of local initiative. Networks may be especially suitable for complex adaptation measures, in cases requiring coordinated action by a variety of actors. Thus networks may function alongside other forms of coordination to enhance their potentials. The adaptive capacity of networks is potentially limited by shortcomings in democratic legitimation. Furthermore, unlike profit-based markets, networks do not in themselves represent any specific motivation for implementation of adaptation measures.

The so-called communicative turn in urban planning signals a shift from the traditional reliance on hierarchical coordination to a renewed focus on interaction and information exchange (Fisher and Forester, 1993; Healey, 1997). Inclusion of non-public actors in planning relies on obtaining community involvement as well as initiating cooperation with market actors such as developers and real-estate holders. Effective adaptation may require a strengthened ability on behalf of public planners and policy-makers to shift between governance modes in order to fully exploit their potentials while circumventing their limits. Herein lies the policy relevance of this theoretically focused chapter.

Acknowledgements

This chapter summarises the theoretical background of a study on adaptation in urban planning and development in Norwegian cities. The study is part of the PLAN project on 'Potentials of and limits to adaptation in Norway', financed by the Research Council of Norway.

References

Berkhout, F., Hertin, J. and Gann, D. M. 2004. *Learning to Adapt: Organizational Adaptation to Climate Change Impacts*, Tyndall Working Paper No. 47. Norwich: Tyndall Centre for Climate Change Research, University of East Anglia.

Berglund, F. and Nergaard, E. 2008. *Utslippsreduksjoner og tilpasninger: Klimatiltak i norske kommuner*, NIBR Working Paper No. 2008:103. Oslo: Norwegian Institute for Urban and Regional Research.

Breheny, M. (ed.) 2002. *Sustainable Development and Urban Form: European Research in Regional Science*. London: Pion.

Brooks, N. and Adger, N. 2005. 'Assessing and enhancing adaptive capacity', in Lim, B., Spanger-Siegfried, E., Burton, I., Malone, E. and Huq, S. (eds.) *Adaptation Policy Frameworks for Climate Change*. New York: Cambridge University Press, pp. 165–182.

Commission of the European Communities (CEC) 1990. *Green Paper on the Urban Environment*. Luxembourg: Directorate-General of Environment, Nuclear Safety and Civil Protection.

Flyvbjerg, B. 1998. *Rationality and Power: Democracy in Practice*. Chicago: University of Chicago Press.

Fisher, F. and Forester, J. (eds.) 1993. *The Argumentative Turn in Policy Analysis and Planning*. Durham: Duke University Press.

Forester, J. 1989. *Planning in the Face of Power*. Berkeley: University of California Press.

Friedman, J. 1987. *Planning in the Public Domain: From Knowledge to Action*. Princeton: Princeton University Press.

Granberg, M. and Elander, I. 2007. 'Local governance and climate change: reflections on the Swedish experience', *Local Environment* **12**: 537–548.

Heinelt, H., Held, G., Kopp-Malek, T., Matthiesen, U., Reisinger, E. and Zimmermann, K. 2006. Working document for the G-FORS project (Governance for Sustainability), an Integrated Project in EU's 6th Framework Programme. August 2006, Darmstadt/Erkner.

Hillier, J. 2002. *Shadows of Power: An Allegory of Prudence in Land-Use Planning.* London: Routledge.

Healey, P. 1997. *Collaborative Planning: Shaping Places in Fragmented Societies.* London: Macmillan.

Hood, C. 2000. *The Art of the State: Culture, Rhetoric and Public Management.* Oxford: Oxford University Press.

Jessop, B. 2002. 'Governance and metagovernance: on reflexivity, requisite variety, and requisite irony', in Heinelt, H., Getimis, P., Kafkalas, G., Smith, R. and Swyngedouw, E. (eds.) *Participatory Governance in Multi-Level Context: Concepts and Experience.* Opladen: Leseke and Budrich, pp. 33–50.

Land Use Consultants 2006. *Adapting to Climate Change Impacts:- A Good Practice Guide for Sustainable Communities*, in association with Oxford Brookes University, CAG Consultants and Gardiner and Theobald. London: Defra.

Mäntysalo, R. 1999. 'Learning from the UK: towards market-oriented land-use planning in Finland', *Housing, Theory and Society* **16**: 179–191.

Newman, P. and Kenworthy, J. 1999. *Sustainability and the Cities: Overcoming Automobile Dependence.* Washington, DC: Island Press.

Næss, P. 2006. *Urban Structure Matters: Residential Location, Car Dependence and Travel Behaviour.* London: Routledge.

Næss, L. O., Bang, G., Eriksen, S. and Vevatne, J. 2005. 'Institutional adaptation to climate change: flood responses at the municipal level in Norway', *Global Environmental Change* **15**: 125–138.

Ouchi, W. G. 1980. 'Markets, bureaucracies and clans', *Administrative Science Quarterly* **25**: 129–141.

Saglie, I. L. and Vabo, S. I. 2004. 'Governing sustainability in Norwegian urban outskirts', in McEldowney, M. (ed.) *European Cities: Governance – Insights on Outskirts.* Brussels: Cost Office Cost C 10, pp. 93–111.

Schmitter, P. C. 2002. 'Participation in governance arrangements: is there any reason to expect it will achieve "Sustainable and innovative policies in multilevel context?"', in Grote, J. R. and Gbikpi, B. (eds.) *Participatory Governance: Political and Societal Implications.* Opladen: Leske and Budrich, pp. 51–79.

St. meld. 34, 2006–2007. *Norsk klimapolitikk.* Oslo: Ministry of Environment.

St. meld. 31, 1992–1993. *Den regionale planleggingen og arealpolitikken.* Oslo: Ministry of Environment.

Stoker, G. 1995. 'Regime theory and urban politics', in Judge, D., Stoker, G. and Wolman, H. (eds.) *Theories of Urban Politics.* London: Sage, pp. 54–71.

United Nations 2007. *World Urbanization Prospects: The 2007 Revision Population Database.* Paris: UN Department of Economic and Social Affairs.

Williamson, O. E. 1985. *The Economic Institutions of Capitalism: Firms, Markets, Relational Contracting.* New York: Free Press.

31

Conclusions: Transforming the world

Donald R. Nelson

Our nature: adaptable

In geological terms, the human presence on the Earth has been exceptionally brief. Yet during that time, biological evolution and cultural innovations have permitted us to colonize most of the Earth's terrestrial regions. We are found in a wide diversity of physical environments and thrive under a range of climate characteristics. If we use the spread of humans as an adaptation metric, we must be considered a successful species. If we increase the focal length of our analytical lens by a couple of magnitudes, however, we begin to see some details that undermine assumptions of overall success. Entire societies disappear. If we increase the focal length even more, we can see that even within societies there are successes and failures across space and through time. Undoubtedly, we are an adaptable species, but adaptation is not synonymous with smooth transition or change. When we consider different scales of analysis we are forced to consider closely the process of adaptation and the consequences of actions and inactions. Our ability to inhabit diverse and changing environments comes at a price – to individuals and entire societies.

Much research in climate adaptation has focused on policy, strategies, technologies and the capacities required to facilitate their successful deployment. While our current level of adaptedness relative to climate and the environment is clearly a product of purposeful action, it certainly is also co-produced by autonomous actions. Autonomous actions occur in response to change, irrespective of policy directives and without external assistance. Our level of adaptedness fluctuates continuously as a function of the entirety of our various activities as individuals and collective entities. As chapters in this book demonstrate, some actions may be in response to perceived changes in the environment, but many are also in response to changes in economic, health, political and cultural environments. Changes in our material behaviour influence our level of adaptedness at any point in time, whether we are conscious of this or not. Changes in agricultural technologies, medical advances and evolving transportation and communication networks all influence our level of adaptedness to the environment.

Climate change adaptation discussions are most often forward-looking. What do we need to do now in order to reduce future risk or develop the capacity to respond effectively? We can define a challenge and measure the success of response actions against a baseline. However, it is only through hindsight that the true complexity of drivers, functional relationships, feedbacks and the range of possible outcomes become apparent. In an analysis of three historical cases of adaptation to climate change, Orlove (2005) finds that the complexity of processes involved makes it very difficult to operationalize a definition of adaptation. This complexity also leads to unforeseen and oftentimes undesirable outcomes from targeted adaptation responses. Orlove (2005) highlights historical unintended consequences, but similar processes are at work today. Responses to changes in water resources for example, frequently depend on engineering solutions. These 'hard' solutions lock in the range of future action and pass on the externalities to future generations (Anderies et al., 2006; Sophie Nicholson-Cole and Tim O'Riordan, Chapter 23, this volume). In addition, just as our actions (whether specific to climate or not) can make us more adapted to our environment they can just as readily increase our vulnerabilities.

Shifting baselines

The proliferation of climate projections – at the global and regional scales – and their coverage in academic and mainstream media have focused the concern of citizens around the world regarding the way that our planet may change throughout the course of the current century. Temperatures and sea levels are projected to rise, ecosystems will be changed – perhaps beyond recognition – and extreme events will be more extreme. We are able to better project future climate trends than we were 50 – 100 years ago. So, is our current concern justified, or is it simply an artefact of increasingly sophisticated modelling capabilities that provide glimpses into a possible future? To be clear, I place more emphasis on the former, rather than the latter. I use the question, however, to illustrate a point. Much of the process of adaptation works are hidden behind the scenes.

Humans are constantly making adjustments in their lives. These adjustments are the result of a complex web of factors including social, environmental and physiological events and feedbacks. These adjustments have influenced our level of adaptedness to our climate. They weren't always attributed as adaptations, but were simply appropriate responses to the pressures of the time. In other words, even before 'adaptation to climate change' became a hot-button topic for international policy-makers, people were adapting to the environment and climate. That they could be considered adaptations often becomes apparent only in reflection. A partial implication of these processes is that what we now consider a normal climate, and to which we consider ourselves adapted, is at least 0.5 °C warmer than the beginning of the last century. The concept of shifting baselines provides insight into this phenomenon.

The capacity to measure or project change implies a baseline from which comparisons can be made. Hulme et al. (2009) discuss what they refer to as unstable climates through an exploration of numbers, culture and psychology. Their analysis demonstrates that baselines themselves undergo frequent change and, as in the case of numbers, may be the result of sometimes arbitrary decisions. The magnitude of projected change is a function of the baseline from which we choose. As baselines change so do our expectations about the implications of what the future will bring. This can most easily be demonstrated using numbers. If we calculate the projected number of very warm days for Central England at the end of the twenty-first century with a 1960–1991 WMO baseline, we will arrive at a number of around 80–100 very warm days. If our calculation is based on a rolling baseline of 10 or 20 years over the course of the next century, the number of very warm days will be only about 25 (Hulme et al., 2009).

It is more difficult to estimate the magnitude of future change using the metric of climate as co-produced through culture or individuals. We will adapt as time moves on. In efforts to maintain some equilibrium in our relationship to the environment, our behaviours will change and new technologies will become available. Our perspectives of 'normal' climate will change accordingly. Thus, even though the climate 100 years from now may be strikingly different from what we now consider normal, the changes in cultural and psychological baselines up until that point will subsume those changes and the co-produced climate will be based on the (then) current perspective.

The manner in which climate projections are frequently communicated – behind a front of scientific methods, scenarios, uncertainties and ranges – conveys the notion that climate change is an empirical and objective phenomenon. But we see that some of the basic assumptions rest upon unquestioned statistical baselines and the belief that climate is and should be stable. In this context, the recognition of the normative aspects of climate is significant and carries with it subtle policy implications.

Slipping and tipping

Cognitive and statistical baselines reflect our perceptions of climate. They are, however, only one part of the story. Our perceptions do not preclude the reality of biophysical properties and interactions that have real world outcomes on the environment. While our cognitive perceptions can be quite elastic, the environment is subject to empirical biophysical thresholds, which when crossed can have significant impacts on the functioning of entire ecosystems.

Climate projections have always included an element of large magnitude change. However, as Schneider (2004) points out, most attention in climate change assessments is placed on events that fit within normal confidence intervals. Less attention has been directed towards low-probability, high-impact events that fall within what he refers to as the 'tails of the distribution'. This is changing however, and there is increased recognition of the potential magnitude and speed of change. Ramanathan and Feng suggest that we are already committed to 2.4 °C change and Anderson and Bows argue that while current political discussion focus on meeting a stabilization target of 550 ppmv CO_2e (CO_2 equivalent – on which much of the Stern analysis is based – meeting even a 650 ppmv CO_2e, is improbable (Anderson and Bows, 2008; Ramanathan and Feng, 2008). More research is pointing to the approaching tipping points for a variety of Earth system elements including the summer arctic sea ice, Himalayan glaciers and the Greenland ice sheet, among others. The increasing attention given to the more extreme projections; sea level rise for example, which is one of the more straightforward projections – is placing increasing urgency on the need for society to develop responses (Hansen, 2007; Risbey, 2008).

Avoiding the tipping points at which system states radically change is a challenge, in part because science has an extremely low predictive skill in these areas. The non-linear nature of ecosystem behaviour in relation to climate change, for example, makes identifying thresholds extremely difficult (Garry Peterson, Chapter 2, this volume). Identifying these thresholds may not be possible, in fact, until after we have passed beyond them. Researchers can, however, predict the approach of tipping points as systems start down slippery slopes and many of the known tipping points could be reached sometime during this century (Lenton et al., 2008). But the complex nature of our planet and the biophysical systems is far beyond the ability of prediction and control (Narasimhan, 2007). For example, Ramanathan and Feng (2008) indicate the possibility that warming may increase more rapidly than currently predicted. They argue that regulation of air pollutants will reduce the effect of aerosol cooling, speeding up global warming.

The way that climate change projections are presented often contains an alarmist tinge (Hulme, 2006). Whereas urgency may be appropriate, the apocalyptic element in much of the public debate is excessive. Novel climate regimes do not signify the end of the world, although they may signify an age in which we have to radically reassess our understanding of the world. Climate change will undoubtedly result in many surprises for our environmental and social systems. Surprises are defined as a disconnect between our expectations and what actually occurs

(Kates and Clark, 1996; Gunderson, 2003). Our expectations are predicated on past experience and while we may be able to anticipate some types of change, there is still significant uncertainty that will contribute to the occurrence of surprise events. Surprise and uncertainty need not always limit our adaptive capacity. For some decision arenas a focus on the robustness of responses permits successful adaptation without accurate understanding of future conditions (Suraje Dessai and co-authors, Chapter 5, this volume). Nevertheless, even when there is high confidence in predicting individual future events, it is significantly more challenging to predict the interactive effects. Streets and Glantz (2000) provide examples of the interactive effects of climate anomalies and biological systems including pest and disease outbreaks. These synergistic processes will contribute to a world where the known and familiar become less so.

Whereas tipping points may be a scientific concept difficult to predict empirically, the processes and changes along a slippery slope towards tipping points have real world implications for humans around the globe, affecting their livelihoods and well-being. These impacts are measurable. Whether we are approaching apocalyptic points of no return or not, the fact remains that most of the change that we document today comes as a result of anthropogenic activities. The most vulnerable, those who are suffering the largest impacts of a changing climate are populations who have contributed the least to the current state of affairs. The urgency of response that many promote must therefore also reflect the moral and ethical obligations that derive from these actions.

Adaptation as transformation

In the way that climate change assessments focus on the most likely range of outcomes, the range of most commonly considered adaptation measures also reflects the middle ground. It is certainly less challenging to imagine adapting to a future world 2 °C rather than 5 °C warmer than our world today. It may also be comforting to think that we can rationally and democratically ease ourselves onto a path of adaptation that accounts for the types of change projected under these scenarios, although Schellnhuber (2008) suggests that meeting 2 °C targets will be a close and difficult race. Nevertheless, there is a disconnect between belief and action. If we believe that changes will continue to occur at a slow and smooth rate, then the urgency regarding climate change may be overdone. If, however, we recognize the presence of approaching thresholds, then a shifting baselines approach will be insufficient. Individuals and society, lulled by a sense of security that we can continue to adapt as we have in the past through incremental steps, will respond only after we have crossed thresholds. If there is genuine agreement that the changes in our climate in the near future will create novel challenges with consequences for human well-being, we must be able to respond in novel ways. This requires a re-evaluation of the way we conceptualize our relationship with the natural environment.

The concept of 'Adaptation to climate change' in many ways reflects the Western nature/culture dichotomy. It frames the issue of climate change as one in which culture stands separate from nature, in which humans degrade the environment and nature needs to be protected from us. As Orlove notes, it also highlights the dominance of culture over nature because it doesn't question the belief that all people will be able to equally adapt (Ben Orlove, Chapter 9, this volume). By separating two parts of a whole, the nature/culture dichotomy constrains the way that we conceptualize climate change and therefore limits our response options. One way it does so is by encouraging us to think about adaptation to climate change as something limited to our impacts on the climate system. If, as a species, we wish to live in a world with changed climate and ecosystems we must expand the way that we think about adaptation.

Narrowly framed, adaptation is focused on reducing risk to specific hazards. Humans are adept at reducing specific risks. We build sea walls and levees, we pump water across deserts, we artificially cool and heat our environments and we modify ecosystems to reduce disease exposure. It is widely recognized that risk-management frameworks run into problems of how to value incommensurable goods (what and where to save, what and where not to save) (e.g. Adger et al., 2009). As Pielke et al. (2007) highlight, most of the projected climate change impacts are only marginal increases in areas where populations already experience significant losses. In addition, narrowly focused responses, while reducing a particular risk, may increase our exposure to other risks or pass the risks onto other populations (Walker et al., 2006; Nelson et al., 2007). All of these factors argue for a more encompassing decision matrix and process-oriented, in addition to outcome-oriented, goals. In other words, the process of adaptation should be designed to make society more resilient to a range of influences – including, but not limited to climate change.

The need to expand our conception of adaptation is a perspective widely held by sustainable development and disaster management practitioners. Some have suggested that responding to climate change is really a question of following sustainable development trajectories. But currently, climate change – not sustainable development – is the topic on the world stage. It is a driver of international policy regimes, generates massive sums of research monies, and the UK now has a Minister of Energy and Climate Change. Too frequently adaptation still reflects a narrow framing, which assumes that climate change is an ultimate, rather than proximate driver of change. Re-evaluating this assumption requires rethinking what we mean by adaptation; moving from an approach of linear, incremental steps, which define the history that brought us to the current point. Transformational change requires reforming the basis on which we think about the world.

The research into adaptation to climate change during the last two decades provides many essential insights into how the process of change occurs. It has provided us with tools and frameworks within which to measure and evaluate actions and outcomes. While they certainly incorporate research from other areas, the concept of vulnerability, multi-level governance, issues of fairness and equity, social learning and the role of adaptive capacity all contribute to our understanding of how we can prepare for change and how it can be strengthened and directed. The force of these ideas begins to converge around the possibility of transformational change.

This volume contributes to this trajectory of understanding, recognizing that adaptation is not limited to decisions about technical and economic responses to our climate. The chapters, which highlight many significant challenges that are often absent in adaptation discussion, make it clear that adapting to climate change will require very hard choices and that the results of those choices are not zero-sum.

Humans are instinctively short-sighted – spatially and temporally. Many of the problems that we see in the world today are a result of this limited vision. The consequences of our actions simply do not affect our daily lives in direct ways. As humans continue to urbanize, and become further removed from direct dependence on natural resources, feedback systems become masked. In this sense, climate change is different from many of the other issues that make up the challenges of our modern world. The reach of climate change impacts will be universal. Undoubtedly, some will suffer more than others and some will come away as winners. But the implication of climate change may become one of the impetuses we need to make radical change. The fact that the problems are coming home to roost in the backyard of those that contributed to the problem, as well as everyone else, may very well mark a shift in the way in which we respond to predictions of change.

The challenge is how to meet our individual and household needs in ways that minimally impact contemporary and future generations. Frankly, this balance is not part of the Western worldview which places humans apart from nature. Within this paradigm, trade-offs are reduced to economic calculations, whose outcomes hinge on how we attribute discount rates and the substitutability of capital. These attributions are rooted in the inviolability of economic growth as an overarching goal. But moving away from this worldview is not a simple process. We live in a complex society and the legacies of complexity include rigidity, fragility and reduced flexibility (Holling and Gunderson, 2002; Redman and Kinzig, 2003). As O'Brien highlights, cultures and values do change through time (Karen O'Brien, Chapter 10, this volume). But culture maintains itself as it transmits information through symbols, meanings and behaviours and naturally resists large-scale change. Once change begins, it can be rapid, but the path to that threshold can be difficult and the process is still not well understood.

For all our well-meaning efforts, governance systems continue to inhibit, as well as facilitate, processes of responding to change (Susanne Moser, Chapter 20, this volume). The implication is that there is no single prescriptive approach that can pave the way for transformational change. Rather, it must be a dynamic process that emerges from many small, individual actions that manage to grow. Research in sustainability, transitions and transformation, environmental movements, moral philosophers, and spiritual and religious movements all have their part to play. Adaptation is not a process that we can fully direct, nor should we wish it to be.

Our deliberate actions are already transforming the world, albeit in many unintentional ways. So the question is whether we begin to be proactive in the process, or whether we will be forced to just respond to a rapidly changing environmental context. Regardless, there will be associated costs in human well-being. Adaptation has both a purposive, directed component as well as an autonomous and emergent component. Future research in adaptation will continue to explore both areas and enhance our understanding of social learning, innovation and how transformational change takes place. This type of research may help overcome some of the adaptation limits and make potential barriers less rigid. In this sense, a diversity of values and cultures may in fact be more beneficial than restricting as humans seek ways to accommodate change.

As has been the case for adaptation and transformation processes throughout time, responding to climate change will require trial and error. The role of learning (Folke et al., 2005) is recognized as a critical part of adaptive capacity. Learning happens as a result of our own actions, but we can also learn from the experiences of others. Similar to biological diversity, diverse cultures and values are an adaptive resource from which we can draw experience and knowledge. This goes beyond recognizing the value of traditional knowledge for resolving local problems. Rather it reflects the idea that the ways we envisage the world and our consequent behaviours have real world outcomes. Atran et al. (2002), for example, show how different cultures demonstrate different behaviours when facing similar conditions. Similarly, Heyd and Brooks suggest that responses to environmental change are conditioned by the way in which societies envision their relationship within the larger environment (Thomas Heyd and Nick Brooks, Chapter 17, this volume).

This volume is centred on the limits *to* adaptation, but it is also useful to reflect on the limits *of* adaptation. Is it possible to respond to the challenges of a changing climate within a framework of adaptation? Adapting to change by reducing vulnerabilities provides significant, measurable benefits. But adaptation does not require reflecting on or addressing the primary drivers of change. Many of the drivers of climate change are anthropogenic in nature, and therefore responsive to our actions. Reflecting on the drivers of change requires moving beyond 'single-loop' learning, which is centred on how we can respond based on the way that things are today. Rather, it will demand 'double-loop' learning where we assess our underlying assumptions (Millbrath, 1989). Re-evaluating our assumptions and values opens the possibility for significant shifts in the way in which we view the world, the way in which we behave and consequently, the outcomes of our actions.

Are we likely to see the world transformed in response to projected climate change? Frankly, I find myself torn between a pessimism born of the past inability of human societies to take proactive measures in response to predicted change and an optimism that the learning that has taken place during the history of humankind is finally providing us with the knowledge, tools, and desire to make necessary change. A critical step, however, will be coming to terms with the fact that what we need may not be more adaptation to nature. Rather, what will be needed is acknowledgement that we co-create 'nature' through our actions. This type of transformational thinking will require recognition that limits do exist and it will require developing the capacity to live within those limits.

References

Adger, W., Dessai, S., Goulden, M., Hulme, M., Lorenzoni, I., Nelson, D.R., Naess, L.-O., Wolf, J. and Wreford, A. 2009. 'Are there social limits to adaptation to climate change?' *Climatic Change* **93**: 335–354.

Anderies, J.M., Ryan, P. and Walker, B. 2006. 'Loss of resilience, crisis, and institutional change: lessons from an intensive agricultural system in Southeastern Australia', *Ecosystems* **9**: 865–878.

Anderson, K. and Bows, A. 2008. 'Reframing the climate change challenge in light of post-2000 emission trends', *Philosophical Transactions of the Royal Society A*: doi:10.1098/rsta.2008.0138.

Atran, S., Medin, D., Ross, N., Lynch, E., Coley, J., Ek' E.U. and Vapnarsky, V. 2002. 'Folk ecology and commons management in the Maya Lowlands', *Proceedings of the National Academy of Sciences of the USA* **96**: 7598–7603.

Folke, C., Hahn, T., Olsson, P. and Norberg, J. 2005. 'Adaptive governance of social–ecological systems', *Annual Review of Environment and Resources* **30**: 441–473.

Gunderson, L.H. 2003. 'Adaptive dancing: interactions between social resilience and ecological crises', in Berkes, F., Colding, J. and Folke, C. (eds) *Navigating Social–Ecological Systems: Building Resilience for Complexity and Change*. Cambridge: Cambridge University Press, pp. 33–52.

Hansen, J. 2007. 'Scientific reticence and sea level rise', *Environmental Research Letter* **2**: 1–6.

Holling, C.S. and Gunderson, L.H. 2002. 'Resilience and adaptive cycles', in Gunderson L.H. and Holling, C.S. (eds.) *Panarchy: Understanding Transformations in Human and Natural Systems*. Washington, DC: Island Press, pp. 25–62.

Hulme, M. 2006. 'Chaotic world of climate truth', in *BBC News*. Available at http://news.bbc.co.uk/1/hi/sci/tech/6115644.stm (accessed 2 November 2008)

Hulme, M., Dessai, S., Lorenzoni, I. and Nelson, D. R. 2009. 'Unstable climates: exploring the statistical and social constructions of 'normal' climate', *Geoforum* **40**: 197–206.

Kates, R. W. and Clark, W. C. 1996. 'Expecting the unexpected', *Environment* **38**(2): 6–18.

Lenton, T. M., Held, H., Kriegler, E., Hall, J. W., Lucht, W., Rahmstorf, S. and Schellnhuber, H. J. 2008. 'Tipping elements in the Earth's climate system', *Proceedings of the National Academy of Sciences of the USA* **105**: 1786–1793.

Millbrath, L. 1989. *Envisioning a Sustainable Society: Learning Our Way Out*. Albany: State University of New York Press.

Narasimhan, T. 2007. 'Limitations of science and adapting to Nature', *Environmental Research Letters* **2**: 1–5.

Nelson, D. R., Adger, W. N. and Brown, K. 2007. 'Adaptation to environmental change: contributions of a resilience framework', *Annual Review of Environment and Resources* **32**: 395–420.

Orlove, B. 2005. 'Human adaptation to climate change: a review of three historical cases and some general perspectives', *Environmental Science and Policy* **8**: 589–600.

Pielke, R. A. Jr, Prins, G., Rayner, S. and Sarewitz, D. 2007. 'Climate change 2007: lifting the taboo on adaptation', *Nature* **445**: 597–598.

Ramanathan, V. and Feng, Y. 2008. 'On avoiding dangerous anthropogenic interference with the climate system: formidable challenges ahead', *Proceedings of the National Academy of Sciences of the USA* **105**: 14245–14250.

Redman, C. L. and Kinzig, A. 2003. 'Resilience of past landscapes: resilience theory, society, and the longue durée', *Conservation Ecology* **7**: 14.

Risbey, J. S., 2008. 'The new climate discourse: alarmist or alarming?' *Global Environmental Change* **18**: 26–37.

Schellnhuber, H. J. 2008. 'Global warming: stop worrying, start panicking?' *Proceedings of the National Academy of Sciences of the USA* **105**: 14239–14240.

Schneider, S. H. 2004. 'Abrupt non-linear climate change, irreversibility and surprise', *Global Environmental Change* **14**: 245–258.

Streets, D. G. and Glantz, M. H. 2000. 'Exploring the concept of climate surprise', *Global Environmental Change* **10**: 97–107.

Walker, B., Gunderson, L. H., Kinzig, A., Folke, C. and Schultz, L. 2006. 'A handful of heuristics and some propositions for understanding resilience in social-ecological systems', *Ecology and Society* **11**: 13.

Index

CPSIA information can be obtained
at www.ICGtesting.com
Printed in the USA
LVHW061951200822
726455LV00005B/55